3rd International Conference on Innovative Technologies for Clean and Sustainable Development

RILEM BOOKSERIES

Volume 29

RILEM, The International Union of Laboratories and Experts in Construction Materials, Systems and Structures, founded in 1947, is a non-governmental scientific association whose goal is to contribute to progress in the construction sciences, techniques and industries, essentially by means of the communication it fosters between research and practice. RILEM's focus is on construction materials and their use in building and civil engineering structures, covering all phases of the building process from manufacture to use and recycling of materials. More information on RILEM and its previous publications can be found on www.RILEM.net.

Indexed in SCOPUS, Google Scholar and SpringerLink.

More information about this series at http://www.springer.com/series/8781

Deepankar Kumar Ashish · Jorge de Brito ·
Sanjay Kumar Sharma
Editors

3rd International Conference on Innovative Technologies for Clean and Sustainable Development

ITCSD 2020

Springer

Editors
Deepankar Kumar Ashish
Department of Civil Engineering
Maharaja Agrasen University
Baddi, Himachal Pradesh, India

Jorge de Brito
Instituto Superior Técnico
University of Lisbon
Lisbon, Portugal

Sanjay Kumar Sharma
Department of Civil Engineering
National Institute of Technical Teachers
Training and Research
Chandigarh, Punjab, India

ISSN 2211-0844 ISSN 2211-0852 (electronic)
RILEM Bookseries
ISBN 978-3-030-51484-6 ISBN 978-3-030-51485-3 (eBook)
https://doi.org/10.1007/978-3-030-51485-3

This Springer imprint is published by the registered company Springer Nature Switzerland AG
The registered company address is: Gewerbestrasse 11, 6330 Cham, Switzerland

Organization

Proceedings Chairs

Deepankar Kumar Ashish, Maharaja Agrasen University, India
Jorge de Brito, Instituto Superior Técnico, University of Lisbon, Portugal
Sanjay Kumar Sharma, National Institute of Technical Teachers Training and Research, India

Conference Chairs

Deepankar Kumar Ashish, Maharaja Agrasen University, India
Sanjay Kumar Sharma, National Institute of Technical Teachers Training and Research, India

Program Committee

Bobby, Federation of Environmental Issues and its Remediation (F-EIR), India
Deepankar Kumar Ashish, Maharaja Agrasen University, India
Jorge de Brito, Instituto Superior Técnico, University of Lisbon, Portugal

Scientific Committee

Josh M. Adam, Polytechnic University of Valencia, Spain
Dimitrios Aggelis, Vrije Universiteit Brussel, Belgium
Shamsad Ahmad, King Fahd University of Petroleum and Minerals, Saudi Arabia
M. Shaheer Akhtar, Chonbuk National University, South Korea
Kursat Esat Alyamac, Firat University, Turkey
Ghazi Gaseem Khaleel Al-Khateeb, University of Sharjah, United Arab Emirates
Deepankar Kumar Ashish, Maharaja Agrasen University, India
Nemkumar Banthia, The University of British Columbia, Canada
Sudhirkumar Barai, Indian Institute of Technology Kharagpur, India
Bishwajit Bhattacharjee, Indian Institute of Technology Delhi, India
Dipendu Bhunia, Birla Institute of Technology and Science, Pilani, India
Tayeb Bouziani, University of Laghouat, Algeria
Paulo Cachim, University of Aveiro, Portugal

Peter Robery, The University of Birmingham, UK
Lukasz Sadowski, Wroclaw University of Science and Technology, Poland
V. Saraswathy, CSIR—Central Electrochemical Research Institute, India
Enrico Sassoni, University of Bologna, Italy
Holmer Savastano Junior, University of São Paulo, Brazil
Nassim Sebaibi, ESITC Caen, France
Marijana Serdar, University of Zagreb, Croatia
Payam Shafigh, University of Malaya, Malaysia
Faiz Shaikh, Curtin University, Australia
Umesh Kumar Sharma, Indian Institute of Technology Roorkee, India
Himanshu Sharma, Maharaja Agrasen University, India
Caijun Shi, Hunan University, China
Kulvinder Singh, Maharaja Agrasen University, India
Harvinder Singh, Guru Nanak Dev Engineering College, India
Shamsher Bahadur Singh, Birla Institute of Technology and Science, Pilani, India
S. P. Singh, National Institute of Technology, Jalandhar, India
Didier Snoeck, Ghent University, Belgium
Anshuman Srivastava, Birla Institute of Technology and Science, Pilani, India
Manju Suthar, Maharaja Agrasen University, India
Zhong Tao, Western Sydney University, Australia
Ahmad Umar, Najran University, Saudi Arabia
Luca Valentini, University of Padova, Italy
Kejin Wang, Iowa State University, USA
Timothy Wangler, ETH Zurich, Switzerland
Frank Winnefeld, Empa—Swiss Federal Laboratories for Materials Science and Technology, Switzerland
Syed Kaleem Afrough Zaidi, Aligarh Muslim University, India

Preface

The 3rd international conference on Innovative Technologies for Clean and Sustainable Development was held in Chandigarh, India, at NITTTR Chandigarh between February 19 and 21, 2020. The event was aimed to bring research professionals from multi-disciplinary fields to cross established sub-disciplinary divides, encouraging the exchange of ideas between scientists, engineering professionals, architects, environmental scientists, academicians, economists, and students.

The conference based on the area of clean technology and sustainable development was attended by more than 250 national and international candidates. Approximately, 500 abstracts were received and reviewed by the scientific committee, out of which nearly 400 abstracts were accepted. Of these, around 380 papers were received and underwent review process by minimum two reviewers. As a result, $\sim$ 108 papers were presented and published at the conference. The proceedings were published under two tracks: Clean Technology published by RILEM Bookseries and Sustainable Materials Science published by Materials Today: Proceedings. A volume of abstracts was published and distributed to the authors at the venue.

This volume contains 38 papers presented at the conference while the other volume contains $\sim$ 70 papers. This clean technology volume issues are discussed on air quality improvement, carbon footprint management, case studies on cleaner production, economics of cleaner production, energy efficiency, energy harvesting, innovative building materials, innovative recycling method, life cycle assessment, pollution prevention techniques and technologies, renewable energy, smart cities and villages, Sustainability, and others.

The editors would like to thank the scientific committee for their unparalleled contribution. Furthermore, we express our sincere appreciation to ACI—American Concrete Institute, RILEM—International Union of Laboratories and Experts in Construction Materials, Systems and Structures, and JCI—Japan Concrete Institute for the informational support provided.

A special thanks to Mr. Suresh Gupta from Maharaja Agrasen University for his constant support. We would also like to thank the local organizing committee from the Maharaja Agrasen University, the National Institute of Technical Teachers

Training and Research, Chandigarh, and the Federation of Environmental Issues and its Remediation (F-EIR). In addition, we appreciate the contribution of all those who helped in organizing the event.

Baddi, India
Lisbon, Portugal
Chandigarh, India
May 2020

Deepankar Kumar Ashish
Jorge de Brito
Sanjay Kumar Sharma

Contents

RILEM Publications

The following list is presenting the global offer of RILEM Publications, sorted by series. Each publication is available in printed version and/or in the online version.

RILEM Proceedings (PRO)

PRO 1: Durability of High Performance Concrete (ISBN: 2-912143-03-9; e-ISBN: 2-351580-12-5; e-ISBN: 2351580125); *Ed. H. Sommer*

PRO 2: Chloride Penetration into Concrete (ISBN: 2-912143-00-04; e-ISBN: 2912143454); *Eds. L.-O. Nilsson and J.-P. Ollivier*

PRO 3: Evaluation and Strengthening of Existing Masonry Structures (ISBN: 2-912143-02-0; e-ISBN: 2351580141); *Eds. L. Binda and C. Modena*

PRO 4: Concrete: From Material to Structure (ISBN: 2-912143-04-7; e-ISBN: 2351580206); *Eds. J.-P. Bournazel and Y. Malier*

PRO 5: The Role of Admixtures in High Performance Concrete (ISBN: 2-912143-05-5; e-ISBN: 2351580214); *Eds. J. G. Cabrera and R. Rivera-Villarreal*

PRO 6: High Performance Fiber Reinforced Cement Composites—HPFRCC 3 (ISBN: 2-912143-06-3; e-ISBN: 2351580222); *Eds. H. W. Reinhardt and A. E. Naaman*

PRO 7: 1st International RILEM Symposium on Self-Compacting Concrete (ISBN: 2-912143-09-8; e-ISBN: 2912143721); *Eds. Å. Skarendahl and Ö. Petersson*

PRO 8: International RILEM Symposium on Timber Engineering (ISBN: 2-912143-10-1; e-ISBN: 2351580230); *Ed. L. Boström*

PRO 9: 2nd International RILEM Symposium on Adhesion between Polymers and Concrete ISAP '99 (ISBN: 2-912143-11-X; e-ISBN: 2351580249); *Eds. Y. Ohama and M. Puterman*

PRO 10: 3rd International RILEM Symposium on Durability of Building and Construction Sealants (ISBN: 2-912143-13-6; e-ISBN: 2351580257); *Ed. A. T. Wolf*

PRO 11: 4th International RILEM Conference on Reflective Cracking in Pavements (ISBN: 2-912143-14-4; e-ISBN: 2351580265); *Eds. A. O. Abd El Halim, D. A. Taylor and El H. H. Mohamed*

PRO 12: International RILEM Workshop on Historic Mortars: Characteristics and Tests (ISBN: 2-912143-15-2; e-ISBN: 2351580273); *Eds. P. Bartos, C. Groot and J. J. Hughes*

PRO 13: 2nd International RILEM Symposium on Hydration and Setting (ISBN: 2-912143-16-0; e-ISBN: 2351580281); *Ed. A. Nonat*

PRO 14: Integrated Life-Cycle Design of Materials and Structures—ILCDES 2000 (ISBN: 951-758-408-3; e-ISBN: 235158029X); (ISSN: 0356-9403); *Ed. S. Sarja*

PRO 15: Fifth RILEM Symposium on Fibre-Reinforced Concretes (FRC)—BEFIB'2000 (ISBN: 2-912143-18-7; e-ISBN: 291214373X); *Eds. P. Rossi and G. Chanvillard*

PRO 16: Life Prediction and Management of Concrete Structures (ISBN: 2-912143-19-5; e-ISBN: 2351580303); *Ed. D. Naus*

PRO 17: Shrinkage of Concrete—Shrinkage 2000 (ISBN: 2-912143-20-9; e-ISBN: 2351580311); *Eds. V. Baroghel-Bouny and P.-C. Aïtcin*

PRO 18: Measurement and Interpretation of the On-Site Corrosion Rate (ISBN: 2-912143-21-7; e-ISBN: 235158032X); *Eds. C. Andrade, C. Alonso, J. Fullea, J. Polimon and J. Rodriguez*

PRO 19: Testing and Modelling the Chloride Ingress into Concrete (ISBN: 2-912143-22-5; e-ISBN: 2351580338); *Eds. C. Andrade and J. Kropp*

PRO 20: 1st International RILEM Workshop on Microbial Impacts on Building Materials (CD 02) (e-ISBN 978-2-35158-013-4); *Ed. M. Ribas Silva*

PRO 21: International RILEM Symposium on Connections between Steel and Concrete (ISBN: 2-912143-25-X; e-ISBN: 2351580346); *Ed. R. Eligehausen*

PRO 22: International RILEM Symposium on Joints in Timber Structures (ISBN: 2-912143-28-4; e-ISBN: 2351580354); *Eds. S. Aicher and H.-W. Reinhardt*

PRO 23: International RILEM Conference on Early Age Cracking in Cementitious Systems (ISBN: 2-912143-29-2; e-ISBN: 2351580362); *Eds. K. Kovler and A. Bentur*

PRO 24: 2nd International RILEM Workshop on Frost Resistance of Concrete (ISBN: 2-912143-30-6; e-ISBN: 2351580370); *Eds. M. J. Setzer, R. Auberg and H.-J. Keck*

PRO 25: International RILEM Workshop on Frost Damage in Concrete (ISBN: 2-912143-31-4; e-ISBN: 2351580389); *Eds. D. J. Janssen, M. J. Setzer and M. B. Snyder*

PRO 26: International RILEM Workshop on On-Site Control and Evaluation of Masonry Structures (ISBN: 2-912143-34-9; e-ISBN: 2351580141); *Eds. L. Binda and R. C. de Vekey*

PRO 27: International RILEM Symposium on Building Joint Sealants (CD03; e-ISBN: 235158015X); *Ed. A. T. Wolf*

PRO 28: 6th International RILEM Symposium on Performance Testing and Evaluation of Bituminous Materials—PTEBM'03 (ISBN: 2-912143-35-7; e-ISBN: 978-2-912143-77-8); *Ed. M. N. Partl*

PRO 29: 2nd International RILEM Workshop on Life Prediction and Ageing Management of Concrete Structures (ISBN: 2-912143-36-5; e-ISBN: 2912143780); *Ed. D. J. Naus*

PRO 30: 4th International RILEM Workshop on High Performance Fiber Reinforced Cement Composites—HPFRCC 4 (ISBN: 2-912143-37-3; e-ISBN: 2912143799); *Eds. A. E. Naaman and H. W. Reinhardt*

PRO 31: International RILEM Workshop on Test and Design Methods for Steel Fibre Reinforced Concrete: Background and Experiences (ISBN: 2-912143-38-1; e-ISBN: 2351580168); *Eds. B. Schnütgen and L. Vandewalle*

PRO 32: International Conference on Advances in Concrete and Structures 2 vol. (ISBN (set): 2-912143-41-1; e-ISBN: 2351580176); *Eds. Ying-shu Yuan, Surendra P. Shah and Heng-lin Lü*

PRO 33: 3rd International Symposium on Self-Compacting Concrete (ISBN: 2-912143-42-X; e-ISBN: 2912143713); *Eds. Ó. Wallevik and I. Níelsson*

PRO 34: International RILEM Conference on Microbial Impact on Building Materials (ISBN: 2-912143-43-8; e-ISBN: 2351580184); *Ed. M. Ribas Silva*

PRO 35: International RILEM TC 186-ISA on Internal Sulfate Attack and Delayed Ettringite Formation (ISBN: 2-912143-44-6; e-ISBN: 2912143802); *Eds. K. Scrivener and J. Skalny*

PRO 36: International RILEM Symposium on Concrete Science and Engineering—A Tribute to Arnon Bentur (ISBN: 2-912143-46-2; e-ISBN: 2912143586); *Eds. K. Kovler, J. Marchand, S. Mindess and J. Weiss*

PRO 37: 5th International RILEM Conference on Cracking in Pavements—Mitigation, Risk Assessment and Prevention (ISBN: 2-912143-47-0; e-ISBN: 2912143764); *Eds. C. Petit, I. Al-Qadi and A. Millien*

PRO 38: 3rd International RILEM Workshop on Testing and Modelling the Chloride Ingress into Concrete (ISBN: 2-912143-48-9; e-ISBN: 2912143578); *Eds. C. Andrade and J. Kropp*

PRO 39: 6th International RILEM Symposium on Fibre-Reinforced Concretes—BEFIB 2004 (ISBN: 2-912143-51-9; e-ISBN: 2912143748); *Eds. M. Di Prisco, R. Felicetti and G. A. Plizzari*

PRO 40: International RILEM Conference on the Use of Recycled Materials in Buildings and Structures (ISBN: 2-912143-52-7; e-ISBN: 2912143756); *Eds. E. Vázquez, Ch. F. Hendriks and G. M. T. Janssen*

PRO 41: RILEM International Symposium on Environment-Conscious Materials and Systems for Sustainable Development (ISBN: 2-912143-55-1; e-ISBN: 2912143640); *Eds. N. Kashino and Y. Ohama*

PRO 42: SCC'2005—China: 1st International Symposium on Design, Performance and Use of Self-Consolidating Concrete (ISBN: 2-912143-61-6; e-ISBN: 2912143624); *Eds. Zhiwu Yu, Caijun Shi, Kamal Henri Khayat and Youjun Xie*

PRO 43: International RILEM Workshop on Bonded Concrete Overlays (e-ISBN: 2-912143-83-7); *Eds. J. L. Granju and J. Silfwerbrand*

PRO 44: 2nd International RILEM Workshop on Microbial Impacts on Building Materials (CD11) (e-ISBN: 2-912143-84-5); *Ed. M. Ribas Silva*

PRO 45: 2nd International Symposium on Nanotechnology in Construction, Bilbao (ISBN: 2-912143-87-X; e-ISBN: 2912143888); *Eds. Peter J. M. Bartos, Yolanda de Miguel and Antonio Porro*

PRO 46: Concrete Life'06—International RILEM-JCI Seminar on Concrete Durability and Service Life Planning: Curing, Crack Control, Performance in Harsh Environments (ISBN: 2-912143-89-6; e-ISBN: 291214390X); *Ed. K. Kovler*

PRO 47: International RILEM Workshop on Performance Based Evaluation and Indicators for Concrete Durability (ISBN: 978-2-912143-95-2; e-ISBN: 9782912143969); *Eds. V. Baroghel-Bouny, C. Andrade, R. Torrent and K. Scrivener*

PRO 48: 1st International RILEM Symposium on Advances in Concrete through Science and Engineering (e-ISBN: 2-912143-92-6); *Eds. J. Weiss, K. Kovler, J. Marchand, and S. Mindess*

PRO 49: International RILEM Workshop on High Performance Fiber Reinforced Cementitious Composites in Structural Applications (ISBN: 2-912143-93-4; e-ISBN: 2912143942); *Eds. G. Fischer and V. C. Li*

PRO 50: 1st International RILEM Symposium on Textile Reinforced Concrete (ISBN: 2-912143-97-7; e-ISBN: 2351580087); *Eds. Josef Hegger, Wolfgang Brameshuber and Norbert Will*

PRO 51: 2nd International Symposium on Advances in Concrete through Science and Engineering (ISBN: 2-35158-003-6; e-ISBN: 2-35158-002-8); *Eds. J. Marchand, B. Bissonnette, R. Gagné, M. Jolin and F. Paradis*

PRO 52: Volume Changes of Hardening Concrete: Testing and Mitigation (ISBN: 2-35158-004-4; e-ISBN: 2-35158-005-2); *Eds. O. M. Jensen, P. Lura and K. Kovler*

PRO 53: High Performance Fiber Reinforced Cement Composites—HPFRCC5 (ISBN: 978-2-35158-046-2; e-ISBN: 978-2-35158-089-9); *Eds. H. W. Reinhardt and A. E. Naaman*

PRO 54: 5th International RILEM Symposium on Self-Compacting Concrete (ISBN: 978-2-35158-047-9; e-ISBN: 978-2-35158-088-2); *Eds. G. De Schutter and V. Boel*

PRO 55: International RILEM Symposium Photocatalysis, Environment and Construction Materials (ISBN: 978-2-35158-056-1; e-ISBN: 978-2-35158-057-8); *Eds. P. Baglioni and L. Cassar*

PRO 56: International RILEM Workshop on Integral Service Life Modelling of Concrete Structures (ISBN 978-2-35158-058-5; e-ISBN: 978-2-35158-090-5); *Eds. R. M. Ferreira, J. Gulikers and C. Andrade*

PRO 57: RILEM Workshop on Performance of cement-based materials in aggressive aqueous environments (e-ISBN: 978-2-35158-059-2); *Ed. N. De Belie*

PRO 58: International RILEM Symposium on Concrete Modelling—CONMOD'08 (ISBN: 978-2-35158-060-8; e-ISBN: 978-2-35158-076-9); *Eds. E. Schlangen and G. De Schutter*

PRO 59: International RILEM Conference on On Site Assessment of Concrete, Masonry and Timber Structures—SACoMaTiS 2008 (ISBN set: 978-2-35158-061-5; e-ISBN: 978-2-35158-075-2); *Eds. L. Binda, M. di Prisco and R. Felicetti*

PRO 60: Seventh RILEM International Symposium on Fibre Reinforced Concrete: Design and Applications—BEFIB 2008 (ISBN: 978-2-35158-064-6; e-ISBN: 978-2-35158-086-8); *Ed. R. Gettu*

PRO 61: 1st International Conference on Microstructure Related Durability of Cementitious Composites 2 vol., (ISBN: 978-2-35158-065-3; e-ISBN: 978-2-35158-084-4); *Eds. W. Sun, K. van Breugel, C. Miao, G. Ye and H. Chen*

PRO 62: NSF/ RILEM Workshop: In-situ Evaluation of Historic Wood and Masonry Structures (e-ISBN: 978-2-35158-068-4); *Eds. B. Kasal, R. Anthony and M. Drdácký*

PRO 63: Concrete in Aggressive Aqueous Environments: Performance, Testing and Modelling, 2 vol., (ISBN: 978-2-35158-071-4; e-ISBN: 978-2-35158-082-0); *Eds. M. G. Alexander and A. Bertron*

PRO 64: Long Term Performance of Cementitious Barriers and Reinforced Concrete in Nuclear Power Plants and Waste Management—NUCPERF 2009 (ISBN: 978-2-35158-072-1; e-ISBN: 978-2-35158-087-5); *Eds. V. L'Hostis, R. Gens and C. Gallé*

PRO 65: Design Performance and Use of Self-consolidating Concrete—SCC'2009 (ISBN: 978-2-35158-073-8; e-ISBN: 978-2-35158-093-6); *Eds. C. Shi, Z. Yu, K. H. Khayat and P. Yan*

PRO 66: 2nd International RILEM Workshop on Concrete Durability and Service Life Planning—ConcreteLife'09 (ISBN: 978-2-35158-074-5; ISBN: 978-2-35158-074-5); *Ed. K. Kovler*

PRO 67: Repairs Mortars for Historic Masonry (e-ISBN: 978-2-35158-083-7); *Ed. C. Groot*

PRO 68: Proceedings of the 3rd International RILEM Symposium on 'Rheology of Cement Suspensions such as Fresh Concrete (ISBN 978-2-35158-091-2; e-ISBN: 978-2-35158-092-9); *Eds. O. H. Wallevik, S. Kubens and S. Oesterheld*

PRO 69: 3rd International PhD Student Workshop on 'Modelling the Durability of Reinforced Concrete (ISBN: 978-2-35158-095-0); *Eds. R. M. Ferreira, J. Gulikers and C. Andrade*

PRO 70: 2nd International Conference on 'Service Life Design for Infrastructure' (ISBN set: 978-2-35158-096-7, e-ISBN: 978-2-35158-097-4); *Eds. K. van Breugel, G. Ye and Y. Yuan*

PRO 71: Advances in Civil Engineering Materials—The 50-year Teaching Anniversary of Prof. Sun Wei' (ISBN: 978-2-35158-098-1; e-ISBN: 978-2-35158-099-8); *Eds. C. Miao, G. Ye and H. Chen*

PRO 72: First International Conference on 'Advances in Chemically-Activated Materials—CAM'2010' (2010), 264 pp., ISBN: 978-2-35158-101-8; e-ISBN: 978-2-35158-115-5; *Eds. Caijun Shi and Xiaodong Shen*

PRO 73: 2nd International Conference on 'Waste Engineering and Management—ICWEM 2010' (2010), 894 pp., ISBN: 978-2-35158-102-5; e-ISBN: 978-2-35158-103-2, *Eds. J. Zh. Xiao, Y. Zhang, M. S. Cheung and R. Chu*

PRO 74: International RILEM Conference on 'Use of Superabsorsorbent Polymers and Other New Addditives in Concrete' (2010) 374 pp., ISBN: 978-2-35158-104-9; e-ISBN: 978-2-35158-105-6; *Eds. O.M. Jensen, M.T. Hasholt, and S. Laustsen*

PRO 75: International Conference on 'Material Science—2nd ICTRC—Textile Reinforced Concrete—Theme 1' (2010) 436 pp., ISBN: 978-2-35158-106-3; e-ISBN: 978-2-35158-107-0; *Ed. W. Brameshuber*

PRO 76: International Conference on 'Material Science—HetMat—Modelling of Heterogeneous Materials—Theme 2' (2010) 255 pp., ISBN: 978-2-35158-108-7; e-ISBN: 978-2-35158-109-4; *Ed. W. Brameshuber*

PRO 77: International Conference on 'Material Science—AdIPoC—Additions Improving Properties of Concrete—Theme 3' (2010) 459 pp., ISBN: 978-2-35158-110-0; e-ISBN: 978-2-35158-111-7; *Ed. W. Brameshuber*

PRO 78: 2nd Historic Mortars Conference and RILEM TC 203-RHM Final Workshop—HMC2010 (2010) 1416 pp., e-ISBN: 978-2-35158-112-4; *Eds. J. Válek, C. Groot and J. J. Hughes*

PRO 79: International RILEM Conference on Advances in Construction Materials Through Science and Engineering (2011) 213 pp., ISBN: 978-2-35158-116-2, e-ISBN: 978-2-35158-117-9; *Eds. Christopher Leung and K.T. Wan*

PRO 80: 2nd International RILEM Conference on Concrete Spalling due to Fire Exposure (2011) 453 pp., ISBN: 978-2-35158-118-6; e-ISBN: 978-2-35158-119-3; *Eds. E.A.B. Koenders and F. Dehn*

PRO 81: 2nd International RILEM Conference on Strain Hardening Cementitious Composites (SHCC2-Rio) (2011) 451 pp., ISBN: 978-2-35158-120-9; e-ISBN: 978-2-35158-121-6; *Eds. R.D. Toledo Filho, F.A. Silva, E.A.B. Koenders and E.M. R. Fairbairn*

PRO 82: 2nd International RILEM Conference on Progress of Recycling in the Built Environment (2011) 507 pp., e-ISBN: 978-2-35158-122-3; *Eds. V.M. John, E. Vazquez, S.C. Angulo and C. Ulsen*

PRO 83: 2nd International Conference on Microstructural-related Durability of Cementitious Composites (2012) 250 pp., ISBN: 978-2-35158-129-2; e-ISBN: 978-2-35158-123-0; *Eds. G. Ye, K. van Breugel, W. Sun and C. Miao*

PRO 84: CONSEC13—Seventh International Conference on Concrete under Severe Conditions—Environment and Loading (2013) 1930 pp., ISBN: 978-2-35158-124-7; e-ISBN: 978-2- 35158-134-6; *Eds. Z.J. Li, W. Sun, C.W. Miao, K. Sakai, O.E. Gjorv and N. Banthia*

PRO 85: RILEM-JCI International Workshop on Crack Control of Mass Concrete and Related issues concerning Early-Age of Concrete Structures—ConCrack 3—Control of Cracking in Concrete Structures 3 (2012) 237 pp., ISBN: 978-2-35158-125-4; e-ISBN: 978-2-35158-126-1; *Eds. F. Toutlemonde and J.-M. Torrenti*

PRO 86: International Symposium on Life Cycle Assessment and Construction (2012) 414 pp., ISBN: 978-2-35158-127-8, e-ISBN: 978-2-35158-128-5; *Eds. A. Ventura and C. de la Roche*

PRO 87: UHPFRC 2013—RILEM-fib-AFGC International Symposium on Ultra-High Performance Fibre-Reinforced Concrete (2013), ISBN: 978-2-35158-130-8, e-ISBN: 978-2-35158-131-5; *Eds. F. Toutlemonde*

PRO 88: 8th RILEM International Symposium on Fibre Reinforced Concrete (2012) 344 pp., ISBN: 978-2-35158-132-2; e-ISBN: 978-2-35158-133-9; *Eds. Joaquim A.O. Barros*

PRO 89: RILEM International workshop on performance-based specification and control of concrete durability (2014) 678 pp., ISBN: 978-2-35158-135-3; e-ISBN: 978-2-35158-136-0; *Eds. D. Bjegović, H. Beushausen and M. Serdar*

PRO 90: 7th RILEM International Conference on Self-Compacting Concrete and of the 1st RILEM International Conference on Rheology and Processing of Construction Materials (2013) 396 pp., ISBN: 978-2-35158-137-7; e-ISBN: 978-2-35158-138-4; *Eds. Nicolas Roussel and Hela Bessaies-Bey*

PRO 91: CONMOD 2014—RILEM International Symposium on Concrete Modelling (2014), ISBN: 978-2-35158-139-1; e-ISBN: 978-2-35158-140-7; *Eds. Kefei Li, Peiyu Yan and Rongwei Yang*

PRO 92: CAM 2014—2nd International Conference on advances in chemically-activated materials (2014) 392 pp., ISBN: 978-2-35158-141-4; e-ISBN: 978-2-35158-142-1; *Eds. Caijun Shi and Xiadong Shen*

PRO 93: SCC 2014—3rd International Symposium on Design, Performance and Use of Self-Consolidating Concrete (2014) 438 pp., ISBN: 978-2-35158-143-8; e-ISBN: 978-2-35158-144-5; *Eds. Caijun Shi, Zhihua Ou and Kamal H. Khayat*

PRO 94 (online version): HPFRCC-7—7th RILEM conference on High performance fiber reinforced cement composites (2015), e-ISBN: 978-2-35158-146-9; *Eds. H.W. Reinhardt, G.J. Parra-Montesinos and H. Garrecht*

PRO 95: International RILEM Conference on Application of superabsorbent polymers and other new admixtures in concrete construction (2014), ISBN: 978-2-35158-147-6; e-ISBN: 978-2-35158-148-3; *Eds. Viktor Mechtcherine and Christof Schroefl*

PRO 96 (online version): XIII DBMC: XIII International Conference on Durability of Building Materials and Components (2015), e-ISBN: 978-2-35158-149-0; *Eds. M. Quattrone and V.M. John*

PRO 97: SHCC3—3rd International RILEM Conference on Strain Hardening Cementitious Composites (2014), ISBN: 978-2-35158-150-6; e-ISBN: 978-2-35158-151-3; *Eds. E. Schlangen, M.G. Sierra Beltran, M. Lukovic and G. Ye*

PRO 98: FERRO-11—11th International Symposium on Ferrocement and 3rd ICTRC—International Conference on Textile Reinforced Concrete (2015), ISBN: 978-2-35158-152-0; e-ISBN: 978-2-35158-153-7; *Ed. W. Brameshuber*

PRO 99 (online version): ICBBM 2015—1st International Conference on Bio-Based Building Materials (2015), e-ISBN: 978-2-35158-154-4; *Eds. S. Amziane and M. Sonebi*

PRO 100: SCC16—RILEM Self-Consolidating Concrete Conference (2016), ISBN: 978-2-35158-156-8; e-ISBN: 978-2-35158-157-5; *Ed. Kamal H. Kayat*

PRO 101 (online version): III Progress of Recycling in the Built Environment (2015), e-ISBN: 978-2-35158-158-2; *Eds I. Martins, C. Ulsen and S. C. Angulo*

PRO 102 (online version): RILEM Conference on Microorganisms-Cementitious Materials Interactions (2016), e-ISBN: 978-2-35158-160-5; *Eds. Alexandra Bertron, Henk Jonkers and Virginie Wiktor*

PRO 103 (online version): ACESC'16—Advances in Civil Engineering and Sustainable Construction (2016), e-ISBN: 978-2-35158-161-2; *Eds. T.Ch. Madhavi, G. Prabhakar, Santhosh Ram and P.M. Rameshwaran*

PRO 104 (online version): SSCS'2015—Numerical Modeling—Strategies for Sustainable Concrete Structures (2015), e-ISBN: 978-2-35158-162-9

PRO 105: 1st International Conference on UHPC Materials and Structures (2016), ISBN: 978-2-35158-164-3; e-ISBN: 978-2-35158-165-0

PRO 106: AFGC-ACI-fib-RILEM International Conference on Ultra-High-Performance Fibre-Reinforced Concrete—UHPFRC 2017 (2017), ISBN: 978-2-35158-166-7; e-ISBN: 978-2-35158-167-4; *Eds. François Toutlemonde and Jacques Resplendino*

PRO 107 (online version): XIV DBMC—14th International Conference on Durability of Building Materials and Components (2017), e-ISBN: 978-2-35158-159-9; *Eds. Geert De Schutter, Nele De Belie, Arnold Janssens and Nathan Van Den Bossche*

PRO 108: MSSCE 2016—Innovation of Teaching in Materials and Structures (2016), ISBN: 978-2-35158-178-0; e-ISBN: 978-2-35158-179-7; *Ed. Per Goltermann*

PRO 109 (2 volumes): MSSCE 2016—Service Life of Cement-Based Materials and Structures (2016), ISBN Vol. 1: 978-2-35158-170-4; Vol. 2: 978-2-35158-171-4; Set Vol. 1&2: 978-2-35158-172-8; e-ISBN : 978-2-35158-173-5; *Eds. Miguel Azenha, Ivan Gabrijel, Dirk Schlicke, Terje Kanstad and Ole Mejlhede Jensen*

PRO 110: MSSCE 2016—Historical Masonry (2016), ISBN: 978-2-35158-178-0; e-ISBN: 978-2-35158-179-7; *Eds. Inge Rörig-Dalgaard and Ioannis Ioannou*

PRO 111: MSSCE 2016—Electrochemistry in Civil Engineering (2016); ISBN: 978-2-35158-176-6; e-ISBN: 978-2-35158-177-3; *Ed. Lisbeth M. Ottosen*

PRO 112: MSSCE 2016—Moisture in Materials and Structures (2016), ISBN: 978-2-35158-178-0; e-ISBN: 978-2-35158-179-7; *Eds. Kurt Kielsgaard Hansen, Carsten Rode and Lars-Olof Nilsson*

PRO 113: MSSCE 2016—Concrete with Supplementary Cementitious Materials (2016), ISBN: 978-2-35158-178-0; e-ISBN: 978-2-35158-179-7; *Eds. Ole Mejlhede Jensen, Konstantin Kovler and Nele De Belie*

PRO 114: MSSCE 2016—Frost Action in Concrete (2016), ISBN: 978-2-35158-182-7; e-ISBN: 978-2-35158-183-4; *Eds. Marianne Tange Hasholt, Katja Fridh and R. Doug Hooton*

PRO 115: MSSCE 2016—Fresh Concrete (2016), ISBN: 978-2-35158-184-1; e-ISBN: 978-2-35158-185-8; *Eds. Lars N. Thrane, Claus Pade, Oldrich Svec and Nicolas Roussel*

PRO 116: BEFIB 2016—9th RILEM International Symposium on Fiber Reinforced Concrete (2016), ISBN: 978-2-35158-187-2; e-ISBN: 978-2-35158-186-5; *Eds. N. Banthia, M. di Prisco and S. Soleimani-Dashtaki*

PRO 117: 3rd International RILEM Conference on Microstructure Related Durability of Cementitious Composites (2016), ISBN: 978-2-35158-188-9; e-ISBN: 978-2-35158-189-6; *Eds. Changwen Miao, Wei Sun, Jiaping Liu, Huisu Chen, Guang Ye and Klaas van Breugel*

PRO 118 (4 volumes): International Conference on Advances in Construction Materials and Systems (2017), ISBN Set: 978-2-35158-190-2; Vol. 1: 978-2-35158-193-3; Vol. 2: 978-2-35158-194-0; Vol. 3: ISBN:978-2-35158-195-7; Vol. 4: ISBN:978-2-35158-196-4; e-ISBN: 978-2-35158-191-9; *Ed. Manu Santhanam*

PRO 119 (online version): ICBBM 2017—Second International RILEM Conference on Bio-based Building Materials, (2017), e-ISBN: 978-2-35158-192-6; *Ed. Sofiane Amziane*

PRO 120 (2 volumes): EAC-02—2nd International RILEM/COST Conference on Early Age Cracking and Serviceability in Cement-based Materials and Structures, (2017), Vol. 1: 978-2-35158-199-5, Vol. 2: 978-2-35158-200-8, Set: 978-2-35158-197-1, e-ISBN: 978-2-35158-198-8; *Eds. Stéphanie Staquet and Dimitrios Aggelis*

PRO 121 (2 volumes): SynerCrete18: Interdisciplinary Approaches for Cementbased Materials and Structural Concrete: Synergizing Expertise and Bridging Scales of Space and Time, (2018), Set: 978-2-35158-202-2, Vol.1: 978-2-35158-211-4, Vol.2: 978-2-35158-212-1, e-ISBN: 978-2-35158-203-9; *Eds. Miguel Azenha, Dirk Schlicke, Farid Benboudjema, Agnieszka Knoppik*

PRO 122: SCC'2018 China—Fourth International Symposium on Design, Performance and Use of Self-Consolidating Concrete, (2018), ISBN: 978-2-35158-204-6, e-ISBN: 978-2-35158-205-3; *Eds. C. Shi, Z. Zhang, K. H. Khayat*

PRO 123: Final Conference of RILEM TC 253-MCI: Microorganisms-Cementitious Materials Interactions (2018), Set: 978-2-35158-207-7, Vol.1: 978-2-35158-209-1, Vol.2: 978-2-35158-210-7, e-ISBN: 978-2-35158-206-0; *Ed. Alexandra Bertron*

PRO 124 (online version): Fourth International Conference Progress of Recycling in the Built Environment (2018), e-ISBN: 978-2-35158-208-4; *Eds. Isabel M. Martins, Carina Ulsen, Yury Villagran*

PRO 125 (online version): SLD4—4th International Conference on Service Life Design for Infrastructures (2018), e-ISBN: 978-2-35158-213-8; *Eds. Guang Ye, Yong Yuan, Claudia Romero Rodriguez, Hongzhi Zhang, Branko Savija*

PRO 126: Workshop on Concrete Modelling and Material Behaviour in honor of Professor Klaas van Breugel (2018), ISBN: 978-2-35158-214-5, e-ISBN: 978-2-35158-215-2; *Ed. Guang Ye*

PRO 127 (online version): CONMOD2018—Symposium on Concrete Modelling (2018), e-ISBN: 978-2-35158-216-9; *Eds. Erik Schlangen, Geert de Schutter, Branko Savija, Hongzhi Zhang, Claudia Romero Rodriguez*

PRO 128: SMSS2019—International Conference on Sustainable Materials, Systems and Structures (2019), ISBN: 978-2-35158-217-6, e-ISBN: 978-2-35158-218-3

PRO 129: 2nd International Conference on UHPC Materials and Structures (UHPC2018-China), ISBN: 978-2-35158-219-0, e-ISBN: 978-2-35158-220-6

PRO 130: 5th Historic Mortars Conference (2019), ISBN: 978-2-35158-221-3, e-ISBN: 978-2-35158-222-0; *Eds. José Ignacio Álvarez, José María Fernández, Íñigo Navarro, Adrián Durán, Rafael Sirera*

PRO 131 (online version): 3rd International Conference on Bio-Based Building Materials (ICBBM2019), e-ISBN: 978-2-35158-229-9; *Eds. Mohammed Sonebi, Sofiane Amziane, Jonathan Page*

PRO 132: IRWRMC'18—International RILEM Workshop on Rheological Measurements of Cement-based Materials (2018), ISBN: 978-2-35158-230-5, e-ISBN: 978-2-35158-231-2; *Eds. Chafika Djelal, Yannick Vanhove*

PRO 133 (online version): CO2STO2019—International Workshop CO2 Storage in Concrete (2019), e-ISBN: 978-2-35158-232-9; *Eds. Assia Djerbi, Othman Omikrine-Metalssi, Teddy Fen-Chong*

RILEM Reports (REP)

Report 19: Considerations for Use in Managing the Aging of Nuclear Power Plant Concrete Structures (ISBN: 2-912143-07-1); *Ed. D. J. Naus*

Report 20: Engineering and Transport Properties of the Interfacial Transition Zone in Cementitious Composites (ISBN: 2-912143-08-X); *Eds. M. G. Alexander, G. Arliguie, G. Ballivy, A. Bentur and J. Marchand*

PRO 21: Durability of Building Sealants (ISBN: 2-912143-12-8); *Ed. A. T. Wolf*

PRO 22: Sustainable Raw Materials—Construction and Demolition Waste (ISBN: 2-912143-17-9); *Eds. C. F. Hendriks and H. S. Pietersen*

PRO 23: Self-Compacting Concrete state-of-the-art report (ISBN: 2-912143-23-3); *Eds. Å. Skarendahl and Ö. Petersson*

PRO 24: Workability and Rheology of Fresh Concrete: Compendium of Tests (ISBN: 2-912143-32-2); *Eds. P. J. M. Bartos, M. Sonebi and A. K. Tamimi*

PRO 25: Early Age Cracking in Cementitious Systems (ISBN: 2-912143-33-0); *Ed. A. Bentur*

PRO 26: Towards Sustainable Roofing (Joint Committee CIB/RILEM) (CD 07) (e-ISBN 978-2-912143-65-5); *Eds. Thomas W. Hutchinson and Keith Roberts*

PRO 27: Condition Assessment of Roofs (Joint Committee CIB/RILEM) (CD 08) (e-ISBN 978-2-912143-66-2); *Ed. CIB W 83/RILEM TC166-RMS*

PRO 28: Final report of RILEM TC 167-COM 'Characterisation of Old Mortars with Respect to Their Repair (ISBN: 978-2-912143-56-3); *Eds. C. Groot, G. Ashall and J. Hughes*

PRO 29: Pavement Performance Prediction and Evaluation (PPPE): Interlaboratory Tests (e-ISBN: 2-912143-68-3); *Eds. M. Partl and H. Piber*

PRO 30: Final Report of RILEM TC 198-URM 'Use of Recycled Materials' (ISBN: 2-912143-82-9; e-ISBN: 2-912143-69-1); *Eds. Ch. F. Hendriks, G. M. T. Janssen and E. Vázquez*

PRO 31: Final Report of RILEM TC 185-ATC 'Advanced testing of cement-based materials during setting and hardening' (ISBN: 2-912143-81-0; e-ISBN: 2-912143-70-5); *Eds. H. W. Reinhardt and C. U. Grosse*

PRO 32: Probabilistic Assessment of Existing Structures. A JCSS publication (ISBN 2-912143-24-1); *Ed. D. Diamantidis*

PRO 33: State-of-the-Art Report of RILEM Technical Committee TC 184-IFE 'Industrial Floors' (ISBN 2-35158-006-0); *Ed. P. Seidler*

PRO 34: Report of RILEM Technical Committee TC 147-FMB 'Fracture mechanics applications to anchorage and bond' Tension of Reinforced Concrete Prisms—Round Robin Analysis and Tests on Bond (e-ISBN 2-912143-91-8); *Eds. L. Elfgren and K. Noghabai*

PRO 35: Final Report of RILEM Technical Committee TC 188-CSC 'Casting of Self Compacting Concrete' (ISBN 2-35158-001-X; e-ISBN: 2-912143-98-5); *Eds. Å. Skarendahl and P. Billberg*

PRO 36: State-of-the-Art Report of RILEM Technical Committee TC 201-TRC 'Textile Reinforced Concrete' (ISBN 2-912143-99-3); *Ed. W. Brameshuber*

PRO 37: State-of-the-Art Report of RILEM Technical Committee TC 192-ECM 'Environment-conscious construction materials and systems' (ISBN: 978-2-35158-053-0); *Eds. N. Kashino, D. Van Gemert and K. Imamoto*

PRO 38: State-of-the-Art Report of RILEM Technical Committee TC 205-DSC 'Durability of Self-Compacting Concrete' (ISBN: 978-2-35158-048-6); *Eds. G. De Schutter and K. Audenaert*

PRO 39: Final Report of RILEM Technical Committee TC 187-SOC 'Experimental determination of the stress-crack opening curve for concrete in tension' (ISBN 978-2-35158-049-3); *Ed. J. Planas*

PRO 40: State-of-the-Art Report of RILEM Technical Committee TC 189-NEC 'Non-Destructive Evaluation of the Penetrability and Thickness of the Concrete Cover' (ISBN 978-2-35158-054-7); *Eds. R. Torrent and L. Fernández Luco*

PRO 41: State-of-the-Art Report of RILEM Technical Committee TC 196-ICC 'Internal Curing of Concrete' (ISBN 978-2-35158-009-7); *Eds. K. Kovler and O. M. Jensen*

PRO 42: 'Acoustic Emission and Related Non-destructive Evaluation Techniques for Crack Detection and Damage Evaluation in Concrete'—Final Report of RILEM Technical Committee 212-ACD (e-ISBN: 978-2-35158-100-1); *Ed. M. Ohtsu*

PRO 45: Repair Mortars for Historic Masonry—State-of-the-Art Report of RILEM Technical Committee TC 203-RHM (e-ISBN: 978-2-35158-163-6); *Eds. Paul Maurenbrecher and Caspar Groot*

PRO 46: Surface delamination of concrete industrial ffioors and other durability related aspects guide—Report of RILEM Technical Committee TC 268-SIF (e-ISBN: 978-2-35158-201-5); *Ed. Valerie Pollet*

Embodied Energy and Cost of Load Bearing Masonry with Alternative Binders and Units—Case Study

B. N. Varsha, P. T. Jitha, and S. Raghunath

Abstract Strength and quality of load bearing masonry is dictated by the properties of masonry unit, mortar and interaction between them. Currently the construction industry is dominated by cement and cement-based products, but parallelly there have been extensive studies in trying to reduce quantity of cement and promote use of alternative cementitious materials. In this regard a masonry unit and mortar combination has been developed which is lime pozzolana cement (LPC) and geopolymer (GP) based, satisfying the strength and serviceability criteria required for the masonry construction. After the experimental investigations, the computation of embodied energy and cost has been carried out which indicate that the designed LPC-GP masonry to be a promising alternative to the conventional cement-based products. The associated advantages of these bricks have also been highlighted through a case study.

Keywords Lime-pozzolana · Geopolymer · Embodied energy · Load bearing masonry

1 Introduction

One of the most efficient and time-tested structural configurations for construction is load bearing masonry [LBM] design. It has proved to be an attractive choice both in terms of aesthetics and economics for the end-user. In this structural system the walling materials accounts for nearly 40% of total material for construction

B. N. Varsha (✉)
ANKANA Construction Consultants, Bangalore, India
e-mail: varsha.bhushana@gmail.com

P. T. Jitha · S. Raghunath
Department of Civil Engineering, BMS College of Engineering, Bangalore, India
e-mail: jithapt@gmail.com

S. Raghunath
e-mail: raghu.civ@bmsce.ac.in

D. K. Ashish et al. (eds.), *3rd International Conference on Innovative Technologies for Clean and Sustainable Development*, RILEM Bookseries 29,
https://doi.org/10.1007/978-3-030-51485-3_1

and hence quality of load bearing masonry design is dictated by the combination of masonry unit and mortar [1]. In the current scenario just like other structural configurations—moment resisting frame [MRF], monolithic reinforced concrete and precast reinforced concrete—the load bearing masonry design is also dominated by cement-based products. Concrete blocks which are solid, hollow or aerated in nature are extensively utilised for construction with cement-based mortars. Soil based hollow clay blocks and table-moulded bricks are also used, but sparingly, due to their cost factor when compared to concrete blocks. Compressed soil cement blocks have low consumer acceptance. The mortar used for majority of these masonry units is cement based. Of late, there have been extensive studies in trying to reduce quantity of cement and promote use of alternative cementitious materials. Traditionally majority of cementitious materials were based on lime and lime-surkhi; modern developments include the use of industrial by-products like ground granulated blast furnace slag (GGBS), fly ash, ultra-fine slag, silica fumes etc.

2 Present Investigation

In view of achieving a lower cement consumption in load bearing masonry without compromising the strength and serviceability criteria a masonry unit and mortar combination has been developed which is lime pozzolana cement (LPC) and geopolymer (GP) based. The constituents of the bricks are soil, crushed stone aggregate (CSA), LPC, GGBS and low molar alkaline solution (2 Molar, 1:1 of Sodium hydroxide (NaOH) and Sodium silicate (Na_2SiO_3)). Following the experimental investigations, the LPC-GP masonry products were found to satisfy the masonry requirements as per IS 1905-1987 [2]. In the present work an attempt has been made to assess the embodied energy (EE) and cost of LPC-GP masonry units based on process method and presented in Table 1 [3–7]. Table 2 shows the details of the LPC-GP masonry units [8].

A real-life case study has been selected to carry out the comparative study of different masonry system on embodied energy and cost. An urban residential building in Bengaluru, comprising of ground floor and two upper floors, constructed as a load-bearing masonry building using stabilised mud blocks [SMB] as masonry units (Fig. 1) has been chosen. The plan of the building is shown in Fig. 2. The walls are 230 mm and 100 mm thick for the load-bearing and partition walls respectively. Table 3 lists the general details of the building. The selection criteria for the particular case study was based on the fact that the block sizes and the masonry finishing of SMB closely matches with table-moulded bricks and LPC-GP masonry [1].

The quantity of material consumed in the construction of the building was estimated to assess the embodied energy and cost of materials of the building—component wise and material wise. It is found that the embodied energy of the building is 1.62 GJ/m^2 and the cost of materials for construction is ₹7302 per m^2 [1]. To work out the different possible combinations that are possible with use of alternatives,

Table 1 EE and cost computation of LPC-GP brick (unit size: 230 × 105 × 75 mm)

S. no	Composition	Specification	Quantity (kg)	Material cost (INR/kg)	EE coefficient (MJ/kg)	Total EE/unit (MJ)	Total cost/unit (₹)
1	Soil	Soil procured from a distance of 1 km	1.04	1.00	0.005	0.0052	1.04
2	CSA	Procured locally	1.04	0.96	0.106	0.11	0.99
3	LPC	LPC produced in a custom-made kiln	0.26	6.00	3.43	0.89	1.56
4	GGBS	Procured locally	0.26	3.00	1.60	0.42	0.78
5	NaOH	Procured locally	0.03	40.0	20.5	0.62	1.20
6	Na_2SiO_3	Procured locally	0.50	25.0	5.36	2.68	12.5
						4.72	18.08

Table 2 General details of LPC-GP brick

S. no.	Properties	Description	Remark
1	Composition	LPC, GGBS, soil, crushed stone aggregates [CSA], alkaline activators [NaOH, Na_2SiO_3]	[8]
2	Dimension	[230 × 105 × 75] mm	Typical table-moulded brick size, made by adobe process
3	Mass	3.2 kg	
4	Density	1767 kg/m^3	
5	Compressive strength	13 MPa	
6	Embodied energy	4.72 MJ/unit	
7	Cost	₹18.0/unit	1 US dollar = ₹70
8	Mortar	LPC-GP based	Same as LPC-GP brick composition
9	Surface finish	Perfect pointed edges and smooth surface	Does not require plastering
10	Post-construction care	Ambient curing	

Fig. 1 Case study building

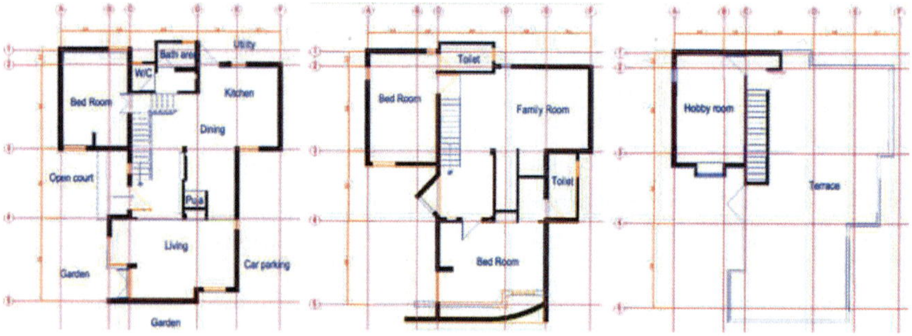

Fig. 2 Plan for the case study building

Table 3 General details of case study building

Plot area	175 m^2
Total built-up area	207 m^2
Structural system	Load-bearing masonry
Type of materials used	
Masonry	Stabilised mud block (7% cement); 6.0 MPa
Flooring	Granite flooring
Wall finishing	No plastering, no painting
Wood work	Teak wood and processed wood
Special features	Jack arch roof system is provided for the ground and first floor. There is a masonry vault constructed in the second floor

the existing building plan was re-worked without changing the structural configuration and building elements. This throws light on the range of embodied energy and cost values that would have been expressed when different masonry units were chosen. Table 4 shows the details of the different masonry units considered and the computations with respect to case study.

The range of values obtained for EE [1.62–2.54 GJ/m^2] and cost [₹6612–9222 per m^2] computations highlight the influence of choice of masonry unit and mortar combination along with the wall finishing required. It turns out that LPC-GP masonry is expensive and high in EE. The contributors for high cost and energy is obviously due to the use of alkaline solution. But these are also accompanied by many benefits. One major benefit that needs mention is the high compressive strength (13 MPa). Listed below are the advantages of utilising these blocks which overweigh their high

Table 4 EE and cost of case study building with alternate masonry systems

S. no.	Type of block	Size (mm)	Mortar type, consumption (%)	Finishing required	EE (GJ/m^2)	Cost (₹/m^2)
1	Autoclaved aerated concrete block [AAC]	600 × 200 × 200	Cement based, 5.23	Plastering and painting	2.54	7638
2	Solid concrete block [SCB]	400 × 200 × 200	Cement based, 7.89	Plastering and painting	1.88	6980
3	Engineered hollow concrete block—7% cement [EHCB]	400 × 200 × 200	Cement based, 5.23	Only painting	1.72	6612
4	Table moulded brick [TMB]	230 × 105 × 75	Cement based, 20.33	Plastering and painting	2.2	7280
5	Stabilised mud block—8% cement [SMB]	230 × 190 × 100	Soil–cement based, 18.77	No plastering and painting	1.62	7302
6	Hollow clay block [HCB]	400 × 200 × 200	Cement based, 7.89	Plastering and painting	1.98	6979
7	LPC-GP Brick	230 × 105 × 75	LPC-GP based, 9.73	No plastering and painting	2.38	9222

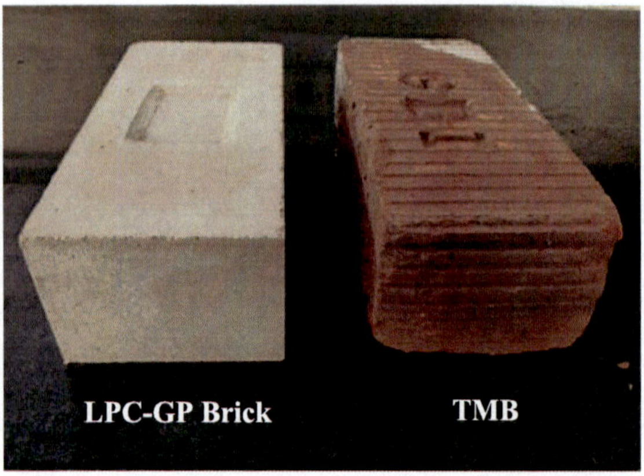

Fig. 3 Typical LPC-GP brick and TMB

EE and cost consumption. Figure 3 shows a typical LPC-GP brick in comparison to table-moulded brick.

- Relatively high compressive strength compared to other available masonry options.
- Homogeneity in construction due to identical composition of masonry unit and mortar, resulting in better masonry efficiency.
- Low mortar consumption due to thin mortar joints.
- Assured masonry quality due to elimination of curing.
- Plastering not required due to excellent surface finish.
- Bricks can be produced with locally available materials and without industrial set-up.
- Produced by 'adobe' process, with very low-capital investment.

It would be instructive to look at the EE and cost when the relatively high compressive strength product is exploited by reducing wall thickness. When compared with TMB (6 MPa) which are of the same dimensions the compressive strength obtained is an order of 53% higher. Also, as per IS 1905-1987 (2002) the required compressive strength for masonry units is in the range of 3.5–10 MPa. Based on these considerations the brick dimensions can be further reduced without compromising the required strength value. This reduction will also lead to reduction in EE and cost of the masonry along with increased carpet area for the building. The modified brick dimension is [150 × 105 × 75] mm. The same building can be constructed with all load bearing walls as 150 mm and partition walls as 105 mm. The EE and cost of the considered case study with LPC-GP modified brick was computed. It is found that a reduction of 13% in EE and 14% in cost can be achieved with respect to the LPC-GP bricks of 230 mm configuration. Figure 4 shows the computed values for the case study with considered masonry units.

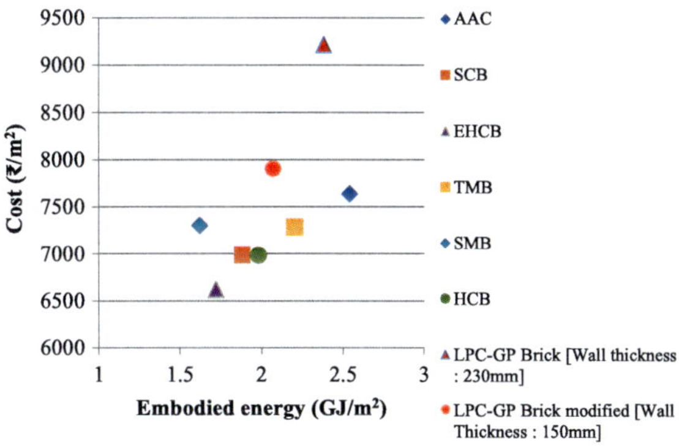

Fig. 4 Range of EE and cost for the case study building with alternate masonry systems

3 Conclusions

From the experimental investigations and computations of embodied energy and cost LPC-GP masonry can be considered to be a promising alternative to the conventional cement-based products. The ease of production of bricks with simple moulds using adobe process, without any machineries and the high compressive strength of the brick makes it best suited for load bearing construction up to 4 storeys. Though the alkaline activators influence the EE and cost of the brick, it can be reduced without compromising the strength by modifying its dimensions.

References

1. Varsha, B.N.: Life cycle energy and cost of urban residential building typologies. Ph.D. thesis, Department of Civil Engineering, B.M.S College of Engineering, Bengaluru, India (2019)
2. IS: 1905-1987: Indian Standard Code of Practice for Structural Use of Unreinforced Masonry. Bureau of Indian Standards, New Delhi, India (2002)
3. Geoff, H., Craig, J.: Inventory of Carbon & Energy (ICE). University of Bath, UK (2008)
4. Jyothi, T.K., Varsha, B.N., Raghunath, S., Jagadish, K.S.: Embodied energy & cost issues of tank-bed-lime based geopolymer adobes. Open J. Energy Effic. **6**, 128–139 (2017). https://doi.org/10.4236/ojee.2017.63010
5. Tempest, B., Sanusi, O., Gergely, J., Ogunro, V., Weggel, D.: Compressive strength and embodied energy optimization of fly ash-based geopolymer concrete. In: World of Coal Ash Conference, Lexington, KY, USA (2009)
6. Fawer, M., Concannon, M., Rieber, W.: Life cycle inventories for the production of sodium silicates. Int. J. Life Cycle Assess. **4**, 207–212 (1999). https://doi.org/10.1007/BF02979498
7. Government of Karnataka: Schedule of Rates for Bengaluru Urban Region 2018–2019
8. Jitha, P.T., Kumar, B.S., Raghunath, S.: Studies on strength development of geopolymer stabilised soil-LPC (lime-pozzolana-cement) mortars. In: Reddy, B., Mani, M., Walker, P. (eds.) Earthen Dwellings and Structures. Springer Transactions in Civil and Environmental Engineering. Springer, Singapore (2019). https://doi.org/10.1007/978-981-13-5883-8_19

Assessing Efficiency of Protective Treatment Materials for Brick Structures

Swathy Manohar and Manu Santhanam

Abstract Protecting the existing brick masonry structures efficiently is a part of sustainable development in the built environment. The effective choice of compatible products in such restoration process as described in this paper reduces the chances of future frequent interventions due to earlier improper repair. In this study, two types of most commonly adopted water-repellant coatings were evaluated for their essential properties like hydrophobicity, breathability, penetration depth and salt weathering resistance in bricks of different inherent properties. Two types of bricks with different pore structure were treated with an acrylic and a silicone-based water-repellents. The breathability and water-absorption of the coatings were measured. Silicone-based treatment was found better with the fundamental requirements—being more water-impervious. However, on studying the performance in accelerated salt crystallization tests, it was found that silicone-based coatings had an adverse effect in the performance of the system, whereas acrylic coatings could improve the salt-resistance. The paper outlines a methodology to be adopted on choosing a protective treatment for a brick structure from the experimental results for obtaining the maximum performance. The methodology follows standard tests for measuring water-repellency, succeeded by the evaluation of breathability and salt weathering resistance of the coating.

Keywords Brick masonry · Protective treatment · Vapour permeability · Salt crystallization

1 Introduction

Choice of the right protective treatment is of paramount importance in conservation of historic structures. The protective treatment should simultaneously ensure optimal, economic and sustainable protection, and also provide protection from any possible aesthetic or physical deterioration of the structure. An ideal surface coating

S. Manohar (✉) · M. Santhanam
Indian Institute of Technology Madras, Chennai 600036, India
e-mail: swathymanohar@yahoo.com

D. K. Ashish et al. (eds.), *3rd International Conference on Innovative Technologies for Clean and Sustainable Development*, RILEM Bookseries 29,
https://doi.org/10.1007/978-3-030-51485-3_2

for stone/brick historic masonry should be oil repellent, transparent, inert towards substrate, durable, chemical and photochemical stability, easy and safe to handle and removable and should be optically stable in order to preserve original stone colour. The coating should be non-hydrophilic and waterproof [1–3]. Nevertheless, the coating layer should allow breathability or water vapour transmission to prevent stress cracking from inside due to the water from rain or capillary rise.

There are several studies that have been conducted for evaluating the performance of different coatings on different materials [2–4]. Most studies show that silicone-based and acrylic-based are the two most common and significant water-repellant coatings which are adapted and effective with masonry structures. Silicone-based coatings had been reported to be better than acrylic siloxane-based. The values of capillary absorption and surface porosity were relatively higher for acrylic-siloxane coated samples [2]. However, better parameters at these tests do not ensure better durability performance. Microscopic studies showed that acrylic siloxane-based coatings had exhibited a cracking tendency, when exposed to salt solutions. Their poor behaviour observed may be attributed to this cracking tendency [2, 4]. But the choice of a particular protective treatment cannot be standardized because the performance or effectiveness of that particular protective treatment need not be depending only on its properties, but also much on the characteristics of the substrate material and the exposure conditions [5].

The current paper shows a systematic methodology of basic tests to look at the most required parameters on selecting the proper water repellent treatment among two common treatments—(one is a silicone-based, and another is an acrylic-siloxane based), for different quality bricks exposed in a saline environment.

2 Materials and Methodology

2.1 Materials

Two types of commercially available fresh brick samples (C1 and C2) were collected from different areas in Chennai, Tamil Nadu, which were of extremely different characteristics expected to show differences in durability performance. C2 was a wire-cut, machine-made brick fired at a higher temperature (>1200 °C) in a modern kiln with a dark red-brown colour whereas C1 was non-wire cut brick fired at 900 °C with a reddish-orange colour. The visual properties and detailed characteristics of the samples are shown in Tables 1 and 2, respectively. Bulk density was measured by Helium gas pycnometer and porosity was measured by Mercury Intrusion Porosimetry (MIP).

Two different types of commercially available and most commonly used water repellent coatings were used in the study, which had different chemical compositions and actions. One was an acrylic siloxane-based water repellent (denoted as 'AS'), and other was a silicone-based water repellent (denoted as 'SC' throughout the paper).

Table 1 Visual properties of samples

Sample	Dimensions (mm)	Colour	Image
C1	220 × 110 × 50	Reddish-orange	
C2	220 × 110 × 50	Dark red-brown	

Table 2 Characteristics of samples

Characteristics	C1	C2
Compressive strength (MPa)	8	12
Bulk density (g/cc)	1.65	2.1
Water absorption (%)	18	10
UPV (m/s)	1400	1950
Porosity (%)	49	31

Sample C2 coated with the silicone-based coating is named as C2-SC, and C2 coated with the acrylic siloxane-based coating is named as C2-AS. The uncoated sample of the set was named as C2 itself.

2.2 *Methods*

1. Water-drop absorption test (RILEM Test Method—Test No. II.8b)

In water-drop test, 1 mL of water was dropped from a height of 1 cm on uncoated and coated samples, dried to a constant mass at 60 °C. To consider the effects of evaporation, an identical drop was placed on a glass surface too. Amount of water absorbed in 24 h in each case was compared and reported.

2. Karstens' pipe test (RILEM Test Method—Test No. II.4, 1980)

In RILEM test for water absorption, Karstens' tubes (graduated glass tubes) were fitted to the samples that have been dried to a constant mass, securely, by using silicone sealant. An equal volume of water was filled in all the tubes, and the amount of water absorbed in regular time intervals was determined and compared.

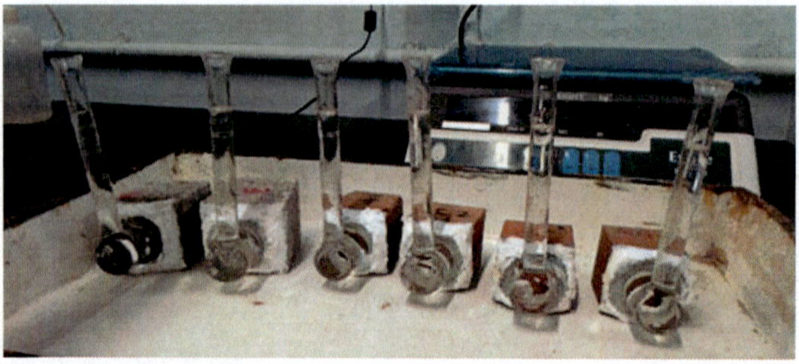

Fig. 1 Illustration of the Karstens' pipe test

3 lithotypes each of uncoated and coated brick samples (both of silicone-based and acrylic-siloxane based treatments) were tested. Figure 1 shows the illustration of the test.

3. Breathability test (RILEM Test No. II.2, 1980)

In this test procedure, rectangular specimens—uncoated and coated on the bottom side with protective treatments were mounted on top of cups having a particular salt (15 ml of saturated potassium nitrate solution in each cup), which constitute a 'measure cell'. The saturated salt solution inside the cup can maintain a constant relative humidity inside the cup. Thus, a gradient is generated by the lower RH lower inside the cup than that of outside, facilitating vapour flow through the specimens. The mass change in the cells was then measured periodically until the rate of change of mass was found constant. From the plots showing mass-changes, the results are then expressed by a parameter called S_d, which is the equivalent air layer thickness (which can provide the same amount of resistance to vapour transmission as that of the specimens).

$$S_d = \frac{\pi^{ar} \cdot A \cdot \Delta P}{G}$$

where π^{ar} is the diffusion coefficient for water vapour in air at atmospheric pressure, A (m^2) is the test area of the specimen, ΔP (Pa) is the vapour pressure differential between the top and bottom surfaces of the specimen, and G (kg/s) is the rate of water vapour flow across the specimen in steady-state conditions.

The test set-up is shown in Fig. 2. The samples were glued to the cups with silicone-sealant ensuring air-tightness. To prevent any further movement of air through the side faces of the samples which were uncoated, additional sealing was provided with silicone-tape as showed in Fig. 2. 6 types of specimens were tested, each had 3 lithotypes.

4. Salt crystallization test (RILEM VI (a)—1994)

In salt weathering test, the samples were subjected to salt crystallization cycles. One cycle is defined as 2 h of immersion in 10% Na_2SO_4 solution, 19 h of drying at 60 °C

Fig. 2 Test setup of vapour permeability test showing the measure cells

and 3 h of cooling at room temperature. The mass of the samples was recorded by the end of every cycle, and the samples were observed for any visible changes. The results are reported as the mass loss by the end of 24 cycles, where the damages and patterns were fully captured.

3 Results and Discussions

3.1 Water-Drop Absorption Test

The uncoated brick samples—both C1 and C2 absorbed the entire drop of water placed on it immediately, i.e. within 2 s. The silicone-coated samples absorb negligible amount of water after 24 h. The acrylic siloxane coated samples had absorbed the entire water after 24 h. But the surface was still wet. The behaviour had no difference between different quality bricks (C1 and C2). Hence, both the considered water repellents were found to have essential liquid impermeability. This is the simplest initial test to screen a good water repellent treatment for brick masonry. Figure 3 shows the water drop over a typical uncoated and coated (with acrylic-siloxane water repellent) captured after 1 h of placing water drop over them.

Fig. 3 Illustration of water drop absorption test

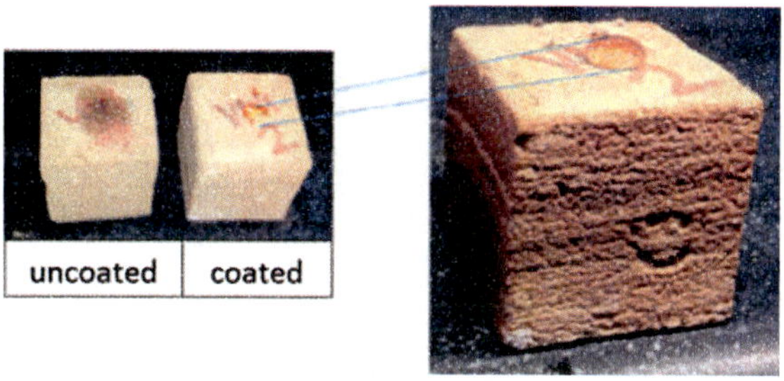

3.2 *Karstens' Pipe Test*

Performances of the coatings were evaluated in Karstens' tube test from the volume of water absorbed by each sample with time. Figure 4 shows the plot between the volume of water absorbed with time elapsed, for C1 and C2 bricks—both uncoated and coated. In C1—which was the highly porous brick, acrylic siloxane-based coating (AS) reduced the rate of water absorption by 1000 times, while the silicone-based coating (SC) for the same brick C1 reduced the rate of absorption by 10,000 times.

In case of C2, which was the wire-cut brick with high density and less porosity, the rate of absorption was already much lesser compared to C1 bricks for uncoated samples. The absorption rate was then reduced roughly by 1000 times by both silicone and acrylic-siloxane coatings. This showed that when the inherent properties of the bricks is not vulnerable to some process (here, because of low density and least porosity and permeability of C2 system, water absorption is controlled by the microstructure of brick itself, unlike the case of C1), then there found no difference

Fig. 4 Karstens' pipe test results showing the volume of absorbed water with time

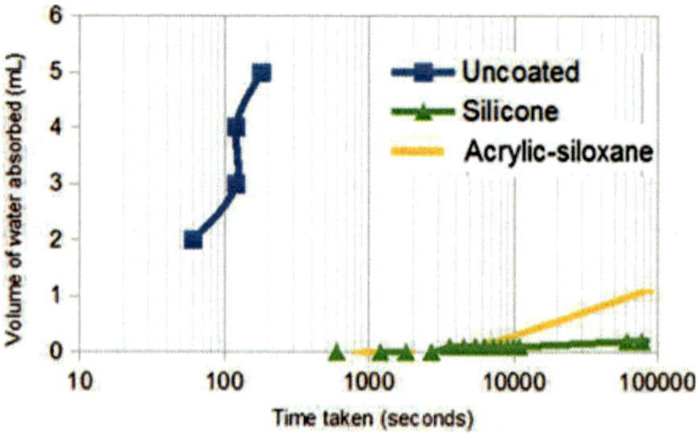

(a) C1

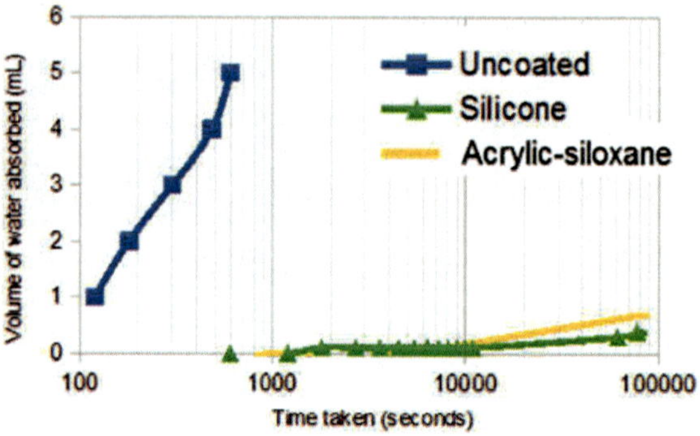

(b) C2

among the performances of the two different coatings. Both water repellents had similar effects on C2.

3.3 Vapour Permeability Tests

The mass of the cell was weighed in every 24 h for 36 days, and the plot is shown in Fig. 5. Equivalent air thickness values were calculated using the plots.

From Fig. 5, it is seen that both SC and AS coatings are providing vapour transmissions through both the systems—C1 and C2. None of the coatings was found to be blocking the movement of vapour through them. S_d values calculated are shown in Table 3. The results show that silicone-based coatings are comparatively providing less breathability than the acrylic-siloxane based treatment, which was not well captured from the graphs.

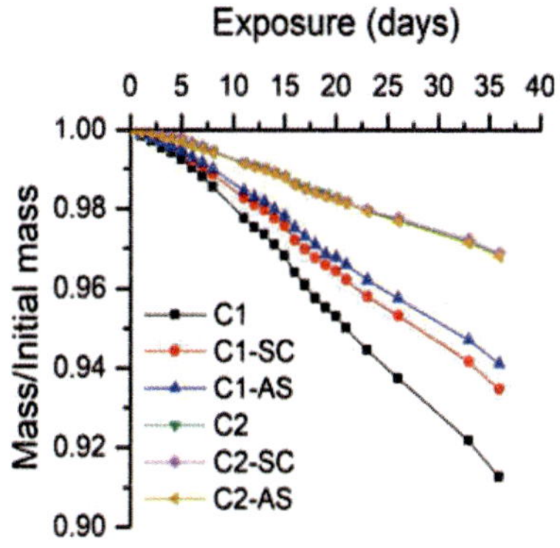

Fig. 5 Mass variation of the measure-cell with time

Table 3 Equivalent air layer thickness values (S_d)

Samples	S_d (m)
C1	0.49
C1-SC	0.67
C1-AS	0.58
C2	1.79
C2-SC	2.25
C2-AS	2.01

3.4 Salt Weathering Tests

The uncoated samples of C2 showed a rough and powdered surface by the end of two cycles. Progressive but predominant mass loss was observed in further cycles. The brick samples coated with silicone-based coatings performed very well initially, with negligible changes in mass. By the end of three cycles, mass deterioration started to show up. For samples C2-SC, deep holes appeared on the bottom of the samples. After six cycles, the brick bulged, spalled and flaked out pieces of roughly 2–3 mm thickness. The brick samples broke into 3–4 pieces at 10–11 cycle for all three samples of C2-SC. C2-AS showed high resistance to salt weathering throughout the cycles. Figures 6 and 7 show the damage pattern observed with C1 and C2 samples respectively for variants of uncoated, silicone-coated (SC) and acrylic-siloxane coated (AS).

The deterioration pattern for C1 was uniform, with progressive smoothening of surface edges, and without any spalling or flaking. Both kinds of specimens coated, which were coated with silicone-based and acrylic-siloxane based (C1-SC and C1-AS respectively) did not show any signs of degradation or mass loss throughout the 15 cycles. Figure 8 shows the mass variation of the samples with the number of cycles. The results clearly show that though the choice of water repellent treatment did not make a difference of brick C1, it was indeed important for the brick C2. For denser brick, choice of improper treatment is destructive than leaving the material uncoated. Acrylic-siloxane treatment could protect denser bricks well in the presence of salt exposure. This could be because of the cracking property of acrylic coatings which was identified in the past, which would help in releasing the internal stresses from pores during salt crystallization [5].

Fig. 6 C1 samples after the salt weathering cycles

Fig. 7 C2 samples after the salt weathering cycles

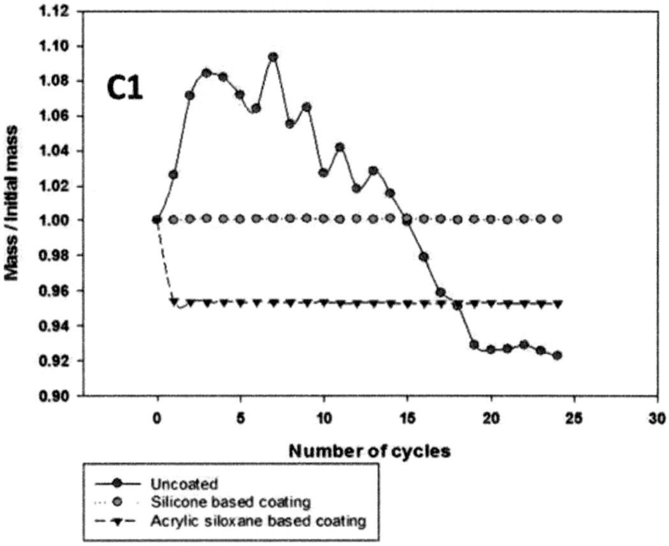

(a) Mass variation of C1 brick

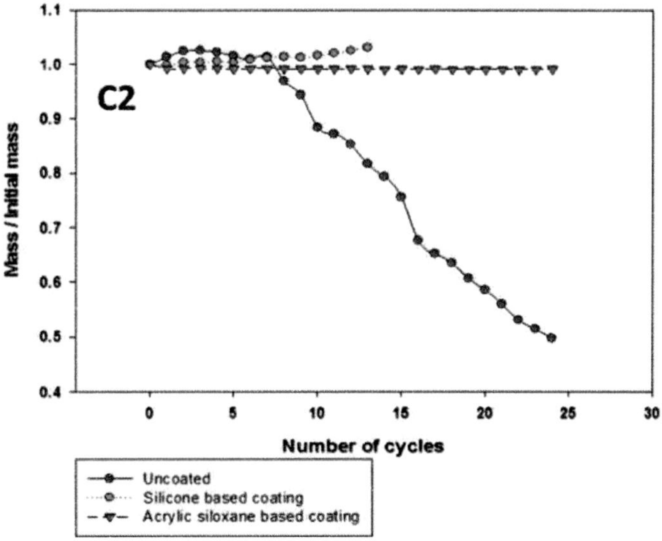

(b) Mass variation of C2 brick

Fig. 8 Mass variation with increasing exposure time for the specimens

4 Conclusions

Basic properties to be ensured on selecting a protective treatment for a brick masonry are water repellency, vapour permeability and salt weathering resistance if the system is expected to perform in a salt exposed location. From water drop absorption and Karstens' pipe test, both acrylic-siloxane and silicone-based treatments were found to have water repellent property, but silicone-based treatment was highly water repellent than acrylic-siloxane. Breathability (vapour permeability) test evaluates the ability of the treatment to allow the movement of water vapour through them. Both the treatments were observed with similar vapour permeability. Measurement of the fundamental properties of water-repellents provides the impression that silicone-based treatments are better than acrylic-based treatments. But the performance should be evaluated based on particular exposure conditions. If the treatment is intended for use in an atmosphere which is exposed to salt, besides the water repellency and breathability, salt weathering resistance has to be tested. Accelerated artificial salt crystallization test with sodium sulfate showed that for bricks which are inherently not resistant towards salt weathering should be protected with treatment, which has good salt weathering resistance. A treatment showing high water repellency and breathability need not perform well in the presence of salt attack. In the current study, acrylic-based treatments were found to have remarkable resistance towards salt crystallization, possibly because of its property to crack itself during the exposure. Cracks over the film would help in releasing internal stresses being generated during salt crystallization. This choice is significant for materials with inherently very low salt weathering resistance (i.e., materials with more micropores) [6].

Hence, water repellent treatments should be tested for basic properties and chose accordingly as per the requirement/exposure conditions. Also, the properties of the substrate (here brick) possess the most important role in allowing the protective treatment to work on its actual efficiency, which was seen from the vapour permeability tests and salt weathering tests.

References

1. Tsakalof, A., Manoudis, P., Karapanagiotis, I.: Assessment of synthetic polymeric coatings for the protection and preservation of stone monuments. J. Cult. Herit. **8**, 69–72 (2007). https://doi.org/10.1016/j.culher.2006.06.007
2. Kronlund, D., Lindén, M., Smått, J.H.: A polydimethylsiloxane coating to minimize weathering effects on granite. Constr. Build. Mater. **124**, 1051–1058 (2016). https://doi.org/10.1016/j.conbuildmat.2016.08.146
3. Alvarez de Buergo, M., Ballester, R.F.G.: Basic methodology for the assessment and selection of water-repellent treatments applied on carbonatic materials. Prog. Org. Coat. **43**, 258–266 (2001)

4. Stefanidou, M., Karozou, A.: Testing the effectiveness of protective coatings on traditional bricks. Constr. Build. Mater. **111**, 482–487 (2016). https://doi.org/10.1016/j.conbuildmat.2016.02.114
5. Manohar, S., Santhanam, M., Chockalingam, N.: Performance and microstructure of bricks with protective coatings subjected to salt weathering. Constr. Build. Mater. **226**, 94–105 (2019). https://doi.org/10.1016/j.conbuildmat.2019.07.180
6. Manohar, S., Bala, K., Santhanam, M., Menon, A.: Characteristics and deterioration mechanisms in coral stones used in a historical monument in a saline environment. Constr. Build. Mat. **241**, 118102 (2020)

Effect of Adhesives on the Structural Performance of Timber-Framed Joints for Sustainable Civil Engineering Construction

S. A. Aejaz, A. R. Dar, and J. A. Bhat

Abstract The need for infrastructural growth in developing countries has increased exponentially consequent to the rapid rise in population. This in turn, has increased the demand for prominent building materials like cement, which is mainly used in concrete construction. With the global cement production crossing 4.1 billion metric tons which releases about 7% of total carbon dioxide to atmosphere, it has become indispensable to promote such a construction material that is sustainable and also, at the same time cost effective. Timber provides a viable alternative in this regard, due to its eco-friendly, energy efficient and light weight qualities, thereby, making it ideal as a building material. Timber framed structures have been used in seismically susceptible regions since times, with connections being the most vulnerable elements that govern the integrity and safety of such systems under diverse loading actions. Although, the behavior of timber framed structures has been studied by many researchers by testing full-scale-connections with the aim to establish consistent design provisions on the same, however, much emphasis in this approach has focused only on a particular connection configuration, with no attempt being made for optimizing the joints through innovative detailing. To address this concern, a comparative study to evaluate the performance of various full-scale timber frame Nailed connections (Bridled Tenon, Cross Halved, Dovetail Halved and Mortise Tenon) supplemented by adhesive with respect to Nailed-Only counterparts, under tensile loading, has been investigated in this paper. The test results have been used to calculate stiffness, load capacity and ductility in both the connection forms (with and without adhesion) which in turn have been compared to other joint profiles along with the observed failure modes. The experimental outcomes showed incorporation of adhesive to be an effective and economical technique in significantly enhancing the performance of different connections considered in this study. Thus, this research is novel as it attempts not only to explore and improve the behavior of various timber framed joints in a logically sequential manner, but also, provides a sustainable alternative to the contemporary building materials, thereby, making it eco-friendly in its approach.

S. A. Aejaz · A. R. Dar (✉) · J. A. Bhat
Department of Civil Engineering, NIT Srinagar, Srinagar, Jammu & Kashmir 190006, India
e-mail: abdulrashid@nitsri.net

D. K. Ashish et al. (eds.), *3rd International Conference on Innovative Technologies for Clean and Sustainable Development*, RILEM Bookseries 29,
https://doi.org/10.1007/978-3-030-51485-3_3

Keywords Timber frame · Tensile capacity · Joint configuration · Failure modes

1 Introduction

The rapid rise in population has exponentially increased the need for infrastructural growth in developing countries. This has resulted in the global demand for cement crossing 4.1 billion metric tons which is expected to rise by 8–10% in the upcoming years. Since one ton of cement production releases about one ton of carbon dioxide into the atmosphere, which by 2020 is estimated to increase by 50% from the current level, therefore, this has imposed pressure on the construction industry to promote such a material that is sustainable and cost-effective. Timber provides a viable alternative due to its eco-friendly, energy efficient and light weight qualities in contrast to contemporary building materials which are also by and large environmentally hazardous. Timber framed structures have been used in seismically susceptible regions since times, with connections being the most vulnerable elements that govern the integrity and safety of such systems under diverse loading actions. The concentration of stresses inherent to material discontinuity makes connections extremely damage susceptible thereby necessitating utmost concern to be divested in their design.

Tensile capacity being an important property of connections in a timber frame under pull-out/uplifting actions has been investigated by many researchers for developing design guidelines and numerical models for strength prediction. However, attention in this area has been confined to pegged mortise-tenon connections wherein several parameters have been studied till date. The first attempt to study the tensile behavior of pegged mortise tenon joints and frames experimentally as well as analytically was made by Brungraber [1] in which the strength and stiffness was seen to enhance as the peg diameter increased. In another study, Schmidt and Daniels [2] identified peg failure to be the most ductile mode when full scale mortise tenon joints were loaded in tension. Similar observations with peg diameter and shape having a significant influence on the failure mode were presented by Shanks et al. [3].

The effect of joint fitness on the performance of pegged mortise tenon connections was explored by Bulleit et al. [4] which revealed that peg suffered less damage in a tightly fit joint compared to a loose connection and that pegged joints can be modelled as pinned connections. Further, the tensile capacity of varying angle pegged mortise tenon connections was evaluated by Walker et al. [5] in which 45° connections exhibited the highest capacity followed by 90° and 67.5° connections respectively with connection stiffness reducing as the angle decreased. This work on angled mortise tenon connections was extended by Judd et al. [6] by developing a theoretical model for evaluating the tensile strength.

Several researchers have also investigated the tensile behavior of mortise tenon joints assembled by fasteners of different materials instead of the conventional wooden pegs. Hassan et al. [7] reported that GFRP dowelled mortise-tenon connections can be considered as a viable alternative to steel and wood dowelled connections because of the comparable performance of former in contrast to latter under tensile action. Similarly, a significant enhancement in the withdrawal capacity of round mortise tenon joints employing steel cross pins instead of wooden pins was observed by Eckelman et al. [8] provided perpendicular to grain failure in the mortised member was prevented by reinforcement. In another study, splitting strength of the mortised member in the cross-grain direction considering the variation of edge distance and timber species by employing steel bolts instead of the wooden pegs was evaluated by Hindman and Milad [9].

In contrast to pegged timber connections little research has been done on the nailed connections in the recent past despite extensive use of nails in timber construction. The load capacity prediction model for dowelled timber connections loaded normal to the fastener axis was developed by Johansen [10] which considered embedded dowels as beam elements. Moller [11] applied this model to symmetrical and unsymmetrical joints in single and double shear, which was later experimentally verified by Siimes et al. [12], Mack [13] and Aune [14]. The application of Johansen's model to nailed joint forms of different embedment strength was developed by Aune and Patton-Mallory [15].

Though a lot of research has been done on pegged mortise joints and dowelled timber connections in the past but none of the work has studied various nailed joint configurations with an effort to refine the joint detailing by adhesive addition to compare the performance for optimization. The primary purpose of the present research work is to conduct a comparative study for experimentally evaluating the tensile behavior of various nailed connection configurations supplemented by adhesive with respect to nailed-only counterparts, in order to ascertain the influence of adhesive addition on the connection performance. Thus, this research is innovative as it attempts not only to explore and improve the behavior of various timber framed joints in a logically sequential manner, but also, provides a sustainable alternative to the environmentally hazardous present-day constructions, thereby, making it eco-friendly in its approach.

2 Experimental Investigation

2.1 Test Specimens

In this study, seasoned Fir, common in traditional timber construction in Jammu & Kashmir, of sectional dimensions 70 mm × 90 mm (2.75″ × 3.5″) has been used for making full scale timber frame T-shaped joints. The timber properties evaluated as per IS-1708 [16] specifications are given in Table 1.

Regular 62.5 mm (2.5″) iron nails and polyvinyl acetate which is a synthetic resin adhesive have been used to construct 12 specimens (3 each of Bridled-Tenon, Cross-Halved, Dovetail-Halved and Mortise-Tenon configurations). Additionally, 12 specimens (3 each) of the said configurations have also been constructed keeping all

Table 1 Material properties of timber

Property	Value
Specific gravity	0.48
Moisture content	12.56%
Modulus of elasticity (MPa)	11,218.41
Poisson's ratio	0.32
Tensile strength parallel to grain (MPa)	128.37
Tensile strength perpendicular to grain (MPa)	2.82

the parameters same except the use of adhesive in order to ascertain its influence on the joint behavior.

2.2 *Testing Programme*

The samples from the four joint types were tested on a Universal Testing Machine in displacement-controlled mode at a constant rate of 2 mm/min for evaluating the tensile capacity. The testing protocol involved testing Nailed-Only connections followed by Adhesive cum Nailed connections from each configuration respectively. The testing of a typical T-joint on U.T.M. has been shown in Fig. 1.

In present research, the four T-connection configurations selected were based upon their common use in timber construction. The similar member dimensions of

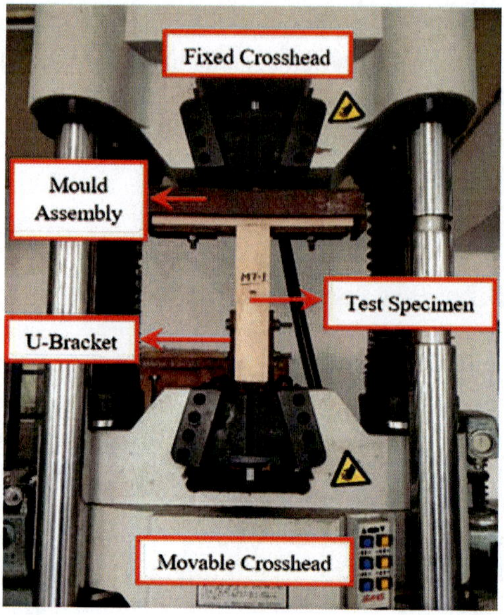

Fig. 1 Typical T-joint on U.T.M.

Fig. 2 Details of the T-joints considered in this research. All dimensions are in mm

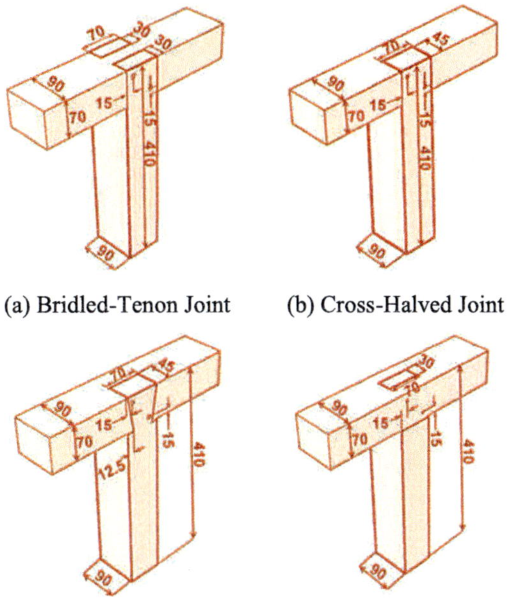

(a) Bridled-Tenon Joint (b) Cross-Halved Joint

(c) Dovetail-Halved Joint (d) Mortise-Tenon Joint

these connections enabled direct comparison, thereby facilitating the performance evaluation of the joint forms (with and without adhesion) individually, in addition to the various connection configurations considered. Thus, the present research is novel for it not only evaluates the structural behavior of Nailed Mortise-Tenon connection on which very little work has been done by the researchers in the past, but it also compares its performance to other configurations along with the refinement in joint detailing by adhesive addition so as to establish quantitatively which connection profile behaves optimally. The details of connections considered in this research are shown in Fig. 2.

For convenience, Nailed-Only connections in this work have been represented by two letters followed by a numeral code wherein letters denote the joint type and the numeral denotes the specimen number whereas Adhesive cum Nailed specimens have been differentiated from the Nailed-Only joints by a (*) symbol.

3 Results and Outcomes

The test results have been presented in the form of Nailed-Only versus Adhesive cum Nailed joints for each joint configuration followed by comparison of various joint types in both forms subsequently. The test results have been used to evaluate stiffness

(K), load capacity (F_{max}) and ductility (μ) of connections in each category. Further, failure modes in various configurations have also been observed and reported.

3.1 Nailed Only Bridled Tenon (BT) Versus Adhesive Cum Nailed Bridled Tenon (BT) Joints*

Both Nailed Only as well as Adhesive cum Nailed Bridled Tenon joints predominantly showed shear bearing failures compounded by complete tear out of the horizontal members in some specimen at large displacements under tensile loading (as shown in Fig. 3). BT* connections resisted very high loads in comparison to BT samples consequent to the largest bound area in adhesion. When loaded, detachment of the binding layer caused BT* to exhibit very sharp load drops after crossing the peak load but the enhancement in load capacity due to adhesive addition being highest when compared to other configurations overshadowed the former effect resulting in high ductility with respect to BT specimen. Additionally, BT* samples showed higher stiffness when referenced to BT. The average values of joint stiffness, peak load and ductility have been reported in Table 2. The comparison of the average values showed that load capacity, stiffness and ductility enhanced by a factor of 2.43, 2.19 and 1.67 respectively by using adhesive in this connection configuration.

Fig. 3 Complete member tear out failure in bridled tenon joint

Table 2 Average test results of BT and BT*

Joint type	K (kN/mm)	F_{max} (kN)	μ
BT average	1.55	8.39	8.13
BT* average	3.40	20.38	13.61

Fig. 4 Shear bearing failure in cross halved joint

Table 3 Average test results of CH and CH*

Joint type	K (kN/mm)	F_{max} (kN)	μ
CH average	0.94	7.74	16.65
CH* average	3.48	14.26	8.93

3.2 Nailed Only Cross Halved (CH) Versus Adhesive Cum Nailed Cross Halved (CH) Joints*

Upon loading, Nailed Only and Adhesive cum Nailed Cross Halved joints primarily demonstrated bearing failure modes (as shown in Fig. 4) in which nails due to higher bearing strength crushed wood along with few samples developing hairline cracks in both horizontal and vertical joint members. CH* connections showed significantly higher stiffness and load capacity with respect to CH connections wherein the enhancement in stiffness due to adhesive incorporation was the largest when compared to other configurations. The CH* connections also exhibited abrupt load drops after crossing the peak value in contrast to CH specimens wherein the load decreased very gradually thereby making ductility values of CH higher in comparison to CH*. Thus, the addition of adhesive was seen to impose adverse effects on the ductility in this joint type. The average connection stiffness, load capacity and the ductility evaluated have been reported in Table 3. On the mean scale, stiffness and load capacity improved by a factor of 3.70 and 1.84 respectively with the use of adhesive in this connection type, however ductility reduced by 46% when compared to CH.

3.3 Nailed Only Dovetail Halved (DH) Versus Adhesive Cum Nailed Dovetail Halved (DH) Joints*

Nailed Only and Adhesive cum Nailed Dovetail Halved specimens exhibited cross grain splitting in the horizontal T-joint members (as shown in Fig. 5) in conjunction

Fig. 5 Cross grain splitting in dovetail halved joint

Table 4 Average test results of DH and DH*

Joint type	K (kN/mm)	F_{max} (kN)	μ
DH average	0.99	11.93	9.83
DH* average	2.98	13.78	15.86

with the shear bearing failure mode consequent to nail induced compression. Despite the sudden load drop post peak point in DH*, considerable improvements in ductility were reported due to nearly plastic post-yield behavior when compared to DH joints. The cross-grain splitting in the horizontal joint members attributed to tenon-taper in this configuration was more pronounced in DH* connections than DH specimens. The average values of stiffness, peak load and ductility of DH and DH*connections are shown in Table 4 respectively. The use of adhesive did not contribute to a notable increment in the load capacity of DH* connections although stiffness showed a substantial increment with respect to DH specimens. The stiffness, ductility and load capacity of DH* exceeded DH by a factor of 3.01, 1.61 and 1.16 respectively, when the average values were considered.

3.4 Nailed Only Mortise Tenon (MT) Versus Adhesive Cum Nailed Mortise Tenon (MT*) Joints

Both Nailed Only and Adhesive cum Nailed Mortise Tenon joints failed in bearing with the eventual tenon tear out in shear besides mortise splitting (as shown in Fig. 6) in a few specimen at large values of displacement. The average joint stiffness, load capacity and ductility are presented in Table 5. In contrast to other Adhesive cum Nailed configurations, incorporation of adhesive in this connection resulted in minimal enhancement in terms of stiffness, load capacity and ductility because of the confinement being imposed by the mortise sidewalls on the tenon thereby resulting in almost similar ductility and load capacity with a marginal improvement in stiffness. On comparing the average curves stiffness, load capacity and ductility rose by a factor of 2.09, 1.12 and 1.01 respectively in MT* when referenced to MT.

Fig. 6 Mortise splitting failure in mortise tenon joint

Table 5 Average test results of MT and MT*

Joint type	K (kN/mm)	F_{max} (kN)	μ
MT average	1.15	10.92	14.58
MT* average	2.41	12.24	14.75

3.5 *Comparison of T-Shaped Nailed-Only Joints*

The mean test results of various Nailed Only joints considered in this study revealed that BT demonstrated the highest stiffness with limited load resistance and minimal ductility. The maximum ductility was shown by CH consequent to very gradual load drop despite exhibiting the lowest stiffness and load capacity in the group. DH noted with the largest load capacity showed lesser stiffness and ductility values. The performance of MT seen to be comparable to DH in terms of load capacity was reported with remarkable ductility and a relatively stiffer response when compared to the latter. Therefore, the optimum performance in the Nailed-Only category was shown by MT. The average stiffness, peak load and ductility of various Nailed Only joints are presented in Figs. 7, 9 and 11 respectively, for comparison.

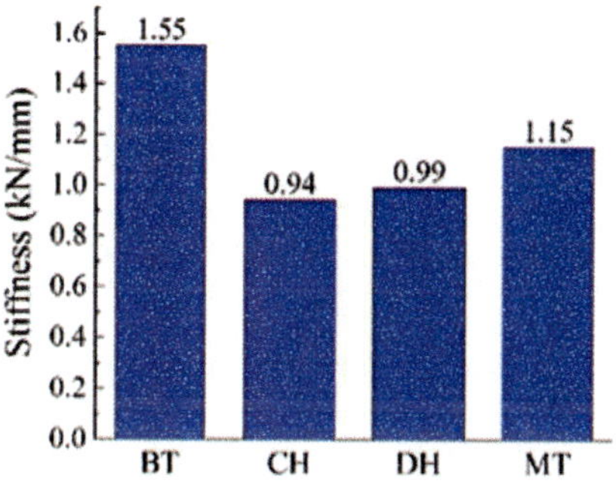

Fig. 7 Comparison of stiffness of various nailed only joints

3.6 Comparison of T-Shaped Adhesive Cum Nailed Joints

The incorporation of adhesive improvised the performance in terms of stiffness, load capacity and ductility in every configuration when referenced to its Nailed-Only counterpart except the ductility in CH* which was noted with adverse effects. It was observed that CH* showed the least ductile behavior with the highest stiffness and appreciable load resistance. Although, adhesive addition led to the highest enhancement in the ductility of BT* when compared to other configurations, yet it did not demonstrate the highest group ductility despite largest load capacity and significant stiffness. The lowest stiffness and load capacity seen in MT* was reported with considerable ductility. DH* exhibited the most ductile behavior with a much stiffer response and better load resistance when compared to MT*. With considerable improvement in stiffness and ductility due to binding layer of adhesion, the optimal performance in this category was shown by DH*. The average stiffness, peak load and ductility of various Adhesive cum Nailed joints are presented in Figs. 8, 10 and 12 respectively, along with the property improvement factors due to adhesive addition in various joints in Table 6, for comparison.

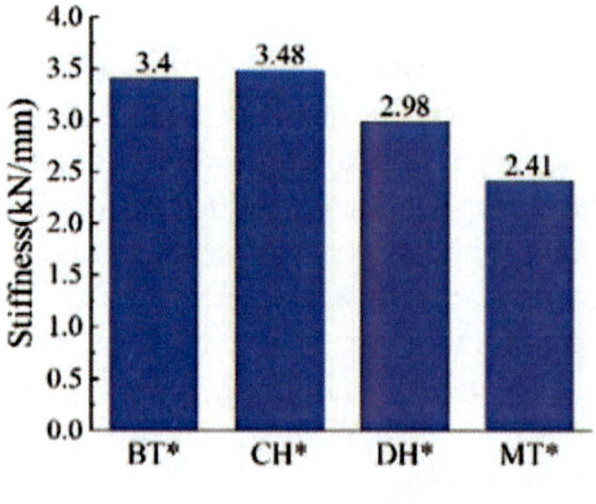

Fig. 8 Comparison of stiffness of various adhesive cum nailed joints

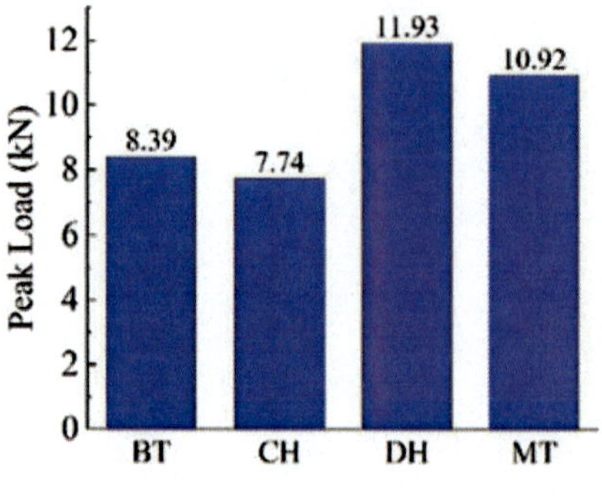

Fig. 9 Comparison of peak load of various nailed only joints

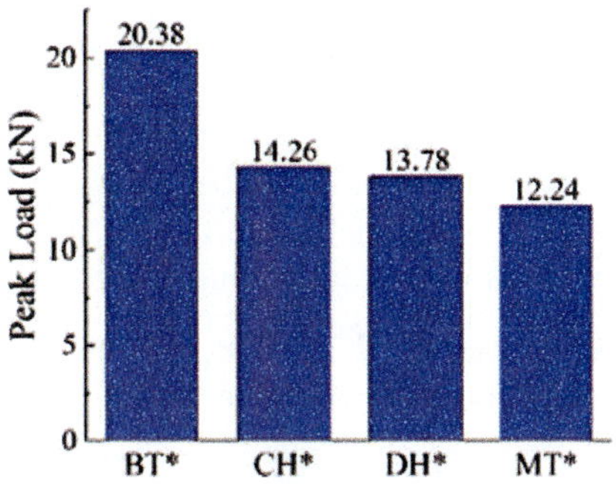

Fig. 10 Comparison of peak load of various adhesive cum nailed joints

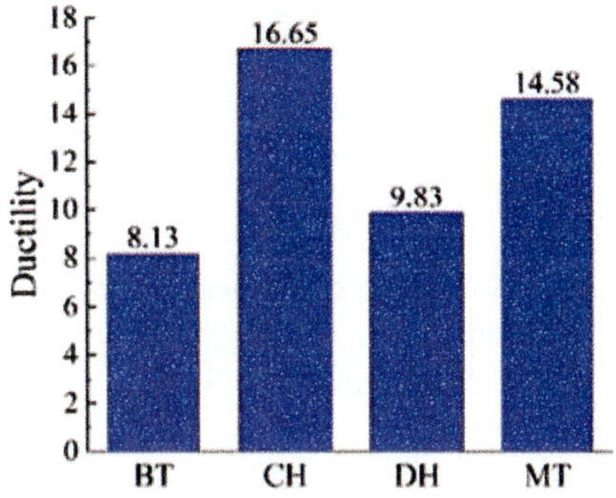

Fig. 11 Comparison of ductility of various nailed only joints

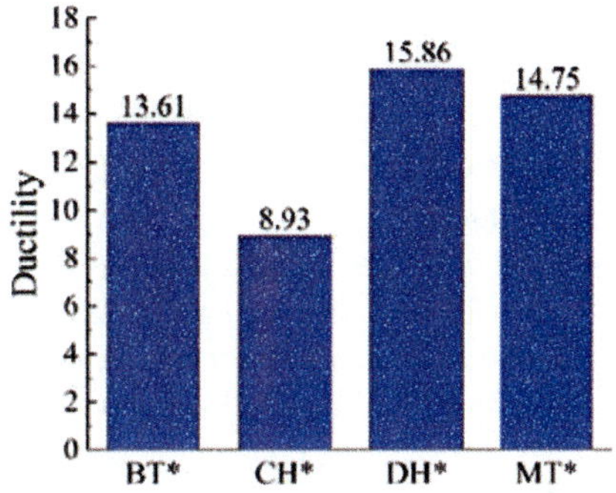

Fig. 12 Comparison of ductility of various adhesive cum nailed joints

Table 6 Property improvement factors due to adhesive addition in various joints

Joint type	K*/K	F*/F	μ*/μ
Bridled tenon	2.19	2.43	1.67
Cross halved	3.70	1.84	0.54
Dovetail halved	3.01	1.16	1.61
Mortise tenon	2.09	1.12	1.01

4 Conclusions

This study involved a comparative study for experimentally evaluating the tensile behavior of various timber joints so as to promote sustainable civil engineering structures, in contrast to the contemporary constructions, which are by and large environmentally hazardous. The following conclusions were drawn from this study:

1. Under tensile loading, all the Nailed-Only as well as Adhesive cum Nailed connections were seen to fail primarily in a shear bearing mode with eventual crushing consequent to the compression induced by pressing of nails against timber.
2. Adhesive cum Nailed connections exhibited abrupt load drops post peak point as a characteristic feature in comparison to Nailed-only connections wherein the load dropped gradually. This sharp load drop was attributable to the detachment of binding layer in Adhesive cum Nailed connections when compared to Nailed Only forms.
3. The incorporation of adhesive was observed to significantly improvise the tensile performance of connections in terms of stiffness, peak load and ductility when compared to Nailed-Only counterparts except the ductility in CH* which was noted with adverse effects.
4. The highest stiffness, load capacity and ductility in the Nailed-Only category was shown by Bridled Tenon, Dovetail Halved and Cross Halved connections respectively. Bridled Tenon displayed the least ductile behavior with the Cross Halved configuration demonstrating the lowest stiffness and load capacity in the group. The optimum performance in this category was shown by Mortise Tenon due to its comparable performance to Dovetail and Cross Halved configurations respectively in terms of load capacity and ductility along with a much stiffer response.
5. The optimum performance in the Adhesive cum Nailed category was exhibited by Dovetail Halved configuration due to its most ductile post yield behavior contrary to Bridled-Tenon, despite highest group capacity and significant stiffness of the latter. The ductile behavior of Mortise Tenon was reported with the lowest stiffness and load capacity values with Cross Halved configuration showing the lowest ductility in this category.

A major limitation of the framed-infill constructions employing these joints is that timber required for their making is obtained at the cost of depletion in green cover which has detrimental environmental consequences. Therefore, a proper balance needs to be struck to ensure that this construction technology is adopted sagaciously, so as to achieve sustainable development without any threat to future generations.

References

1. Brungraber, R.L.: Traditional timber joining: a modern analysis. Ph.D. dissertation, Stanford University, California (1985)
2. Schmidt, R.J., Daniels, C.E.: Design Considerations for Mortise and Tenon Connections. Report No. 9702896. USDA, NRI/CGP, Washington, DC and Timber Frame Business Council, New Hampshire, U.S. (1999)
3. Shanks, J.D., Chang, W.S., Komatsu, K.: Experimental study on mechanical performance of all-softwood pegged mortice and tenon connections. Biosyst. Eng. **100**, 562–570 (2008)
4. Bulleit, W.M., Sandberg, L.B., Drewek, M.W., O'Bryant, T.L.: Behavior and modelling of wood-pegged timber frames. J. Struct. Eng. **125**(1), 18707 (1999)
5. Walker, C.R., Fonseca, F.S., Judd, J.P., Thorley, P.R.: Tensile capacity of timber-framed mortise and tenon connections. M.S. dissertation, Brigham Young University, Utah (2008)
6. Judd, J.P., Fonseca, F.S., Walker, C.R., Thorley, P.R.: Tensile strength of varied-angle mortise and tenon connections in timber frames. J. Struct. Eng. **138**(5), 636–644 (2012)
7. Hassan, R., Ibrahim, A., Ahmad, Z.: Experimental performance of mortice and tenon joint strengthened with glass fibre reinforced polymer under tensile load. In: IEEE Symposium on Business, Engineering & Industrial Applications (ISBEIA), Indonesia, Sept 2012
8. Eckelman, C.A., Haviarova, E., Akcay, H.: Exploratory study of the withdrawal resistance of round mortise and tenon joints with steel pipe cross pins. For. Prod. J. **56**(11/12) (2006)
9. Hindman, D., Milad, M.: Splitting strength of mortise members in timber frame joints. J. Mater. Civ. Eng. **28**(12) (2016)
10. Johansen, K.W.: Theory of timber connections. Int. Assoc. Bridge Struct. Eng. **9**, 249–262 (1949)
11. Moller, T.: New Method of Estimating the Bearing Strength of Nailed Wood Connections. Report No. 117. Gothenburg, Sweden (1950)
12. Siimes, F.E., Johanson, P.E., Niskanen, E.: Investigations on the Ultimate Embedding Stress and Nail Holding Power of Finish Pine. The State Institute for Technical Research, Helsinki, Finland (1954)
13. Mack, J.J.: The Strength of Nailed Timber Joints, Tech. Pap. No. 9. CSIRO Division of Forest Products (1960)
14. Aune, P.: The Load Carrying Capacity of Nailed Joints. Calculations and Experiments. The Norwegian Institute of Technology, Norway (1966)
15. Aune, P., Patton-Mallory, M.: Lateral Load-Bearing Capacity of Nailed Joints Based on the Yield Theory: Theoretical Development, Paper No. FPL 469. Forest Products Lab (1986)
16. IS:1708: Methods of Testing Clear Specimens of Timber. Indian Standards Institution, New Delhi, India (2005)

Enhancement of Sub-grade Soil Strength with Additives: Cement and Molasses

Ankit Bansal, Tripta Goyal, and Umesh Sharma

Abstract Increased vehicular traffic has tremendously increased the stresses on roads especially sub-grade. In order to make the roads more sustainable to the higher stresses, the stabilization of sub-grade is required. Therefore, the present study aims to enhance the subgrade soil strength by stabilization using cement and molasses (in combination with the lime). The results of the study propose using either 9% cement or 9% molasses (with 4.5% lime) for stabilizing the poorly graded sand. With the addition of cement, the liquid limit has decreased from 24 to 20%, however, with molasses it decreases to 19%. The maximum dry density of the soil has increased from 1.86 to 1.98 g/cm^3 with the cement and increased to 2 g/cm^3 by the addition of molasses. Moreover, the addition of cement increases the un-soaked CBR value from 1.75 to 3.5% whereas molasses increases the value to 2.83%. Likewise, soaked CBR value has augmented from 1.17 to 2.92% with cement and 2.54% with molasses. Therefore, the authors suggest using molasses for stabilization in case of low volume roads whereas for heavy traffic, cement should be preferred as it yield better results in enhancing the bearing strength of sub-grade as compared to molasses.

Keywords Stabilization · Sub-grade · Molasses · Cement · Soil

1 Introduction

The integral part of the pavement is sub-grade which provides the adequate support for the pavement. The sub-grade should have adequate support or stability under worst climatic and loading conditions. The formation of waves, rutting, and corrugations

A. Bansal (✉) · T. Goyal · U. Sharma
Civil Engineering Department, Punjab Engineering College, Chandigarh, India
e-mail: ankitbansal.phdcivil16@pec.edu.in

T. Goyal
e-mail: triptagoyal@pec.ac.in

U. Sharma
e-mail: umeshsharma@pec.ac.in

D. K. Ashish et al. (eds.), *3rd International Conference on Innovative Technologies for Clean and Sustainable Development*, RILEM Bookseries 29,
https://doi.org/10.1007/978-3-030-51485-3_4

are some of the failures due to the poor sub-grade soil. Sometimes, the edges of the pavement also break due to the instability of the soil. Stability, incompressibility, permanency of strength, good drainage and ease of compaction are the important properties of good sub-grade. The pressure transmitted to the bottom should be good enough to bear them. The top 50 cm of the sub-grade is compacted with optimum moisture content to attain certain density, however, when the required strength is not met, the soil stabilization needs to be carried out.

2 Literature Review

Road stabilization is the way of improving the natural sub-grade and enhancing the strength against the adverse traffic and climatic conditions with the help of admixtures or by compaction and densification of the soil strata [1]. Admixture can be chemical binders, industrial wastes, cement, fly ash and rice husk ash. In laymen terns, Soil Stabilization is a practice of improving the soil properties and augmenting the performance for engineering purpose [2]. Selection of stabilizer for increasing strength of sub-grade mainly depends on the type of soil, construction materials and technology to be adopted [3]. Stabilization improves the engineering performance and makes the soil suitable for construction of pavement. Modification or improvement of a soil can be done in following ways: (a) Stabilization by applying drainage facilities and compaction; (b) Mechanical stabilization by using low cost admixtures like lime, rice husk ash, molasses, cement etc.; (c) Chemical Stabilization [4]. The stabilization by using admixtures only is the most common among all the available techniques.

The problematic soil for any construction activity is black cotton soil. Due to the presence of montmorillonite mineral, it shows swelling of soil and does not provide the adequate bonding stability. So, every soil should have limited amount of clay only and it needs to be stabilized using lime and rice husk ash [5]. Also lime stabilization is suggested for soils in waterlogged areas as lime slows down the pozzolanic reaction [6]. In addition to lime, rice husk ash is also used for stabilization process which decreases the differential free swell and improves the index properties. Rice husk ash up to 12% can be used for strengthen the expansive soils in sub-grade [7]. Besides stabilization, sub-grade can also be strengthened by reinforcing it with geotextiles, improvement by compaction, grouting and electro-osmosis technique, but these all are costly as compared to mechanical stabilization [8].

The main purpose for the stabilization of the road sub-grade soil is increasing the strength, reduction in the pavement construction cost and making it economical. It is evident from the previous researches that about 43% reduction in construction cost can be achieved using mechanical stabilization technique using admixtures such as Lime and Cement which are found to save the construction cost by 46.2% and 27.56% respectively [9, 10].

In India, repeated failure of the pavement occurs because of soil problem which can be rectified with stabilization. Stabilized sub-grade not only reduces thickness of pavement but also increases load bearing capacity. Providing overlays for strengthening at later stages can be decreased and the material can be saved with stabilization. Therefore, in the present study the weak sub-grade soil is stabilized with the help of readily available materials such as cement and molasses.

3 Material and Methods

Lime Lime is the one of the admixtures used for stabilization, generally in clay soils and some granular soils. Hydrated lime is to be used instead of quick lime because quick lime will react with the water only. It will not alter the properties of soil but also decreases the moisture content in the soil whereas hydrated lime reacts with the soil and improves it for further construction. In soil after adding adequate amounts of lime and water, the clay and alumina present will react with the calcium and form calcium-aluminate-hydrates (CAH) and calcium-silica-hydrates (CSH). Addition of lime results in breaking the soil and granular material into hard impermeable layer decreasing the plasticity index and having significant bearing capacity. The process of stabilization can continue for years and years and matrix formed will be durable, impermeable, and hard for both the flexible and rigid pavements.

Lime stabilization is of three phases; with first phase as hydration of hydrated lime, second phase as flocculation (immediate reducing of plasticity making the clay more workable and mixable) and the last phase is the pozzolanic action that slowly strengthens. When the pH is greater than the 12.4 the calcium combines with silica, and this reaction is based upon availability of silica "that's the reason the lime stabilization is more effective for montmorillonite soils than for kaolinite soil". Lime stabilization mechanism constitutes scarification and initial pulverization, Lime spreading, preliminary mixing and watering, final mixing and compaction, and final curing. Factors affecting the lime-soil stabilization are soil type, lime content, types of lime, compaction, curing and additives. In Lime stabilization, the immediate effect is that compaction properties of soil are improved, bearing capacity is increased and plasticity index (PI) is reduced [6]. The decrease in MDD and increase in OMC makes the soil compact easily as the soil moves into humidity range.

Cement Cement is the best suitable material for stabilization because of its easy availability and providing the soil with the great strength and durability. Portland cement is widely used as a soil stabilizer, because of its quality control properties and easy handling. The hydrated cement binds with soil to form the strong cement-stabilized base or cement treated aggregate base. The strength of the stabilised base is directly proportional to quantity of cement used in the soil that further depends on the type of soil [11]. The literature reveals that cement stabilization is preferable for granular soils and clay soils having low Plasticity Index [12]. The mechanism involves the formation of (C-A-H) and (C-S-H) as well as calcium-hydroxide due to reaction of calcium-silicate and calcium-aluminates present in Portland cement with water.

This reduces the permeability and results in moisture-resistant material. The steps involved in the preparation of sub-grade are pulverization, application of cement and mixing without water, spraying of water and again mixing, spreading, grading and compaction, curing, field control tests. Factors influencing properties of cement stabilized soil are soil-cement pulverization and mixing, degree of compaction (should not be very high), curing, and additives [13, 14].

Molasses Molasses is dark viscous syrup obtained during the processing of sugarcane. Molasses depends on the amount of sugar, method of extraction, and age of plant. Molasses could be hazardous to environment if not properly disposed off. It could also degrade water streams if enters into river streams. There are three grades of molasses available: light molasses, also known as first molasses; dark or second molasses; and blackstrap. The molasses usually has a density of 1.6 g/cm^3 and viscosity of 5000 centipoise [15, 16]. Sugars in molasses are principally sucrose and reducing sugars that consists of organic non-sugar matter (nitrogen-containing materials), inorganic constituents (pozzolanic materials) and water [17]. These pozzolans in presence of water at ordinary temperature conditions react chemically to form compounds possessing cementitious properties that are essential in bonding with soil during the stabilization process [18]. This potentially was the scientific motivation for the use of molasses and not any other readily available and cheap agricultural materials.

For the present study, molasses sample is collected from the Wahid Sandar Sugar factory, Phagwara, Punjab. Ordinary Portland cement of grade 33 is used for cement-soil stabilization. Soil sample is collected from the excavation site at Phagwara, Punjab where the road construction process is going on.

4 Experimental Results

In the present study, the tests to be carried out are sieve analysis, liquid limit, plastic limit, standard proctor and California bearing ratio test in order to obtain the optimum content for soil-cement and soil-molasses stabilization.

4.1 Sieve Analysis

The first step adopted is oven drying of the soil sample and classifying according to Unified Soil Classification (UCS) as shown in Table 1.

After calculating the weight retained on each individual sieve, the graph is plotted as shown in Fig. 1.

From Fig. 1, Cu and Cc are calculated using values of D_{10}, D_{30} and D_{60}. Cu comes out to be 3.3 and Cc is 0.78. Acc. to UCS classification table, soil is classified as poorly graded sand with the specific gravity of 1.86.

Table 1 Sieve analysis (IS: 2720-Part 4)

Sieve size (mm)	Mass retained on each sieve (g)	% Retained	Cumulative % retained	% Passing
4.75	78	7.85	7.85	92.15
2	80	8.06	15.91	84.09
1	82	8.26	24.17	75.83
0.425	84	8.46	32.63	67.37
0.212	214	21.55	54.18	45.82
0.09	405	40.79	94.97	5.03
0.075	1	0.10	95.07	4.93
0.01	49	4.93	100	0
Total	993			

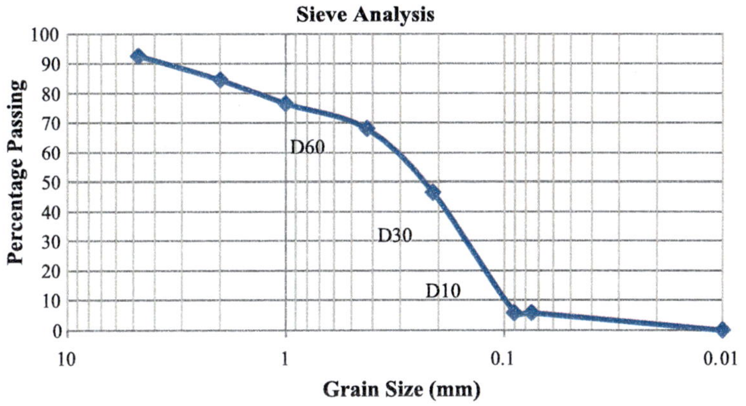

Fig. 1 Semi log graph to effective size values

4.2 Liquid Limit and Plastic Limit

After classifying the soil, the liquid limit and plastic limit for normal soil is found out by using the ASTM apparatus (IS: 2720 Part 5). Table 2 and Fig. 2 depict the liquid limit for normal soil.

Table 2 Liquid limit for normal soil

Wt. of wet soil (g)	Dry weight of soil (g)	Wet wt. – dry wt. of soil (g)	Wt. of water/dry wt. of soil	MC (%)	Blows count
26	22	4	0.18	18.18	36
42	34	8	0.23	23.5	29
30	24	6	0.25	25	19

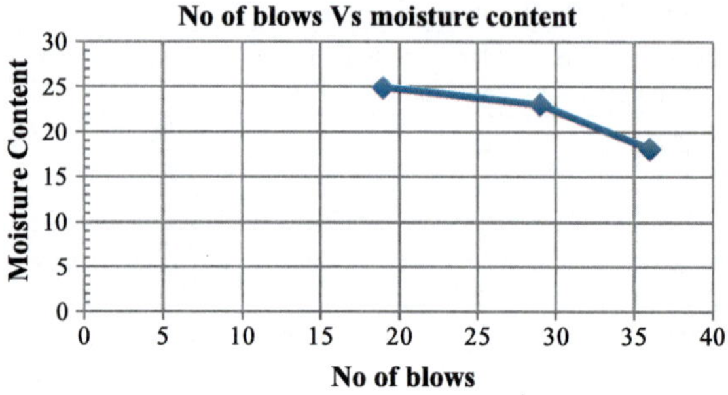

Fig. 2 Liquid limit for normal soil

From Fig. 2, it is observed that the liquid limit corresponding to 25 no. of blows is 24%. As far as plastic limit is concerned, it is difficult to obtain in sandy soils because the sandy soils cannot be easily rolled into 3 mm threads. Therefore, the soil used in the present study is non-plastic in nature.

4.3 *Soil Stabilization Using Cement*

For the determination of optimum content of cement, standard proctor test and California bearing ratio tests are performed as per IS:2720 Part 7 and IS:2720 Part 16 respectively.

4.3.1 Standard Proctor Test

The main purpose of standard proctor test is to determine the optimum moisture content (OMC) and maximum dry density (MDD). The reason for not using the modified proctor test is that vibrations are also to be considered in modified proctor. The vibrations are observed only in airports. So, for normal roads, preference is given to Standard Proctor Test. The variation in cement content are taken as 0, 3, 6, 9, 12 and 15 and for each trial mix, OMC and MDD values are obtained. The results of OMC and MDD for cement stabilized soil are further summarized in Table 3 and presented in Figs. 3 and 4.

Figure 4 shows that the maximum dry density of the cement stabilized soil (with 9% cement content) increases from 1.86 to 1.98 g/cm^3 and the optimum moisture content decreases from 15.25 to 9.3%.

Table 3 Optimum moisture content (OMC) and maximum dry density (MDD) for cement stabilized soil

Cement content (%)	Optimum moisture content (%)	Maximum dry density (g/cm^3)	Graph b/w moisture content (%) and dry density (g/cm^3)
0	15.25	1.86	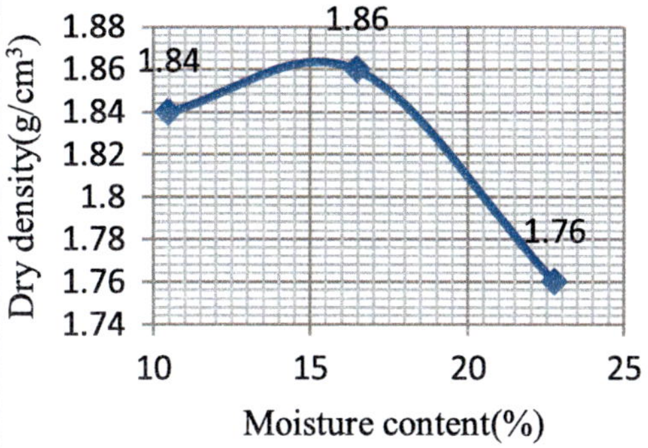
3	14.4	1.87	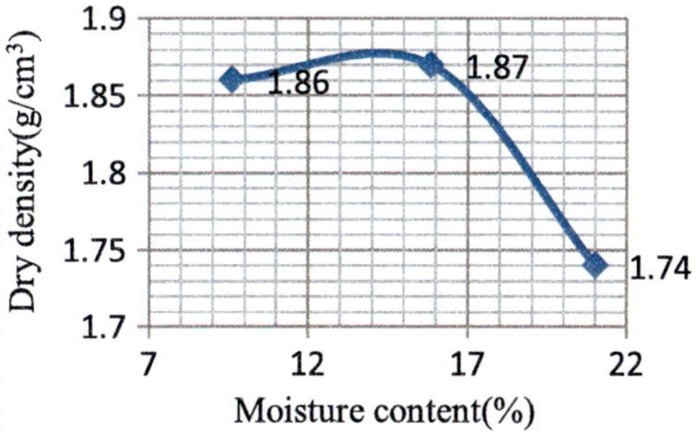
6	13.5	1.88	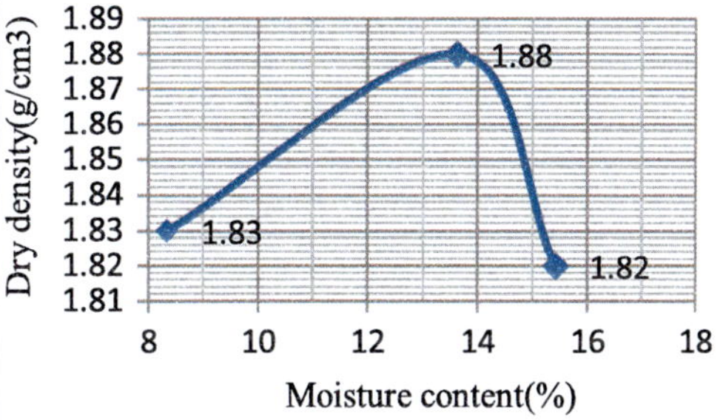

(continued)

Table 3 (continued)

Cement content (%)	Optimum moisture content (%)	Maximum dry density (g/cm^3)	Graph b/w moisture content (%) and dry density (g/cm^3)
9	9.3	1.98	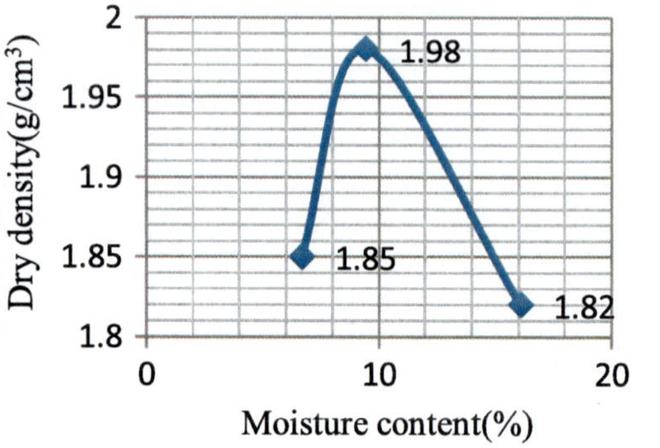
12	14.2	1.91	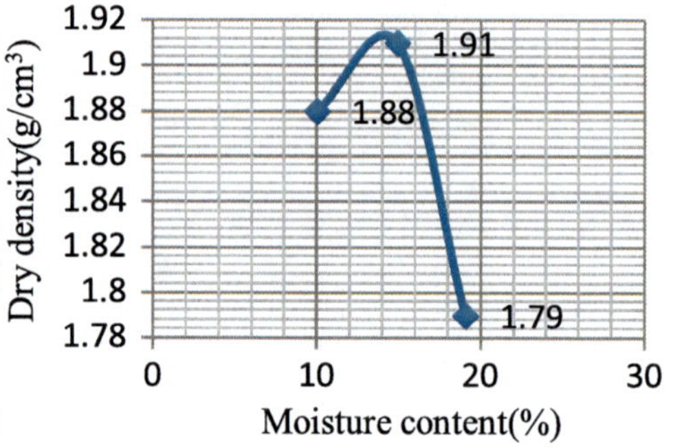
15	16.2	1.85	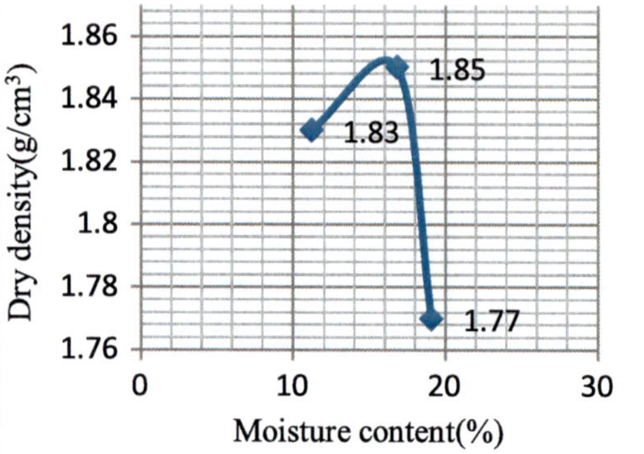

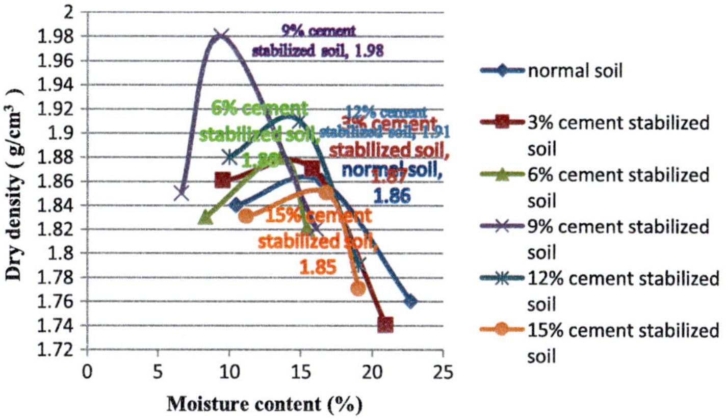

Fig. 3 Dry density versus moisture content for various percentages in cement stabilization

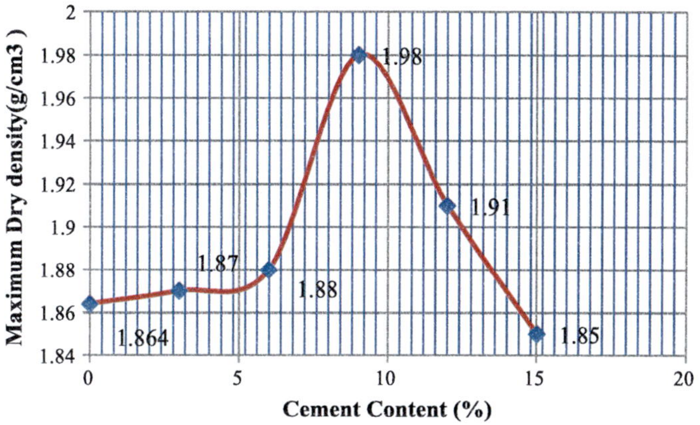

Fig. 4 Maximum dry density versus cement content

4.3.2 California Bearing Ratio Test

After obtaining the OMC values using the standard proctor test, both the un-soaked and soaked CBR values tests are carried out according to IS:2720 Part 16. Necessary corrections are also applied. The values obtained for light compaction (No. of layers—3, Weight of Hammer—2.6 kg, Height of Fall—31 cm, No. of blows—56) are depicted in Table 4. The soaked CBR is done as according to MET, the annual rainfall in the Punjab state is around 69 cm. Therefore, to stimulate the field conditions, 4-days soaked CBR tests are carried out.

Table 4 depicts that both the un-soaked CBR (3.50) and soaked CBR values (2.92) are maximum corresponding to the 9% cement content. This implies that 9% cement

Table 4 CBR values for cement stabilized soil

% of cement	Un-soaked CBR (%)	Soaked CBR (%)
0	1.75	1.17
3	2.34	1.75
6	2.73	2.33
9	3.50	2.92
12	2.63	2.43
15	2.41	2.35

is the optimum content for carrying out the stabilization of poorly graded soil with the cement.

4.4 Soil Stabilization Using Molasses

Molasses are to be used in combination with the lime. The purpose of mixing the lime with molasses is to make the molasses-stabilized-soil as the molasses has high affinity towards water. Lime reduces the affinity and helps in making the mix workable. The lime content to be used is half of the content of molasses. The variation in molasses content similar to that of cement i.e. 0, 3, 6, 9, 12 and 15% and for each trial mix, OMC and MDD values are obtained. The results of OMC and MDD for molasses stabilized soil are further summarized in Table 5 and represented in Fig. 5.

Figure 5 shows that the maximum dry density of the molasses stabilized soil (with 9% molasses content) increases from 1.864 to 2.0 g/cm^3 and the optimum moisture content decreases from 15.25 to 13.8%. Due to chemical reaction between the lime and molasses with soil, the moisture content has not reduced considerably.

After proctor test, California bearing ratio is performed and both the un-soaked and soaked results are obtained which are summarized in Table 6.

Table 6 depicts that both the un-soaked CBR (2.83) and soaked CBR values (2.54) are maximum corresponding to the 9% molasses content. This implies that

Table 5 OMC and MDD for molasses stabilized soil

% of molasses	Optimum moisture content (%)	Maximum dry density (g/cm^3)
0	15.25	1.86
3	15.1	1.88
6	14.5	1.89
9	13.8	2.0
12	15.39	1.84
15	15.1	1.75

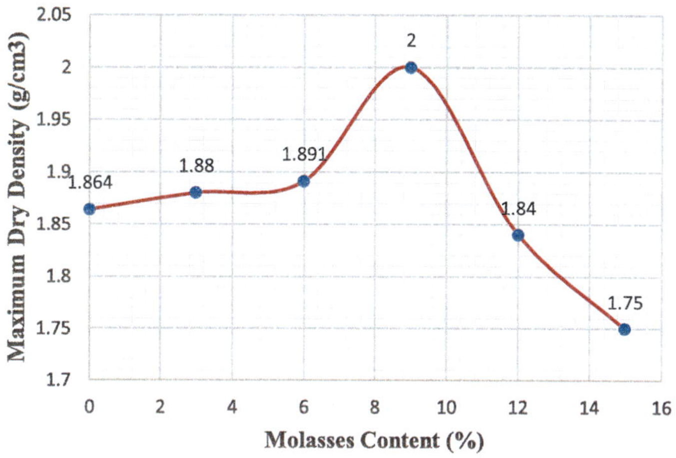

Fig. 5 Maximum dry density versus molasses content

Table 6 CBR values for molasses stabilized soil

% of molasses	Un-soaked CBR (%)	Soaked CBR (%)
0	1.75	1.17
3	2.25	1.63
6	2.53	2.34
9	2.83	2.54
12	2.81	2.51
15	2.73	2.14

9% molasses (with 4.5% lime content) is the optimum content for carrying out the stabilization of poorly graded soil with the molasses.

Therefore, the optimum content for both cement and molasses stabilized sub-grade soil is 9%. In the end, the liquid limit values are determined for both cement stabilized soil and molasses stabilized soil. Liquid limit of cement stabilized soil (with 9% cement as admixture) is 20% and for molasses stabilized soil (with 9% molasses as admixtures) is 19%.

5 Discussions

For the present study, soil selected is a poorly graded sand as per UCS classification which is mechanically stabilized using locally available admixtures like cement and molasses (in addition of lime) so as to enhance the soil sub-grade strength. Further, the liquid limit, standard proctor test and CBR (both soaked and un-soaked) test

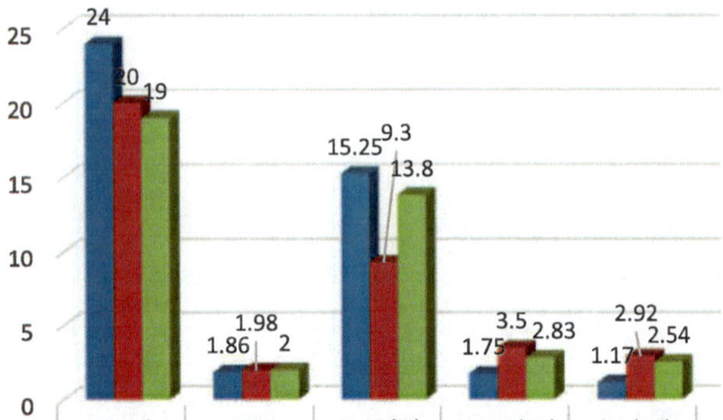

	Liquid Limit (%)	MDD (gm/cm3)	OMC (%)	Unsoaked CBR (%)	Soaked CBR (%)
Normal Soil	24	1.86	15.25	1.75	1.17
Cement Stabilised Soil	20	1.98	9.3	3.5	2.92
Molasses Stabilised Soil	19	2	13.8	2.83	2.54

Fig. 6 Different results for normal, cement and molasses stabilized soils

results are compiled for comparing the competency of cement (9%) and molasses (9%) stabilized soil against the normal soil. The results are presented in Fig. 6.

Figure 6 indicates that the liquid limit of the both the soil-cement (20%) and soil-molasses (19%) after stabilization process has fallen below the normal soil (24%). It is observed that there is 4% more reduction in liquid limit for the molasses stabilized soil than the cement stabilized soil. The MDD for soil-cement has increased to 1.98 g/cm^3 and for soil-molasses has increased to 2 g/cm^3. There is 1.07% more increase in soil-molasses than soil-cement. Also, OMC has reduced to 9.3% and 13.8% for soil-cement and soil-molasses respectively from 15.25% which is OMC for the normal soil. The data indicates 29.51% less reduction in case of soil-molasses. This is due to chemical action of lime which is added for getting the stabilized mix with molasses. Both the un-soaked CBR and soaked CBR values have improved significantly. However, there is 38.29% and 32.48% more increase in both un-soaked and soaked CBR values respectively for soil-cement than soil-molasses. This indicates that soil-cement has high bearing capacity of soil. But as far as strength is concerned, soil-molasses has more strength for the similar content (9%) of cement and molasses during stabilization.

6 Conclusion

The research shows that the stabilization of sub-grade soil with cement and molasses (with the addition of lime) has increased the bearing capacity of the soil (reduced the liquid limit, increased the maximum dry density and enhanced the un-soaked and soaked CBR values). This will result in the decrease of thickness of the pavement, thus, eventually reducing the cost of construction. It is also concluded that molasses is more competent stabilizer than cement in terms of improving engineering properties of the soil but CBR results are much better in case of cement. The soil stabilization will also prevent the future maintenance works of sub-grade soil. The authors further recommend to try different other waste materials in addition of molasses and cement to enhance the properties of soil and carry out the design of the pavement as per IRC:37 using the stabilized soils for estimating the actual reduction in cost.

References

1. Winterkorn, H.F.: Soil stabilization, Chap. 8. In: Winterkorn, H.F., Fang, H.Y. (eds.) Foundation Engineering Handbook, pp. 312–336. Van Nostrand Reinhold, New York (1975)
2. Saitoh, S.Y., Suzuki, S., Shirai, K.: Hardening of soil improvement by deep mixing method. In: Proceedings of the 11th ISCMFE, pp. 1745–1748, San Francisco (1985)
3. Broms, B.B.: Stabilization of Soft Clay with Lime and Cement Columns in Southeast Asia Applied Research Project. Nanyang Technical Institute, Singapore (1986)
4. Holt, C.: Chemical stabilization of inherently weak sub-grade soils for road construction–applicability in Canada. In: Proceedings of Annual Conference of the Transportation Association of Canada (2014)
5. Otoko, G.R.: A review of the stabilization of problematic soils. Int. J. Eng. Technol. Res. **2**(5), 1–6 (2014)
6. Negi, A.S., Faizan, M.: Soil stabilization using lime. Int. J. Innov. Res. Eng. Technol. **2**(2), 9–16 (2013)
7. Shrivastava, D.: Effect of lime and rice husk ash on engineering properties of black cotton soils. Int. J. Eng. Res. Sci. Technol. **3**(2), 24–28 (2014)
8. Ramaji, A.E.: A review on the soil stabilization using low-cost methods. J. Appl. Sci. Res. **3**(2), 15–20 (2014)
9. Malhotra, M.: Stabilization of expansion soils using low cost materials. Int. J. Eng. Innov. Technol. **2**(11), 13–16 (2013)
10. Currin, D.D., Allen, J.J., Little, D.N.: Validation of Soil Stabilization Index System with Manual Development. Frank J. Seisler Research Laboratory, United States Air Force Academy, Colorado (1976)
11. Derucher, K.N., Korfitatis, G.P., Ezeldin, A.S.: Materials for Civil and Highway Engineers. Prentice-Hall (1989)
12. Khanna, S.K., Justo, C.E.G.: Highway Engineering. Nem Chand & Bros, Roorkee, India (2010)
13. White, J.D.: Cement stabilization of sub base and sub-grade. In: Boone County Expo Test Sections, pp. 45–53 (2013)
14. Baviskar, K.R., Bacchav, S.S., Ksirsagar, D.Y.: Stabilization of local expansive soil with fly ash and cement. Int. J. Eng. Innov. Technol. **5**(11), 19–23 (2016)
15. M'Ndegwa, J.K.: The effect of cane molasses on strength of expansive clay soil. Emerg. Trends Eng. Appl. Sci. **2**(6), 1034–1041 (2011)
16. Ravi, E., Animesh, S., Manikandan, A.T., Karthick, G., Jameel, A.A.: Study on effect of molasses on strength of soil. Int. J. Adv. Res. Trends Eng. Technol. **2**(2), 57–61 (2015)
17. Baikow, V.E.: Manufacture and Refining of Raw Cane Sugar, Revised edn. s.l.: Elsevier Science (2013)
18. Vinodhkumar, S., Kulanthaivel, P., Kabilan, A., Lokeshkanna, K.M.: Study of black cotton soil characteristics with molasses. Asian J. Eng. Appl. Technol. **7**(1), 73–77 (2018)

Development of Innovative Green Self-compacting Concrete with Partial Replacement of Fine and Coarse Aggregate by Using Slag

S. Girish, N. Ajay, N. Ishani, D. M. Chaitra, and M. Hrushikesh

Abstract There is a need to minimize the use of conventional natural materials in concrete and the use of alternative marginal materials could be the answer for sustainable development. In this study, Self-Compacting Concrete (SCC) was developed using slag as cement, filler, fine aggregate and coarse aggregate. A new mix proportioning method was adopted starting with a volume of paste (Vp) based on absolute volume concept. Vp chosen was 0.37, 0.39 and 0.41 with maximum size of the aggregate restricted to 20 mm. Cement and water contents of 300, 375 and 450 kg/m^3 and 180 and 190 l/m^3 respectively were chosen. For each of the mix, slag is used as cement, filler, fine and coarse aggregates and aggregates are replaced in different proportions to the natural aggregate by volume. Compressive strength tests were conducted on cubes at 28 days. The results show for a given w/c ratio, Vp and water content, an optimal replacement level of 50% slag as fine and coarse aggregate gives higher results than at higher percentage and opens a new beginning of using slag as fine as well as coarse aggregate. Interestingly SCC mixes with slag showed higher compressive strength than normal concrete.

Keywords Compressive strength of concrete · Self-compacting concrete · Slag aggregates · Sustainability · Volume fraction method

1 Introduction

Present day industry faces the challenge of disposing huge amount of waste/marginal materials that is generated due to many activities. In addition, it is well known that the manufacture of Ordinary Portland Cement will have impact on the environment due to its emission of carbon dioxide. Also the use of natural aggregates poses the

S. Girish (✉) · N. Ishani · D. M. Chaitra · M. Hrushikesh
Department of Civil Engineering, BMSCE, VTU, Bangalore, Karnataka 560019, India
e-mail: girish.civ@bmsce.ac.in

N. Ajay
CIM, RASTA, VTU, Bangalore, Karnataka 560058, India

D. K. Ashish et al. (eds.), *3rd International Conference on Innovative Technologies for Clean and Sustainable Development*, RILEM Bookseries 29,
https://doi.org/10.1007/978-3-030-51485-3_5

Fig. 1 Slag sand dumped in open field

problem of sustainability in the backdrop of depletion of the natural resources. To balance the need and supply in concrete industry, without any negative impact either on the environment or in the depletion of natural resources, it is necessary to use sustainable solutions. There is an urgent need to reduce the carbon emission and to bring down the carbon footprints for greenhouse effect and to find suitable alternatives to replace natural aggregates in concrete. Huge quantity of Ground Granulated Blast Slag (GGBS), which is a by-product in the manufacture of pig-iron and considered as waste material, has found a way to land fill and unfortunately piled up without much benefits. Figure 1 shows the piled up slag in an open field near the place of the pig iron plant. There is a quest among engineers for using technological advances and to relay on energy efficient, low carbon or carbon negative cements or geo-polymers and many other alternatives for sustainability.

The use of Self-compacting concrete (SCC) is very advantageous with its numerous advantages over Normal concrete (NC) especially when unfavorable aggregates are used. SCC mix design as developed and proposed by Girish [1, 2] is used in this study, which is based on absolute volume concept. The proposed method starts by choosing a volume of paste (Vp) and water based on fresh property requirements. Further proportioning of cement and filler is done by considering the requirement of compressive strength.

This study is conducted with the objective of replacing the natural ingredients of NC by alternative marginal material, keeping in mind the sustainability. SCC was developed starting with Vp using slag as cement, filler, fine aggregate and coarse aggregate and proceeds to find the optimum re-placement level of natural aggregates by assessing the compressive strength of SCC and comparing the same with NC.

2 Literature Review

In 1986, Okamura developed the first prototype SCC [3]. Generally, SCC is defined as a "highly flowable, non-segregating concrete that spreads and fills the formwork and passes through congested reinforcement without any external energy". This provides better solution for durable concrete structure with better compaction using minimum number of skilled labors [3]. Study conducted on volume fraction of SCC by Girish et al. [4] have reported a mean value of 0.38 with a variation of 0.03, as the paste content for achieving robust SCC based on successfully developed SCC mixes world over. This indicates that SCC can be achieved with the Vp ranging from 0.35 to 0.41 and generally requires nearly 40% of paste volume and about 60% of aggregate volume. Cement and water can have the same range as that of NC. Higher value of paste and water can be considered for crushed angular aggregates or unfavorable aggregates. Alyamac et al. [5] studied SCC mixes made with three types of marble powder as fillers. Among the three marble powders cherry and white marble powders were coarser than the gold marble powder. These marble powders differ in their fineness, with cherry having the least fineness of 3924 cm^2/g. The scanning electron microscope (SEM) photographs of all the marble powders indicated angular shape with rough textured surface. SCC mixes were developed for different w/c ratio ranging from 0.38 to 0.70, water content 180–210 l/m^3 and powder content 400–800 kg/m^3 with cement content 300, 400 and 500 kg/m^3. Khrapko [6] examined the use of quarry rock dust along with fly ash for producing SCC. W/C ratio was kept constant at 0.43 and the total paste content for SCC mixes were taken as 0.33, 0.38 and 0.39. The study conducted by Kou et al., Limbachiya et al., and Padmini et al. [7–9] indicate a lower compressive strength with different proportion of recycled aggregates. However the study conducted by Girish [10, 11] and Kothai et al., and Xiao et al. [12, 13], using recycled aggregates and powder type SCC indicate better performance compared to NC in terms of compressive strength. There is always quest among engineers to look for alternative materials for the natural ingredients used for SCC.

3 Experimental Program

3.1 Materials

The material properties are tabulated in Table 1. The slag as cement, filler and aggregates are procured from the same plant nearby to the place of work. The aggregates were then processed in the laboratory and separated based on their size for fine and coarse, using the appropriate sieves. The coarse aggregate were sieved through 20 mm sieve and retained on 4.75 mm sieve and fine aggregate were sieved through 4.75 mm sieve and retained on 0.15 mm sieve.

Table 1 Material properties

Materials	Specific gravity	Specific surface (m^2/kg)	Water absorption (%)	Remarks
Slag cement	2.93	389	–	PSC-53 grade, conforming to IS: 455-1989
Fine aggregate	2.64	–	1.5	Crushed natural river sand used as fine aggregates passing 4.75 mm sieve and confirming to zone-II as per IS 383-2016 (see Fig. 2)
	2.7	–	5.0	Slag sand used as fine aggregates passing 4.75 mm sieve and confirming to zone-II as per IS-383-2016 (see Fig. 2)
Coarse aggregates	2.6	–	0.5	Crushed natural coarse aggregate passing 20 mm was used
	2.5	–	2.0	Slag coarse aggregate maximum nominal size of 20 mm was used
GGBS	2.8	315	–	–
Super plasticizer (SP)	1.09	–	–	Base: polycarboxylic ether (PCE)
Water	1.00	–	–	Potable water—pH 7.7 [conforming to IS 456-2000]

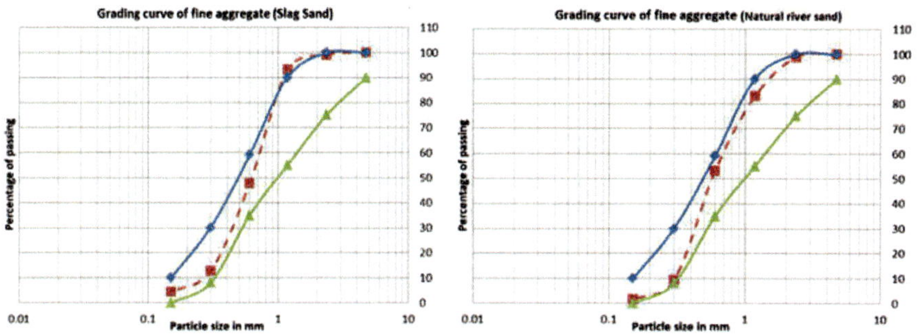

Fig. 2 Grading curve for fine aggregates (slag and natural river sand)

3.2 Methodology

In the present study, powder type SCC based on absolute volume concept starting with a Vp as proposed and developed by Girish [1, 2, 14] was used. The advantage of this method is that Vp, which plays a very important role and an important parameter in achieving a robust SCC, can take into different types of aggregates and if the aggregate shape are unfavorable like angular, flaky, the same can be accommodated with higher Vp and the method is simple and easy and it requires less number of trials to develop SCC. The Vp can be 0.38 with a variation of ±0.03 for different works and is based on the type of aggregates used. Once the Vp is fixed the aggregates can be calculated by considering the concrete as one unit. The proportioning of fine aggregate and coarse aggregate can be based on equal proportion or in some cases with fine aggregate at higher percentage of 5–10%. Generally powder type SCC makes the mix more robust and the cement is supplemented with filler like GGBS which is being produced every year in huge quantity and is unutilized, can be considered in concrete.

SCC with varying Vp (0.37, 0.39 and 0.41); water contents (180 and 190 l/m^3) and cement contents (300, 375 and 450 kg/m^3) were developed. The optimum dosage of SP was fixed for all the experiments based on the results of Marsh cone test for each mix. The optimum dosage depends on cement, filler and water in a mix. A modified mixing procedure was used and the details are discussed elsewhere [14]. EFNARC guidelines were used to assess the fresh properties of SCC [15]. All the mixes satisfied the guidelines as per EFNARC [15] barring minor variations. For the sake of discussion slump flow value and Visual Stability Index (VSI) are given for fresh properties and results of the cubes of size 150 mm are given for compressive strength (IS 516) [16] values. The curing of the specimen is done by immersion in water and tested at 28 days.

The experimental program are carried out in 3 phases. The details are discussed in the subsequent paragraphs.

Phase 1

In this phase, SCC trial mixes were developed with constant Vp of 0.39, water content of 190 l/m^3 and cement content of 375 kg/m^3. Different proportions of natural fine and coarse aggregates were replaced by slag fine and coarse aggregates. Based on published literature [11] and trials, a replacement level of 50% was found to be optimal. In addition 100% replacement was also done in order to know the effect of complete replacement on fresh and compressive strength. For flowability the minimum flow was to achieve 550 mm. SCC mixes were tested for compressive strength at 28 days. Table 2 show the mix proportion, fresh and compressive strength of SCC considered in this phase.

Phase 2

Based on the encouraging results of phase 1, SCC mixes were developed with constant Vp 0.39 for varying cement contents of 300, 375 and 450 kg/m^3, water content 180 l/m^3 along with replacement of natural aggregates by slag aggregates by 50 and 100% and tested for compressive strength at 28 days. Table 3 show the mix proportion, fresh and compressive strength of SCC. The results of SCC mixes were compared with the results of NC. For NC, in the mix proportion used for SCC, only the filler was replaced with coarse aggregate on volume basis. The comparison of NC with SCC has necessitated in understanding the influence of filler (GGBS) content on the strength of SCC.

Phase 3

In order to understand the influence of Vp, paste contents, 0.37, 0.39 and 0.41 were considered. For each paste content, different percentage replacement of natural aggregates was done with slag aggregate. Based on phase 2 results natural fine and coarse aggregate were replaced by 50% and in some cases with 100%, whereas for coarse aggregate it was restricted to 50%. The cement was taken as 300, 375, 450 kg/m^3 with constant water content of 190 l/m^3. In this phase also minimum flowability to be achieved was 550 mm. Further SCC was tested for compressive strength at 28 days. For comparing the results, NC mixes were also developed and the results are shown in Table 5.

4 Results and Discussion

4.1 Fresh Properties

Tests were conducted to assess the properties like flowability, passing ability and segregation resistance of the SCC. All the tests satisfied the acceptance values as per EFNARC guidelines [15] barring few mixes. In this section discussion is done only for flowability. As can be seen from Tables 3, 4 and 5, flowability of SCC were assessed by slump flow and the secondary measurement of T500 were also recorded

Table 2 Mix proportion, fresh properties and compressive strength of SCC trial mixes (Vp = 0.39, cement 375 kg/m^3 and water 190 l/m^3)

Mix	Cement (kg/m^3)	Water (l/m^3)	GGBS (kg/m^3)	Fa (natural) (kg/m^3)		Fa (slag) (kg/m^3)		Ca (natural) (kg/m^3)		Ca (slag) (kg/m^3)		SP (%)	Slump flow (mm)	T500 (s)	VSI	Compressive strength (MPa)
				100%	50%	100%	50%	100%	50%	100%	50%					
M1	375	190	209	805	–	–	–	793	–	–	–	0.6	755	5.0	0	55
M2	375	190	209	–	403	–	143	793	–	–	–	0.6	740	5.0	0	49
M3	375	190	209	–	–	827	–	793	–	–	–	0.6	725	4.0	0	41
M4	375	190	209	–	403	–	413	–	396	–	380	0.5	675	4.0	0	49
M5	375	190	209	–	403	–	413	–	–	760	–	0.5	690	3.5	0	39
M6	375	190	209	805	–	–	–	–	397	–	380	0.5	670	4.0	0	52
M7	375	190	209	805	–	–	–	–	–	760	–	0.6	725	3.5	0	44
M8	375	190	209	–	–	827	–	–	–	760	–	0.5	665	3.5	1	41
M9	375	190	209	–	–	827	–	–	397	–	380	0.5	720	3.5	0	46

Table 3 Mix proportion, fresh properties and compressive strength of SCC mixes for varying proportion of aggregates with Vp = 0.39 and water 180 l/m^3

Mix	Cement (kg/m^3)	Water (l/m^3)	GGBS (kg/m^3)	Fa (natural) (kg/m^3)		Fa (slag) (kg/m^3)		Ca (natural) (kg/m^3)		Ca (slag) (kg/m^3)		SP (%)	Slump flow (mm)	T500 (s)	VSI	Compressive strength (MPa)
				100%	50%	100%	50%	100%	50%	100%	50%					
M10	300	180	312	805	–	–	–	793	–	–	–	0.50	550	5.5	0	47
M11	300	180	312	–	403	–	413	–	396	380	–	0.55	700	3.3	0	38
M12	300	180	312	–	–	827	–	–	396	–	380	0.65	695	5.0	0	51
M13	375	180	238	805	–	–	–	793	–	–	–	0.50	600	3.5	0	51
M14	375	180	238	–	403	–	413	–	396	–	380	0.55	690	4.0	0	47
M15	375	180	238	–	–	827	–	–	396	–	380	0.65	630	5.0	0	49
M16	450	180	164	805	–	–	–	793	–	–	–	0.45	695	2.5	0	56
M17	450	180	164	–	403	–	413	–	396	–	380	0.40	670	4.0	0	55
M18	450	180	164	–	–	827	–	–	396	–	380	0.45	640	5.0	0	47

Table 4 Mix proportion, fresh properties and compressive strength of SCC mixes for varying proportion of aggregates with Vp = 0.37, 0.39 and 0.41 and water = 190 l/m^3

Mix	Cement (kg/m^3)	Water (l/m^3)	GGBS (kg/m^3)	Vp	Fa (natural) (kg/m^3)		Fa (slag) (kg/m^3)		Ca (natural) (kg/m^3)		Ca (slag) (kg/m^3)		SP (%)	Slump flow (mm)	T500 (s)	VSI	Compressive strength (MPa)
					100%	50%	100%	50%	100%	50%	100%	50%					
M19	300	190	255	0.37	832	–	–	–	819	–	–	–	0.40	535	4.0	0	38
M20	300	190	255		–	416	–	427	–	410	–	392	0.40	620	4.0	0	36
M21	375	190	151		832	–	–	–	820	–	–	–	0.40	705	3.0	0	50
M22	375	190	151		–	416	–	427	–	410	–	392	0.40	640	5.0	0	41
M23	450	190	77		832	–	–	–	820	–	–	–	0.30	570	5.0	0	45
M24	450	190	77		–	416	–	427	–	410	–	392	0.30	605	4.0	0	44
M25	300	190	312	0.39	805	–	–	–	793	–	–	–	0.40	600	4.0	0	43
M26	300	190	284		–	403	–	413	–	396	–	380	0.45	735	3.4	0	41
M27	300	190	284		–	–	827	–	–	396	–	380	0.50	600	2.3	0	39
M1	375	190	209		805	–	–	–	793	–	–	–	0.40	755	5.0	0	55
M4	375	190	209		–	403	–	413	–	396	–	380	0.45	675	4.0	0	49
M9	375	190	209		–	–	827	–	–	397	–	380	0.50	720	3.5	0	46
M28	450	190	136		805	–	–	–	793	–	–	–	0.40	725	2.3	0	57
M29	450	190	136		–	403	–	413	–	396	–	380	0.45	645	2.7	0	53
M30	450	190	136		–	–	827	–	–	396	–	380	0.50	650	3.0	0	52

(continued)

Table 4 (continued)

Mix	Cement (kg/m^3)	Water (l/m^3)	GGBS (kg/m^3)	Vp	Fa (natural) (kg/m^3)		Fa (slag) (kg/m^3)		Ca (natural) (kg/m^3)		Ca (slag) (kg/m^3)		SP (%)	Slump flow (mm)	T500 (s)	VSI	Compressive strength (MPa)
					100%	50%	100%	50%	100%	50%	100%	50%					
M31	300	190	341	0.41	779	–	–	–	767	–	–	–	0.40	550	0.5	4.0	43
M32	300	190	341		–	389	–	400	–	384	–	367	0.50	600	4.0	1	41
M33	375	190	267		779	–	–	–	767	–	–	–	0.40	685	3.0	0	55
M34	375	190	267		–	389	–	400		384	–	367	0.40	620	5.0	0	49
M35	450	190	193		779	–	–		767		–		0.30	590	5.0	0	57
M36	450	190	193		–	389	–	400		384	–	367	0.30	600	4.0	0	53

Table 5 Mix proportion, fresh property and compressive strength of NC mixes for varying proportion of aggregates with equivalent Vp = 0.37, 0.39 and 0.41 and water = 190 l/m^3

Mix	Cement (kg/m^3)	Water (l/m^3)	W/C ratio	Vp	Fa (kg/m^3)	Ca (kg/m^3)	Slump (mm)	Compressive strength (MPa)
M_{NC1}	300	190	0.63	0.37	964	896	39	29
M_{NC2}	375	190	0.50			855	36	42
M_{NC3}	450	190	0.42			769	41	51
M_{NC4}	300	180	0.60	0.39	917	943	24	32
M_{NC5}	375	180	0.48			893	25	43
M_{NC6}	450	180	0.40			866	30	45
M_{NC7}	300	190	0.63	0.39	964	896	39	29
M_{NC8}	375	190	0.50			855	36	42
M_{NC9}	450	190	0.42			769	41	51
M_{NC10}	300	190	0.63	0.41	964	896	39	29
M_{NC11}	375	190	0.50			855	36	42
M_{NC12}	450	190	0.42			769	41	51

along with VSI. The values of all the mixes were well within the acceptance values as per EFNARC guidelines [15]. The slump flow value varies from 550 to 755 mm, T500 from 2.3 to 5.5 s and VSI varies from 0 to 1. It may be observed that the slump flow is generally enhanced when Vp or the powder is increased. For a constant water and SP content, the flowability is increased with increase with powder or Vp or both. This is due to better coating of the aggregate surface with higher powder or Vp thereby reducing inter-particle friction between the aggregates which facilitated the increased flowability or lower yield stress.

4.2 *Compressive Strength*

Tables 2, 3 and 4 show the rounded off values of compressive strength of SCC and Fig. 3 shows the strength values plotted against Vp and w/c ratio for SCC made with natural sand and slag sand of 0 and 50% replacement for 28 days compressive strengths with water contents of 190 l/m^3. It is evident from Fig. 3 that the 28 days compressive strength increases with increase in Vp for a particular w/c ratio and is true for different w/c ratios. This observation is true even when the slag is replaced by 50% in SCC. The importance of the paste is also observed in other studies [17–20]. The results of SCC made with 50% slag aggregate show 30% higher compressive strength when compared with NC at 28 days. However the increase is marginal and this may be due to porous nature of slag aggregates compared to compact or dense nature of natural aggregates. The interesting observation is that 0 and 50% replacement of slag, the values of compressive strength is generally higher when

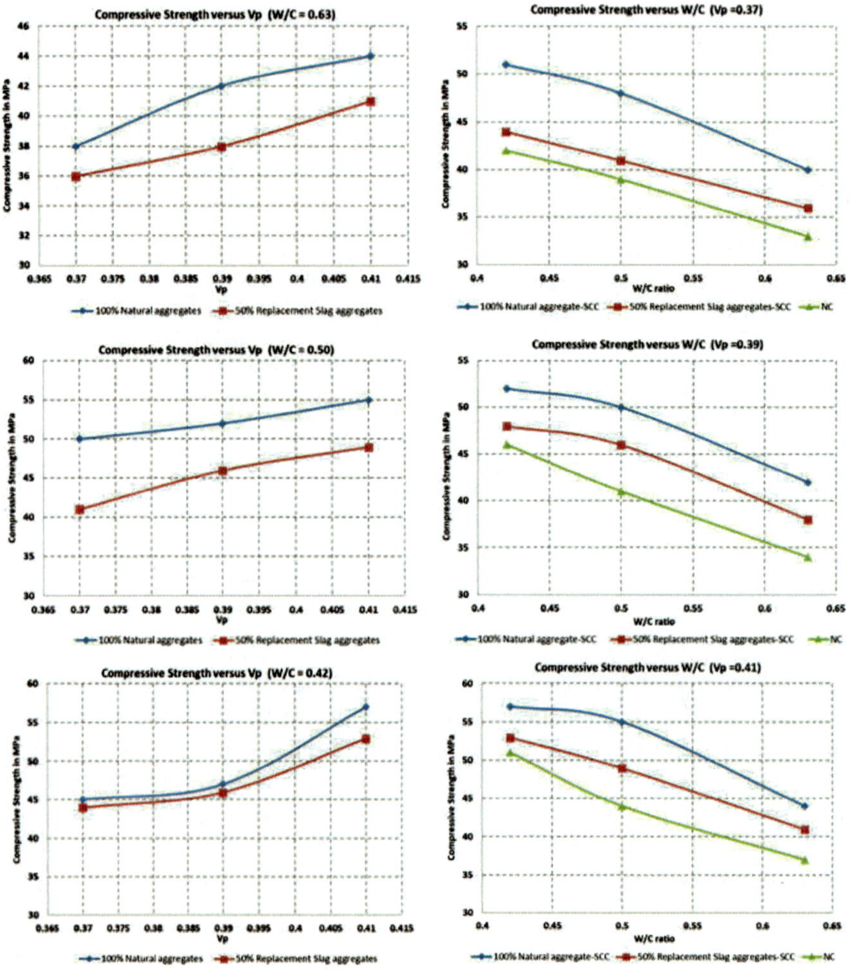

Fig. 3 Compressive strength versus Vp and W/C ratio for SCC for 0% slag and 50% slag for Vp = 0.37, 0.39 and 0.41 and water 190 l/m^3

compared with the results of NC as given in Table 5, for all values of Vp. This can be attributed to better coating of aggregates in SCC due to higher powder or paste thereby reducing the inter-particle friction with better physical packing resulting in improved microstructure in addition to chemical action.

Similarly when the results of compressive strength were compared with w/c ratio it is evident that SCC also follows the Abram's hypothesis and the compressive strength is higher at lower w/c ratio. This observation is true for different percentage replacement of slag and the interesting observation is that for the same w/c ratio, SCC mixes with natural aggregates and replacement level of 50% slag aggregate has shown higher compressive strength than NC as is evident from the relation between compressive strength versus w/c ratio.

It is clearly evident that SCC is much superior to NC even with marginal materials like slag through their physical and chemical action and one can take advantage by considering SCC in place of NC and most importantly use marginal materials for sustainability. The increase in compressive strength due to higher Vp is a factor to be considered in the mix design. It is also important to note that results of compressive strength are shown only for 28 days and the long term increase in compressive due to the use of filler slag is not considered here. It is also clearly evident from the results that the compressive strength of SCC mixes with 50% replacement of slag shows higher compressive strength than 100% replacement and possibly 50% replacement can be considered as the optimum level. Even though it is well known that Vp play important role in enhancing the flowability and compressive strength, the present work reinforces the importance of Vp for marginal material like slag as compared to natural aggregate. It is needless to say the Vp is one of the factor to be considered in mix design.

5 Concluding Remarks

Based on the study conducted following observation can be made:

SCC can be successfully developed replacing natural aggregates with slag aggregates following the mix design method proposed by Girish [14] which is simple, easy and involves less number of trials. The Vp plays an important role in mix design for the development of SCC and in this study it varies from 0.37 to 0.41. SCC made with 50% slag aggregate show 30% higher compressive strength when compared with NC at 28 days for the materials used in this study. The optimal replacement of slag aggregates can be 50% and the use of slag cement, filler, fine aggregate and coarse aggregate can lead to green concrete with sustainability.

Acknowledgements The authors wish to thank the College Management and Government of Karnataka for extending the financial support under TEQIP-II for the research work and M/s JSW Cements Limited for sponsoring the slag. Also the authors thank Ms. Arpitha and Ms. Kavya for their help during the experimental stage.

References

1. Girish, S., Arjun, H.: R, Development of SCC with crusher rock fines as filler. J. Struct. Eng. **38**(1), 60–66 (2011)
2. Girish, S., Ranganath, R.V., Jagadish, V.: Influence of powder and paste on flow properties of SCC. Constr. Build. Mater. **24**, 2481–2488 (2010)
3. Okamura, H., Ochi, M.: Self-compacting concrete (invited paper). J. Adv. Concr. Technol. **1**(1), 5–15 (2003)

4. Girish, S., Vengala, J., Ranganath, R.V.: Volume fractions in self-compacting concrete—a review. In: Proceedings of the 5th International RILEM Symposium on Self-Compacting Concrete, pp. 73–79, Ghent (2007). ISBN 978-2-35158-050-9
5. Alyamac, K.E., Ince, R.: A preliminary concrete mix design for SCC with marble powders. Constr. Build. Mater. **23**, 1201–1210 (2009)
6. Khrapko, M.: Development of SCC containing quarry rock dust. In: Proceedings of SCC, Chicago, IL (2005)
7. Kou, S.C., Poon, C.S., Chan, D.: Influence of fly ash as a cement addition on the hardened properties of recycled aggregate concrete. Mater. Struct. **41**, 1191–1201 (2008)
8. Limbachiya, M.C., Leelawat, T., Dhir, R.K.: Use of recycled concrete aggregate in high-strength concrete. Mater. Struct. **33**, 574–580 (2000)
9. Padmini, A.K., Ramamurthy, K., Mathews, M.S.: Influence of parent concrete on the properties of recycled aggregate concrete. Constr. Build. Mater. **23**, 829–836 (2009)
10. Girish, S.: Influence of powder and paste on SCC using recycled concrete aggregates. Bonfring Int. J. Ind. Eng. Manag. Sci. **2**(4), 68–75 (2012). ISSN 2277-5056
11. Girish, S., Karthik Pattaje, S.: Experimental study on the use of blast furnace slag sand as an alternative to fine aggregate in self-compacting concrete with emphasis on sustainability. Indian Concr. J. **91**(1), 53–61 (2017)
12. Kothai, P.S., Malathy, R.: Utilization of slag sand in concrete as a partial replacement material for fine aggregates. Int. J. Innov. Res. Sci. Eng. Technol. **3**(4), 11585–11592 (2014)
13. Xiao, J., Li, J., Zhang, C.: On relationship between the mechanical properties of recycled aggregate concrete: an overview. Mater. Struct. **39**, 655–664 (2006)
14. Girish, S.: Importance of volume of paste on the compressive strength of SCC—a parameter to be considered in mix design. Indian Concr. J. **91**(4), 51–62 (2017)
15. EFNARC Guideline: Specification and Guidelines for Self-Compacting Concrete. UK (2005). https://www.efnarc.org/
16. IS 516: Method of Test for Strength of Concrete. Bureau of Indian Standards, New Delhi, India (1959)
17. Zsigovics, I.: Effect of limestone powder on the consistency and compressive strength of SCC. In: Proceedings of SCC, Chicago, IL. ACBM (2005)
18. Roziere, E., Granger, S., Turcry, P., Loukili, A.: Influence of paste volume on shrinkage cracking and fracture properties of self-compacting concrete. Cem. Concr. Compos. **22**, 626–636 (2007)
19. Pineaud, A., Cabrillac, R., Remond, S., Pimienta, P., Rivillon, P.: Mechanical properties of self-compacting concrete—influence of composition parameters. In: Proceedings of SCC, Chicago, IL. ACBM (2005)
20. Heirman, G., Vandewalle, L.: The influence of fillers on the properties of self-compacting concrete in fresh and hardened state. In: Proceedings of the 3rd International Symposium on Self-Compacting Concrete, pp. 606–618, Reykjavik, Iceland (2003)

Treatment of Nutrient Laden Wastewater Using Simultaneous Nitrification and Denitrification

Susan N. James and Arya Vijayanandan

Abstract The excess discharge of the untreated wastewater into the natural water bodies stresses the biotic components in it. Discharge of nutrient-rich wastewater into water bodies leads to eutrophication. Simultaneous Nitrification and Denitrification (SND) is a process in which nitrification and denitrification occur simultaneously under microaerobic conditions in a single reactor. Depending upon the microbial community and floc size, the mechanism behind the SND process can change. The creation of stable aerobic and anoxic conditions within the floc is the most tedious process in SND. Slight variation in the dissolved Oxygen (DO) concentration can affect the SND process. Compared to conventional nitrification and denitrification processes, SND is cost-effective due to the decrease in structural footprint, low oxygen requirement, and low energy requirement. This review focuses on the applicability of SND as a clean technology for nutrient removal by investigating the mechanism in detail, factors that affect the process efficiency such as microbial population, DO concentration, Carbon/Nitrogen (C/N) ratio, reactor configurations and thermophysical parameters.

Keywords Nutrient removal,biological wastewater treatment · Simultaneous nitrification and denitrification · Mechanism and pathway of SND · Microbial community · DO concentration · Reactor configuration · Carbon/nitrogen ratio

1 Introduction

Natural water bodies such as river basins, lakes, etc. are surrounded by a sensitive ecosystem. Most of the water bodies face the risk of pollution due to population growth, industrial, agriculture, and commercial activities. United Nations World

S. N. James (✉) · A. Vijayanandan
Department of Civil Engineering, Indian Institute of Technology Delhi, New Delhi 110016, India
e-mail: susannjames@gmail.com

A. Vijayanandan
e-mail: aryav@civil.iitd.ac.in

D. K. Ashish et al. (eds.), *3rd International Conference on Innovative Technologies for Clean and Sustainable Development*, RILEM Bookseries 29,
https://doi.org/10.1007/978-3-030-51485-3_6

Water Development Reports of 2017 published that overall, 80% of wastewater in the world is disposed into the environment without sufficient treatment [1]. The treatment capacity available for domestic sewage in India is 6,000MLD, which means that only 21% of the wastewater generated is being treated [2]. Wastewater treatment plants (WWTPs) designed for the removal of the nutrients are less in number. The inadequate treatment facility for nutrient removal in treatment plant increases the amount of nutrient content in treated wastewater. Finally, untreated and partially treated wastewater with high nitrogen content is discharged into the water bodies. Other sources of nutrients in the aquatic system are fertilizer use, crop nitrogen fixation, agriculture waste, poultry farm, atmospheric nitrogen fixation, and fossil fuel combustion, etc. [3–7]. Excess nutrient content in water bodies accelerates the plant growth and leads to eutrophication [7, 8]. Eutrophication is an ecological state of an aquatic system where biological growth speed-up due to nutrient enrichment in water bodies [7, 9]. Other adverse environmental impacts due to eutrophication are depletion of oxygen in water bodies due to bacterial oxidation of ammonia to nitrate, algae bloom, toxicity to aquatic organisms, etc. [7, 10].

Nowadays, discharge limits are getting stringent due to the scarcity of drinking water and deterioration of the quality of available freshwater. Biological nutrient removal is an economical and sustainable treatment technique to meet new stringent discharge limits recommended by the Central Pollution Control Board (CPCB). Conventional nitrification and denitrification are two distinct processes that occur in two different conditions and reactors [11, 12]. The disadvantages of the traditional biological nitrogen removal methods are large structural footprints and high processing costs. For overcoming these limitations, and for enhancing nutrient removal, various kinds of biological processes have been studied, and SND is found to be effective in achieving this objective [13]. In the SND process, nitrification and denitrification occur concurrently in one reactor under the aerobic condition with low dissolved oxygen (DO) concentration [12]. The main advantages of SND compared with conventional biological nutrient removal processes are cost-effectiveness due to the decrease in structural footprint, less external carbon sources, no internal recirculation, low oxygen requirement and diversity in the microbial population results [14]

The present study reviews our existing knowledge related to the SND process and its adaptability in WWTPs. The review paper focuses on the suitability of the SND process in nitrogen removal and presents the mechanism and factors affecting the SND process. The microbial community and physical characterization of floc structure are controlled by DO concentration, reactor configuration, C/N ratio, and thermophysical parameters of the wastewater. This review paper is focused on different mechanisms, microbial populations in the SND system, and factors such as DO concentration, reactor configuration, C/N ratio and thermophysical parameters affecting the SND process.

2 Mechanism and Pathway of the SND

In any biological process, the microbial community is an important parameter. The mechanism behind the biological process is complicated due to the existence of a large number of different microorganisms in microbial flocs. Each microorganism has its optimum environmental conditions. The ecological environment conditions for nitrifiers are aerobic, whereas for denitrifiers are anoxic. Therefore, creating an environment that favors the co-existence of nitrifiers and denitrifiers is a major challenge in the SND process. The mechanisms of the SND process is based on the diverse microbial populations and physical characteristics of flocs, and it could be explained in terms of physical and biological nature [13].

2.1 Physical Mechanism Behind SND

The physical mechanism is based on the distribution of DO, organic matter, and substrate concentration within the sludge flocs [10]. The aerobic and anoxic conditions within flocs are developed due to the diffusional limitation of DO. The creation of an oxygen concentration gradient across the agglomerates is the major challenge in the SND process. Nitrifiers are located on the periphery of the flocs because of the aerobic state, which favors the nitrification. Denitrifiers are distributed inside the flocs because of the anoxic condition and result in the denitrification. Organic matter requirement is the other essential element for denitrification [15, 16]. For better heterotrophic denitrification, the penetration of organic matter into the flocs should be more, but oxygen penetration should be less. The diffusions of oxygen and organic matter are affected by the size and density of flocs [16]. He et al. [10] stated that large and dense flocs were the best flocs for developing anoxic conditions [10]. The large diameter of flocs (>125 μm) limits the oxygen diffusion to the inner ring of flocs resulted in an anoxic environment [11]. In dense flocs, more heterotrophic bacteria could be present in the outer ring of the flocs, leading to an increase in oxygen utilization and limited diffusion of oxygen into the flocs. Pochana and Keller, 1999 stated that nitrogen removal through the SND process decreased from 52 to 21% when flocs size reduced from 80 to 40 μm [16]. Denitrification rate was affected by the decrease in the size of flocs because the internal volume of the anoxic zone has been reduced, while the nitrification rate was independent of the size of the flocs [16]. "Fig. 1 shows the distribution of DO, organic matter (BOD), and substrate concentration within the floc".

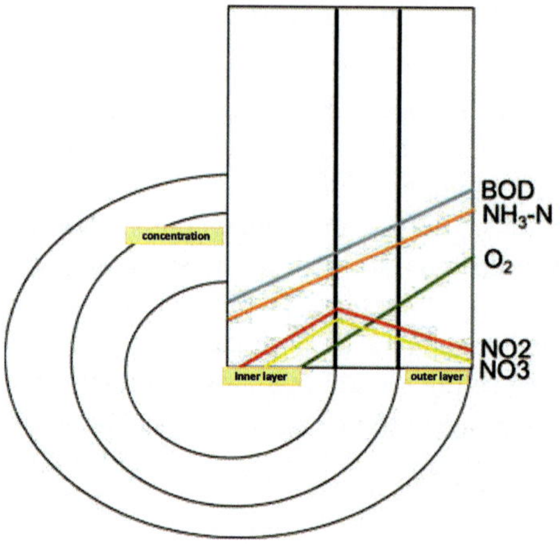

Fig. 1 The distribution of DO, organic matter (BOD), and substrate concentration within the floc. Adapted from He et al. [10]

2.2 *Mechanism of SND Process Based on Bacterial Population*

Depending on the bacterial community, the mechanisms behind the SND process are conventional autotrophic nitrification and heterotrophic denitrification, heterotrophic nitrification and aerobic denitrification and direct conversion of ammonia into di-nitrogen (N_2) gas [17]. Each of these processes are explained below in detail.

Conventional Autotrophic Nitrification and Heterotrophic Denitrification
The conventional nitrification is a two-step aerobic process which is performed by autotrophic bacteria. In former, the ammonia (NH_3) is converted into nitrite (NO_2^-) by ammonia-oxidizing bacteria (AOB) in the presence of ammonia monooxygenase (AMO), and NO_2^- later is converted into nitrate (NO_3^-) by nitrite-oxidizing bacteria (NOB) which is driven by nitrites reductase [11, 14]. NH_3 is the electron donor, and oxygen is the electron acceptor in the autotrophic nitrification process. Conventional denitrification is the reduction of NO_3^- into N_2 gas by heterotrophic bacteria in an anoxic environment. There are four steps ($NO_3^- \rightarrow NO_2^- \rightarrow NO \rightarrow N_2O \rightarrow N_2$) involved in the heterotrophic denitrification process. Nitrite reductase, nitrate reductase, nitrous-oxide reductase and nitric oxide reductase which are produced by denitrifiers during the heterotrophic denitrification. In the heterotrophic denitrification process, the electron donor is organic carbon, and the electron acceptor is NO_3^- or NO_2^-. "Eq. 1 represents the overall autotrophic nitrification equation, and Eqs. 2 to 4 define the each reactions involved in autotrophic nitrification". "Eqs. 5 represents the overall heterotrophic denitrification and, Eqs. 6 and 7 describe reactions involved in the heterotrophic denitrification".

$$NH_3 \rightarrow NO_2^- \rightarrow NO_3^- \tag{1}$$

$$NH_3 + O_2 + 2H^+ \rightarrow NH_2OH + H_2O \quad (2)$$

$$NH_2OH + 0.5O_2 \rightarrow NHO_2 + 2H^+ + 2e^- \quad (3)$$

$$NO_2^- + 0.5O_2 \rightarrow NO_3^- \quad (4)$$

$$NO_3^- \rightarrow NO_2^- \rightarrow NO \rightarrow N_2O \rightarrow N_2 \quad (5)$$

$$NO_2^- + 3H^+ \rightarrow 0.5N_2 + H_2 + OH \quad (6)$$

$$NO_3^- + 5H^+ \rightarrow 0.5N_2 + 2H_2 + OH \quad (7)$$

In SND, the nitrification pathway can happen through NO_3^- or NO_2^-. The traditional path of nutrient removal is through NO_3^-. Benefits of nitrification via NO_2^- are the reduction in oxygen consumption upto 25% during nitrification and 40% reduction in carbon requirement during denitrification [18]. Accumulation of NO_2^- is accomplished by inhibiting the NOB. The main factor which affects the accumulation of NO_2^- was DO concentration. When DO concentration is decreased, NO_2^- accumulation increased. The other factors which help in the accumulation of nitrite are temperature, pH, free ammonia (FA), free nitrite (FNA), shortened hydraulic residence times (HRT), etc. [14].

Heterotrophic Nitrification and Aerobic Denitrification

The second mechanism for the removal of nitrogen in SND is heterotrophic nitrification and aerobic denitrification. Aerobic denitrification is the conversion of NO_3^- into N_2 gas under the aerobic condition [19]. The aerobic denitrifiers are heterotrophic bacteria that simultaneously use NO_3^- and oxygen as an electron acceptor, which is termed as a co-respiration mechanism [20]. The periplasmic nitrate reductase is the vital enzyme that takes part in the aerobic denitrification [19]. The presence of NapA gene may be a proof of aerobic denitrification [19]. The enzymes involved in aerobic denitrification are same as those responsible for conventional denitrification [19]. Some of the aerobic denitrifiers could be performed nitrification by utilizing organic carbon, which leads to heterotrophic nitrification [21]. The mechanism of heterotrophic nitrification resembles that of autotrophic nitrification. Ammonia is oxidized to hydroxylamine by AOM, and then hydroxylamine is oxidized to NO_2^- by hydroxylamine oxidase [20]. Chai et al. [22] studied the different pathways of SND in SBBR (sequencing biofilm batch reactor) [22]. He reported that 55.4% of total ammonia was removed through heterotrophic nitrification and aerobic denitrification. The mechanism was studied by using inhibits such as allylthiourea (ATU) and sodium chlorate ($NaClO_3$). ATU inhibited the enzyme activity of AMO (which helps in conversion of NH_3 into NO_2^-), and $NaClO_3$ inhibited the enzyme activity of nitrite oxide reductase (which helps in the denitrification).

Direct Conversion of Ammonia into Di-Nitrogen Gas
A particular group of heterotrophic bacteria (e.g., *Cupriavidus sp.S1*, *Alcaligenes faecalis*, and *Thiosphaera pantotropha*) could be produced N_2 gas directly from ammonia without accumulation of NO_3^-or NO_2^-. The mechanism behind this process was reported as the hydroxylation of ammonium to hydroxylamine (NH_2OH) by AMO under aerobic conditions. Gradually hydroxylamine is oxidized to NO_2^- by hydroxylamine oxidase. Finally, NO_2^- is directly transferred to N_2 [23].

3 Factors Affecting SND Process

3.1 Microbial Community

Microbial community is a key factor in biological nutrient removal. Changes in the operational parameters and chemical parameters are affected by the microbial community in a system. DO concentration is an imperative factor affecting the microbial community. A small change in the DO concentration is reflected in the transformation of the microbial community. Chemolithoautotrophic bacteria perform conventional nitrification. Both AOB and NOB are involved in nitrification. Autotrophic bacteria are slow-growing bacteria compared to heterotrophic bacteria. The presence of organic matter inhibits the growth of nitrifiers. Denitrifiers are heterotrophic bacteria, so for denitrification, organic matter is an important parameter. The other factors which are affected by the physiology and biochemistry of the microbial community are growth rate, C/N ratio, NO_2^-, and NO_3^- accumulation. Bacteria species which are abundantly present in the group of AOB and NOB are *Nitrosomonas* and *Nitrobacter, Nitrospira*, etc. respectively. Yan et al. [14] studied microbial communities involved in the process of SND via NO_2^- in an oxygen-limited Sequencing batch reactor (SBR) [14]. Under low DO concentration, in the genus level, *Nitrosomonas* was more abundant than *Nitrobacter* because it was more susceptible to low oxygen concentration. *Candidatus Nitrotoga* was present in this system, which was a NOB bacteria present in low DO concentration, and also it was suitable for low-temperature conditions. Also, this study identified 15 genera of denitrifying bacteria. *Thauera, zoogloea, flavobacterium* showed the highest population among denitrifiers. When DO concentration is decreased from 1.2 to 0.7 mg/l, the relative abundance of the *Thauera* also increased from 1.53% to 9.85%. In the phylum level dominating community in SBR is *Saccharibacteria* (32.54%). The presence of AOB in *β-Proteobacteria* and *γ-Proteobacteria* are responsible for the conversion of NH_3 into NO_2^-.

Chai et al. [22], studied about the microbial communities of the seed sludge and biofilm in SBBR system [22]. Seed sludge was collected from an aeration tank of the wastewater treatment plant. *Zoogloea* was the dominating bacteria found in biofilm and, also it had a sticky exterior layer that enables the bacteria to adheres to surfaces and provide a protective layer on the biofilm. The presence of *Nitrospira defluvii* was reported as the second dominating species in biofilm, which is in the group of NOB. The population of *Nitrospira defluvii* was increased in the biofilm compared to seed sludge, while the population of *Candidatus Microthrix* was decreased in the biofilm. *Proteobacteria* was the dominating community in the phylum level, followed by *Actinobacteria, chloroflexi, Bacteroides,* and *Nitrospirae* in the seed sludge, but in biofilm *Proteobacteria, Nitrospirae* and *Saccharibacteria* were dominated.

Chang et al. [17] studied the microbial community in the biological folded non-aerated filter reactor (BFNAF), and the samples were taken from the upper part, middle part, and lower part of the reactor [17]. This study identified seven genera of denitrifying bacteria, such as *Azoarcus, Arcobacter, Flavobacterium, unclassified Comamonadaceae,* and *Thurea.* The *Azoarcus* was reported as the highest population at the genus level and it had denitrification capacity. The presence of denitrifiers was more in the middle part compare to the upper portion because of high COD, which inhibited the growth of nitrifiers and denitrifiers. The co-existence of ABO, NOB and denitrifiers were higher in the lower part, which indicated the SND process.

Padhi et al. [24] studied the heterotrophic nitrification-aerobic denitrification using bacteria strain *Klebsiella* sp. (KSND), which had the capability of reducing inorganic nitrogen sources into N_2 gas using organic matter as a substrate [24]. *Pseudomonas, Rhodococcus,* and *Acinetobacter* were the bacteria, which performed both heterotrophic nitrification and aerobic denitrification in nutrient removal system [22, 25, 26]. *Arcobacter* is an autotrophic denitrifying bacteria that could be grown in both conditions aerobic and anaerobic [17]. Yan et al. [27] reported *Comammonas* and *Paracoccus* are aerobic denitrifiers [27]. *Zoologea* is the most common denitrifiers and phylogenetically similar to the common aerobic denitrifiers, which helped in the complete denitrification in the SND system [28].

3.2 DO Concentration

DO concentration is an important parameter affecting the SND. Microbes are susceptible to DO concentration. Denitrification process was more efficient at zero DO concentration, and efficiency was decreased when DO concentration is more than 0.2 mg/l. DO concentration of more than 2 mg/l was the most favorable condition for nitrification [16]. Optimization of DO concentration is a critical process in SND because nitrifiers and denitrifiers are existed in different DO concentrations. The optimized DO concentration for the SND process is the DO concentration at which nitrifiers and denitrifiers are co-existing in a single system. DO concentration could be affected by the microbial community and volume of the anoxic zone in flocs

He et al. [10] reported the effect of DO on NH_3 and Total Nitrogen (TN) removal [10]. When DO was 0.8 mg/l, the removal rate of NH_3 was in the range of 92–94%. When DO concentration was above 1.5 mg/l, the removal rate of NH_3 was independent of DO, but TN removal efficiency was decreased. The decrease in the TN removal was due to the change in the volume of the anoxic zone in the sludge flocs. The DO and concentration of sludge were increased in the system, the penetration of oxygen into the flocs was increased, and thereby, the volume of the anoxic zone reduces [16].

Jun et al. [29] reported that the SND process was very efficient in the DO range of 2.5–4 mg/L [29]. The volumetric removal rate of TN decreased when the DO concentration was above 4 mg/L because of the reduction in the denitrification rate. DO concentration below 2.5 mg/L had been resulted in the low effluent of TN. At low DO concentration, aeration rate could be reduced, but this condition increases the aeration time, and thereby volumetric removal rate of TN became small. The increase in DO concentration had resulted in the rise of NH_3 removal and a decrease in denitrification. At low DO, nitrification was less, and denitrification was more. At high DO, the denitrification rate was decreased not only due to the less anoxic area but also due to less organic carbon. Heterotrophic aerobic bacteria is grown fastly at high DO by utilizing more organic carbon, which results in less penetration of organic carbon into flocs [10]. At high DO concentration, electron acceptor for denitrifiers will change from nitrate/nitrite to oxygen. This condition could cease the activity of heterotrophic denitrifiers [15].

In the study of Yan et al. [14], nitrite accumulation was increased when DO concentration was decreased from 1.5 to 0.7 mg/l [14]. The Monod oxygen saturation constant for AOB and NOB is 0.3 and 1.1.mg/L. At the lower value of DO concentration, NOB is more sensitive to oxygen compare to AOB. Thereby controlling the DO concentration gradient in the reactor enhanced nitrite accumulation [30].

Yan et al. [30] study had been shown that the metabolic activity of the microbial population for amino acids, nucleotides, coenzymes, and inorganic ions was decreased, but the activity of carbohydrates and lipids was increased as the DO concentration decreased [30]. The DO concentration was reduced from 1.5 to 0.7 mg/; the population of denitrifiers was increased from 11.22 to 26.09% [14].

3.3 C/N Ratio

Imbalance in nitrification and denitrification process can happen due to the carbon deficiency or NO_3 deficiency in the wastewater, which had resulted in an inefficient SND process [10, 16, 23]. The optimum value of the C/N ratio is required for the efficient SND process [23].

When the C/N ratio is too high, heterotrophic bacteria could be accumulated, and it affects the growth of autotrophic bacteria. It is due to the slow-growth rate of autotrophs bacteria. One of the reasons for the decrease in TN removal was the lack of carbon sources [31]. Chang et al. [17] studied the effect of C/N at a different C/N ratio, such as 4, 7, 10 [17]. The NH_3–N removal was highest at the lowest value of the C/N ratio of 4.TN, and NH_3–N removal was smaller (around 42%) at a high C/N ratio of 10. It might be due to high organic load shock, and it leads to insufficient NO_3^- as the reactant in the denitrification process.

Chiu et al. [23] studied in three different C/N ratios 6.3, 11.1, and 19.7 [23]. At the lowest C/N ratio (6.3), autotrophic nitrification had taken place, but it inhibited

Table 1 Studies conducted on SND using different reactor configurations

Reactor configuration	DO range (mg/l)	C/N ratio	Removal efficiency NH_3–N (%)	Removal efficiency TN (%)	References
SBBR	2	3.8	72.45	97.9	[22]
SBR	(a) 0.7	6.7	96.89	71.43	[14]
	(b) 1.2	6.3	91.14	63.32	
Folded non-aerated filter	–	(a) 4	90	40	[17]
		(b) 7	75	66	
		(c) 10	42	42	
Aerated SPD	1.50	–	83.85	93.88	[28]
SNDPR-AOA-(aerobic granular)	(a) 3.5–4.5	10.7	97.03	63.71	[34]
	(b) 1.7–2.3			76.87	
	(c) 0.8–1.2			94.12	
SND bioreactor	0.30–0.80	8	98	–	[30]
Oxidation ditch	0.2–3	4–9	100	60	[15]
A/O-MBR	0.3	(a) 17.7	94.74	91.8	[12]
	0.4–2	(b) 11.87	95.75	91.9	
		(c) 9.3	85.91	84.6	
MBR	(a) 0.8	12.4	93	86.58	[10]
	(b) 1.5		97	77.53	
	(c) 3		97	65.33	
	(d) 5		97	54.21	

heterotrophic denitrification. "From Table 1 clear that at a low C/N ratio decreased the TN removal". In the study of Chiu et al. [23], the nitrification rate was more at a high C/N ratio of 19.7; these results were indicated the chances of heterotrophic nitrification [23]. A similar result was found in the study of Pochana and Keller [16]. In this study, a maximum SND efficiency of 80% was attained at an initial C/N ratio of 14.5 [16].

3.4 Reactor Configuration

SND process can be carried out in the attached growth system and suspended growth system. Many studies were conducted on both systems. Reactor configuration is a critical factor for the enrichment of both nitrifying and denitrifying bacteria in a single reactor. "Table 1 shows the different reactor configuration and removal efficiency".

Any reactor configuration performing SND aims to form a DO concentration gradient across flocs and to create an anoxic microenvironment inside the flocs [14].

Various reactor types have been used in suspended growth systems are ASP (Activated sludge process) [13, 16], SBR [11, 23], oxidation ditch [15] etc. The size and density of sludge flocs are essential factors in the suspended growth system. In the study of Yan et al. [14], the reaction zone in the reactor was divided into an aerobic zone, and an anoxic zone using a piece of clapboard for creating a DO concentration gradient across the SBR system [14]. Aeration was provided at one side of the reactor only. This aerobic zone was beneficial for the enrichment of aerobic microorganisms such as nitrifiers, aerobic denitrifiers, and the anoxic zone was suitable for enriching the denitrifying bacteria population. Lack of organic carbon for the denitrification process was one of the most common problems in any reactor configuration [11]. Carbon sources could be external carbon sources (such as Glucose, sucrose, etc.) or organic matter present in wastewater and internal carbon sources such as PHB, polymers, etc. The denitrification that happens in lower COD value was a clear indication of the use of internal carbon sources as an electron donor. The alternation of anaerobic and aerobic conditions helped in the storage of carbon sources as polymers (e.g., PHB) in the anaerobic zone. Later, it could be utilized as an electron donor for denitrification in the aerobic stage.

The reactor configuration that has been studied in the attached growth systems are SBBR [22], MBBR (moving bed biofilm reactor) [19], FBBR (fluidized-bed biofilm reactor) [32], etc. Biofilm formation and its thickness are the main factors affecting the SND process in the attached growth system. Jun et al. [29] studied SND through the SBBR system [29]. Research had shown that the thicker biofilm contributes to better efficiency in the removal of TN. When the thickness of biofilm was increased, TN removal also increased because of the stable anoxic environment inside the biofilm. When the thickness of biofilm increased, the penetration of oxygen into the biofilm could be decreased, which results in better denitrification. In the attached growth system, denitrifiers and nitrifiers are the present lower part of the system because of less COD in the lower part [17].

Fu et al. [12] studied in an anoxic/oxic MBR system without any circulation [12]. NH_3 was oxidized without accumulation of NO_x–N during operation, which indicated the occurrence of the SND process in the aerobic zone using internal carbon sources. The exclusion property of the membrane, which helped in reducing the competition between autotrophic nitrifiers and heterotrophs [10]. SND had studied in some innovated reactors such as aerated solid-phase denitrification system, aerobic granular sludge, biological folded non-aerated filter (BFNAF), and achieved high NH_3–N removal of and NO_3^- removal [17, 27, 28]. The advantages of the BFNAF reactor are the frame with folded structure could enhance the contact area between the substrate, air, and biomass carrier, and also does not require a separate aeration system [17].

3.5 *Thermophysical Parameters*

Thermophysical parameters such as pH,temperature can also affect the SND process. Effect of pH in SND was studied by He et al. [10]. The study reported, the optimum pH for SND was 7.5.The pH value above or below the optimum value, the removal of NH_3–N and TN was decreased. At pH 7.5, the removal of NH_3–N and TN was 99% and 91% respectively. Liu et al. [15] studied effect of temperature in SND process [15]. The results indicate that, below 11 °C, nitrification was decreased even though enough DO present. It was due to the decrease in the activity of nitrifiers. AOB has a higher specific growth rate at a temperature above 15 °C compare with NOB. Zhang et al. [33] reported the highest ammonia oxidation rate and TN removal were obtained at a temperature of 31 °C, and the lowest value was observed at a temperature of 15 °C [33]. Temperature more than 35 °C, ammonia oxidation rate was decreased due to enzymatic proteins denature. At the lower temperature, oxygen transfer into the biofilm was decreased and therby long aeration time was required [33].

4 Conclusion

Many researchers reported the SND has a better nitrogen removal efficiency than the conventional nitrification and denitrification process. The SND process occurs when the nitrification and denitrification rates are in a balanced equilibrium. The stable equilibrium is based on factors such as DO, C/N ratio, reactor configuration, and thermophysical parameters. The size of the flocs is an important parameter that affects the physical mechanism of the SND process. Depending upon the microbial community present in the reactor, the mechanism can shift from conventional autotrophic nitrification and heterotrophic denitrification to heterotrophic nitrification and aerobic denitrification. Studies have reported that both the DO concentration more than 1 mg/L and less than 1 mg/L as a favorable condition for the SND process. Because of that, the optimization of DO concentration is very much essential for a stable SND process. The microbial community shift can happen because of improper DO concentration and C/N ratio. Proper C/N ratio is essential for the penetration of the organic carbon into the flocs. The reactor configuration helps in the creation of an anoxic zone and aerobic zone within the flocs. A major challenge facing the SND process is the maintenance of both denitrifiers and nitrifiers in the same reaction. In terms of structural footprint and savings in oxygen requirement, the SND is the best process for the treatment of nutrient-laden wastewater.

Acknowledgement This research was supported by the Science and Engineering Research Board (SERB).

References

1. Weerasekara, P.: The United Nations world water development report 2017 wastewater. Future Food J. Food Agric. Soc. **5**(2), 80–81 (2017)
2. Suneethi, S., Keerthiga, G., Soundhar, R., Kanmani, M., Boobalan, T., Krithika, D., Philip, L.: Qualitative evaluation of small scale municipal wastewater treatment plants (WWTPs) in South India. Water Pract. Technol. **10**(4), 711–719 (2015)
3. Finlay, J.C., Small, G.E., Sterner, R.W.: Human influences on nitrogen removal in lakes. Science **342**(6155), 247–250 (2013)
4. Galloway, J.N., Dentener, F.J., Capone, D.G., Boyer, E.W., Howarth, R.W., Seitzinger, S.P., Asner, G.P., Cleveland, C.C., Green, P.A., Holland, E.A., Karl, D.M.: Nitrogen cycles: past, present, and future. Biogeochemistry **70**(2), 153–226 (2004)
5. Mulholland, P.J., Helton, A.M., Poole, G.C., Hall, R.O., Hamilton, S.K., Peterson, B.J., Tank, J.L., Ashkenas, L.R., Cooper, L.W., Dahm, C.N., Dodds, W.K.: Stream denitrification across biomes and its response to anthropogenic nitrate loading. Nature **452**(7184), 202 (2008)
6. Paerl, H.W.: Mitigating harmful cyanobacterial blooms in a human-and climatically-impacted world. Life **4**(4), 988–1012 (2014)
7. Yao, X., Zhang, Y., Zhang, L., Zhou, Y.: A bibliometric review of nitrogen research in eutrophic lakes and reservoirs. J. Environ. Sci. **66**, 274–285 (2018)
8. Seitzinger, S.: Nitrogen cycle: out of reach. Nature **452**(7184), 162 (2008)
9. Vinçon-Leite, B., Casenave, C.: Modelling eutrophication in lake ecosystems: a review. Sci. Total Environ. **651**, 2985–3001 (2019)
10. He, S.B., Xue, G., Wang, B.Z.: Factors affecting simultaneous nitrification and de-nitrification (SND) and its kinetics model in membrane bioreactor. J. Hazard. Mater. **168**(2–3), 704–710 (2009)
11. Marin, J.C.A., Caravelli, A.H., Zaritzky, N.E.: Nitrification and aerobic denitrification in anoxic–aerobic sequencing batch reactor. Biores. Technol. **200**, 380–387 (2016)
12. Fu, Z., Yang, F., An, Y., Xue, Y.: Simultaneous nitrification and denitrification coupled with phosphorus removal in an modified anoxic/oxic-membrane bioreactor (A/O-MBR). Biochem. Eng. J. **43**(2), 191–196 (2009)
13. Bueno, R.F., Piveli, R.P., Campos, F., Sobrinho, P.A.: Simultaneous nitrification and denitrification in the activated sludge systems of continuous flow. Environ. Technol. **39**(20), 2641–2652 (2018)
14. Yan, L., Liu, S., Liu, Q., Zhang, M., Liu, Y., Wen, Y., Chen, Z., Yang, Q.: Improved performance of simultaneous nitrification and denitrification via nitrite in an oxygen-limited SBR by alternating the DO. Biores. Technol. **275**, 153–162 (2019)
15. Liu, Y., Shi, H., Xia, L., Shi, H., Shen, T., Wang, Z., Wang, G., Wang, Y.: Study of operational conditions of simultaneous nitrification and denitrification in a Carrousel oxidation ditch for domestic wastewater treatment. Biores. Technol. **101**(3), 901–906 (2010)
16. Pochana, K., Keller, J.: Study of factors affecting simultaneous nitrification and denitrification (SND). Water Sci. Technol. **39**(6), 61–68 (1999)
17. Chang, M., Wang, Y., Pan, Y., Zhang, K., Lyu, L., Wang, M., Zhu, T.: Nitrogen removal from wastewater via simultaneous nitrification and denitrification using a biological folded non-aerated filter. Biores. Technol. **289**, 121696 (2019)
18. Kornaros, M.S.N.D., Dokianakis, S.N., Lyberatos, G.: Partial nitrification/denitrification can be attributed to the slow response of nitrite oxidizing bacteria to periodic anoxic disturbances. Environ. Sci. Technol. **44**(19), 7245–7253 (2010)
19. Ji, B., Yang, K., Zhu, L., Jiang, Y., Wang, H., Zhou, J., Zhang, H.: Aerobic denitrification: a review of important advances of the last 30 years. Biotechnol. Bioprocess Eng. **20**(4), 643–651 (2015)
20. Chen, Q., Ni, J.: Heterotrophic nitrification–aerobic denitrification by novel isolated bacteria. J. Ind. Microbiol. Biotechnol. **38**(9), 1305–1310 (2011)

21. Rout, P.R., Bhunia, P., Dash, R.R.: Simultaneous removal of nitrogen and phosphorous from domestic wastewater using Bacillus cereus GS-5 strain exhibiting heterotrophic nitrification, aerobic denitrification and denitrifying phosphorous removal. Biores. Technol. **244**, 484–495 (2017)
22. Chai, H., Xiang, Y., Chen, R., Shao, Z., Gu, L., Li, L., He, Q.: Enhanced simultaneous nitrification and denitrification in treating low carbon-to-nitrogen ratio wastewater: treatment performance and nitrogen removal pathway. Biores. Technol. **280**, 51–58 (2019)
23. Chiu, Y.C., Lee, L.L., Chang, C.N., Chao, A.C.: Control of carbon and ammonium ratio for simultaneous nitrification and denitrification in a sequencing batch bioreactor. Int. Biodeterior. Biodegradation **59**(1), 1–7 (2007)
24. Padhi, S.K., Tripathy, S., Sen, R., Mahapatra, A.S., Mohanty, S., Maiti, N.K.: Characterisation of heterotrophic nitrifying and aerobic denitrifying Klebsiella pneumoniae CF-S9 strain for bioremediation of wastewater. Int. Biodeterior. Biodegradation **78**, 67–73 (2013)
25. Chen, P., Li, J., Li, Q.X., Wang, Y., Li, S., Ren, T., Wang, L.: Simultaneous heterotrophic nitrification and aerobic denitrification by bacterium Rhodococcus sp. CPZ24. Biores. Technol. **116**, 266–270 (2012)
26. Zhao, B., He, Y.L., Hughes, J., Zhang, X.F.: Heterotrophic nitrogen removal by a newly isolated Acinetobacter calcoaceticus HNR. Biores. Technol. **101**(14), 5194–5252 (2010)
27. Yan, L., Zhang, S., Hao, G., Zhang, X., Ren, Y., Wen, Y., Guo, Y., Zhang, Y.: Simultaneous nitrification and denitrification by EPSs in aerobic granular sludge enhanced nitrogen removal of ammonium-nitrogen-rich wastewater. Biores. Technol. **202**, 101–106 (2016)
28. Sun, H., Wang, T., Yang, Z., Yu, C., Wu, W.: Simultaneous removal of nitrogen and pharmaceutical and personal care products from the effluent of waste water treatment plants using aerated solid-phase denitrification system. Biores. Technol. **287**, 121389 (2019)
29. Li, J., Peng, Y., Gu, G., Wei, S.: Factors affecting simultaneous nitrification and denitrification in an SBBR treating domestic wastewater. Front. Environ. Sci. Eng. China **1**(2), 246–250 (2007)
30. Liu, J., Wang, J., Zhao, C., Liu, J., Xie, H., Wang, S., Zhang, J., Hu, Z.: Performance and mechanism of triclosan removal in simultaneous nitrification and denitrification (SND) process under low-oxygen condition. Appl. Microbiol. Biotechnol. **101**(4), 1653–1660 (2017)
31. Wang, H., Sun, Y., Wu, G., Guan, Y.: Effect of anoxic to aerobic duration ratios on nitrogen removal and nitrous oxide emission in the multiple anoxic/aerobic process. Environ. Technol. **40**(13), 1676–1685 (2019)
32. Seifi, M., Fazaelipoor, M.H.: Modeling simultaneous nitrification and denitrification (SND) in a fluidized bed biofilm reactor. Appl. Math. Modell. **36**(11), 5603–5613 (2012). Environ. Sci. **66**, 274–285
33. Zhang, L., Wei, C., Zhang, K., Zhang, C., Fang, Q., Li, S.: Effects of temperature on simultaneous nitrification and denitrification via nitrite in a sequencing batch biofilm reactor. Bioprocess Biosyst. Eng. **32**(2), 175–182 (2009)
34. He, Q., Zhang, W., Zhang, S., Wang, H.: Enhanced nitrogen removal in an aerobic granular sequencing batch reactor performing simultaneous nitrification, endogenous denitrification and phosphorus removal with low superficial gas velocity. Chem. Eng. J. **326**, 1223–1231 (2017)

Utilization of Industrial Waste in Concrete Mixes—A Review

Rajwinder Singh, Vaibhav Chaturvedi, Ankit Kumar Chaurasiya, and Mahesh Patel

Abstract In the current scenarios, the employment of various varieties of by-products in pozzolanic material has become a typical practice in concrete mixes. In this epoch of industry, innovation in technology for the utilization of fabric with higher potency and specifically reusing identical material with equal effectiveness and productivity is in high demand to avoid wasting natural resources. In this study, the feasibility of adding various industrial waste materials at discrete levels in construction material has been suggested. For Example, waste rubber tire, electric arc furnace dust, induction furnace dust used foundry sand, welding slag and others are the by-products of the industry which have been characterized as perilous because of containing some heavy metals such as zinc, cobalt, lead, copper and some other extraneous material. A significant assessment has been carried out to explore the physical characteristics and business potential of scrap tires, which can be used as an alternative to natural aggregates in concrete. Similarly, replacing rubber for natural aggregates can provide plain rubberized concrete (PRC) and it can be used for non-structural applications. Welding slag can also be a substitute of fine aggregates in plain cement concrete which shows an impactable effect by increasing the strength of concrete. This paper summarizes and provides extensive conclusions from the outcomes of the previous studies in terms of the contemporary and mechanical properties of concrete. It has been concluded that the outcome is favorable for solving socio-environmental problems with the effective use of these waste in concrete mixes in different forms.

Keywords Industrial wastes · Hazardous/toxic · Cement replacement · Concrete · Mechanical properties

R. Singh · V. Chaturvedi (✉) · A. K. Chaurasiya · M. Patel
Dr BR Ambedkar National Institute of Technology, Jalandhar, Punjab, India
e-mail: vaibhavraja104@gmail.com

D. K. Ashish et al. (eds.), *3rd International Conference on Innovative Technologies for Clean and Sustainable Development*, RILEM Bookseries 29,
https://doi.org/10.1007/978-3-030-51485-3_7

1 Introduction

While increasing the population and industrialization, the constructional activities have also been increased in the past 30 years [1]. By following that uses of natural resources like fine and coarse aggregates are decreasing at a higher rate to meet the pace of construction. In this regard, cement is an essential constructional material which has been used in almost every manufacturing activity of construction [2]. According to the ACC 2018 [3], it was observed that the consumption of cement increased by 6–7% in 2018. Due to the significant use of cement and its by-products are enhancing the environmental problems [4]. Therefore, various authors have suggested the alternate material in place of cement and aggregate in order to fulfill the demand for construction activities, consequently, it may help to reduce the impact on environment and natural resources [5–7]. Further, there are different fields such as agricultural and industrial produce waste materials, which are having the cementitious properties. These waste materials can be utilized in the concrete as a replacement material of cement or fine aggregates [5, 8–11]. Apart from that waste rubber tire, electric arc, and induction furnace dust, waste foundry, and fly ash are the main by-products that are generated in large quantities in the various industries [11–14]. Some of these wastes have higher toxicity than the permissible limits as has been suggested by EPA. As per the available literature, it has observed that such materials can be used for the cost-effective production of concrete through safe disposal and stabilization [5, 8–10, 12, 15]. Thus, this paper aims to review the available studies on the utilization of industrial waste as a construction material for concrete mixes as well as to provide the research possibilities to be carried out in near future.

2 Types of Industrial Waste Reviewed

In this section, the various types of industrial wastes are taken into consideration to evaluate the benefits provided by these materials when integrated into the concrete through the replacement of cement, coarse, or fine aggregates. The different wastes are described below:

2.1 Waste Foundry Sand (WFS)

WFS is a type of waste material that is produced in large quantities in the foundries of industries. It is generally obtained from the industries which involve the usage of ferrous and non-ferrous metals. This waste is directly disposed in the landfill sites, causing the environment and health hazard as it consists of heavy metals, and it produces leachate in the form of Cd, Cr, Mn, Zn, and Pb, etc., which contaminates the ground water. As per the available literature, the addition of WFS waste in the concrete mixes resulting in improvement of the mechanical properties of concrete [8, 16–20].

2.2 *Electric Arc Furnace Dust (EAFD)*

This is the waste that is generated from the Electric Arc Furnace during the steel-making process. In the past few decades, with the increase in the usage of electric arc furnaces in the industries from 14 to 34% from 1970 to 1998 has resulted in increasing the production of steel [21]. Because of this increment, the creation of by-products has also increased. Generally, these by-products are in powdered form and are discharged to the environment. Electric arc furnace dust is one of those pollutants which are generated from the electric arc furnace industries. It is estimated that these industries produce 3.7 million tons of dust per year [22]. It has been stated that apart from disposing of Electric arc furnace dust in the landfills, it can be used to recover the essential metals such as Pb, Zn and Fe [23]. This material can also be integrated in the cement-based industries [24–26], glass-ceramic industry for making materials out of it [26, 27] and as extra admixture in the asphalt-based cement [28].

2.3 *Crumb Rubber (CR)*

It is a type of rubber that is recycled from automobile companies and used truck tires. Meanwhile, for recycling of rubber, the steel part, and fluff are generally taken out, which is soft, small in size, and has a granulated shape. During the further processing of this rubber, its size is further reduced with the help of mechanical equipment. The utilization of this rubber as aggregates in the concrete as a replacement material has gained popularity in the past two decades. Because of the increase in the number of four-wheel vehicles around the globe, scrap tires are being generated in large quantities and have resulted in causing major environmental and health-based problems [29]. According to the United States Environmental Protection Agency (EPA), it is stated that in the year 2003, 290 million waste tires were produced [30]. Out of which, a large portion of around 45 million was used to make the tires of trucks and other vehicles. However, in Europe, from 90 plants, about 355 million numbers of tires are generated, which makes the overall 24% of worldwide production [31]. Whereas, in the United States of America, the number of tires produced is nearby 275 million/year [32]. A large number of scrap tires are also used to provide the shock protection at the banks of rives in order to reduce the impact of ships and waves, whereas, in some areas, the remaining amount of tires are open burnt which cause the numerous types environmental pollution beyond the unacceptable limits [33]. A large number of countries choose to find an alternative way to recycle and to use the scrap tires instead of disposing of in the landfill. Out of such alternatives, the addition of scrap tires in the concrete as a replacement material of cement or coarse aggregates is done, which causes a reduction in the problems associated with scrap tires and cost-effective production of concrete [21].

2.4 Fly Ash

The combustion of coal in the power plants and coal-fires boilers, a very fine powdered form of material has generated that causes the problems after disposing it into the environment. This waste is generally coming out with the flue gasses which are emitted during the combustion process, are settled down with the application of electrostatic precipitator. The composition of fly ash is generally depended upon the composition of coal, which is being burnt. Fly ash mainly consists of higher quantities of silicon dioxide (SiO_2), Aluminum dioxide (Al_2O_3), and calcium oxide (CaO) in a large amount than other compounds. In accordance with the American coal ash Association (ACAA), 43% of fly ash is recycled from the power plants and is being added to the cement-based construction as a mineral admixture to reduce the problems related to its disposal [34]. The particle size of ash is normally ranging from 0.5 microns to 300 microns. According to the Charlotte observer 2014 [35], it has been noticed that about 65% of total fly ash is disposed of directly in the landfills and water bodies without any treatment. The integration of fly ash as a partial replacement in the cement and concrete is done by various researchers, and its positive impact on the performance of concrete is observed [36–38].

2.5 Welding and Induction Furnace Slag

Welding slag is a type of waste that is toxic and is produced during the welding process. During the welding, various types of fumes and gasses are emitted in the atmospheres. Due to the higher temperature during the welding process, the materials in the arc get volatilized in the environment, which pollutes the air in surrounding area. The particle of flues gasses is so small that it can easily go into the respiratory system, creating some serious health issues like cancer [39, 40]. The size of the particle of fumes is normally below 2.5 microns [41]. The use of welding slag as a flux in concrete has been done as a partial replacement of coarse aggregates, and it was found that its addition in the concrete results in reducing the performance of concrete in terms of compressive, split tensile and flexural strength [42]. The Submerged arc welding flux (SWF) waste was added in the mortar by crushing it in very fine particle. Consequently, when it was used in the mortar specimens and it has shown the increases in the improvement of physical and mechanical properties of mortar specimens at all curing periods than ordinary mortar specimens [43]. However, industries involved in the usage of induction furnace produce various kinds of wastes such as slag, dust, and oils. In general, these waste materials are produced in greater quantities and are responsible for the air, land, and water pollution when disposed of directly without any treatment. The particle size of induction furnace dust (IFD) ranges from 1–100 microns, and 12–15 kg of slag is produced per ton of processing the raw materials [44–46]. Due to the cost-effective production of steel and other alloys, an induction furnace is normally used. Because of this reason, the generation of dust and slag also gets increased. The generated dust usually involves the higher concentrations of heavy metals such as Ni, Zn, Pb, and others, which are responsible for polluting the environment. The generated dust is also responsible for the formation of smog [47].

3 Other Industrial Waste Materials

3.1 Copper Slag (CS)

It is an industrial waste material that is broadly utilized within the sandblasting industries for the production of rough materials that are supposed to be used for with in due course of time. Copper slag is being utilized to manufacture the materials such as tiles and glasses. The main compound of this material is SiO_2, Fe_2O_3, Fe_3O_4, CaO, MgO, Al_2O_3, Na_2O, K_2O, and ZnO. The amount of SiO_2 and Fe_2O_3 is higher as compared to cement. Where SiO_2 is about 25–40%, and Fe_2O_3 is about 50–60% [48].

3.2 Rice Husk (RH)

It is a waste created in large amounts from rice-processing units around the world. The rice husk is smaller in size and normally trashed directly burnet or disposed of in the landfills. Due to the presence of a higher amount of silica than the other oxides, it is burnt in a controlled manner and then added to the cement for the production of mortar and concrete [49, 50]. These properties of rice husk ash obtained after the burning depending upon the temperature, time, and type of rice husk burnt. All these parameters affect the reactivity of rice husk ash which makes its usage in the concrete specimens [51].

3.3 Ground Granulated Blast Furnace Slag (GGBFS)

It is a special type of industrial waste that is obtained from industries where scrap iron is used for the production of steel. This waste normally falls under the category of mineral admixture, which is added to the cement to influence the properties of concrete. GGBFS is a toxic industrial waste, which consists of heavy metals. The concentration of heavy metals depends upon the type of scrap material and process used for the melting of scrap materials. When this material is added in the cement-based matrix, due to the smaller size of its particles, its hydration reactions gets started in the presence of water. After the completion of the hydration reaction, the secondary reaction gets started, and it produced the denser structure because of the consumption of the $Ca(OH)_2$ and formation of CSH gel in the concrete microstructure [52]. The chemical composition of this waste is quite similar to the composition of the cement composition, which has large amount of oxides such as Fe_2O_3, Al_2O_3 and SiO_2. The amount of lime in this waste is ranges from 30–40% [53, 54]. As per the available literature on the addition of this waste in the concrete, it has concluded that GGBFS has the capability to enhance the durability and mechanical properties of concrete [52]. It has been reported that approximately 300 kg of this waste is being produced during the per ton of pig iron manufactured [48].

The industrial wastes which are taken reviewed in this paper are represented in Fig. 1.

Fig. 1 Pictures of the different types of industrial wastes used

4 Environment Concern

The pH of waste foundry sand depends upon the metal is to be casted. The range of pH of waste foundry sand ranges from 4 to 8, which is comparable to the normal range of atmospheric rainfall [55]. During the heavy rainfalls, the heavy metals present in the waste foundry sand started leaching towards the groundwater. Thus, it may cause the contamination of drinkable groundwater [56]. It has also been reported that there are some of the foundry sands which have caused the corrosion to the metals, leading to reducing the strength of those metals [57].

As reported earlier, Electric arc furnace dust (EAFD) is one of the most toxic by-products which are being generated from the steel industries during the production of steel and iron-based materials. This waste falls under the hazardous category suggested by the Environmental Protection Agency (US). The concentration of heavy metals is generally higher than the acceptable limits. The heavy metals such as Zn, Pb, Cr, Cd, Ni, Co are present in the electric arc furnace dust. It has also been reported that many toxic products are recovered from being used again in the industries. During the production of one ton of steel, around 15–20 kg of dust is generated, which is a higher content in terms of handling. For easy disposal of dust, it is sent to open landfills, which may also be the reason to cause the contamination of groundwater [7, 58, 59].

Waste tires are a burden on landfills when they are not reused or recycled properly. As reported by the already conducted studies, unused tires have the capacity to produce acid rain when they are burned in the atmosphere. After the vapourization of burned tires, some of the carbon portions remain unburned and cause the ill-effect on the soil. The ash portion of the rubber also has a tendency to show leaching [21, 32]. A large amount of steel and rubber can be recycled from the waste tires, which further can be used in the manufacturing process at very low or negligible changes in the normal production processes [33].

Fly ashes also have a higher amount of toxic/heavy metals as they are originated from the thermal power plant. Even the board of environment ministry's has discussed the effect of mixing fly ash with the water. The studies have been reported on the detection of heavy metals from fly ash in the contaminated groundwater. It was observed that the high content of heavy metals was detected. Due to the smaller size of fly ash, it easily gets mixed to the air and may easily get into the respiratory system and may cause serious health problems associated with the kidney and lungs [6]. According to a study conducted by the Centre for Science and Environment (CSE) [60], the disposal of fly ash is still a major issue. It has reported that the fly ash has caused the reduction in the recharging of groundwater.

Welding slag normally contains the different set of elements which are hazardous to the environment. Along with the generation of slag, flue gasses are also originated during the arc process. Arc process involves the generation of higher temperatures between the materials being adjoined together due to this small quantity of powdered form, and fine white-colored smoke is generated. Basically, the slag, which is non-ferrous, involves a higher risk of causing hard to the environment. Only a few amounts

Table 1 Concentration of heavy metals present in the electric arc furnace dust [62–66]

Heavy metal elements	Percentage (%)
Zn	7–41
Pb	3.9–9.0
Fe	23.8–27.1
Cd	0.1–0.3
Mn	1–1.5
Si	1–1.5
Cu	0.1–0.5
Cr	0.16–0.26
Cl	2–3.9
F	0.2–0.9

of welding slag is recycled and recovered. The nature of slags is normally alkaline and has the tendency to produce the leachate, which usually has pH around 11 [61]. The agricultural soils are also gets affected by the welding slag when run-off water transports the sediments of slag along with it. In industries, welding slag is being used as a discharge neutralizing agent [41].

The concentration of heavy metals of EAF dust used in various studies is given in Table 1.

5 Physical and Chemical Properties

5.1 Physical Properties

The physical properties of these industrial wastes are written in this section with appropriate discussion.

Shape and Appearance In general, the particles of waste foundry sand are in the round shape. However, the dark color is noticed of sand, which is obtained from green foundries, the greyish color is observed when the sand particles are chemically bonded [16]. Electric Arc Furnace Dust (EAFD) is a waste product that has a dark yellow color in appearance [62]. The color of fly ash may vary, which depends upon the type time of combustion of coal. Generally, fly ash possesses the grey, or light black color, and size is spherical in nature [34, 36, 37].

Particle Gradation The size of crumb rubber ranges from 75 μm to 9.5 mm, the large-sized crumb rubber is used to replace coarse aggregates in the concrete. Crumb rubber is in granular, in nature, dark black polished in nature [67]. The size of fly ash ranges from 10 to 100 μm, i.e. the particles are of the size the same as silt [36, 37]. The waste foundry sand has a minimum particle size of 75 μm and a maximum of

600 μm. Being finer in nature, sometimes in the presence of moisture, its particles get attached to one another to form lumps. A large share of dust, i.e. 85–95% has a particle size between 150 and 600 μm, whereas a lesser amount around, i.e. 5–15% has particles lesser than 75 μm [16, 17, 68, 69]. The EAF dust has the majority of particles in range between 0.5 μm to 600 μm. In most cases, a large share of dust is collected in the 300 μm sized sieve during mechanical sieving of dust [7, 58, 59]. The minimum and maximum particle size of fly ash is 0.5 and 300 μm, respectively [36–38]. The induction furnace slag has the normal particle size 1.18 mm, but may go up to 03 mm when ground through the application of mechanical equipment [70].

Specific Gravity The waste foundry sand has specific gravity around 2.18 [19], whereas another study evaluated that the WFS can also exhibit the specific gravity in the range between 2.39–2.79 [71]. Apart from that, it was reported that the steel slag has a specific gravity nearby 3.0–3.15 [69, 72]. The crumb rubber, which was free from the impurities and soft in nature, was used to measure the specific gravity, and it was found to be around 1.15. The specific gravity of fly ash obtained from the coal-based water boiler normally between 2.1 and 3.0 [73].

Density The bulk density of waste foundry sand is normally around 1690–1890 kg/m^3 [16]. A study has reported that the normal bulk density of industrial steel industrial slag is around the range between 1475 and 2395 kg/m^3 [74]. It was observed that the density of bottom ash (BA) ranges between 620 kg/m^3 [75, 76]. The density of GGBFS has reported nearly 1052–1236 kg/m^3 [76]. The bulk density varies from 2.31 to 2.36 for the various combinations of Crumb Rubber [77]. The loose density of fly ash loose ranges from 540 to 860 kg/m^3, whereas the compacted density of this waste is varying from 1120 to 1500 kg/m^3 [78, 79]. For EAFD, has the bulk density between 1.1 and 2.5 g/cm^3 [80].

5.2 *Chemical Properties*

The chemical properties and composition of various type of wastes are explained which influence the properties of concrete mixes.

Composition of Oxides

The chemical composition of industrial waste foundry sand is rich in the amount of silica content [16]. Because of this, it has hydrophilic properties, which helps it to attract the water on the surface [56]. It has also observed that the amount of silica in the waste foundry sand can be lower than the usual, due to the presence of impurities in it [81]. Electric Arc Furnace Dust (EAFD) contains Fe_2O_3, which is about 32% as major components and Zinc Oxide (ZnO) 29% and other components such as silica (4%), NaCl (5.79%), MgO (4.66%) [82]. It has been observed that SiO_2, Fe_2O_3, and CaO are the major compounds that are present in higher amounts than the other materials such as magnesium oxide. However, the chemical composition of electric

Table 2 Chemical composition of industrial by-products [66, 33, 86–96]

Oxides	Electric arc furnace dust	Induction furnace slag	Fly ash	Waste foundry sand	Cement
Oxides	% Dry wt				
ZnO	6.47–46.4	–	–	–	–
Fe_2O_3	24.2–39.56	20.05	3.70–9.30	0.25–4.83	2.40–3.90
PbO	4–5.9	–	–	–	–
SiO_2	5.3–5.76	40.59	39.90–59	78.81–95.10	21.9–22.60
Al_2O_3	0.74–1.6	13.54	16–25.80	0.8–6.62	4.30–6.9
CaO	3.5–6.59	7.12	3.30–24.30	0.035–1.88	63.0–64.40
K_2O	0.48–2.2	–	0.090–1.30	0.04–0.68	0.60–1.00
Cl	0.6–1.8	–	–	–	–
MgO	1.8–4.25	–	1.40–4.60	0.19–0.30	2.10–2.5
SO_3	1–1.6	–	0.60–3.30	0.01–0.09	1.70–2.30
Na_2O	–	–	–	0.04–0.26	0.20–0.27
MnO	1.8–5.88	11.68	–	–	–

arc furnace dust totally depends upon its origin, manufacturing process, quality of scrap used [83].

The chemical composition of industrial waste (i.e. steel slag) gets changed when different types of furnaces used during the melting process. This industrial waste from the steel industry mainly has a high content of SiO_2, CaO, Fe_2O_3, Al_2O_3, MgO, MnO, P_2O_5 [84]. The chemical composition of fly ash determined by ASTM C 311 and it was found two major quantities SiO_2 (55.3%) and Al_2O_3 (25.7%) [74, 85].

The chemical compositions of all these wastes used in previous studies are given in Table 2.

6 Fresh Properties of Concrete

6.1 Slump Test

Slump is the property of fresh concrete mix, which provides finishing ability, mobility, and stability [20]. According to Indian Standard 456, 2000, it has been defined the values for low workability (25–75 mm), medium workability (50–100 mm) and high workability (100–150 mm) [97]. Different researchers have performed slump tests on the concrete mix, which are prepared by partial replacement of aggregates with industrial wastes. They are defined as follows-

Waste Foundry Sand An examination of the slump value of concrete having the varying percentage of waste foundry sand at w/c content of 0.44. After carrying out the experimental work, it was taken into account that WFS does not have much effect on workability [98]. Other research have been carried out with the same experimental setup to examine the value of slump with the replacement of normal sand by foundry sand at different percentages with w/c ratio of 0.50. It was concluded that the value of slump reduces as percentages of WFS is increasing in the concrete mix [99]. For determining the value of slump with partial replacement of WFS with aggregates, the sand that passes through 3/8-inch sieve, was taken for the replacement at different percentages 0, 20, 60%, 80%. It has been observed that value of slump decreases as the amount of WFS increased [17].

Crumb Rubber The slump value of concrete containing crumb rubber results in decreasing the value [100]. A similar pattern has been seen when the recycled aggrades and crumb rubber, along with some fibers, were used in the concrete mix by various researchers. It was obtained that there is a significant reduction in slump value when crumb rubber is introduced to the concrete mix [101–103]

Fly Ash Research work carried out to form a lightweight concrete by adding 10, 20, 30% of fly ash. It was concluded that the value of slump was in good working level for the concrete mix at 10%. Although, as the value of fly ash increased to 20 and 30% there is a slight reduction in the value of slump [104].

In addition to that, the study was carried out to examine the effect of fly ash in the concrete reveal the reduction in slump value. The test was performed by adding 60% of fly ash, and it was noticed the 5 mm low slump value. The results of study suggests that increment in fly ash content would increase the water demand in fresh concrete, and it was not worthy to carry out workability if the value of slump is less than 10 mm [105]. The workability of Portland Fly Ash and PC control mix for water/binder of 0.56, showing high slump for Portland Fly Ash than PC mix [106].

6.2 Compressive Strength-

Various physical and mechanical tests are being performed on concrete specimens. However, compressive strength test is very important to determine all properties of the concrete mix. Performing a compressive strength test can illustrate concrete work in different states. Their work define that all parameters of concrete are based on compressive strength results [107].

Waste Foundry Sand The decrease in the compressive strength of concrete specimens made with the introduction of WFS and BA at different proportions has been noticed. Experimental results obtained at 7, 28, 90, and 365 days are illustrated in Fig. 2 [20]. The results of the compressive strength of concrete specimens found at different curing days are presented in Fig. 2

Values of compressive strength after introducing WFS as fine aggregates at 25 and 35% were decreased at 28 days [108]. While there was an increase in the value of

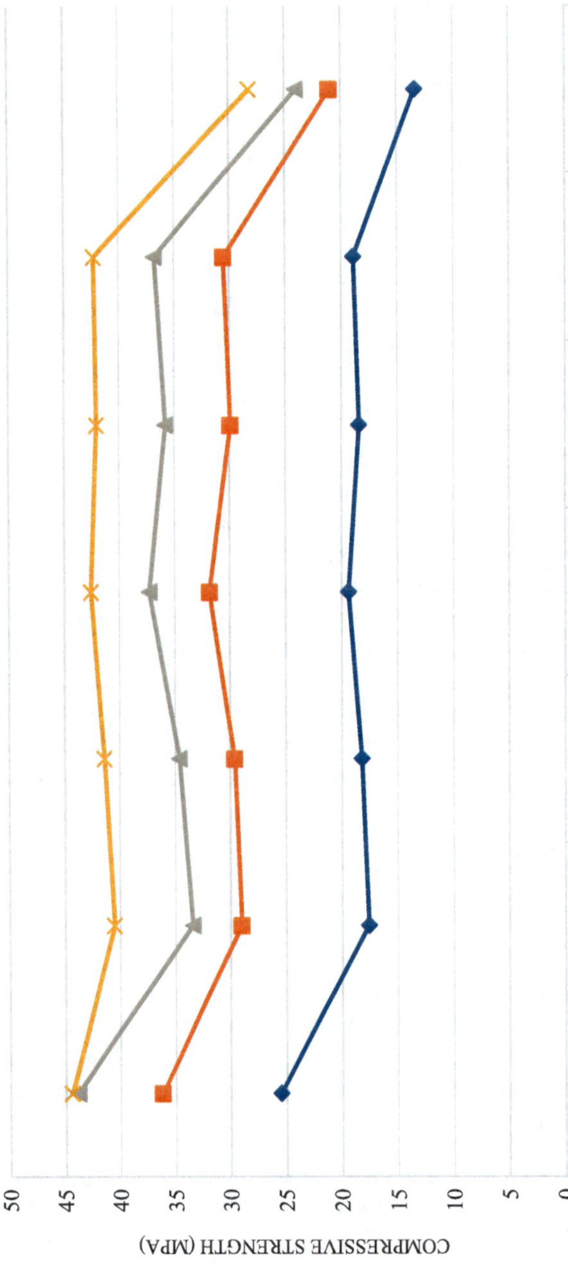

	Control Mix (0% WFS + 0% BA)	Mix 1 (5% WFS + 5% BA)	Mix 2 (10% WFS + 10% BA)	Mix 3 (15% WFS + 15% BA)	Mix 4 (20% WFS + 20% BA)	Mix 5 (25% WFS + 25% BA)	Mix 6 (30% WFS + 30% BA)
7 days	25.5	17.6	18.2	19.36	18.4	18.9	13.4
28 days	36.2	29	29.6	31.8	29.9	30.5	21
90 days	43.9	33.4	34.6	37.3	35.8	36.8	24
365 days	44.42	40.5	41.4	42.6	42.03	42.27	28.2

Fig. 2 Pictures of the different types of industrial wastes used

strength to 8.25–17% after 28 days of curing, when WFS is introduced as a natural sand replacement [19]. There was an increase in compressive strength at 28 days when sand is replaced with WFS [8]. For WFS passing through the 3/8-in. sieve, was taken for the replacement in the concrete, resulting in the reduction in the value of the compressive strength of the concrete mix [21].

Fly Ash It was found that the strength increases as there is an increase in the percentage of fly ash at different w/c ratios with different proportions of fly ash [74]. An significant increase in fly ash content shows a decline in the value of compressive strength [104, 106]. It can be revealed that the addition of fly ash in concrete command better packing which results in filling voids between cement particles. However, it was found out that if fly ash value increased beyond the 20%, there was a reduction in compressive strength [109].

Crumb Rubber As the percentage of crumb rubber in the mix increases, causing declination in the value of compressive strength. Researchers have also suggested the adhesion property between rubber and cement particles causes the reduction in the strength [100]. Various researchers observed that the reduction in compressive strength value at 20% replacement of crumb rubber [101, 110]. Similar results were observed that there is a decrease in strength as an increase in the percentage of crumb rubber when specimens are cured up to 90 days. It has been noticed that a significant effect of w/c ratio on compressive strength because of amount of crumb rubber [111].

Induction Furnace Slag Slag was replaced with recycled aggregates at different proportions for the formation of mix concrete. It was found that the compressive strength of concrete made with 50% replacement of recycled aggregates was highest among all of the other mixes [33].

EAFD The compressive strength of concrete specimen's manufactured with the addition of dust after 90 days of curing ranges of 44–50 Mpa. However, researchers have also reported that the addition of EAF dust with the ordinary Portland cement shows almost equal results to the concrete made without the addition of EAF dust after 90 days of curing [112]. The addition of dust in concrete with varying percentages (0–3%) and different curing time periods, resulting in the addition of 2% of dust to the concrete yields more compressive strength than the other percentages [10]. It has also been investigated that when concrete containing EAF dust cured at elevated temperatures for a longer time period, a significant reduction in the compressive strength of specimen is observed [113]. Experimental work was done by adding a microorganism with a combination of EAFD and Rice straw. Their experimental research carried out with a comparison of compressive strength with and without microorganism. The compressive strength shows better results with the addition of microorganism [114].

6.3 Flexural Strength

Waste Foundry Sand A study was carried out with the addition of waste foundry sand in the concrete, it was observed that the flexural strength of the concrete mix containing 20% of WFS was approximately equal to the flexural strength of reference mix [98]. Experimental works carried out with the result that there is a marginal increase in flexural strength value with an increment of WFS [99]. There was a decrease in flexural strength in the concrete mix at different percentages. It is estimated that concrete mixes varied marginally with the increase in waste foundry sand [20].

Fly Ash As same as the compressive strength, flexural strength increases. Results reported that there is an increase in flexural strength with fly ash percentage increment [74, 115].

Crumb Rubber The concrete containing rubber ash in it reported that the flexural strength of concrete decrease with an increase in the amount of rubber ash. This is mainly due to the shape of the rubber particle. The results of flexural strength of concrete, having waste rubber in it, show that when the amount of crumb rubber in concrete increased the flexural strength gets decreased [100].

7 Split Tensile Strength

Waste Foundry Sand Various researchers suggested that the tensile strength decreases with an increase in foundry sand substitution rate. They have found that at 28 days of curing, there was a decrement in the value of split tensile strength concrete as the percentage of WFS increases [68, 98]. In case of sand replacement with equally bottom ash and foundry sand, it was observed that split tensile decreases as the amount of replacement is increased [20].

Crumb Rubber Split tensile strength was higher at 10% recycled coarse aggregates (RCA) replacement, 0% crumb rubber (CR) replacement along with 2% fiber and decreases when there is an increase in RCA content and CR content [102]. Further, it was found that when the replacement of fine aggregate with crumb rubber along with high volume fly ash for the production of roller compacted pavement, causing reduction in split tensile strength. However, it was increased with the addition of nano-silica content [116].

8 Conclusion and Discussion

This review paper presents the studies on the addition of toxic and hazardous industrial by-products in the concrete, and their observations are made. The appropriate discussion has been carried out in accordance with the results of the previous studies to explore the future possibilities of using such wastes in other construction materials. The conclusions of the various studies have been made as follows:

Waste foundry sand

- The slump value of the concrete made with the addition of waste foundry sand is generally lower than the reference mix (0%). As the content of WFS is increased, a significant decrease in the slump value was observed.
- By the addition of WFS in the concrete mixes, improvement in the compressive strength was observed. Even after the treatment of WFS with the microbes, the increase in the strength is reported.
- The flexural strength concrete gets decrease with increase in the amount of waste foundry sand in the concrete. At 10% of replacement, the obtained results are marginal to the control mix.
- Split tensile strength of concrete gets decreased as the proportion of WFS is increased.

Crumb Rubber

- The addition of crumb rubber has resulted in decreasing the slump values. The same pattern is observed when recycled aggregates were used along with crumb rubber.
- With the increase in the percentage of crumb rubber, the compressive strength was reduced by a significant amount. The reason behind was the lack of adhesion between the particles of cement and rubber.
- The flexural strength of rubber ash concrete decrease with an increase in the percentage of rubber ash in the matrix.
- The split tensile strength of specimens was reduced when crumb rubber was added in them.

Electric Arc Furnace Dust

- The addition of EAF dust in concretes provides the best results when optimum content of 5% is added as a replacement material of cement. Beyond this percentage, the negative impact on setting times and compressive strength are observed.

Induction Furnace Dust and Slag

- The compressive strength of IF slag was improved when added up to 50% as a replacement of recycled aggregates.

Fly Ash

- At the addition of 10% (optimum concentration), the slump value is improved. Beyond this percentage, a noticeable reduction in the slump value was observed.
- The addition of fly ash in the concrete mixes caused the enhancement in strength. However, the more integration of fly ash results in a reduction in the strength.
- The integration of fly ash in concrete specimens has caused the improvement of the flexural strength after 28 days of curing.

All the available literatures based on the addition of these materials were examined thoroughly. It was suggested that these materials can be added in the combinations with one another in the various proportions to check their effect on the performance of concrete. Microbes have the ability to improve the properties of concrete. Thus, these waste materials can also be added along with different concentrations of microbes for concrete production. As the stabilization of these materials in the concrete results in the safe disposal, their addition in the clayey soil after the treatment is suggested be done in order to evaluate the change in the properties.

References

1. US Geological Survey: Major countries in worldwide cement production from 2012 to 2017 (in million metric tons). https://www.statista.com/statistics/267364/world-cement-productionby-country/. Last accessed 19 Dec 2014 (2018)
2. Gibbs, M., Soyka, P.: CO_2 emissions from cement production. In: Good practice guidance and uncertainty management in National Greenhouse gas inventories. 175–182 (2001)
3. ACC: Cement demand. In: Economic Times. https://economictimes.indiatimes.com/industry/indl-goods/svs/cement/cement-demand-to-grow-by-7-in-2018-excess-capacity-a-concern-cc/articleshow/64621120.cms. Last accessed 19 Dec 2014
4. Karstensen, K.H., Engelsen, C.J., Ng, S., Saha, P.K., Malmedal, M.N.: Cement manufacturing and air quality. Comprehens. Anal. Chem. **73**, 683–705 (2016)
5. Manso, J.M., Gonzalez, J.J., Polanco, J.A.: Electric arc furnace slag in concrete. ASCE J. Mater. Civil Eng. **16**, 639–645 (2004)
6. Hutchinson, T.C., Meema, K.M.: Human health concerns of lead, mercury, cadmium and arsenic. In: Hutchinson, T.C. (ed.) Lead, Mercury, Cadmium and Arsenic in the Environment, 1st edn, pp. 53–68. Wiley, UK (1987)
7. Guézennec, A.G., Huber, J.C., Patisson, F.: Dust formation in electric arc furnace: birth of the particles. Powder Technol. **157**(1–3), 2–11 (2005)
8. Kaur, G., Siddique, R., Rajor, A.: Micro-structural and metal leachate analysis of concrete made with fungal treated waste foundry sand. Constr. Build. Mater. **38**, 94–100 (2013)
9. Filipponi, P., Polettini, A., Pomi, R., Sirini, P.: Physical and mechanical properties of cement-based products containing incineration bottom ash. Waste Manag **23**(2), 145–156 (2003)
10. Al-Zaid, R.Z., Al-Sugair, F.H., Al-Negheimish, A.I.: Investigation of potential uses of electric-arc furnace dust (EAFD) in concrete. Cement Concr. Res. **27**(2), 267–278 (1997)
11. Mujedu, K.A., Adebara, S.A., Lamidi, l.: The use of corn cob ash and saw dust ash as cement replacement in concrete works. Int. J. Eng. Sci. (IJES) **3**(4), 22–28 (2014)
12. Alizadeh, R., Chini, M.,: Utilization of electric arc furnace slag as aggregates in concrete—environmental issue. In: Proceedings of the 6th CANMET/ACI International Conference on Recent Advances in Concrete Technology, pp. 451–464, Bucharest, Romania(2003)
13. Mehta, P.K.: Pozzolanic and cementitious byproducts as mineral admixtures for concrete—A critical review. Spec. Publ. **79**, 1–46 (1983)

14. Sikalidis, C., Mitrakas, M.: Utilization of electric arc furnace dust as raw material for the production of ceramic and concrete building products. J. Environ. Sci. Health., Part A **41**(9), 1943–1954 (2006)
15. da Silva Magalhães, M., Faleschini, F., Pellegrino, C., Brunelli, K.: Cementing efficiency of electric arc furnace dust in mortars. Const. Building Mater. **157**, 141–150 (2017)
16. Singh, G.: Strength and durability studies of concrete containing waste foundry sand. Ph.D. thesis, pp. 1–153. Thapar University, Patiala-147004, Punjab (India) 2012)
17. Khatib, J.M., Herki, B.A., Kenai, S.: Capillarity of concrete incorporating waste foundry sand. Constr. Build. Mater. **47**, 867–871 (2013)
18. Basar, HM., Aksoy, ND.: The effect of waste foundry sand (WFS) as partial replacement of sand on the mechanical, leaching and micro-structural characteristics of ready-mixed concrete. Constr. Build. Mater. **35**, 508–515 (2012)
19. Singh, G., Siddique, R.: Effect of waste foundry sand (WFS) as partial replacement of sand on the strength, ultrasonic pulse velocity and permeability of concrete. Constr. Build. Mater. **26**, 416–422 (2012)
20. Aggarwal, Y., Siddique, R.: Microstructure and properties of concrete using bottom ash and waste foundry sand as partial replacement of fine aggregates. Constr. Build. Mater. **54**, 210–223 (2014)
21. Sofilic, T., Rastovcan-Mioc, A., Cerjan-Stefanovic, S., Novosel-Radovic, V., Jenko, M.: Characterization of steel mill electric arc furnace dust. J. Hazardous Mater. **B109**, 59–70 (2004). ISSN 0304-3894
22. Néstor, G.G., Borja, G.E.V.: The situation of EAF dust in Europe and the upgrading of the Waelz process. Waste Treat. Clean Technol. **99**(2), 1511–20 (2003). ISSN 0210-5621
23. MacRay, D.R.: Electric arc furnace dust disposal recycle and recovery. CMP Report 85-2. Center for Metals Production, Pittsburgh, PA (1985)
24. Vargas, A.S., Masuero, A.B., Vileia, A.C.F.: Investigations on the use of electric-are furnace dust (EAFD) in Pozzolan-modified Portland cement 1 (MP) pastes. Cement Concr. Res. **36** (2006)
25. Holter, M., Malinowska, A.: The reduction of harmfulness of heavy metals from electric arc furnace dust. Hutnik-Wiadomosci hutnicze 7–8 (2005)
26. LIST, Paluchewicz, Z.: The utilization of steel-making dusts in the glass-making industry. Hutnik-Wiadomosci hunticze **5** (2007)
27. Kavouras, P., Kehagias, T., Tsilika, I., Kaimakamis, G., Chrissafis, K., Kokkou, S.: Glass-ceramic materials from electric arc furnace dust. J. Hazard. Mater. A139 (2007)
28. Sturm, T., Murko, S., Vahčič, M., Mladenovič, A., Suput, J.S. Scancar, J.: The use of EAF dust in cement composites: assessment of environmental impact. J. Hazard. Mater. **166** 277–83 (2009)
29. Najim, K,B., Hall, M.R.: A review of the fresh/hardened properties and applications for plain-(PRC) and self-compacting rubberized concrete (SCRC). J. Constr. Build. Mater. **24**, 2043–2051 (2010) (Elsevier)
30. US Environment Protection Agency: Wastes—Resource conservation—Common wastes & materials—Scrap tires. https://archive.epa.gov/epawaste/conserve/materials/tires/web/html/basic.html. Last accessed 14 Dec 2019
31. Davidelo, P.: Recycled tyre rubber modified bitumens for road asphalt mixtures: a literature review. Const. Build. Mater. **30**, 863–881 (2013)
32. Kumar, A., Yadav, S.: A Review on use of crumb rubber as fine aggregate in concrete. JETIR **4**(11) (2017). ISSN-2349-5162
33. Najim, K.B.: Modulus of elasticity and impact resistance of chopped worn-out tires concrete. Iraqi J. Civil Eng. **6**, 93–111 (2005)
34. American coal ash association, best coal ash management practices: integrating strategies for disporsal and beneficial use. https://www.acaa-usa.org/Portals/9/Files/PDFs/ASH02-2016.pdf. Last accessed 14 Dec 2019
35. Duke energy plant reports coal-ash spill. The Charlotte Observer. https://www.charlotteobserver.com/news/business/article9094658.html. Last accessed 14 Dec 2019

36. Malhotra, V.M.: High-performance high-volume fly ash concrete. Concr. Int. **24**(7), 30–34 (2002)
37. Chindaprasirt, P., Jaturapitakkul, C., Sinsiri, T.: Effect of fly ash fineness on compressive strength and pore size of blended cement paste. Cement Concr. Compos. **27**(4), 425–428 (2005)
38. Kurama, H., Kaya, M.: Usage of coal combustion bottom ash in concrete mixture. Constr. Build. Mater. **22**(9), 1922–1928 (2008)
39. International Agency for Research on Cancer (IARC): Welding fumes and gases. lARC Monogr **49** (1987)
40. National Occupational Health and Safety Commission: Welding: fumes and gases. In: Commonwealth of Australia. Ambassador Press Pty Ltd. (1990)
41. Golbabaei, F., Khadem, M.: Air pollution in welding processes—Assessment and control methods. Curr. Air Qual. Issues 33–63 (2015)
42. Ganga, V., Rajkohila, A.: Experimental studies on concrete using SAW flux waste. SSRG Int. J. Civil Eng. (SSRG-IJCE) **12**, 1–4 (2015). http://www.internationaljournalssrg.org/IJCE/2015/Volume2-Issue12/IJCE-V2I12P101.pdf
43. Viana, C.E., Dias, D.P., Holanda, J.N.F.D., Paranhos, R.P.D.R.: The use of submerged-arc welding flux slag as raw material for the fabrication of multiple-use mortars and bricks. Soldagem Inspeção **14**(3), 257–262 (2009)
44. Saggu, T.S.: Power quality improvement by harmonic reduction with special reference to induction furnaces. Int. J. Emerging Electr. Power Syst. **17**(3) (2016)
45. Central Pollution Control Board Comprehensive Industry Document on Electric Arc And Induction Furnaces (2010)
46. Kishore, K.: Sand for concrete from steel mills induction furnace waste slag. https://economictimes.indiatimes.com/industry/indl-goods/svs/cement/cement-demand-to-grow-by-7-in-2018-excess-capacity-a-concern-acc/articleshow/64621120.cms. Last accessed 2019 Dec 2014
47. Geddes, J.A., Murphy, J.G.: The science of smog: a chemical understanding of ground level ozone and fine particulate matter. Metropolit. Sustain 205–230 (2012)
48. Government of India Ministry of Mines Indian Bureau of Mines. Indian Minerals Yearbook 2017 (Part-II: Metals & Alloys) 56th edn. NAGPUR (2018)
49. Krishnarao, R.V., Mahajan, Y.R.: Effect of acid treatment on the formation of SiC whiskers from raw rice husks. J. Eur. Ceram. Soc. **15**, 1229–1234 (1995)
50. Gastaldini, A.l.G., Isaia, G.C., Hoppe, T.F., Missau, F., Saciloto, AP.: Influence of the use of RHA on the electrical resistivity of concrete. A technical and economic feasibility study. Constr. Build. Mat. **23**, 3411–3419 (2009)
51. Pande, A.M., Makarande, S.G.: Effect of rice husk ash on mortar. Int. J. Eng. Res. Appl. (IJERA) **3**, 1710–1717 (2013). ISSN 2248-9622
52. Yeau, K.Y., Kim, E.K.: An experimental study on corrosion resistance of concrete with ground granulate blast-furnace slag. Cement Concr. Res. **35**, 1391–1399 (2005)
53. Shariq, M.: Studies in ultrasonic pulse velocity of concrete containing GGBFS. Constr. Build. Mater. **40**, 944–950 (2013)
54. Chidiac, S.E., Panesar, D.K.: Evolution of mechanical properties of concrete containing ground granulated blast furnace slag and effects on the scaling resistance test at 28 days. Cement Concr. Compos. **30**, 63–71 (2008)
55. Johnson, C.K.: Phenols in foundry waste sand modern casting. American Foundry Men's Society (1981)
56. Olutoge, F.A., Olawale, S.O.A., Gbadamosi, M.A.: Strength behavior of concrete produced with foundry sand as fine aggregate replacement. Int. J. Emerg. Technol. Adv. Eng. **5**, 35–38 (2015)
57. John, E.: Mineral aggregate conservation reuse and recycling. The Ministry (1992)
58. Cubukcuoglu, B., Ouki, S.K.: The use of alternative constituents in cement-based solidification/stabilization of electric arc furnace dust. In: Proceedings of the 7th International Conference on Sustainable Management of Waste and Recycled Materials in Construction. WASCON (2009)

59. Pavao, B., De Vargas, A.S., Masuero, A.B., Dal Molin, D.C., Vilela, A.C.: Influence of electric arc furnace dust (EAFD) in alkali-activated fly ash binder. In: Proceedings of the 11th International Conference on Non-conventional Materials and Technologies (NOCMAT'09) (2009)
60. Centre for Science and Environment, Environment Ministry Panel recommends 10-year study on fly ash disposal, Livemint. 21 Mar 2016. https://www.livemint.com/Politics/5MKRRdsTDwpFW84GDFtX5J/Environment-ministry-panel-recommends-10year-study-on-fly-a.html. Last accessed 14 Dec 2019
61. Proctor, D.M., Fehling, K.A., Shay, E.C., Wittenborn, J.L., Green, J.J., Avent, C., Bigham, R.D., Connolly, M., Lee, B., Shepker, T.O., Zak, M.: APhysical and chemical characteristics of blast furnace, basic oxygen furnace, and electric arc furnace steel industry slags. Environ. Sci. Technol. **34**(8), 1576–1582 (2000)
62. Orhan, G.: Leaching and cementation of heavy metals from electric arc furnace dust in alkaline medium. Hydrometallugy **78**, 236–245 (2005)
63. Bayraktar, C.: Solidification and stabilization of electric arc furnace dust originated from steel production by using low grade MgO additive. Master thesis, ITU Graduate School of Institute of Science Engineering and Technology, Turkey (2011) (in Turkish)
64. Bayraktar, A.C., Avşar, E., Toröz, İ., Alp, K.: Hanedar,:AStabilization and solidification of electric arc furnace dust originating from steel industry by using low grade MgO. Arch. Environ. Protect. **41**(4), 62–66 (2015)
65. De Vargas, A.S., Masuero, Â.B., Vilela, A.C.: Investigations on the use of electric-arc furnace dust (EAFD) in Pozzolan-modified Portland cement I (MP) pastes. Cem. Concr. Res. **36**(10), 1833–1841 (2006)
66. Sohal, K.S., Kaur, I., Singh, R.: Use of electric arc furnace dust in concrete: a review. In: International Conference on Sustainable Waste Management through Design, pp. 464–472 (2018)
67. Sunthonpagasit, N., Duffey, M.R.: Scrap tires to crumb rubber: feasibility analysis for processing facilities. Resour. Conserv. Recycl. **40**(4), 281–299 (2004)
68. Basar, H.M., Aksoy, N.D.: The effect of waste foundry sand (WFS) as partial replacement of sand on the mechanical, leaching and micro-structural characteristics of ready-mixed concrete. Constr. Build. Mater. **35**, 508–515 (2012)
69. Qasrawi, H., Shalabi, F., Asi, I.: Use of low CaO unprocessed steel slag in concrete as fine aggregate. Constr. Buildi. Mater. **23**, 1118–1125 (2009)
70. Morrison, C., Hooper, R., Lardner, K.: The use of ferro-silicate slag from ISF zinc production as a sand replacement in concrete. Cement Concr. Res. **33**, 2085–2089 (2003)
71. Siddique, R.: Utilization of industrial by-products in concrete. Procedia Eng. Struct. Constr. Mater. **95**, 335–347 (2014)
72. Devi, V.S., Gnanavel, B.K.: Properties of concrete manufactured using steel slag. Procedia Eng. **97**, 95–104 (2014)
73. Siddique, R.: Effect of fine aggregate replacement with Class F fly ash on the mechanical properties of concrete. Cement Concr. Res. **33**, 539–547 (2003)
74. Chang-long, W., Yan-ming, Q.I., Jin-yun, H.: Experimental study on steel slag and slag replacing sand in concrete. Int. Workshop Model. Simul. Optim. **13**, 451–455 (2008)
75. Yuksel, I., Genc, A.: Properties of concrete containing non ground ash and slag as fine aggregate. ACI Mater. J. **104**(4), 397–403 (2007)
76. Bilir, T.: Effects of non-ground slag and bottom ash as fine aggregate on concrete permeability properties. Const. Build. Mater. **26**, 730–734 (2012)
77. Malarvizhi, G., Senthul, N., Kamaraj, C.A.: Study on recycling of crumb rubber and low density polyethylene blend on stone matrix asphalt. Int. J. Sci. Res. **2**(10) (2012)
78. Malhotra, V.M.: Fly ash, silica fume, slag, and natural pozzolans in concrete (No. CONF-890607–). American Concrete Institute, Detroit, MI, USA (1989)
79. Fly ash, slag, silica fume, and natural pozzolans, In: Chapter 3, Design and Control of Concrete Mixtures EB001. The University Of Memphis

80. Nyirenda, R.L.: The processing of steelmaking flue-dust: a review. Miner. Eng. **4**(7–11), 1003–1025 (1991)
81. Basar, H.M., Aksoy, N.D.: The effect of waste foundry sand (WFS) as partial replacement of sand on the mechanical, leaching and micro-structural characteristics of ready-mixed concrete. Constr. Buil. Mater. **35**, 508–515 (2012)
82. Alsheyab, M.A.T., Khedaywi, T.S.: Effect of electric arc furnace dust (EAFD) on properties of asphalt cement mixture. Resour. Conserv. Recycl. (2013)
83. Rojas, M.F., Rojas, D.M.S.: Chemical assessment of the electric arc furnace slag as construction material: expansive compounds. Cement Concr. Res. **34**(10), 1881–1888 (2004)
84. Yi, H., Xu, G., Cheng, H., Wang, J., Wan, Y., Chen, H.: An overview of utilization of steel slag. Procedia Environ. Sci. **16**, 791–801 (2012)
85. ASTM C311/C311M-13: Standard test methods for sampling and testing fly ash or natural pozzolans for use in Portland-cement concrete
86. Fernández, A.I., Chimenos, J.M., Raventós, N., Miralles, L., Espiell, F.: Stabilization of electrical arc furnace dust with low-grade MgO prior to landfill. J. Environ. Eng. **129**(3), 275–279 (2003)
87. Laforest, G., Duchesne, J.: Stabilization of electric arc furnace dust by the use of cementitious materials: Ionic competition and long-term leachability. Cement Concr. Res. **36**(9), 1628–1634 (2006)
88. Guney, Y., Sari, Y.D., Yalcin, M., Tuncan, A., Donmez, S.: Re-usage of waste foundry sand in high-strength concrete. Waste Manag **30**(8–9), 1705–1713 (2010)
89. Siddique, R., Aggarwal, Y., Aggarwal, P., Kadri, E.H., Bennacer, R.: Strength, durability, and micro-structural properties of concrete made with used-foundry sand (UFS). Constr. Build. Mater. **25**(4), 1916–1925 (2011)
90. American Foundrymen's Society: Alternative utilization of foundry waste sand. Final Report (Phase I) Prepared by American Foundrymen's Society Inc. for Illinois Department of Commerce and Community Affairs, Des Plaines, IL. https://www.odot.org/materials/tech-reports/FOUNDRY_SAND.pdf. Last accessed 2019 Dec 14
91. Etxeberria, M., Pacheco, C., Meneses, J.M., Berridi, I.: Properties of concrete using metallurgical industrial by-products as aggregates. Constr. Build. Mater. **24**, 1594–1600 (2010)
92. Vegas, I., Urreta, J., Frías, M., Rodríguez, O., Ferreiro, S., Nebreda, B., Vigil, R.: Engineering properties of cement mortars containing thermally activated paper sludge. In: WASCON, Proceedings of the 7th International Conference on Sustainable Management of Waste and Recycled Materials in Construction, p. 225–232 (2009)
93. Dunuweera, S.P., Rajapakse, R.G.: Cement types, composition, uses and advantages of nanocement, environmental impact on cement production, and possible solutions. Adv. Mater. Sci. Eng. 1–11 (2018)
94. Gamage, N., Liyanage, K., Fragomeni, S., Setunge, S.: Overview of different types of fly ash and their use as a building and construction material (2011)
95. Ismail, K.N., Hussin, K., Idris, M.S.: Physical, chemical and mineralogical properties of fly ash. J. Nucl. Relat. Technol. **4**, 47–51 (2007)
96. Khairul, N., Al-Bakri, M.M., Rafiza, A.R., Kamarudin, H., Abdullah, A., Yahya, Z.: Study on physical and chemical properties of fly ash from different area in Malaysia. Key Eng. Mater. **594**, 985–989 (2014)
97. IS 456: Indian Standard Plain and Reinforced Concrete—Code of Practice, 4th revision. Bureau of Indian Standards 2000
98. Prabhu, G.G., Hyun, J.H., Kim, Y.Y.: Effects of foundry sand as a fine aggregate in concrete production. Constr. Build. Mater. **70**, 514–521 (2014)
99. Siddique, R., Schutter, G.D., Noumowe, A.: Effect of used-foundry sand on the mechanical properties of concrete. Constr. Build. Mater. **23**, 976–980
100. Bisht, K., Ramana, P.V.: Evaluation of mechanical and durability properties of crumb rubber concrete. Const. Build. Mater. **155**, 811–817 (2017)

101. Eshmaiel, G., Khorami, M., Maghsoudi, A.A.: Scrap tyre rubber replacement for aggregate and filler in concrete. Constr. Build. Mater. **23**, 1828–1836 (2009)
102. Hossain, F.M.: Mechanical properties of recycled aggregate concrete contain crumb rubber and polypropylene fiber. Constr. Build. Mater. **225**, 983–996 (2019)
103. Adeboje, A.O.: Experiment investigation of modified bentonite clay-crumb rubber concrete. Constr. Build. Mater. **233** (2019)
104. Muhammad, A.Z., Akmal, N.: Utilization of fly ash as partial sand replacement in oil palm shell lightweight aggregate concrete. In: IOP Conference: Matereial Science and Engineering, vol 271 (2017)
105. Mohammed, B.S., Fang, O.C.: Assessing the properties of freshly mixed concrete containing paper-mill residuals and class F fly ash. J. Civil Eng. Constr. Technol. **2**(2), 17–26 (2011)
106. Nochaiya, T., Wongkeo, W., Chaipanich, : A utilization of fly ash with silica fume and properties of Portland cement–fly ash–silica fume concrete. Fuel **89**(3), 768–774 (2010)
107. Dash, M.K., Patro, S.K., Rath, A.K.: Sustainable use of industrial-waste as partial replacement of fine aggregate for preparation of concrete—A review. Int. J. Sustain. Built Environ. **5**(2), 484–516 (2016)
108. Naik, T.R., Rudolph, K.N., Yoon-moon. C., Ramme, Bruce, W., Siddique, R.: Precast concrete products using industrial by-products. ACI Mater. J. **101**(3), 199–206 (2004)
109. Shirish, Arun: Parametric study for replacement of sand by fly ash for better packing and internal curing. Open J. Civil Eng. **5**, 118–130 (2015)
110. Gupta, T., Chaudhary, S., Sharma, R.K.: Assessment of mechanical and durability properties of concrete containing waste rubber tire as fine aggregate. Constr. Build. Mater. **73**, 562–57 (2014)
111. Al-Akhras, N.M., Smadi, M.M.: Properties of tire rubber ash mortar. Cement Concr. Compos. **26**(82), 1–6 (2004)
112. Maslehuddin, M., Awan, F.R., Shameem, M., Ibrahim, M., Ali, M.R.: Effect of electric arc furnace dust on the properties of OPC and blended cement concretes. Constr. Build. Mater. **25**(1), 308–312 (2011)
113. Malone, P.G., Kirkpatrick, T., Randall, C.A.: Potential applications of alkali-activated alumi-nosilicate binders in military operations. Report WES/MP/GL-85-15. US Army, Corps of Engineers, Vicksburg, MS (1986)
114. Singh, R., Sodhi, A.K., Bhanot, N.: Sustainable concrete production by integrating wastes: a comparative study with and without bacillus megaterium. In: International Conference on Sustainable Waste Management through Design, pp. 377–385. Springer, Cham. (2018)
115. Bouzoubaa, N., Zhang M.H., Malhotra, V.M.: Mechanical properties and durability of concrete made with high-volume fly ash blended cements using a coarse fly ash. Cement Concr. Res. **31**, 1393–1402 (2001)
116. Adamu, M., Mohammed, B.S., Shafiq, N., Liew, M.S.: Durability performance of high volume fly ash roller compacted concrete pavement containing crumb rubber and nano silica. Int. J. Pavement Eng. 1–8 (2018)

Precious Recycling of Reclaimed Asphalt as Hot Mix Asphalt by Use of Rejuvenator

Pahirangan Sivapatham and Norbert Simmleit

Abstract Asphalt is one of the most frequent and effective recyclable construction materials. The recycling of reclaimed asphalt (RAP) makes both environmental and economical, sense. In this study, asphalt mixes with four different recycling agents (rejuvenators) in different content were produced and tested in the laboratory of the TPA in Cologne. A great number of tests on asphalt mixes and extracted binders were carried out to produce a green asphalt mix with similar or better performance than asphalt mixes prepared with virgin material. The viscosity of the resulting bitumen was effectively decreased with the increase in recycling agent content. Due to that, the mixing and compaction temperature of asphalt mix with RAP could be decreased from 170 to 140 °C. By means of optimierung the contents of recycling agents, asphalt mixes with RAP content can be produced in any quality comparable to the asphalt mixes prepared with virgin bitumen and aggregates. Finally, asphalt base course mixes as warm mix asphalt with a RAP content of 40% in combination with four-selected rejuvenator were produced at asphalt mix plant and constructed in test fields. The determined test results of asphalt mixes produced at asphalt mix plant confirm the test results determined on samples produced in the laboratory. In addition, the determined test results show also that the mechanical and performance properties of resulting bitumen of asphalt mix with RAP content cannot be described by means of a single test method and several test methods are necessary. But, the respective characteristics on low, medium and high temperatures can be determined by means of adequate test methods according to well known European Standards EN 12697.

Keywords Reclaimed asphalt pavement · RAP · Rejuvenator · Warm mix asphalt · Recycling agent

---

P. Sivapatham (✉)
Road Construction and Maintenance, University of Wuppertal, Wuppertal, Germany
e-mail: psivapatham@uni-wuppertal.de

N. Simmleit
Competence Center of Construction Material of Strabag AG, TPA GmbH, Cologne, Germany

D. K. Ashish et al. (eds.), *3rd International Conference on Innovative Technologies for Clean and Sustainable Development*, RILEM Bookseries 29,
https://doi.org/10.1007/978-3-030-51485-3_8

1 Introduction

The milled material of deteriorated pavement is called reclaimed asphalt pavement (RAP). In Germany, approximately more than 11 million tons of RAPs are reused every year, while more than 40 million tons of new asphalt mixes were produced yearly in the last 25 years. In average, a hot recycling rate of more than 80% of RAP material was realized. Due to the experience on that field, a big compilation of long-term experience is available [1, 2]. The remains of 20% of RAP material are used in cold recycling [2]. This study contains only the hot recycling of RAP Material. According to the Specification, the use of recycling agent to influence the binder properties of RAP is forbidden in Germany [2]. But, in the case of re-recycling (multiple recycling) of RAP material, fulfill of the requirement of SP R&B ($\leq$71 °C) according to the German Specification will not be possible. Due to that, new approaches are necessary to influence the properties of hardened bitumen in the RAP. Therefore, in this study, asphalt base course and asphalt wearing courses mixes with a very stiffen RAP material in combination with four different recycling agents were produced and investigated. The gained test results of bitumen and asphalt mixes were compared with the test results of reference variants with virgin material to study the effect of recycling agents on resulting bitumen and asphalt properties.

2 Recycling of Asphalt

RAP material can be reused for construction of new asphalt pavement. Due to that, the recycling of RAP become of crucial interest because of economic and environmental benefits. The performance of recycled asphalt is highly affected by the properties of the RAP used, because the bitumen in RAP is brittle and susceptible to cracks. In addition, the low temperature resistance, the properties of bonding and adhesion decrease, too. In Germany, a lot of experience is available for recycled asphalt [3, 4]. The Specification TL AG-StB [5] contains the criteria for using of RAP in Germany. The maximum dosage of RAP material can be calculated by a range of characteristics of RAP composition and bitumen properties. These characteristics are

- Softening point ring and ball (SP R&B) (°C)
- Binder content (% by mass)
- Filler content <0.063 mm (% by mass)
- Fine aggregate 0.063–2.0 mm (% by mass)
- Coarse aggregate >2.0 mm (% by mass).

The average value of SP R&B is a crucial criterion and the maximum average value of SP R&B shall be $\leq$71 °C. In most cases, this criterion can be fulfilled by means of blending of soft and stiffen RAP materials. Due to that, the available experiences in Germany are based on RAP material with SP R&B $\leq$71 °C. In any case, recycling agent is not allowed to use according to the Specification TL Asphalt-StB 08/13 [6].

European Standard (EN 13108-1) suggests using only 10% for wearing course and 20% for binder and base layers in hot-mix recycling. Exceeding this amount of RAP, additional tests are needed to determine the effect of the aged bitumen in RAP on the mechanical and performance characteristics of resulting asphalt mix.

3 Ageing of Bitumen

Ageing behavior of bitumen is very complex and dependent upon provenience of crude oil and method of production. The bitumen is an organic material and that is affected by the oxygen, ozone, ultraviolet radiation and structural changes during the service time [7]. As a consequence, the structure of bitumen molecules and proportion of bitumen constituents vary and influence the behavior of asphalt performance [8]. Worldwide, researchers have developed several methods to analyze chemical composition of bitumen in virgin status and with increasing the ageing process [9]. In his study, Mr. Kallas determined that the colloidal stability of asphaltene of bitumen changes due to the aging process during the lifetime and the bitumen alters rheologicaly [10]. In addition, the mix composition, binder thickness, temperature of asphalt production and air voids in the compacted asphalt affect also the ageing process of bitumen in situ [11]. Due to that, the ageing of bitumen cannot be avoided.

The simple method to influence the viscosity of bitumen of RAP is use of softer bitumen. However, this method is not effective in the case of re-recycling because of very stiffen bitumen in RAP. Due to that, new effective methods, to restore the rheological properties of aged bitumen or to improve the performance behavior of asphalt pavements produced with RAP, have to be developed. The literature study shows that a modification of aged bitumen in the RAP by means of recycling agent promise a lot, but the efficiency of the recycling agent differs significantly regarding to the type of new asphalt, bitumen and RAP used [12]. Thus, in this study the effect of recycling agent on aged bitumen in RAP gained in Germany will be investigated.

4 Research Objectives

Currently, only the physical characteristics SP R&B and grade of bitumen describes the aged bitumen in RAP. This parameter describes only the viscosity of bitumen and does not give information about the mechanical or performance properties of asphalt mix. Few studies have been found to evaluate the fatigue resistance and low-temperature properties of the extracted bitumen. The effect on the performance of the asphalt mix produced with RAP and recycling agent compared to the reference variant needs to be thoroughly studied and must be emphasized. Therefore, in this study asphalt mixes for asphalt base course (ABC mix) and stone mastic asphalt (SMA) with several recycling agents (rejuvenators) were produced and tested regarding to mechanical properties and performance behavior at low, medium and high temperatures. However, it is important that asphalt mixes containing RAP should perform similar or better than asphalt mixes without it. Finally, five different ABC mixes as warm mix asphalt and SMA as hot mix asphalt with different recycling agents were produced at the asphalt mixing plant and paved as test fields to determine the workability and the compactibility.

4.1 Bitumen and Asphalt Tests (Asphalt Mix Produced in Laboratory)

To determine the influence of the recycling agents on aged bitumen in RAP, ABC mixes with two different recycling agents (agent 1 and agent 2) with three different ratios of 6, 9 and 12% (by weight of bitumen obtained from RAP) were investigated. A pen grade bitumen 70/100 has been chosen as virgin bitumen. A great number of tests on extracted and PAV aged bitumen were carried out to evaluate the physical and performance properties of the resulting bitumen. Particularly, thermal stress restrained specimen test (TSRST) was utilized to evaluate the low temperature behavior of the asphalt mixes. To find out the long-term behavior of recycling agents under the thermal loading, the extracted bitumen of asphalt mixes were aged in PAV and analyzed. The bitumen tests were conducted according to the respectively European Standards.

4.2 Bitumen and Asphalt Tests (Asphalt Mix Produced at Mix Plant)

Four ABC warm mixes with recycling agents and RAP content of 40% by mass were produced for test fields. The selected production temperature of warm asphalt mixes was 140 °C. The conventional production temperature of hot mix is between 160 and 180 °C. The equipment of central mixing plant permits only an addition of maximum 40% of RAP material with respect to the German Standard [6]. The dosages of recycling agents were selected with respect to the bitumen and asphalt test results gained and the recommendation of the recycling agents' producers. In addition, following asphalt tests were carried out:

- Compactibility in the laboratory (EN 12697-10A)
- Stiffness modulus (at different temperatures) (ITT, EN 12697-26)
- Fatigue behavior (ITT, EN 12697-24).

Table 1 Characteristics of RAP material

	Binder content in RAP (% by mass)	Binder content in 40% RAP (kg)	SP R&B (°C)	PEN (1/10 mm)
RAP	4.8	19.2	73.3	10

4.3 Properties of RAP

The gradation, binder content and bitumen properties were determined and the mix design was established upon the determined test results. Table 1 contains the determined test results of binder content and bitumen characteristics. The determined SP R&B (73.3) overruns the permitted values of 71 °C according to the Specification TL AG-StB. In addition, the PEN value of 10 (1/10 mm) indicates on very harden bitumen in the RAP.

4.4 Recycling Agents

Worldwide several recycling agents are available in the form of lake asphalt (natural bitumen), mineral oils, flux oil (recycled fuel oil/REOB), biogenous products (vegetable oils, fatty acid resin etc.) and chemical additive. The influence of recycling agent plays a key role because they need to restore the performance of asphalt mix produced with RAP. In this study, four types of commercial recycling agents from four different sources have been used. The recycling agents are described below respectively as supplied by manufactures.

- Recycling agent 1: This liquid additive is made of renewable raw materials and designed to fully restore the binder properties of RAP, ensuring longer lifetime and lower maintenance.
- Recycling agent 2: This liquid additive restore also the properties of the aged bitumen in RAP and improve the compactibility and low temperature behavior. The rutting resistance will not be affected.
- Recycling agent 3: This additive is a synthetic hard paraffin wax based on Fischer–Tropsch-Synthesis. This product improves the workability of mastic asphalt and rutting resistance.
- Recycling agent 4: This additive is also a wax based on non-oxidized and oxidized polyethylene- and EVA-copolymer-waxes. This additive modifies the bitumen, improves rutting resistance and workability.

The recycling agents 3 and 4 were used only for the asphalt mix production at the asphalt mixing plant. The recycling agents will be denoted as agent 1, 2, 3 and 4.

5 Analyzing of the Test Results

5.1 Bitumen and Asphalt Tests Results (Asphalt Mix Produced in the Laboratory)

The characteristics of each component must meet the requirement of the asphalt specification [6] to ensure that the performance behavior are equivalent to those of the reference material without RAP content. Due to that, in the first stage of the study the investigation focuses the conventional and rheological properties of extracted bitumen.

Figures 1, 2, 3 and 4 show the results of the conventional and rheological bitumen properties for the extracted and PAV aged bitumen with different types and contents of recycling agent. The PEN values increase and SP R&B decrease with the increase in recycling agent content. Nevertheless, the gained SP R&B for the selected recycling contents do not meet the value of the reference variant. However, higher PEN values on extracted and PAV aged bitumen were determined compared to the reference variant. The gained test results of PAV aged bitumen show similar behavior as extracted bitumen without recycling agent (Fig. 1). This imply that PEN value of the reference variant with the selected recycling agent content can be reached, but the SP R&B of reference variant cannot be met. Due to that, the viscosity of resulting bitumen at high temperature will be higher and hardness at the medium temperature will be lower compared to the reference variant.

The results of bending beam rheometer (BBR) at a temperature t = (−12 °C) is shown in Fig. 2. The creep stiffness decreases and the m-value increases with an increase in the content of recycling agents. The relaxations and creep stiffness of the resulting bitumen can be improved significantly compared to the reference

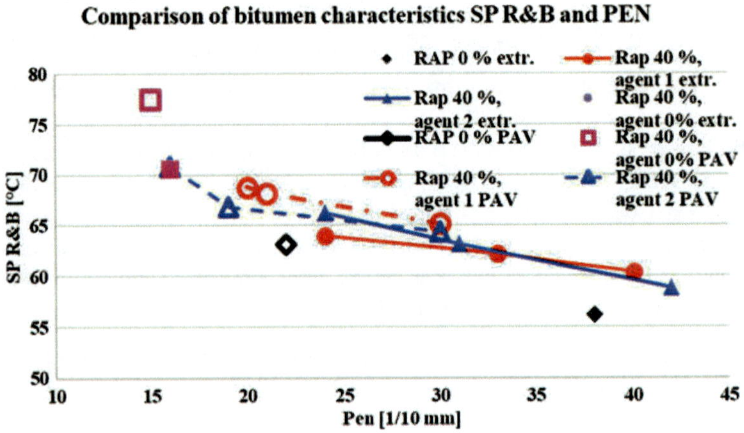

Fig. 1 Relationship of the PEN- and SP R&B values

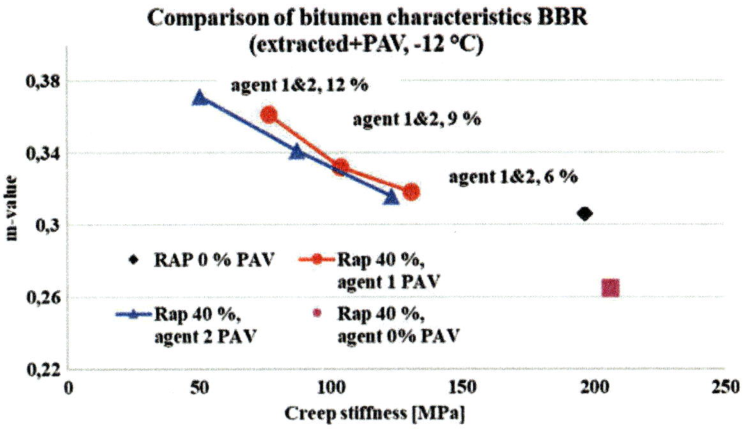

Fig. 2 Low temperature behavior (BBR), Lab mix

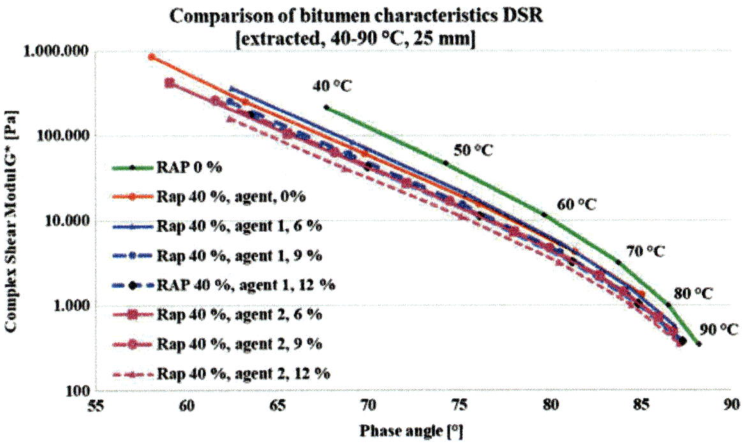

Fig. 3 High temperature behavior (DSR), Lab mix

variant. With 6% of recycling agent, the flexibility is better than that of the reference variant. The recycling agent 2 improves the low-temperature behavior more effectively compared to the recycling agent 1.

Figure 3 show the influence of the recycling agent contents on the complex shear moduli and phase angles at high temperatures. The variant without recycling agent shows higher complex shear moduli and lower phase angles. But, the variant without RAP shows the lowest complex shear moduli and highest phase angles. In common, the complex shear moduli of variants with recycling agents decrease and phase angles increase compared to the variant without recycling agent. However, the phase angle values of reference variant cannot be reached with the selected content of recycling agents. It means that the elastic properties of resulting bitumen with recycling agent

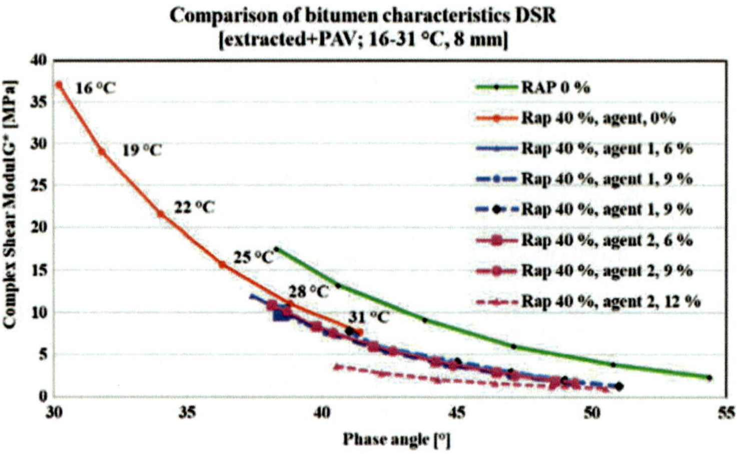

Fig. 4 Medium temperature behavior (DSR), Lab mix

is higher compared to the reference variant. Due to that, these variants will show high rutting resistance compared to the reference variant with similar complex shear moduli.

The test results of dynamic shear rheometer (DSR) after PAV ageing at medium temperature are shown in Fig. 4. The variants with recycling agents show significantly lower complex shear moduli compared to the both other variants. The phase angles of the variants with recycling agents are also lower compared to the reference variant. It means that in medium temperature, the extracted bitumen of the variants with recycling agents meet the properties of the reference variants and show better fatigue resistance with low phase angles and complex shear moduli.

Figure 5 illustrates the (cryonic) tensile stresses determined at temperatures between +20 and −20 °C and the failure temperatures. The tensile stress and the

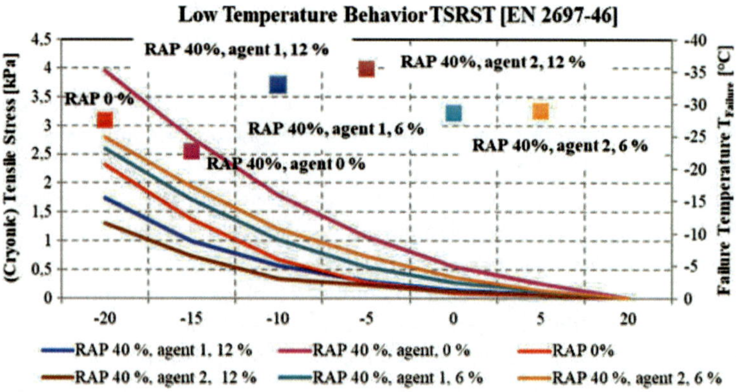

Fig. 5 Low temperature behavior of asphalt specimen

failure temperature greatly decrease with the increase in recycling agents. With decreasing tensile stresses and the failure temperatures, the resistance against low temperature cracks increases. The variant without recycling agent shows the highest tensile stress and the highest failure temperature compared to the other variants. Due to that, this variant shows the lowest resistance against the low-temperature cracks. The variants with recycling agent of 6% range between the test results of reference variant and the variant without recycling agents. The resistance against low temperature of the variants with the recycling agents of 6% is higher to those of the reference variants and the chance for thermal cracking of these variants at low temperatures is significantly reduced. With precious selection of content of recycling agent, the low temperature behavior of aged bitumen in RAP material can be restored or improved compared to the reference asphalt used in this study.

5.2 Bitumen and Asphalt Tests Results (Asphalt Mix Produced at Asphalt Mix Plant)

The test results of bitumen investigation are presented in Table 2. The determined test results of variants with recycling agents 1 and 2 confirm mostly the test results determined on asphalt mixes produced in the laboratory. Test results of the variant with the recycling agent 2 show that this recycling agent has a big influence on the aged bitumen compared to the recycling agent 1. The recycling agent 2 with a content of only 6.3% can decreases more effectively the SP R&B, creep stiffness, BP after Fraass, complex modulus and BTSV and increases the PEN value, phase angles and m-value compared to the variant with a content of 9% of recycling agent 1. Variants 3 and 4 increase the viscosity and decrease the flexibility of bitumen compared to the variants without recycling agent. The rutting resistance and the stiffness of these variants increase, but a restoration of the properties of aged bitumen cannot be determined. As expected, the compactibility of all variants tested improves with increasing compaction temperature. At the compaction temperature of 135 °C, similar compactibility has been determined compared to the compaction temperature of 150 °C. Due to that, the compaction temperature can be reduced up to 135 °C. However, the recycling agent 2 causes a better compactibility. The additives 3 and 4 do not improve the compactibility (Fig. 6).

Recycling agents 1 and 2 decrease and the recycling agents 3 and 4 increase the mechanical parameter: stiffness modulus compared to the variant without recycling agent. However, the recycling agent 2 reduce more effectively the stiffness modulus at low temperatures (Fig. 7). With decreasing stiffness modulus, the bearing capacity decreases and as consequence, the strain and deformation in the asphalt pavement will increase.

The test results of fatigue behavior are displayed in Fig. 8. The determined load cycles to failure of variants with recycling agents 1 and 2 decrease compared to the variant without recycling agent. The recycling agents 3 and 4 increase the load cycle

Table 2 Test results of bitumen investigation (extracted and extracted + PAV)

		Agent, 0%		Agent 1, 9%		Agent 2, 6.3%		Agent 3, 6.2%		Agent 4, 6.2%	
		Extracted	PAV	Extracted	PAV	Extracted	PAV	Extracted	PAV	Extracted	PAV
PEN	(1/10 mm)	27	15	34	22	40	29	23	15	24	14
SP R&B	(°C)	62.4	70.7	56.9	65	55.6	64.5	83.3	85.9	65.1	75.3
BP after Fraass	(°C)	−4.6	+ 0.2	−9.7	−6.2	−11.7	−5.6	−4.6	0.1	−6.2	−3.7
BBR creep stiffness	Extr. at −16 °C and PAV at −10 °C	170.993	112.438	118.935	75.326	115.194	71.579	188.706	116.311	169.211	89.244
BBR m-value		0.319	0.3	0.358	0.348	0.36	0.356	0.295	0.294	0.31	0.306
BTSV	T_{BTSV} (°C)	61.4	(−)	55.9	(−)	54.4	(−)	69.1	(−)	62.8	(−)
	δ_{BTSV}	76	(−)	77.2	(−)	78.3	(−)	68.6	(−)	74.3	(−)

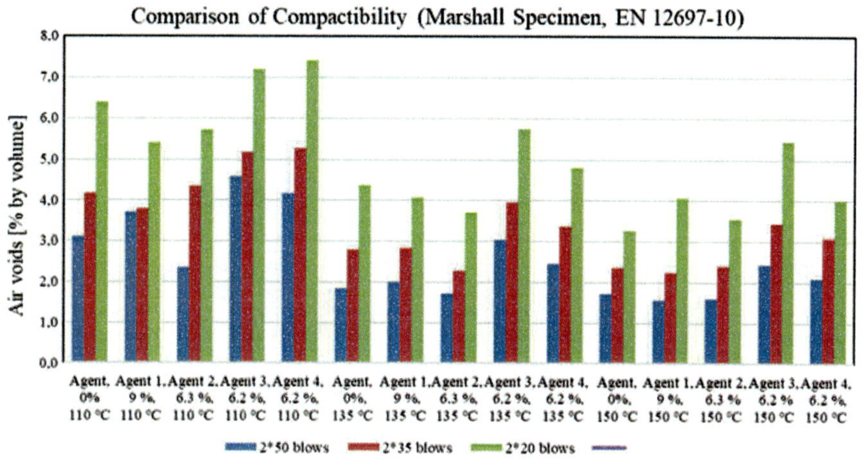

Fig. 6 Comparison of compactibility

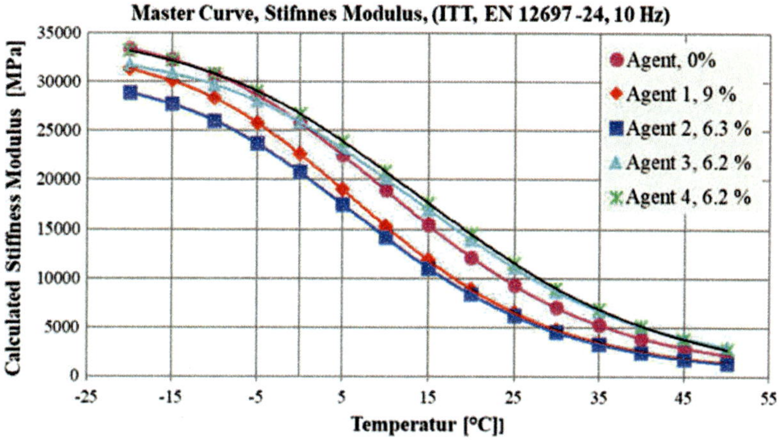

Fig. 7 Stiffness modulus

to failure. For better comparison, load cycles to failure for an initial strain of 0.08 ‰ were calculated. The variant with the recycling agent 3 shows the highest and the variant with the recycling agent 2 shows the lowest calculated load cycle. The variant without recycling agent ranges between the variants with recycling agents 1 and 4. The determined test results indicate that the fatigue resistance is depending on the stiffness modulus and hardness of resulting bitumen in medium service temperature.

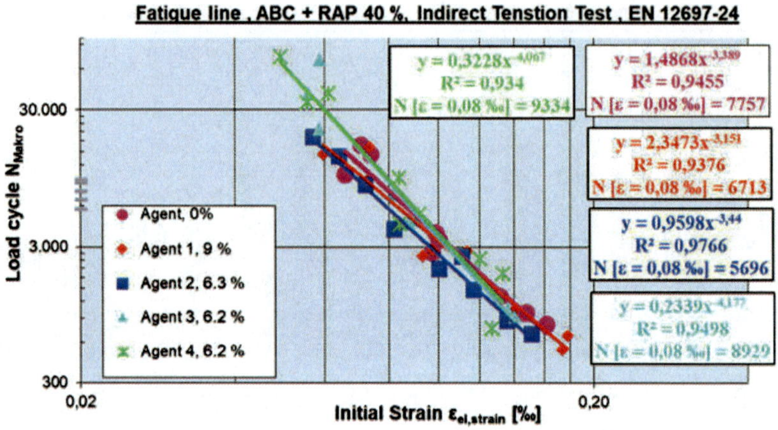

Fig. 8 Fatigue test results and calculated load cycle for ε 0.08 ‰

6 Conclusions

Several asphalt mixes with RAP and four different recycling agents were investigated in this study. The recycling agent 1 and 2 increase the PEN value, phase angle and m-value and the decrease the complex modulus, SP R&B, creep stiffness, failure low temperature behaviour, stiffness modulus and load cycles to failure. The recycling agent 3 and 4 improve the rutting and fatigue behavior, but the hardness and viscosity of bitumen increase and the flexibility at low temperature decreases. A restoration of the properties of aged bitumen cannot be determine and the recycling agents act much more as additives used to improve the asphalt performances. The recycling agents 3 and 4 are not suitable to restore the aged bitumen in RAP and cannot be used as recycling agents for constructing a standard asphalt.

The recycling agent 1 and 2 act as a recycling agent. These reduce more effectively the hardness and improve the flexibility of aged bitumen of RAP. In a quantity between 6 and 12% by mass of bitumen in RAP can be used to restore the physical properties of aged bitumen. The determined mechanical and performance properties show that asphalt mixes with RAP and recycling content can be produced with similar or better performance than asphalt mixes prepared with virgin material.

References

1. Willis, X.P.: Analysis of the Use of Reclaimed Asphalt Pavement (RAP) in Europe. Milano (2016)
2. German Asphalt Association: Asphalt Production in Germany, Bonn (2019)
3. Hiersche, E.-L.: Effect of the RAP on long-term behavior of asphalt base course material. Report of AIF Research Project Nr. 7026, University of Karlsruhe, Germany (1988)

4. Charid, K.: Recycling of RAP in the asphalt binder and asphalt wearing course mix—investigation of long-term behavior. J. Bitumen **3**, 92 (1992) (Hamburg, Germany)
5. Technical Delivery Conditions for Reclaimed Asphalt Pavement, TL AG-StB 09 (in German Language). Road and Transportation Research Association (FGSV), Bonn (2009)
6. Technical Delivery Conditions for Asphalt Mix, TL Asphalt-StB 07/13 (in German Language). Road and Transportation Research Association (FGSV), Bonn (2013)
7. Zhang, Q., Sun, H., Geng, J.: Chemical effects of environmental factors on asphalt aging. In: Advances in Civil Engineering and Architecture, Haikou, China (2011)
8. Jung, S.D: The Effects of Asphalt Binder Oxidation on Hot Mix Asphalt Concrete Mixture Rheology and Fatigue Performance. A&M University Texas, Aug 2006
9. Van den Bergh, W.: The Effect of Ageing on the Fatigue and Healing Properties of Bituminous Mortars. TU Delft, NL (2011)
10. Kallas, B.F.: Flexible pavement mixture design using reclaimed asphalt concrete. FHWA-RD-84-088, RR-84-2. National Technical Information Service, Alexandria, VA (1984)
11. Radenberg, M., Louis, H.: Change of bitumen properties during the service time of asphalt pavements. Research Report 774, Bundesministerium fuer Verkehr, Bau- und Wohnungswesen, Bonn, Germany (1987)
12. Zaumanis, M., Mallick, R.B., Frank, R.: Evaluation of different recycling agents for restoring aged asphalt binder and performance of 100 % recycled asphalt. Mater. Struct. (2015)

Thermal Properties of Foamed Concrete: A Review

Chandrashekhar D. Wagh, Gandhi Indu Siva Ranjani, and Abhishek Kamisetty

Abstract Among various types of building insulation materials developed to solve the problem of energy crisis, foam concrete is particularly interesting for its special attributes such as excellent low density, high flowability and excellent thermal insulation. The focus of this paper is to classify the literature on thermal behaviour of foamed concrete which includes the major factors affecting thermal properties and available methods of measuring it along with its pros and cons. Based on the review conducted it is observed that among the various factors studied, the microstructural parameters such as porosity, pore size and pore shape influences the thermal conductivity of foamed concrete to a great extent. Further, the literature evidence indicates that the constituent materials affects the microstructure of foamed concrete, which eventually affects its thermal behaviour. Also, studies indicate that use of foamed concrete for different structural and non-structural building applications is a viable method for reducing the heat transfer owning to its lower thermal conductivity value.

Keywords Foamed concrete · Thermal conductivity · Thermal insulation · Low density · Thermal diffusivity

1 Introduction

Considering the tremendous population boom in the recent decades and limited sources of natural energy, energy conservation becomes a prima facie solution that needs to be addressed. As per report of united nations environment program (UNEP)

C. D. Wagh · G. Indu Siva Ranjani (✉) · A. Kamisetty
Department of Civil Engineering, Indian Institute of Technology Guwahati, Guwahati 781039, Assam, India
e-mail: gindu@iitg.ac.in

C. D. Wagh
e-mail: c.wagh@iitg.ac.in

A. Kamisetty
e-mail: abhishekkamishetty@iitg.ac.in

D. K. Ashish et al. (eds.), *3rd International Conference on Innovative Technologies for Clean and Sustainable Development*, RILEM Bookseries 29,
https://doi.org/10.1007/978-3-030-51485-3_9

published in 2009, about one-third of the total energy consumption and 30% of greenhouse gas emissions are attributed to construction sector in most countries [1]. Since most people spend 90% of time living indoor for which 40% of the energy is consumed to overcome the heat flow and acquisition via surface coating and building wall units, hence prioritizing energy conservation over thermal comfort becomes rather a debatable issue [2, 3]. Using materials with low thermal conductivity (TC) during design and construction of building will help to reduce this energy consumption to a great extent. Foamed concrete which is classified as lightweight concrete (Density 400–1850 kg/m^3) can be considered as one of the most suitable material in modern building industry from thermal insulation point of view. It is basically a cement paste or mortar with air voids entrained by suitable foaming agent and possess special attributes such as low self-weight, high flowability, minimal consumption of aggregates, and excellent thermal insulation properties [4, 5]. Foamed concrete can be prepared by two techniques, pre-foaming method where a stabilized foam is prepared separately with the help of air compressor and then later mixed with base mix (consisting of cement, sand and water) while in other, mix foaming method, foam agent solution is practically mixed along with base mix in an high speed mixer [4, 6]. Air being poorest conductor of heat, light weight foamed concrete (LFC) which signifies greater porosity due to air entrainment possess lower thermal conductivity which varies according to the degree of porosity. Although the material was first patented in 1923 [7], its application in construction as lightweight non- and semi-structural material have increased in the last few years. Because of its wide spread use in countries like Netherlands, Sweden, Germany, USA, Switzerland and UK, investigators have been considering its engineering properties [8–10]. From the first comprehensive review on cellular concrete presented by Valore in 1954, there have been several reviews focusing on its applications and properties, but a detailed review on thermal properties of foamed concrete is still not available [7, 11–13]. Hence the objective of this paper is to present a state of art review of the available literature on thermal properties of foamed concrete, which includes the major factors influencing the thermal properties. The available literature on thermal properties of foamed concrete being limited, at some instances thermal performance of other types of lightweight concrete and normal weight concrete is also discussed for comparative analysis.

Thermal conductivity, thermal diffusivity and specific heat are considered as thermo-physical properties of foamed concrete. Thermal diffusivity (α-value) expresses the rate of heat spread through materials while specific heat (C-value) represents the heat storage capability of material [14]. The thermal conductivity (k-value) which refers to heat transfer by conduction through material is considered as most important property of all [15]. Though there are various steady state and transient methods that are used to measure the thermal conductivity of materials but it was observed that different methods implemented for the same specimens may result in different values of thermal conductivity [16, 17]. For calculating the energy consumption of buildings it is deemed necessary to have accurate values of thermal conductivity, however insufficient literature on suitable measurement methods compels to have a more systematic research in this direction. Hence, this paper also reviews pros and cons of various measurement methods for the thermal conductivity of foamed concrete.

2 Thermal Conductivity Measurement Methods for Foamed Concrete

Out of all the thermos-physical properties of foamed concrete, thermal conductivity (k-Value) is commonly used to check the energy performance of buildings and for conducting energy audits of the existing buildings [18, 19]. Considering this most of the researchers tried to address factors affecting Thermal conductivity (TC). TC which demonstrates heat conduction capability of material is measured primarily by two approaches namely steady state methods and transient methods which have different heat transfer conditions across materials [20–22]. A constant heat transfer approach is adopted by steady state methods, where in the temperature or heat flow is not dependent on time while transient method is dependent on time and temperature changes over time. A graphical representation of different available TC measurement methods of foamed concrete is shown in Fig. 1, while Table 1 summarizes the different TC measurement techniques, specimen sizes, testing ages, dry density range, corresponding TC range and scientific standard reported by researchers in previous studies. Also based on literature review the percentage distribution of usage of different TC measurement methods employed for foamed concrete over 32 studies is shown in Fig. 2.

2.1 *Steady State Method*

In steady state method the thermal conductivity is measured in just one Cartesian direction. From Table 1 it can be noted that the sample size required for instruments based on steady state method is much larger than instruments based on transient

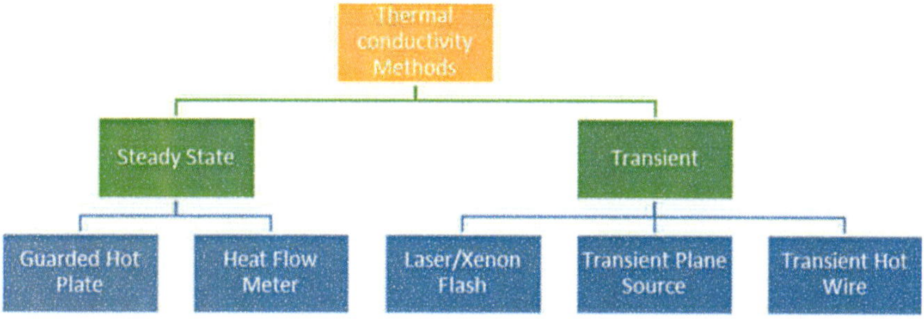

Fig. 1 Different thermal conductivity measurement methods

Table 1 Summary of thermal conductivity measurement techniques

	Measurement technique	Description	Specimen size (mm)	Testing age (days)	Standards reported by researchers	Dry density (kg/m^3)	T.C range measured (W/mK)	Refs.
Steady state method	GHP	Effect of porosity, density and pore size on K-value of FC	300 × 300 × 50	14	ASTM C 177	650–1200	0.23–0.39	[23]
	GHP	Effect of OPS as coarse aggregate on K-value	300 × 300 × 50	28	NA	1100–1600	0.4–0.57	[24]
	GHP	Effect of LS waste on K-value	300 × 300 × 50	28	IS 3346	800–1000	0.3–0.33	[17]
	GHP	Effect of SF, slag and FA on K-value	NA	NA	ASTM C 177	1490–430	0.72–0.16	[25]
	GHP	Effect of OPS on K-value of GFC	300 × 300 × 55	28	BS EN 12664	1300–1700	0.47–0.54	[26]
	GHP	Effect of temperature on K-value of FC	430 × 415 × 150	NA	ASTM C 177	650–1850	0.226–0.484	[27]
	GHP	Effect of pore size on the K-value of FC	300 × 300 × 30	28	NA	252–1870	0.065–0.5	[28]
	GHP	Effect of newspaper sandwiched on K-value of FC	300 × 300 × 50	28	BS EN 12664	1100–1700	0621–0391	[29]
	GHP	Effect of POFA on	300 × 300 × 100	28	BS EN 12664	1300±50	0.65–0.74	[30]
	GHP	Effect of W/C ratio and HPMC on K-value	300 × 300 × 30	NA	GB/T 10294	181–409	0.059–0.078	[31]
						151–180	0.050–0.055	

(continued)

Table 1 (continued)

	Measurement technique	Description	Specimen size (mm)	Testing age (days)	Standards reported by researchers	Dry density (kg/m^3)	T.C range measured (W/mK)	Refs.
	HFM	K-value for super low density FC	300 × 300 × 30	28	ASTM C 518	150–300	0.050–0.071	[32]
	HFM	K-value for very low density FC	300 × 300 × 30	28	GB/T 10294-2008	110–270	0.036–0.063	[33]
	HFM	Effect of FA and SF replacement on K-value	305 × 305 × 50	28	ISO 8301:1996	1280–1870	0.475–0.962	[34]
	HFM	Effect of MPC as binder and hydrogen peroxide as foaming agent on K-value	300 × 300 × 30	28	ASTM C518	300–1000	0.08–0.23	[35]
Transient Method	TPS	Effects of foamed content, QL and SF on the K-value of soil-based FC	200 × 200 × 50	28	NA	800–1800	0.19–0.755	[36]
	TPS	Effect of LS waste on K-value	30 × 30 × 20	28	NA	800–1000	0.32–0.34	[17]
	TPS	Effect of Fine aggregate proportion on K-value of FC	70 × 70 × 70	60	NA	916–1070	0.132–0.230	[37]
	TPS	Effect of varying FA % on K-value	25 × 50 × 10	NA	NA	600–1400	0.19–0.59	[38, 39]

(continued)

Table 1 (continued)

	Measurement technique	Description	Specimen size (mm)	Testing age (days)	Standards reported by researchers	Dry density (kg/m^3)	T.C range measured (W/mK)	Refs.
	TPS	Effect of Mineral additives—PFA, POFA, SF, WA Fibres—Steel, Polypropylene, coconut on K-value	Ø 75 mm × 40 mm	NA	NA	700–1400	0.24–0.74	[40]
	TPS	Effect of alkali activated class C FA on K-value	Ø 50 mm × 100 mm	28	ISO 22007–2	940–1310	0.23–0.31	[41]
	TPS	Effect of EPS beads on K-value of FC	200 × 200 × 50	NA	NA	400–800	0.07–0.15	[42]
	TPS	Effect of aero gel on K-value of ultra-lightweight FC	NA	NA	NA	700–200	0.164–0.049	[43]
	TPS	Effect of slag substitution on K-value of GFC.	Ø 53 mm × 15 mm	28	NA	585–1370	0.15–.048	[44]
	THW	Effect of water repellents on K-value	NA	28	NA	550	0.149–0.159	[45]
	THW	Effect of geopolymerisation on K- value	Ø 50 mm × 110 mm	56	NA	978	0.28	[46]
	THW	Effect of FA, SF and foamed % on K-value	40 × 40 × 160	NA	EN 993-15	873–1998	0.2385–0.942	[47]

FC Foamed concrete, *OPS* Oil palm shells, *LS* Limestone slurry, *FA* Flyash, *SF* Silica fume, *QL* Quick lime, *GFC* Geopolymer foamed concrete, *POFA* Palm oil fuel ash, *HPMC* Hydroxypropyl methyl cellulose, *MPC* Magnesium phosphate cement, *PFA* Pulverized fly ash, *WA* Wood ash, *EPS* Expanded polystyrene, *TC* Thermal conductivity (k-value), *NA* Not available in literature

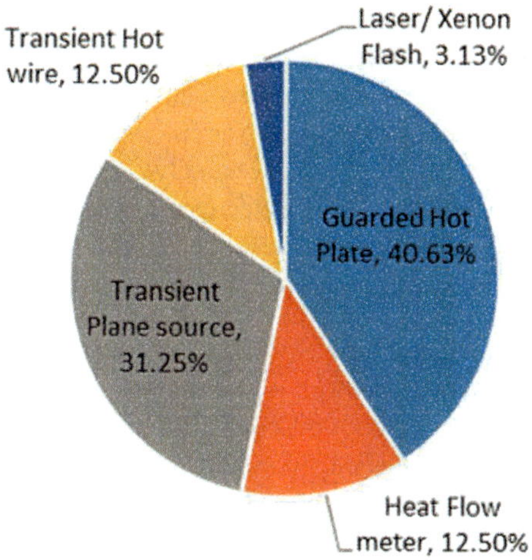

Fig. 2 The distribution of usage of different TC measurement methods employed for foamed concrete

method [17]. In contrast to transient methods, it is possible to determine the overall thermal conductivity of multi-layered specimens by using steady state method [29]. Though, the time taken by the steady state method is more but the k-value measured is more precise than that of transient method.

Guarded Hot Plate Method (GHP) The TC of a foamed concrete can be measured using GHP method by placing the specimen in between two plates cold and hot as mentioned in ASTM C177 [48]. In most of the cases the temperature of cold and hot plate were set as 18 and 40 °C to represent the interior and exterior of building structure respectively [24, 26, 30]. In case of GHP method the major advantage is its sophisticated nature and straight forward evaluation of results However long run time which is around 3 h can be considered as disadvantage [28]. Usually the specimens that were used for testing with GHP method were oven dried at above 100 °C for 24 h [26, 30, 31].

Heat Flow Meter (HFM) The HFM is an axial type steady state instrument based on measurement of density of heat flow rate. The test is conducted by placing the specimen in between the heat source at top and the sink at the bottom with the heat-flow transducer (flux gauge) placed between sample and sink. The merits and demerits of HFM are similar to that of guarded hot plate. Researchers have used this method for determining the TC of very low density foamed concrete as this method gives more accurate results at very low TC values [32, 33].

2.2 *Transient Method*

As compared to steady state method for which four different functional units (heat sink, heat source, thermometer, and sample) are required, the transient method requires only two (heat source and sample). Here the heat source also acts as thermometer while the specimen samples act as heat sink. TC of heterogeneous materials with high moisture content can be calculated by using transient method, though repetitive trials are required for precise results [49, 50]. The transient instruments come in two different classes namely contact and non-contact. Transient plane source and transient hot wire comes in contact class where there needs to be contact between sensor and specimen, while the non-contact class consists of the optical laser/xenon flash techniques.

Transient Plane Source (TPS) Which comes under contact class can be used to measure TC as well as thermal diffusivity and specific heat of foamed concrete [38–41, 44]. The hot disk sensor measuring the current and voltage drop over a period of time is embedded in between the two halves of the specimens forms a key part of the TPS technique [51]. The main advantages of this method are flexible specimen size and very less testing time which is in range of few seconds [17, 36, 41–43]. Since it comes in contact class, for better and precise results and minimizing the influence of contact resistance, the surfaces of the foamed concrete specimens need to be polished beforehand [36, 42, 44].

Transient Hot Wire (THW) is another contact class technique is method, which is usually used for measuring the TC of liquids but can also be used for solids. For measuring TC of solids two wires are required namely heater wire and a temperature sensor wire which normally have different functions.

Laser/Xenon Flash (LFA/XFA) Which is a non-contact class technique, is the most commonly used technique for determining the thermal diffusivity (directly) and TC (indirectly) of foamed concrete. While the major advantage of this method can be attributed to its ability of working over a broad temperature range and its simplicity in handling, however the requirement of a very small and homogeneous sample can be considered as disadvantage [3].

3 Factors Affecting Thermal Properties of Foamed Concrete

Table 2 summarizes the literature addressing various factors influencing the thermal properties of foamed concrete. Also it should be noted that, most researchers have investigated the TC rather than focusing on all thermal properties for foamed concrete as represented in Fig. 3 and observed in Table 2. The proportion of 39 studies evaluating the different thermal properties of foamed concrete are presented in Fig. 3. The subsequent section reviews the influence of various factors related to materials, mix composition and microstructure on thermal properties of foamed concrete.

Table 2 Factors affecting the Thermal conductivity

Factor	Type of additive	Foaming agent used	Observation	Density (kg/m^3)	Thermal properties	Other properties studied	Refs.
Air void system and density	None	Noraite PA-1 (protein based)	Lower density FC indicates greater porosity and hence lower TC.	650–1200	k = 0.23–0.39	Porosity and pore size	[23]
	Ultrafine GGBS and PF	Hydrogen peroxide	TC achieved was significantly lower than that of normal weight concrete (1.7 W/mK)	150–300	k = 0.05–0.07	CS, water absorption	[32]
	Foam replaced with EPS	Protein based	Increase in EPS % not only increased the strength but also decreased the T.C	400–800	k = 0.07–0.15	Porosity CS, TS	[42]
	None	NA	Super low density concrete with porosity ranging from 88–95% was achieved with extremely lower TC values	110–270	k = 0.036–0.063	Rheological tests, porosity, CS	[33]

(continued)

Table 2 (continued)

Factor	Type of additive	Foaming agent used	Observation	Density (kg/m^3)	Thermal properties	Other properties studied	Refs.
	None	Protein based	In case of high porosity FC radiation heat transfer is a considerable factor affecting TC. However the radiation influence starts diminishing as the porosity decreases.	252–1870	k = 0.065–0.5	Microstructure study, porosity	[28]
Moisture content and temperature	None	Noraite PA-1 (protein based)	The TC values of FC were less for high temperatures due to lower moisture content	650–1850	k = 0.226–0.484	Moisture content, Fire resistance.	[27]
	S.F as cement and F.A as filler replacement	EABASSOC (chemical)	T.C of prepared FC was more in saturated state compared to specimens in dry state due to presence of moisture.	1280–1870	k = 0.475–0.962	CS, TS	[34]
	GGBS	Synthetic	FC specimen with air dried curing showed the best thermal performance.	1167–1293	K = 0.62–0.72	CT, Ultrasonic pulse velocity	[52]

(continued)

Table 2 (continued)

Factor	Type of additive	Foaming agent used	Observation	Density (kg/m^3)	Thermal properties	Other properties studied	Refs.
Aggregate/ Filler Substitution	RTC	NA	Addition of RTC made concrete more thermal resistive due to improvement in porosity.	600 ± 25	k = 0.15–0.2	Porosity, Acoustic. Water absorption and CS	[53]
	OPS	Naphthalene sulfonated (Synthetic)	The T.C was on par with FC prepared with expanded perlite aggregate and very less compared to clay brick and blocks	1100–1600	k = 0.4–0.57	CS	[24]
	LS	Natural Protein based	T.C values were way less than normal brick and concrete	800–1000	k = 0.32–0.34 C = 1.025–1.07 α = 0.32–0.35	CS, Porosity	[17]
	OPS	Sika AER 50/50 (Synthetic)	The porous structure of OPS aggregates resulted in better thermal resistivity of OPS GFC compared to conventional materials	1300–1700	k = 0.47–0.54	Sorpitivity, porosity, CS TS and UPV	[26]

(continued)

Table 2 (continued)

Factor	Type of additive	Foaming agent used	Observation	Density (kg/m^3)	Thermal properties	Other properties studied	Refs.
	AP	NA	With the porosity maintained constant, the TC could be efficiently decreased by increasing the AP content and decreasing the cement content.	700–200	k = 0.164–0.049	CS, porosity	[43]
	POFA	Synthetic foaming agent	Replacement of sand with POFA as filler increased the TC due to densification of FC.	1300±50	k = 0.65–0.74	CS, TS	[30]
Binder/Cement and pozzolanic materials	PFA, POFA, SF, WA	Noraite PA-1 (protein based)	Use of S.F as mineral admixture resulted in less T.C than PFA. But when PFA mixed with optimum WA is used it exhibited better heat resistant behaviour.	700–1400	k = 0.24–0.74 C = 0.879–0.794 α = 0.39-0.69	None	[40]
	S.F and F.A	Hydrogen peroxide	TC of FC has much relevance with dry density	190–470	k = 0.05–0.085	CS	[54]
	FA, SF and slag	Protein based	TC depends mainly on porosity rather than mix constituents.	1490–430	k = 0.72–0.16	CS	[25]

(continued)

Table 2 (continued)

Factor	Type of additive	Foaming agent used	Observation	Density (kg/m^3)	Thermal properties	Other properties studied	Refs.
	FA, SF	Synthetic based	SF introduction resulted in superior compressive strength/TC ratios than FA introduction.	873–1998	k = 0.238–0.942	CS, water absorption, porosity	[47]
	MPC	Hydrogen peroxide	TC of the MPC foamed concrete is higher than that of OPC for a given dry density	300–1000	k = 0.08–0.23	Porosity, CS, water absorption	[35]
Mix Cement + Filler replacement	Soil as filler and SF and QL as cement replacement	Protein based	Addition of SF decreased TC while quick lime increased the T.C	800–1800	k = 0.19–0.755	Hygroscopic properties, CS,	[36]
	FA and lime as filler replacement, PF	Noraite PA-1 (protein based)	Addition of FA and PF decreased T.C	600–1400	k = 0.19–0.59 C = 0.54–0.98 α = 0.35–0.60	CS, flexural strength, porosity	[38, 39]
	S.F as cement and F.A as filler replacement	EABASSOC (chemical)	Addition of S.F and F.A leads to slightly increased TC in the dry state. However TC in the saturated state was slightly lower due to better microstructure	1280–1870	k = 0.475–0.962	CS, TS	[34]

(continued)

Table 2 (continued)

Factor	Type of additive	Foaming agent used	Observation	Density (kg/m^3)	Thermal properties	Other properties studied	Refs.
Influence of fibers and admixtures	Fibres- Steel, Polypropylene, coconut.	Noraite PA-1 (protein based)	Among various types of fibres studied coconut fibres showed lowest TC.	700–1400	k = 0.24–0.74	None.	[40]
					C = 0.879–0.794		
					α = 0.39–0.69		
	HPMC	Modified sodium alcohol ether sulfate	Addition of HPMC decreases TC to a great extent.	181–409	k = 0.059–0.078	Porosity	[31]
				151–180	k = 0.050–0.055		
	Newspaper	Synthetic	Addition of 0.05 g/cm^2 newspaper, sandwiched in foamed concrete panels were able to reduce the TC by 20%	1100–1700	k = 0.303–0.621	CS.	[29]

(continued)

Table 2 (continued)

Factor	Type of additive	Foaming agent used	Observation	Density (kg/m^3)	Thermal properties	Other properties studied	Refs.
	Water repellents	Protein based	The addition of potassium trimethylsilanolate (PT) and calcium stearate (CS) showed an increasing trend while siloxane-polymer (SP) showed irregular trend in case of TC	550	k = 0.149–0.159	CS, Sorpitivity and Hygroscopicity	[45]
Geopolymer Foam concrete	GGBS, FA	Protein based	Type of activator used in GFC does not affect the TC,	325–492	k = 0.088–0.129	Porosity, CS	[55]
	FA	Protein based	TC performance of alkali activated system was comparable with normal OPC foamed concrete system	940–1310	k = 0.23–0.31 C = 1.18–1.69 α = 0.18–0.20	None.	[41]
	Class F-FA, GGBS	NA	GFC exhibits better thermal insulation property than normal FC with similar density range.	585–1370	k = 0.15-.048 α = 0.27–0.34	CS, acoustic	[44]

K—Thermal conductivity (W/mK), C—Specific heat (J/g K), α—Thermal diffusivity (mm^2/s)
FC Foamed concrete, *OPS* Oil palm shells, *LS* Limestone slurry, *FA* Flyash, *SF* Silica fume, *QL* Quick lime, *GFC* Geopolymer foamed concrete, *POFA* Palm oil fuel ash, *HPMC* Hydroxypropyl methyl cellulose, *MPC* Magnesium phosphate cement, *PFA* Pulverized fly ash, *WA* Wood ash, *EPS* Expanded polystyrene, *PF* Polypropylene fibre, *RTC* Recycled tyre crumb, *AP* Aerogel powder, *CS* Compressive strength, *TS* Tensile strength

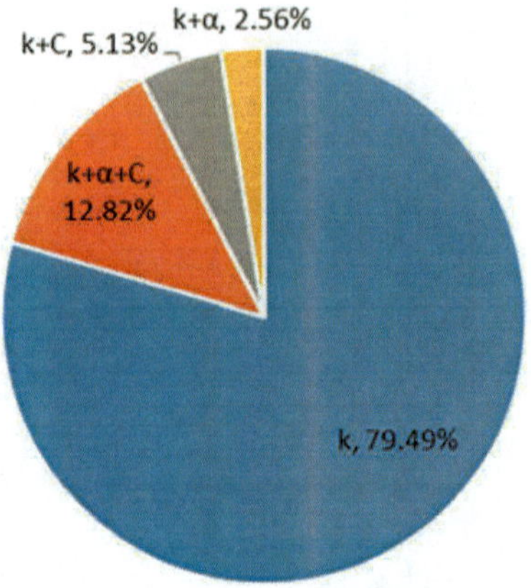

Fig. 3 Distribution of reported different thermal properties of foamed concrete (k—Thermal conductivity, C—Specific heat, α—Thermal diffusivity

3.1 Influence of Air Void System and Density

One of the most substantial factors influencing thermal properties is the density and subsequently porosity. Many of the earlier studies have established the correlation that thermal resistance is indirectly proportional to density of foamed concrete [23, 56, 57]. In this line, Pan et al. [32] prepared a super low density foamed concrete with density in range of 150–300 kg/m^3 and hence was able to achieve a very low TC of 0.05–0.07 W/mK. This was possible due to use of rapid hardening cement and accelerators which prevented the foam collapse by quick setting. Similar research was carried out by Jiang et al. [33] on very low density foamed concrete. Adding to above Chen and Liu [42] in their research replaced foam with EPS (expanded polystyrene) beads in foamed concrete. It was observed that there was a significant increase in T.C as well as compressive strength of concrete, though the density of the mixes remained almost similar. Hence the above finding proved that apart from density, properties of mix constituents also plays an important role in T.C of foamed concrete. Wei et al. [28] explored the effect of porosity level on thermal properties and established that at high porosity levels radiation heat transfer becomes a considerable factor while with decrease in porosity the radiation influence starts to diminish. TC is also predominantly influenced by directional homogeneity of pore distribution. Pores when arranged perpendicular to heat flow allowed more heat to pass through thus improving thermal resistance. However layer of pores were populated in parallel to heat flow direction offered lower thermal resistance [58].

3.2 *Influence of Moisture Content, Temperature and Curing Condition*

Studies on influence of moisture content on TC of foamed concrete proved that TC increases with increase in moisture content [11]. For instance every 1% increment in moisture content lead to increase in TC value by 6% [34]. This can be ascribed to the fact that TC of water is 25 times higher than that of air [59]. While investigating the influence of temperature variation on thermal properties of foamed concrete, Richard et al. [60] noted that rate of increase in TC with increase in density is lesser at lower temperature. Hence the thermal insulation property of foamed concrete was improved with reduction in temperature. On similar lines Othuman and Wang [27] investigated the thermal properties of foamed concrete at different temperatures ranging from 20 to 250 °C. Their experimental outcomes indicated that the TC remained constant with gradual increase in temperature till 90 °C after which it started to decrease rapidly till 170 °C. This can be attributed to evaporation of the available moisture in the pores of specimen. Beyond 170 °C the increase in temperature resulted in gradual increase in TC due to the radiation in pores. Hence moisture content in pores of foamed concrete plays an important role in thermal properties. It was also noted that specific heat was very less sensitive to temperature changes as compared to TC.

While studying the influence of curing conditions on TC, Zhao et al. [52] subjected foamed concrete specimens to four different curing conditions namely 28-day air, 7-day water + 21-day air, 28-day natural weather and 7-day water + 21-day natural weather. The 28-day air cured foamed concrete exhibited the lowest TC. This was due the lowest amount of moisture present in 28-day air cured specimens where in curing temperature and humidity was kept constant at 30 °C and 65% respectively. Further Ji et al. [61] reported that foamed concrete specimens cured at higher temperature showed lower TC. This could be attributed to the increased porosity resulting from increase in curing temperature.

3.3 *Influence of Aggregate/Filler Substitution*

Since the coarse aggregate (CA) phase is mostly absent in the foamed concrete, the filler mainly consists of only fine aggregate (FA) and the replacement of which also affects the thermal properties to a great extent. Researchers have tried to incorporate different waste materials such as recycled tyre crumbs [53], waste limestone slurry [17], granulated blast furnace slag [62], palm oil fuel ash (POFA) [30] etc. as replacement for filler in foam concrete. Their experimental outcomes indicated that the thermal insulation properties were improved in most cases due to increase in porosity. However, use of POFA as filler replacement resulted in increase in TC due to densification of microstructure. Li et al. [43] were able to prepare a high performance aerogel foamed concrete by using aerogel powder of extreme low density as filler which resulted in dry density of 198 kg/m^3 and TC of 0.049 W/Mk with better compressive strength and decreased porosity. Salvini et al. [63] proposed that use of porous ceramics as filler material for preparation of foamed concrete could result in insulating refractory composite with an acceptable thermal conductivity.

Though natural coarse aggregate due to their high specific gravity and angular shape are not preferred in foamed concrete but experimental evidences have showed that in order to improve the properties of foamed concrete, researchers have used lightweight aggregate composed of expanded clays and expanded shales as CA providing a filler effect [57]. This not only resulted in better mechanical properties but also improved the thermal resistivity of foamed concrete. On similar lines oil palm shells (OPS) were used as CA for preparing foamed concrete. The physical and thermal properties were found to be on par with concrete prepared with expanded perlite aggregates as OPS itself are porous aggregates having large number of micropores [24, 26, 64]. Wang et al. [65] prepared a new type of foamed concrete with a density of 600 kg/m^3 by incorporating ceramsite as lightweight coarse aggregate which resulted in concrete with best thermal and mechanical behaviour.

3.4 Influence of Binder/Cement and Pozzolanic Materials

The chemistry of cement plays a significant role on the thermal performance of foamed concrete. To date very few research works have been undertaken on influence of type of binder on thermal properties of foamed concrete. Based on the extensive studies, researchers have established that higher heat of hydration of cement also have a significant role on thermal performance of foamed concrete [66, 67]. Hence the above problem can be addressed by optimum use of cement and its replacement with pozzolanic fly ash with lower heat of hydration. In a study by Li et al. [35] magnesium phosphate cement (MPC) was used instead of OPC and the prepared foamed concrete was found to have higher TC along with higher compressive strength for a given dry density.

Studies on use of flyash and silica fume as additive in foamed concrete showed slight increase in TC. However a significant increase in the mechanical properties was observed and this could be attributed to the densification of microstructure due to addition of additives [25, 47]. The above increment in TC was more significant for silica fume concrete when compared to flyash blended concrete. However the TC range reported for silica fume concrete was on par with similar density range of non-autoclaved cellular gas concrete produced with perlite [68]. Ganesan et al. [40] based on their comparative studies on the various possible binder replacements such as Pulverized fuel ash (PFA), Silica Fume (SF), Palm oil fuel ash (POFA) and Wood ash (WA), found out that replacement with combination of PFA and wood ash provided the best thermal performance of all.

3.5 Influence of Cement + Filler Replacement

In order to optimize the performance as well as cost of foamed concrete, researchers have tried to replace the cement as well as filler with various substitutes. Similar such attempt was made when silica fume (SF) was used as cement replacement and flyash (FA) as fine sand replacement [34]. This lead to better mechanical properties but a slight increase in TC was observed for oven dried specimens due to dense microstructure. On contrary, in case of saturated specimens the TC observed was lower and this could be attributed to reduced water absorption resulting from improved microstructure of SF-FA concrete. In same line, Cong and Bing [36] based on comparative studies on use of SF and quick lime as cement replacement with soil as filler in foamed concrete, proved that addition of quick lime resulted in poor thermal performance. This was because in mixes with quick lime foam collapse occurred due to consumption of water from foam surface by quick lime while mixing and eventually leading to increase in concrete density. Awang and Mydin [38, 39] replaced the filler with lime and cement with flyash and found out that lime as aggregate replacement did not contribute much on mechanical behaviour but good in thermal performance by improving all three thermo-physical properties.

3.6 Influence of Addition of Fibers and Admixtures

Generally, fibers and admixtures though added in small quantities leave a good impact on the properties of foamed concrete. Use of fibres in foamed concrete resulted in better thermal as well as shrinkage resistance performance. This was due to ability of fibres to form small and uniform pores by swelling and shrinking in a concrete during mixing and drying [38, 39]. Researchers have tried the incorporation of various synthetic fibers such as polypropylene (PP), alkaline resistance glass, kenaf, steel, basalt, chrysotile asbestos and also natural fibers like oil palm fiber, coir (coconut) fibres etc. in foamed concrete. Ganesan et al. [40] studied the relative performance of different fibres such as steel, PP and coir and their experimental outcomes proved that coir fibres showed the best performance of all. Based on a similar comparative study of fibres such as PP, alkaline resistance glass, kenaf, steel and oil palm fibre Ahmad and Awang [69] reported that PP fibres turned out to be better. On the same note, Li et al. [70] studied the seismic and thermal behaviour of foam concrete precast self-insulation shear wall panels with polypropylene fibres incorporated in the mix and reported that the TC was very much lesser than previous related research. On contrary another study on synthetic fibers showed that thermal performance of polypropylene was relatively poor than basalt and chrysotile asbestos fibre [71].

Experimental results presented by Ng and Low [29] showed that mere addition of 0.05 g/cm^2 newspaper, sandwiched in foamed concrete panels were able to reduce the TC by 20% for same density. Sang et al. [31] found that the mere inclusion of 4% of hydroxypropyl methyl cellulose (HPMC) in foamed concrete lead to decrease in TC to a great extent due to increase in porosity. Ma and Chen [45] used water repellents in foamed concrete and found out that the addition of potassium trimethylsilanolate (PT) and calcium stearate (CS) showed an increasing trend while siloxane-based polymer (SP) showed constant trend in case of thermal conductivity.

3.7 Geopolymer Foamed Concrete (GFC)

Yang et al. [55] studied the properties of alkali activated GGBS foamed concrete prepared with three different types of activators. Based on their studies it was concluded that TC of alkali activated GGBS foamed concrete was found to be more dependent on density rather than on type of activator used. Liu et al. [26] evaluated the thermal properties of geopolymer foamed concrete (GFC) in which crushed oil palm shells were used as coarse aggregates. The obtained results established that the foamed geopolymer concrete was superior in thermal resistivity compared to conventional blocks and bricks available in market. While investigating the thermal properties of GFC made from class F fly ash along with partial slag substitution researchers reported. better thermal insulation properties than normal Portland cement foamed concrete at the same density [41, 44]. This was due to the low proportion of chemically bound water in the geopolymer gel, providing a more discontinuous gel structure in these material. The thermal diffusivity and specific heat values were also on the lower side, thus recommending GFC as excellent thermal insulation building material. A commercially viable and environmentally-friendly geopolymer foamed concrete was synthesized with a mix composed of a hollow cenospheres exhibiting high strength/density ratio. The material thus prepared reported a strength of 17.5 MPa at density of 978 kg/m^3 with TC value to be around 0.28 W/mK which is comparable to normal foamed concrete [46].

4 Thermal Insulating Applications of Foamed Concrete

Properties of foamed concrete like low self-weight, low thermal conductivity, high flowability, self-compacting nature, ease of production etc., have made possible the use of foam concrete over a wide range of applications in structural as well as nonstructural areas [4, 72]. On reviewing the history of use of foamed concrete it was found that in the 1930s, the Soviet Union started extensive manufacturing of thermally insulating cellular concrete blocks, wall boards, and floor slab for commercial and residential structures. later in the mid-1940 and 1950s, equipment and standards were developed, some of which are still used today [73]. Due to superior thermal insulation property, foamed concrete was used for roof insulation in middle east and South Africa. In the middle far east, almost 3000 houses were built between 1948 and 1958 using 1100–1500 kg/m^3 density of foamed concrete [13, 19]. The conditions of these dwellings assessed later were found to have performed better than contemporary timber or brick and concrete houses. In India a successful trial has been attempted in Delhi by G. B. Singh, who had constructed a G + 3 storey building with foamed concrete [74]. Approximately 250,000 m^3 of foamed concrete is used annually in Korea as an essential component in floor heating system [55]. Hence based upon the above instances it is evident that foam concrete can be a promising thermally insulating building material.

5 Summary

Based on the review conducted, the salient observations pertaining to the thermal properties of foamed concrete are summarized below. In recent years, considering the importance of energy efficiency in building sector, researchers have prioritized studies on thermal properties of foamed concrete. Out of all three thermos-physical properties, it is observed 80% of researchers had studied thermal conductivity of foamed concrete.

Steady state or transient methods are the most commonly used methods for evaluating the thermal properties of foamed concrete. it should be noted that though transient method does not prove its greater accuracy over steady state methods, however very less experiment measurement time, ease of use, compactability, and the ability to measure TC of moist specimens can be considered as some factors compelling the researchers for preferring transient methods over steady state.

Review on factors affecting thermal properties of foamed concrete indicates that density and the air void system are the most substantial factors. Studies indicated that thermal resistance is indirectly proportional to density and directly proportional to porosity of foamed concrete. Moisture content, temperature and curing conditions also have a considerable impact on thermal properties of foamed concrete. Foamed concrete prepared with lightweight aggregates not only improves mechanical properties but also significantly improve the thermal resistivity of foamed concrete. Selection of type of filler also has significant influence on TC of foamed concrete as it modifies the microstructure of concrete to a great extent. Further selection of binder also alters the thermal performance as the heat of hydration is dependent on it. Fibres and admixtures though added in small quantities but affects the micro structure and porosity to a great extent in foamed concrete thus reducing the thermal conductivity. Geopolymer foamed concrete not only on par but in some cases resulted in better insulating properties than normal foamed concrete. Hence the above extensive review indicates that foamed concrete has great potential to improve the energy performance of buildings for structural as well as non-structural applications.

In summary, it appears that availability of literature on the thermal properties of foamed concrete is very limited when compared to normal weight concrete. Hence the review highlights the need for more systematic studies to evaluate the thermal behaviour of foamed concrete and energy consumption of buildings in different weather conditions.

References

1. Hirst, N,: Buildings and climate change. Des. Manage. Sustain. Built Environ. **9781447147**, 23–30 (2013)
2. De Giuli, V., Da Pos, O., De Carli, M.: Indoor environmental quality and pupil perception in Italian primary schools. Build. Environ. **56**, 335–345 (2012)
3. Cha, J., Seo, J., Kim, S.: Building materials thermal conductivity measurement and correlation with heat flow meter, laser flash analysis and TCi. J. Therm. Anal. Calorim. **109**(1), 295–300 (2012)
4. Ramamurthy, K., Kunhanandan Nambiar, E.K., Indu Siva Ranjani, G.: A classification of studies on properties of foam concrete. Cement Concr. Compos. 31(6):388–396 (2009)
5. Sahu, S.S., Gandhi, I.S.R., Khwairakpam, S.: State-of-the-art review on the characteristics of surfactants and foam from foam concrete perspective. J. Inst. (India): Ser. A **99**(2), 391–405, 2018 (Springer, India)
6. Amran, Y.H.M., Farzadnia, N., Ali, A.A.A.: Properties and applications of foamed concrete; a review. Constr. Build. Mater. **101**, 990–1005 (2015)
7. Valore, R.C.: Cellular concretes part 1 composition and methods of preparation. ACI J. Proc. **50**(5), 773–796 (1954)
8. Wimpenny, D.E.: Some aspects of the design and production of foamed concrete. In: Concrete in the Service of Mankind: Appropriate Concrete Technology, pp. 243–252 (1996)
9. Cox, L.S.: Major road and bridge projects with foam concrete. In: Proceedings of the International Conference on the Use of Foamed Concrete in Construction, pp. 105–112 (2005)
10. Pavan Ranmale, M.: Feasibility study on conventional concrete and cellular light weight concrete (foamed concrete). Novat. Publ. Int. J. Innov. Eng. Res. Technol. 3(11): 2394–3696 (2016)
11. Valore, R.C.: Cellular concretes part 2 physical properties. ACI J. Proc. **50**(6), 817–836 (1954)
12. Short A., Kinniburgh, W.: Lightweight concretes. 1–2 (1963)
13. Jones, M.R., McCarthy, A.: Behaviour and assessment of foamed concrete for construction applications. In: Proceedings of the International Conference on the Use of Foamed Concrete in Construction, pp. 61–88 (2005)
14. Bejan, A.: Heat Transfer, 2nd edn. (1995)
15. Silva, M.A.G.: Influence of environmental aging on properties of polymeric mortars. J. Mater. Civ. Eng. **16**(5), 461–468 (2004)
16. Gomes, M.G., Flores-Colen, I., Manga, L.M., Soares, A., de Brito, J.: The influence of moisture content on the thermal conductivity of external thermal mortars. Constr. Build. Mater. **135**, 279–286 (2017)
17. Kumar, R., Lakhani, R., Tomar, P.: A simple novel mix design method and properties assessment of foamed concretes with limestone slurry waste. J. Clean. Prod. **171**, 1650–1663 (2018)
18. Ficco, G., Iannetta, F., Ianniello, E., D'Ambrosio Alfano, F.R., Dell'Isola, M.: U-value in situ measurement for energy diagnosis of existing buildings. Energy Build. **104**, 108–121 (2015)
19. Giannakou A., Jones, M.R.: Potential of foamed concrete to enhance the thermal performance of low-rise dwellings. In: Innovations and Developments in Concrete Materials and Construction. Proceedings of International Conference held at the University of Dundee, Scotland, UK 9–11 Sept 2002, pp. 533–544 (2002)
20. Bindiganavile, V., Batool, F., Suresh, N.: Effect of fly ash on thermal properties of cement based foams evaluated by transient plane heat source. Ind. Concr. J. **86**(11), 7–14 (2012)
21. Tong, X.C.: Characterization methodologies of thermal management materials. In: Advanced Materials for Thermal Management of Electronic Packaging, pp. 59–129. Springer, New York, NY (2011)
22. Zhang, W., Min, H., Gu, X., Xi, Y., Xing, Y.: Mesoscale model for thermal conductivity of concrete. Constr. Build. Mater. **98**, 8–16 (2015)
23. Mydin, A.O.: Effective thermal conductivity of foamcrete of different densities. Concr. Res. Lett. **2**(March), 181–189 (2011)

24. Johnson Alengaram, U., Al Muhit, B.A., Bin Jumaat, M.Z., Jing, M.L.Y.: A comparison of the thermal conductivity of oil palm shell foamed concrete with conventional materials. Mater Des. **51**(2013), 522–529 (2013)
25. Pan, Z., Hiromi, F., Wee, T.: Preparation of high performance foamed concrete from cement, sand and mineral admixtures. J. Wuhan Univ. Technol. Mater. Sci. Ed. **22**(2), 295–298 (2007)
26. Liu, M.Y.J., Alengaram, U.J., Jumaat, M.Z., Mo, K.H.: Evaluation of thermal conductivity, mechanical and transport properties of lightweight aggregate foamed geopolymer concrete. Energy Build. **72**, 238–245 (2014)
27. Othuman, M.A., Wang, Y.C.: Elevated-temperature thermal properties of lightweight foamed concrete. Constr. Build. Mater. **25**(2), 705–716 (2011)
28. Wei, S., Yiqiang, C., Yunsheng, Z., Jones, M.R.: Characterization and simulation of microstructure and thermal properties of foamed concrete. Constr. Build. Mater. **47**, 1278–1291 (2013)
29. Ng, S.C., Low, K.S.: Thermal conductivity of newspaper sandwiched aerated lightweight concrete panel. Energy Build. **42**(12), 2452–2456 (2010)
30. Lim, S.K., Tan, C.S., Lim, O.Y., Lee, Y.L.: Fresh and hardened properties of lightweight foamed concrete with palm oil fuel ash as filler. Constr. Build. Mater. **46**, 39–47 (2013)
31. Sang, G., Zhu, Y., Yang, G., Zhang, H.: Preparation and characterization of high porosity cement-based foam material. Constr. Build. Mater. **91**, 133–137 (2015)
32. Pan, Z., Li, H., Liu, W.: Preparation and characterization of super low density foamed concrete from Portland cement and admixtures. Constr. Build. Mater. **72**, 256–261 (2014)
33. Jiang, J., Lu, Z., Niu, Y., Li, J., Zhang, Y.: Study on the preparation and properties of high-porosity foamed concretes based on ordinary Portland cement. Mater. Des. **92**, 949–959 (2016)
34. Hilal, A.A., Thom, N.H., Dawson, A.R.: The use of additives to enhance properties of pre-formed foamed concrete. Int. J. Eng. Technol. **7**(4), 286–293 (2015)
35. Li, T., Wang, Z., Zhou, T., He, Y., Huang, F.: Preparation and properties of magnesium phosphate cement foam concrete with H_2O_2 as foaming agent. Constr. Build. Mater. **205**, 566–573 (2019)
36. Cong, M., Bing, C.: Properties of a foamed concrete with soil as filler. Constr. Build. Mater. **76**, 61–69 (2015)
37. Mahawan, J., Maneewan, S., Patanin, T., Thongtha, A.: Investigation of physical, mechanical and thermal properties of building wall materials. In: Key Engineering Materials, vol. 751, pp. 521–526. KEM (2017)
38. Mydin, A.O., Awang, H., Roslan, A.F.: Determination of lightweight foamed concrete thermal properties integrating various additives. Elixir Cement Concr. Compos. **48**(2012), 9286–9291 (2012)
39. Awang, H., Mydin, A.O., Roslan, A.F.: Effect of additives on mechanical and thermal properties of lightweight foamed concrete. Adv. Appl. Sci. Res. **3**(5), 3326–3338 (2012)
40. Ganesan, S., Othuman Mydin, M.A., Mohd Yunos, M.Y., Mohd Nawi, M.N.: Thermal properties of foamed concrete with various densities and additives at ambient temperature. Appl. Mech. Mater. **747**, 230–233 (2015)
41. Stolz, J., Boluk, Y., Bindiganavile, V.: Mechanical, thermal and acoustic properties of cellular alkali activated fly ash concrete. Cement Concr. Compos. **94**(August), 24–32 (2018)
42. Chen, B., Liu, N.: A novel lightweight concrete-fabrication and its thermal and mechanical properties. Constr. Build. Mater. **44**, 691–698 (2013)
43. Li, P., Wu, H., Liu, Y., Yang, J., Fang, Z., Lin, B.: Preparation and optimization of ultra-light and thermal insulative aerogel foam concrete. Constr. Build. Mater. **205**, 529–542 (2019)
44. Zhang, Z., Provis, J.L., Reid, A., Wang, H.: Mechanical, thermal insulation, thermal resistance and acoustic absorption properties of geopolymer foam concrete. Cement Concr. Compos. **62**, 97–105 (2015)
45. Ma, C., Chen, B.: Properties of foamed concrete containing water repellents. Constr. Build. Mater. **123**, 106–114 (2016)
46. Hajimohammadi, A., Ngo, T., Provis, J.L., Kim, T., Vongsvivut, J.: High strength/density ratio in a syntactic foam made from one-part mix geopolymer and cenospheres. Compos Part B Eng. **173**, 106908 (2019)

47. Gökçe, H.S., Hatungimana, D., Ramyar, K.: Effect of fly ash and silica fume on hardened properties of foam concrete. Constr. Build. Mater. **194**, 1–11 (2019)
48. ASTM C177-13 (2013) Standard test method for steady-state heat flux measurements and thermal transmission properties by means of guarded-hot-plate apparatus. Am. Soc. Test. Mater., **2001**, 1–13 (2013)
49. Gandage, A.S., Rao, V.R.V., Sivakumar, M.V.N., Vasan, A., Venu, M., Yaswanth, A.B.: Effect of perlite on thermal conductivity of self compacting concrete. Procedia Soc. Behav. Sci. **104**, 188–197 (2013)
50. Hladik, J.: Métrologie des propriétés thermophysiques des matériaux. Masson (1990)
51. Log, T., Gustafsson, S.E.: Transient plane source (TPS) technique for measuring thermal transport properties of building materials. Fire Mater. **19**(1), 43–49 (1995)
52. Zhao, X., et al.: Properties of foamed mortar prepared with granulated blast-furnace slag. Materials (Basel) **8**(2), 462–473 (2015)
53. Kashani, A., Ngo, T.D., Mendis, P., Black, J.R., Hajimohammadi, A.: A sustainable application of recycled tyre crumbs as insulator in lightweight cellular concrete. J. Clean. Prod. **149**, 925–935 (2017)
54. Cui, Y.L., Qian, F.G., Liu, S.X., Yin, H.T.: Effects of hydrogen peroxide on foam concrete performances. Appl. Mech. Mater. **584–586**, 1746–1749 (2014)
55. Yang, K.H., Lee, K.H., Song, J.K., Gong, M.H.: Properties and sustainability of alkali-activated slag foamed concrete. J. Clean. Prod. **68**, 226–233 (2014)
56. Shrivastava, O.P.: Lightweight aerated concrete—a review. Ind. Concr. J. **51**, 10–23 (1977)
57. Weigler, H., Karl, S.: Structural lightweight aggregate concrete with reduced density-lightweight aggregate foamed concrete. Int. J. Cem. Compos. Light. Concr. **2**(2), 101–104 (1980)
58. Hajimohammadi, A., Ngo, T., Mendis, P., Kashani, A., van Deventer, J.S.J.: Alkali activated slag foams: the effect of the alkali reaction on foam characteristics. J. Clean. Prod. **147**, 330–339 (2017)
59. Bessenouci, M.Z., Bibi-Triki, N.E., Bendimerad, S., Nakoul, Z., Khelladi, S., Hakem, A.: Influence of humidity on the apparent thermal conductivity of concrete pozzolan. Phys. Procedia **55**, 150–156 (2014)
60. Richard, T.G., Dobogai, J.A., Gerhardt, T.D., Young, W.C.: Cellular concrete—a potential load-bearing insulation for cryogenic applications? IEEE Trans. Magn. **11**(2), 500–503 (1975)
61. Ji, J., Liu, X., Tan, S., Ni, W.: Preparation and properties of waste-solid based foam concrete energy-saving materials. In: IOP Conference Series: Earth and Environmental Science, vol. 295, no. 4 (2019)
62. Oren, O.H., Gholampour, A., Gencel, O., Ozbakkaloglu, T.: Physical and mechanical properties of foam concretes containing granulated blast furnace slag as fine aggregate. Constr. Build. Mater. **238**, 117774 (2020)
63. Salvini, V.R., Luz, A.P., Pandolfelli, V.C.: Foam sprayed porous insulating refractories. Refract. World Forum **4**(April), 93–97 (2012)
64. Demirboğa, R., Gül, R.: Thermal conductivity and compressive strength of expanded perlite aggregate concrete with mineral admixtures. Energy Build. **35**(11), 1155–1159 (2003)
65. Wang*, W.Q.X., Shi*, Y.X., Shi, J.B., Zhang, Y.G.: An experimental study on thermal conductivity of ceramsite cellular concrete. In: International Conference on Structural, Mechanical and Materials Engineering, pp. 64–69 (2015)
66. Jones, M.R., McCarthy, A.: Heat of hydration in foamed concrete: effect of mix constituents and plastic density. Cem. Concr. Res. **36**(6), 1032–1041 (2006)
67. Tarasov, A.S., Kearsley, E.P., Kolomatskiy, A.S., Mostert, H.F.: Heat evolution due to cement hydration in foamed concrete. Mag. Concr. Res. **62**(12), 895–906 (2010)
68. Fabien, A., Sebaibi, N., Boutouil, M.: Effect of several parameters on non-autoclaved aerated concrete: use of recycling waste perlite. Eur. J. Environ. Civ. Eng. **0**(0), 1–18 (2019)
69. Ahmad, M.H., Awang, H.: Effect of steel and alkaline-resistance glass fibre on mechanical and durability properties of lightweight foamed concrete. Adv. Mater. Res. **626**(7), 404–410 (2013)

70. Li, J., Chen, Z., Chen, W., Xu, Z.: Seismic performance of pre-cast self-insulation shear walls made by a new type of foam concrete with high strength and low thermal conductivity. Structures **24**(2019), 124–136 (2020)
71. Kudyakov, A.I., Steshenko, A.B.: Heat insulating reinforced air hardened foamed concrete. Vestn. TSUAB 60–65 (2014)
72. Just, A., Middendorf, B.: Microstructure of high-strength foam concrete. Mater. Charact. **60**(7), 741–748 (2009)
73. Panesar, D.K.: Cellular concrete properties and the effect of synthetic and protein foaming agents. Constr. Build. Mater. **44**, 575–584 (2013)
74. Singh, G.B.: Site produced cellular lightweight concrete—A boon for housing. https://eco-web.com/edi/050113.html (2005)

Influence of Bacillus Megaterium on Crack Healing and Mechanical Properties of Concrete

Rishab Attri, Abhilash Shukla, Vinayak Sharma, and Ayush Thakur

Abstract Inclusion of calcite precipitating bacteria during mixing is one of the promising techniques of improving the concrete durability. Concrete cracks can occur due to shrinkage, creep or due to different loading conditions. Bacteria in presence of moisture, precipitates calcite in the microcracks and repair the concrete. *Bacillus megaterium* was isolated from the soil and introduced in concrete at 10^5 cells/ml concentration. This paper reports the effect of the inclusion of *Bacillus megaterium* on the strength, porosity and concrete crack healing under two different curing techniques (standard water curing and air curing). Six different concrete mixtures were made, two control, two with standard bacteria and two with bacteria isolated from the soil for standard water curing and air curing techniques. *Bacillus megaterium* in concrete resulted in increased strength and decreased porosity of concrete. These results were obtained due to the deposition of bacteria on pores.

Keywords Crack healing · Biomineralization · Calcite precipitation · Bacteria · Concrete

1 Introduction

In recent years concrete has come out as the material of choice in the construction industry. Due to low cost and easy availability of its ingredients, extensive use of concrete all over the world was observed in the past as well as in the present.

R. Attri · A. Shukla (✉) · V. Sharma · A. Thakur
Civil Engineering Department, Jaypee University of Information Technology, Waknaghat, India
e-mail: abhilash.shukla29@gmail.com

R. Attri
e-mail: rishabattri1@gmail.com

V. Sharma
e-mail: vinayaksharma203@gmail.com

A. Thakur
e-mail: ayushthakur92@gmail.com

D. K. Ashish et al. (eds.), *3rd International Conference on Innovative Technologies for Clean and Sustainable Development*, RILEM Bookseries 29,
https://doi.org/10.1007/978-3-030-51485-3_10

Rapid urbanization and industrialization majorly increase demand of concrete which directly increases the demand of cement. Therefore, continuously driving concrete industry to develop. Extensive use of cement also give rise to problems such as resource and energy conservation and production cost. Cement is also responsible for 10% of total CO_2 emissions globally. Moreover, the biggest problem is the tendency of concrete to form cracks even by non-load factors. Due to shrinkage, creep and adverse loadings conditions, surface cracking occurs in concrete as these produce tensile stresses. Cracks developed in concrete can be filled only by hydrated products or precipitation of calcite over cracks. Presence of portlandite and amorphous silica is necessary for formation of hydrated products. Concrete can heal microcracks by itself if there is presence of unhydrated products and in presence of moisture. In absence of Portlandite and amorphous silica in required amount concrete will be unable to heal its microcracks.

Degree of durability of concrete depends upon the type of exposure and properties it is exposed to. Capacity of concrete to withstand against weathering actions, abrasion and chemical attacks without compromising with its engineering properties like compressive strength, modulus of elasticity etc. defines the durability of concrete. Conventionally, a variety of repair agents like polymer modified concrete and Polymer concrete can be used to repair concrete [1]. Use of repair agents have their own limitations of high cost, unstable molecular structure, incompatible surfaces and susceptibility to Ultra-violet radiations. Complexity of mix design while using SCM's, high cost of repair materials are enough proofs to indicate need of another type of concrete which can cure itself.

Use of self-healing concrete might be the solution, in which concrete will be able to heal its cracks automatically using biomineralization. Biomineralization is process in which precipitation is induced biologically with the help of microorganism. Organisms creates a microenvironment having optimal conditions for extracellular chemical precipitation. [2–6]. Microorganism are small living organism which cannot be seen with the naked eye. Micro-organism can be found in variety of natural environments, geographical conditions, freshwater, lakes, soil. Numerous of microorganism are available which can precipitate mineral carbonates. The role of microbes in precipitation of calcium carbonate is still not defined and precipitation of calcium carbonate occurs as the by-product of metabolic process of bacteria. Primarily, Biomineralization helps to decrease the permeability of concrete by filling the pores depending upon various factors such as, pH, temperature and pressure. Concrete being extremely alkaline in nature and it is mixed under high mechanical stresses. Hence the microbe to be used should be alkaliphilic in nature and also have the property to withstand against high mechanical stresses [7]. One of the important aspects to investigate is the ways in which bacteria can be added into concrete. There are primarily four different approaches for bacterial addition in concrete. These four methods are (1) Direct addition of microbial broth or it can be added in form of spores, (2) immobilized form of bacteria onto activated carbon gel or silica gel, (3) by encapsulation and (4) by vascular networks. Out of these four ways, first 3 methods of bacteria addition tend to show higher workability and addition of bacteria by using vascular networks shows slightly less workability when compared to concrete without bacteria [8–11]. *Bacillus* genus have the property to survive in high alkaline environment, high mechanical stresses and in presence of moisture and calcium source it can precipitate calcium carbonate. *Bacillus* are rod shaped, Gram Positive, endospore forming bacteria which is found commonly in soil [12]. *Bacillus megaterium,* member of *Bacillus* family can convert ammonium carbonate from urea and further to precipitation of calcium carbonate occurs. Various investigations were done on incorporation of bacteria in concrete as an effective method of crack healing of concrete. In studies it is proven that *Bacillus Megaterium* on mixed with mixing water in concrete resulted in increased compressive strength, stiffness, decreased permeability in comparison to the concrete made without bacteria [13–21]. Therefore, in this study *Bacillus Megaterium* is incorporated to check its influence over the mechanical properties of concrete. Based on literature, behavior of concrete with bacteria in different curing conditions is not mentioned. In this study, three different concrete samples were made normal concrete which will act as control concrete, Concrete with standard *Bacillus Megaterium* MTCC 1684, procured from MTCC, Chandigarh, India. Third concrete samples were made with *Bacillus Megaterium* isolated from soil. The concentration of bacteria in mixing water was kept 10^5 cells/ml [13]. After performing the study, the authors suggests that the process of biomineralization in concrete can be a promising technique to improve characteristics of concrete.

2 Experimental Programme

2.1 Materials Used

2.1.1 Cement

For this study, OPC meeting guidelines of IS 8112:2013 was used. The physical properties and chemical composition of cement used are in Tables 1 and 2 respectively.

Table 1 Physical properties of OPC

Physical properties	Obtained values
Standard consistency	30%
IST (min)	40
FST (min)	230
Specific gravity	3.1

Table 2 Chemical composition of OPC

Constituent	Percentage (%)
SiO_2	21.10
Al_2O_3	5.24
Fe_2O_3	3.10
CaO	64.39
MgO	1.10
K_2O	0.57
Na_2O	0.23
SO_3	2.52
TiO_2	0.17

2.1.2 Fine and Coarse Aggregates

The results of sieve analysis of Coarse aggregate and Fine aggregate are shown Tables 3 and 4 respectively. River sand having maximum size of 4.75 mm was used as fine aggregate. All the tests on sand were conducted as per IS 383-2016. Fineness modulus and specific gravity of sand was 2.75 and 2.62 respectively. Fine aggregate conforming to Zond II according to IS 383—2016 was used. This type of sand can be considered as Medium Sand.

Crushed stone were used as coarse aggregates. The maximum size of coarse aggregate used was 20 mm. Specific gravity of coarse aggregate was 2.73 and water

Table 3 Sieve analysis of Fine aggregate

Size/Sieve No.	Weight of aggregate retained (grams)	Percentage of aggregate retained (%)	Cumulative percentage of aggregate retained (%)	Cumulative percentage of aggregate passing (%)
10 mm (3.8″)	0	0	0	100
4.75 mm (No. 4)	13	1.3	1.3	98.7
2.36 mm (No. 8)	34	3.4	4.7	95.3
1.18 mm (No. 16)	136	13.6	18.3	81.7
600 μm (No. 30)	364	36.4	54.7	45.3
300 μm (No. 50)	293	29.3	84.0	16.0
150 μm (No. 100)	131	13.1	97.1	2.9

Table 4 Sieve analysis of Coarse aggregate

Sieve size (mm)	Weight of aggregate retained (grams)	Percentage of aggregate retained (%)	Cumulative percentage of aggregate retained (%)	Cumulative percentage of aggregate passing (%)
20 mm	00	00	00	100
12.5 mm	2212	73.67	73.67	26.33
10 mm	642	21.4	95.07	4.93
4.75 mm	131	4.37	99.44	0.56
Pan	9	0.3	–	–

Table 5 Water analysis of mixing water

Temperature	DO	pH	Conductivity
8 °C	8 mg/L	7.1	292.3 μm hos/cm

absorption was 1.925% respectively. Normal tap water was used during the preparation of concrete specimens. Test were conducted on water and their results are mentioned in Table 5.

2.2 *Concrete Mix Design and Testing*

For this study, mix design for M20 concrete was made according to IS 10262-1989. Water cement ratio of 0.47 was adopted. The mix proportion of concrete used are given in Table 6. Bacteria was added at 10^5 cells/ml concentration. For compressive strength test cubes of 150 mm were casted. Similarly, for Split Tensile Strength test concrete cylinder were casted with 200 mm height and 100 mm diameter and beams of size (100 mm × 100 mm × 500 mm) were made for flexural strength test. Two sets of three mix proportions were made. The first mix was made as control mix (without bacteria), second mix was the standard mix (with standard bacteria obtained from MTCC Chandigarh) and third mix was the isolate mix (with bacteria isolated from the soil). After casting, both sets of mixes could kept in molds at room temperature (25 ± 2 °C) for 24 h. After 1 day all the samples were demolded and first set was placed in water tank for next 27 days at room temperature. After demolding the

Table 6 Mix design proportions

Material	Quantity (kg/m^3)
Cement	394.32
Water	186
Coarse aggregate	1258.6
Fine aggregate	658.8

second set of mixes was placed in water for next 6 days in water and in air for the next 21 days at room temperature. Test for compressive strength, flexural strength and split tensile strength was done in accordance with IS 516:2018.

2.3 Bacterial Isolation and Identification

A total of 15 soil samples from various locations in Himachal Pradesh, India were collected. Collected samples were kept in sterile bags at 4 °C until their use. Luria broth (LB) was prepared with hydrolysate of casein enzyme 10.0 g/L, was prepared with casein enzymic hydrolysate 10.0 g/l, Yeast extract 5.0 g/l, NaCl 5.0 g/l, pH 7 ± 0.2. This nutrient media was autoclaved at 121 °C for 15 min at 15 psi pressure. 1 g of each soil sample was inoculated in their respective flasks under laminar airflow and then incubated at 37 °C for 3–5 days in a shaker. Urea Agar (UA) and Urea Broth (UB) were prepared separately. All the samples after incubation were inoculated from Lurea Broth flasks and 1 ml was transferred to respective UB flasks under Laminar air flow by micropipette and then flasks of Urea Broth were incubated at 37 °C, for 3–5 days at 150 rpm. Subsequently frozen autoclaved UA was melted and poured under Laminar air flow into petri-dishes. Serial dilution was carried out to avoid the dense concentration of bacterial cells. After dilution, 1 ml of diluted sample was transferred on petri plates and spread evenly with the help of cell spreader. Petri-plates were incubated for 5 days at 37 °C. Streaking of all the visible colonies on UA petri plates was done and incubation for 5 days at 37 °C was done in Static incubator. Grown colonies were subculture two more times in order to obtain pure colonies. For each different colony 5 ml of UB was prepared. Individual colony from petri plates was inoculated and incubated for 3–4 days at 37 °C. Morphological and biological studies were performed on the selected colonies. [22] Tests like Gram Straining Test, Endospore staining test, Calcite precipitation test, Urease Assay test were conducted. Growth curves of the selected isolates were also observed.

2.4 Crack Healing Quantification

Cracks in concrete was induced with thin copper plate, with thickness ranging from 0.3 to 0.5 mm. The plates were introduced in fresh concrete paste. The plates were induced up to depth of 10–15 mm. The plates were removed after 24 h, during demolding. Visual inspection of concrete specimens was done at regular intervals to check the amount of crack healed in defined time.

3 Results and Discussion

3.1 Bacteria Identification

For bacterial isolation, soils of various types were collected. After cultivating bacteria in nutrient medium we found that colonies appeared from almost every soil sample. The data for type of soil along with time taken to obtain colonies and their size is provided in Table 7.

For sub culturing, sample no 1, 2, 5, 6, 7, 10 and 14 along with *Bacillus megaterium* MTCC 1684 (BM 1684) were selected for obtaining best bacterial isolates. A total of 26 different isolates were obtained. Morphology od selected bacterial isolates was checked. Different Biochemical tests such as Gram Staining test, Calcite precipitation test, Urease assay test were performed on the selected isolates. Gram staining test is done to check if bacteria is Gram positive or negative. It also Tells us about the shape of bacteria under microscope. Gram Staining test indicated That most of the isolates belongs to genus *Bacillus*. Figure 1 shows images obtained during gram staining test with microscope at 100× zoom.

Bacteria is gram positive when color of bacteria becomes purple. The color changes to purple in presence of thick cell wall known as Peptidoglycan (90% of the cell wall). Pink color is obtained during gram staining test when the bacteria is gram

Table 7 Size of colonies and time taken to obtain colonies

Sample No.	Type	Colonies size	Time (days)
1	Alkaline soil	Small	3
2	Crusher Soil	Medium	4
3	Natural soil	Small	4
4	Vegetative soil	Very small	4
5	Natural soil	Medium–Large	4
6	Poultry Farm Soil	Very Small	5
7	Limestone sample	Small	3
8	Cement sample	Small	5
9	Fine sand	No colony	–
10	Crushed brick sample	Medium	5
11	Natural soil	No colony	–
12	Sandy soil	No colony	–
13	Natural soil	Very small	5
14	Vegetative soil	Medium	4
15	Sandy soil	No colony	–

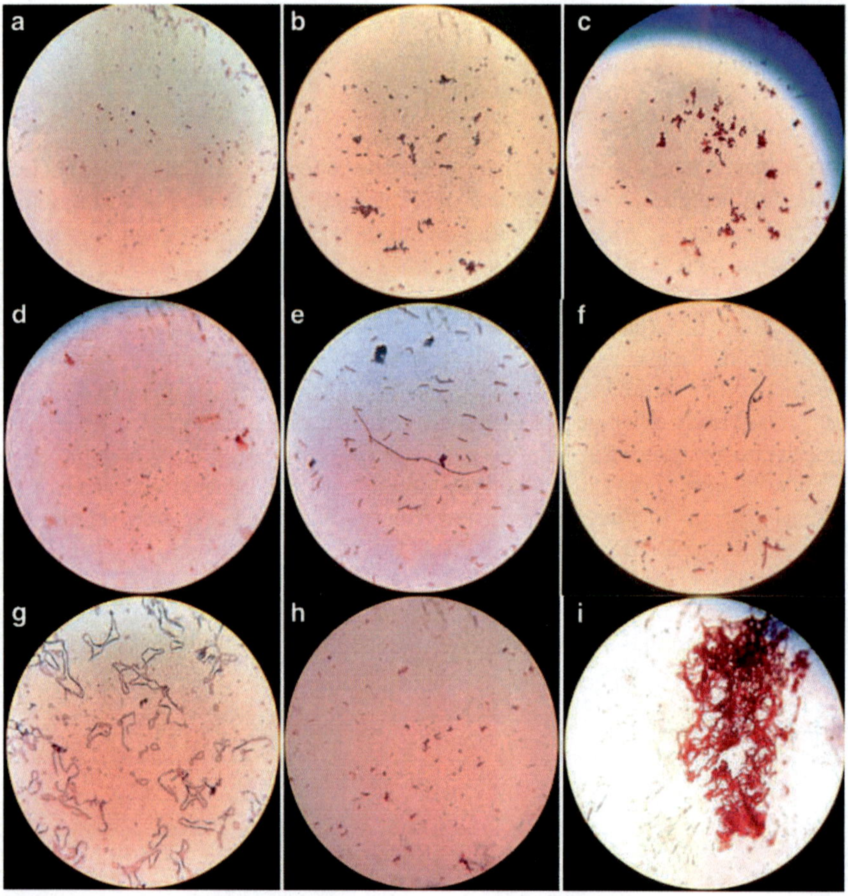

Fig. 1 Images of results obtained in gram Staining of different isolates (**a**—Isolate 1, **b**—Isolate 3, **c**—Isolate 4, **d**—Isolate 6, **e**—Isolate 12, **f**—Isolate 13, **g**—Isolate 18, **h**—Isolate 24, **i**—Isolate 26)

negative (thin layer of Peptidoglycan). Calcite precipitation test was conducted to check the calcite precipitation by different isolates. Urease activity test was conducted on selected isolates to check whether bacteria is showing positive or negative urease activity. Table 8 shows the results obtained during various biochemical test conducted on 26 isolates.

The growth parameters were determined after performing growth curve on all the selected bacterial isolates (Isolate 3 and Isolate 26). For growth curve test the cells were cultured in UB medium. TO determine the Optical Density of isolates, Spectrophotometer was used for 0–64 h at 600 nm. Figure 2 shows the growth curve for Isolate 3. Similarly, Fig. 3 shows the growth curve for 26 (Standard Bacteria Isolate).

Table 8 Cell morphology results along with results obtained during biochemical tests

Isolate No.	Sample No.	Colony Size	Shape	Colour	Gram	Shape	Weight of Calcite precipitated (mg)	Urease activity	Colour change
1	1	Small	Round	Greyish white	+	Bacilli	79.6	+	Yes
2	1	Small	Perfectly round	Greyish white	+	Streptococci	68.9	+	Yes
3	1	Small	Round	White	+	Bacilli	117.5	+	Yes
4	1	Very Small	Round	White	+	Bacilli	86.4	+	Yes
5	1	Small	Round	Pale white	+	Bacilli	68.9	+	Yes
6	1	Small	Irregular round	Pale white	+	Bacilli	92	+	Yes
7	1	Small	Round	White	+	Bacilli	45.1	+	Yes
8	1	Small	Round	Pale white	+	Cocci	39	+	Yes
9	1	Very Small	Perfectly round	Greyish white	Not found	–	16	+	Yes
10	1	Very Small	Round	White turned safranin	+	Bacilli	47.6	+	Yes
11	1	Small	Irregular round	White turned green	+	Bacilli	51.1	+	Yes
12	2	Large	Irregular round	White	+	Bacilli	116.8	+	Yes
13	2	Large	Irregular round	Faded white	+	Bacilli	112.1	+	Yes
14	2	Medium	Irregular round	Faded Yellow	+	Cocci	41.3	+	Yes
15	7	Small–Medium	Round	Greyish white	+	Bacilli	64.4	+	Yes

(continued)

Table 8 (continued)

Isolate No.	Sample No.	Colony Size	Shape	Colour	Gram	Shape	Weight of Calcite precipitated (mg)	Urease activity	Colour change
16	5	Medium	Perfectly round	White	+	Bacilli	74.5	+	Yes
17	5	Medium–Large	Irregular round	Greyish white	+	Bacilli	62	+	Yes
18	5	Medium	Perfectly round	White	+	Bacilli	76.2	+	Yes
19	5	Small	Round	White	+	Bacilli	97.3	+	Yes
20	6	Very Small	Round	Yellow	+	Bacilli	56.5	+	Yes
21	10	Medium	Irregular round	Greyish white	+	Bacilli	53.8	–	No
22	14	Medium	Irregular round	Greyish white	+	Bacilli	22.6	+	Yes
23	14	Medium	Irregular round	Greyish white	+	Streptobacilli	19.8	+	Yes
24	14	Small	Round	Pale white	+	Cocci	117.9	–	No
25	14	Small	Round	Pale white	+	Cocci	96.6	–	No
26	BM 1684	Medium - Large	Perfectly round	White	+	Bacilli	109.2	+	Yes

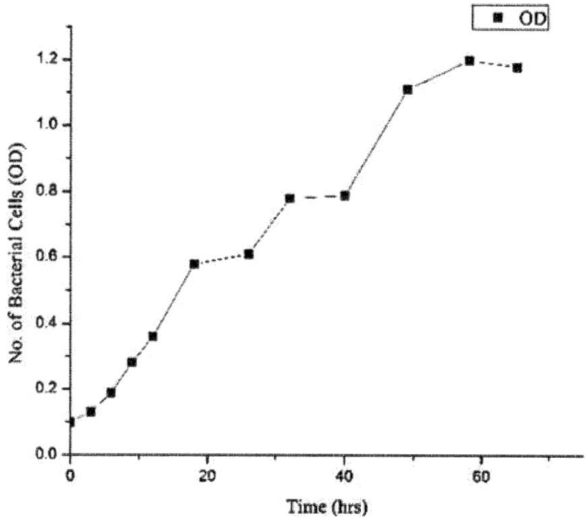

Fig. 2 Isolate 3 growth curve

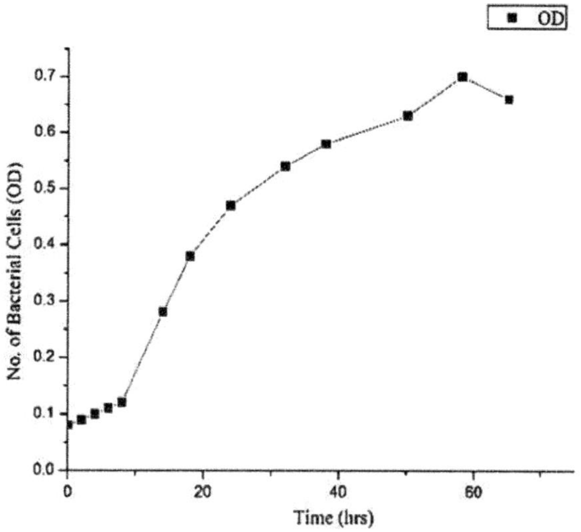

Fig. 3 Isolate 26 growth curve

From curves, higher and quicker growth was observed in Isolate 3 when compared to Isolate 26 (Standard bacterial isolate). After performing all biochemical test Isolate 3 was selected and the concrete made with it was termed as Isolate Concrete and the concrete made with Standard Bacteria was termed as Standard Concrete.

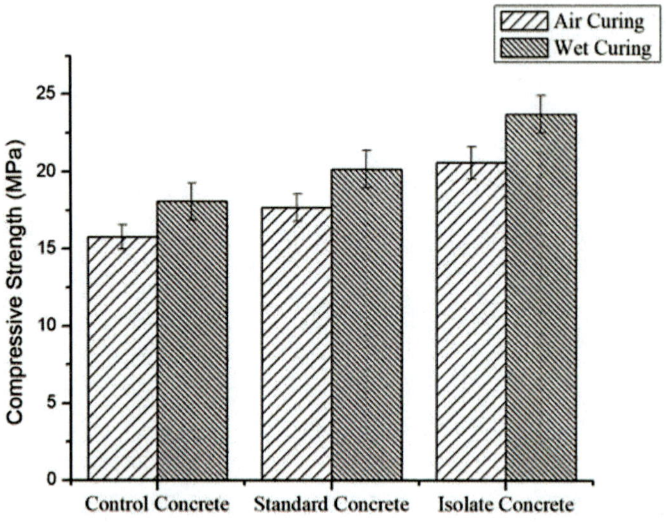

Fig. 4 Compressive strength at 7 days of curing

3.2 *Strength Tests*

3.2.1 Compressive Strength

Compressive strength of concrete increased with incorporation of bacteria. Isolate concrete showed maximum strength between the three, Minimum strength was shown by control concrete in both water and air curing. Air cured samples showed 10–14% decrease in strength with respect to water cured specimens. Highest strength (28.64 MPa in 28 days) achieved during the study was in case of Isolate concrete cured in water. Increase of strength in case of Standard and Isolate concrete was due to precipitation of bacterial cells on pores. It made the concrete matrix denser which in turn increased the concrete compressive strength. Strength was increased by 17.28% in case of Isolate concrete and 10.44% in case of standard concrete than control concrete. Comparison of compressive strength between water cured concrete and air concrete at 7 days, 14 days and 28 days are shown in Figs. 4, 5 and 6 respectively.

3.2.2 Split Tensile and Flexural Strength Test

Split Tensile sand Flexural strength test also shows that the increment in strength was highest in case of Isolate Concrete. Maximum increase of 27.76% was observed in Isolate concrete and 16.73% in Standard Concrete during Split tensile strength test. Formation of additional calcium silicate hydrate lead to increase in strength. The pore space of concrete mix was also reduced. Flexural Strength test also revealed maximum strength in case of Isolate concrete. Strength increase of Isolate concrete

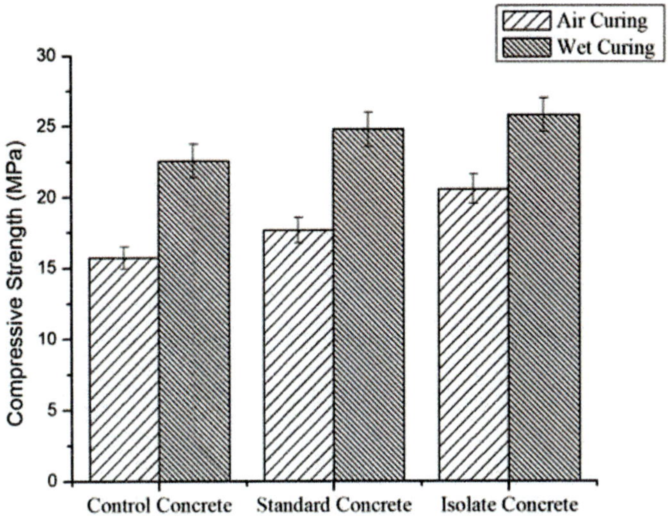

Fig. 5 Compressive strength at 14 days of curing

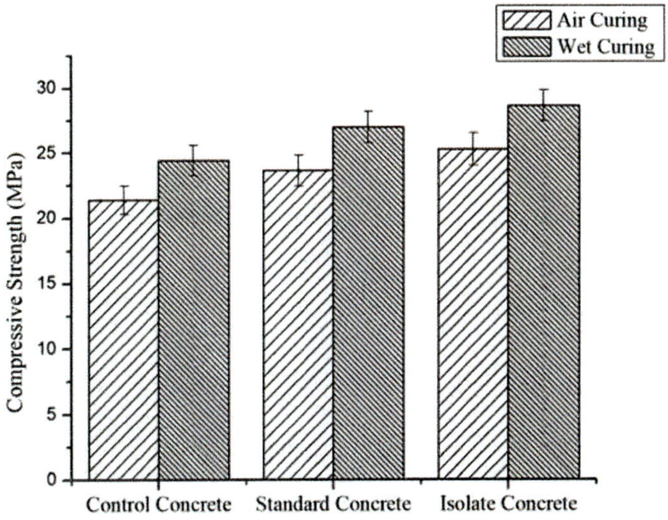

Fig. 6 Compressive strength at 28 days of curing

was 35.03% and 18.47% of standard concrete. Figures 7 and 8 gives the comparison of strength between water cured samples and air cured sample in split tensile and flexural strength test respectively.

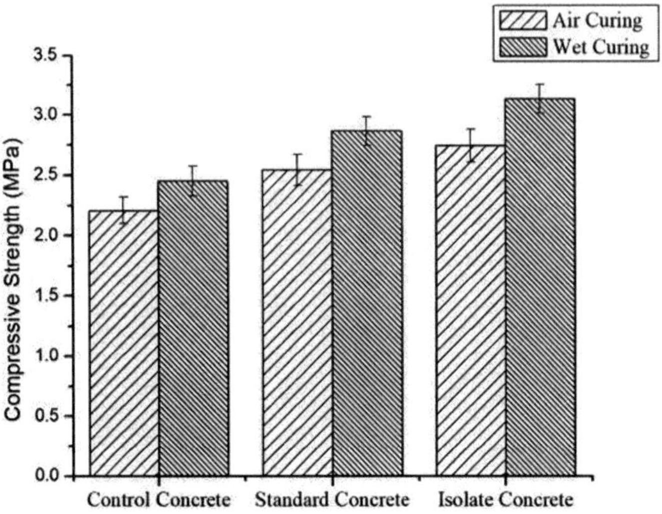

Fig. 7 Split tensile strength test

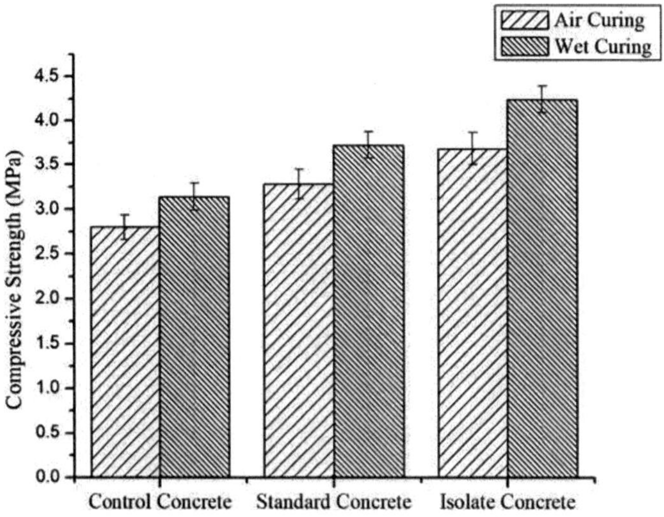

Fig. 8 Flexural strength test

3.3 Quantification of Crack Healing

Crack healing quantification was done by visual inspection at regular intervals. Autogenous healing was observed in control concrete as cracks introduced in them was healed by 0.1–0.2 mm width. Maximum crack healing was observed in isolate concrete where healing of 0.4 mm was observed in 28 days. From inspection we concluded that due to abundance of moisture in water cured concrete, bacteria were able to heal larger cracks. Crack healing in concrete took place for only 0.2–0.3 mm but precipitation of calcium was observed near crack edges. Figure 9a–d shows crack healing of control concrete cured in water and air respectively. Similarly, Fig. 9e–h shows crack healing of concrete with standard bacterial isolate and Fig. 9i–l show crack healing of isolate concrete.

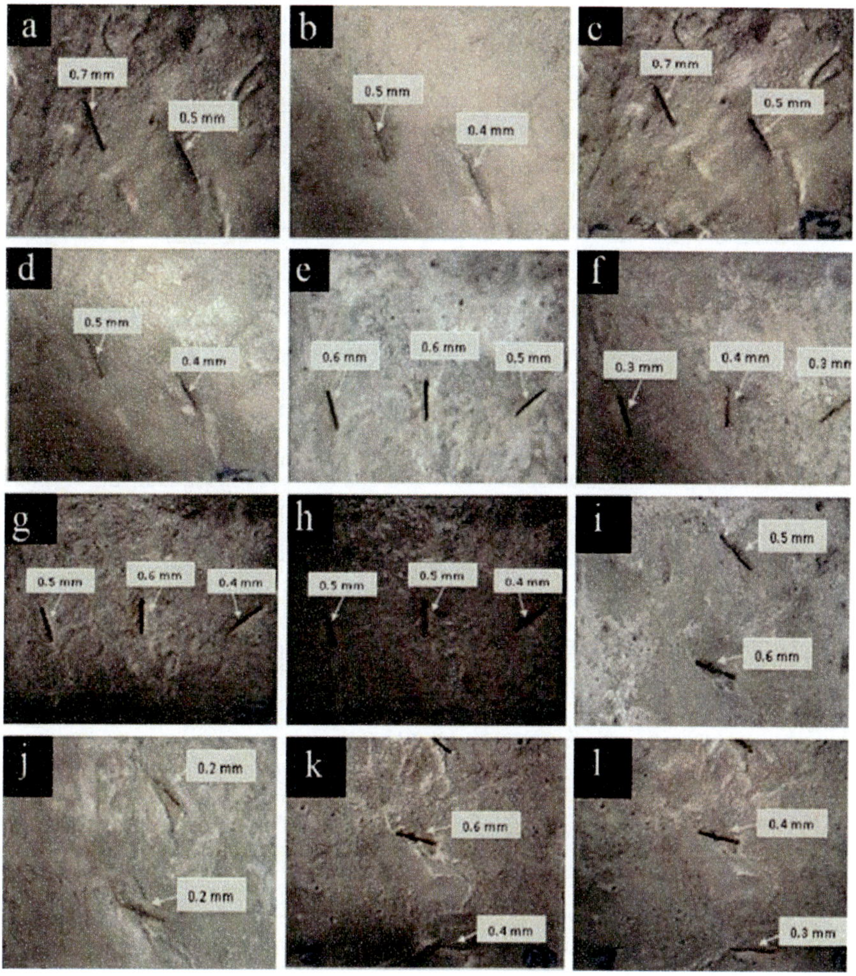

Fig. 9 **a**, **c**, **e**, **g**, **i**, **k** Crack width at the end of day 1 curing and **b**, **d**, **f**, **h**, **j**, **l** Crack width at the end of 28 days curing

4 Conclusions

The present study helps to understand a new biological method to repair concrete with calcium precipitation through biomineralization. Through study, it is clear that by inducing bacteria in concrete enhance the resistance of concrete towards degradation. Following conclusions can be drawn according to the finding in the study.

1. Alkaline soil which is rich in iron oxide and lime should be used for isolating bacteria capable of precipitating calcite.
2. Bacterial isolates capable of forming endospores and having positive urease activity should be considered for crack healing in concrete samples. Bacteria without endospore forming capability will not be able to survive in high alkaline environment present in fresh concrete.
3. The highest increase in different strength tests was observed in concrete with Isolate 3 when compared to the control concrete specimens with water curing. While air cured sample gave low strength between 10-14% when compared to water cured concrete.
4. The maximum crack width healed by Isolate 3 in 28 days was 0.3–0.4 mm. Similarly crack of width 0.2–0.3 mm was healed by Isolate 26 in 28 days. For control concrete maximum crack width healed was only 0.1–0.2 mm. However, for cracks having width greater than 0.5 mm, the inclusion of bacteria will be of little use, as also confirmed in various studies.
5. From study it was concluded that Isolate 3 can be used for inducing autogenous healing of concrete structures.

References

1. Fowler, D.W.: Repair Materials for Concrete Structures. Woodhead Publishing Limited (2009)
2. Achal, V., Mukherjee, A., Reddy, M.S.: Microbial concrete: a way to enhance durability of building structures. In: 2nd International Conference on Sustainable Construction Materials and Technologies (2010)
3. Bundur, Z.B., Kirisits, M.J., Ferron, R.D.: Cement and concrete Research Biomineralized cement-based materials: impact of inoculating vegetative bacterial cells on hydration and strength. Cement Concr. Res. **67**, 237–245 (2015)
4. Van Tittelboom, K., De Belie, N., De Muynck, W., Verstraete, W.: Use of bacteria to repair cracks in concrete. Cement Concr. Res. (2010)
5. Ramachandran, S.K., Ramakrishnan, V., Bang, S.S.: Remediation of concrete using micro-organisms. ACI Mater. J. (2001)
6. Luo, M., Qian, C., Li, R., Rong, H.: Efficiency of concrete crack-healing based on biological carbonate precipitation. J. Wuhan Univ. Technol. Mater. Sci. Ed. (2015)
7. Erşan, Y.Ç., Da Silva, F.B., Boon, N., Verstraete, W., De Belie, N.: (2015) Screening of bacteria and concrete compatible protection materials. Constr. Build. Mater. (2015)
8. Wu, M., Johannesson, B., Geiker, M.: A review: self-healing in cementitious materials and engineered cementitious composite as a self-healing material. Constr. Build. Mater. (2012)
9. Toohey, K.S., Sottos, N.R., Lewis, J.A., Moore, J.S., White, S.R.: Self-healing materials with microvascular networks. Nat. Mater. **6**(8), 581–585 (2007)

10. Wang, J., Van Tittelboom, K., De Belie, N., Verstraete, W.: Use of silica gel or polyurethane immobilized bacteria for self-healing concrete. Constr. Build. Mater. (2012)
11. Thakur, A., Phogat, A., Singh, K.: Bacterial concrete and effect of different bacteria on the strength and water absorption characteristics of concrete: a review. Int. J. Civil Eng. Technol. (2016)
12. Krishnapriya, S., Venkatesh Babu, D.L., Prince Arulraj, G.: Isolation and identification of bacteria to improve the strength of concrete. Microbiol. Res. (2015)
13. Andalib R., et al.: Optimum concentration of Bacillus megaterium for strengthening structural concrete. Constr. Build. Mater. (2016)
14. Reinhardt, H., Jooss, M.: Permeability and self-healing of cracked concrete as a function of temperature and crack width. vol. 33, pp. 981–985 (2003)
15. Talaiekhozan, A., Keyvanfar, A., Shafaghat, A., Andalib, R., Majid, M.Z.A., Fulazzaky, M.A.: A review of self-healing concrete research development. J. Environ. Treat. Tech. **2**(1), 1–11 (2014)
16. Li, V.C., Yang, E.H.: Self healing in concrete materials. In: Springer Series in Materials Science, vol. 100, pp. 161–193 (2007)
17. Anne, S., Rozenbaum, O., Andreazza, P., Rouet, J.: Evidence of a bacterial carbonate coating on plaster samples subjected to the Calcite Bioconcept biomineralization technique. Constr. Build. Mater. **24**(6), 1036–1042 (2010)
18. Van Tittelboom, K., De Belie, N., De Muynck, W., Verstraete, W.: Cement and Concrete research use of bacteria to repair cracks in concrete. Cement Concr. Res. **40**(1), 157–166 (2010)
19. Wang, J.Y., Soens, H., Verstraete, W., De Belie, N.: Self-healing concrete by use of microencapsulated bacterial spores. Cement Concr. Res. (2014)
20. Carino N.J., Clifton, J.R.: Prediction of cracking in reinforced concrete structures. Natl Inst. Stand. Technol. (1995)
21. Van Tittelboom K., De Belie, N.: Self-healing in cementitious materials—a review (2013)
22. Vashisht, R., Attri, S., Sharma, D., Shukla, A., Goel, G.: Monitoring biocalcification potential of *Lysinibacillus* sp. isolated from alluvial soils for improved compressive strength of concrete. Microbiol. Res. (2018)
23. Jonkers, H.M., Thijssen, A., Muyzer, G., Copuroglu, O., Schlangen, E.: Application of bacteria as self-healing agent for the development of sustainable concrete. Ecol. Eng. (2010)
24. Luo, M., Qian, C.: Influences of bacteria-based self-healing agents on cementitious materials hydration kinetics and compressive strength. Constr. Build. Mater. **121**, 659–663 (2016)
25. Tziviloglou, E., Wiktor, V., Jonkers, H.M., Schlangen, E.: Bacteria-based self-healing concrete to increase liquid tightness of cracks. Constr. Build. Mater. (2016)

Re-interpreting and Adapting the Site Specific Vernacular Passive House Architectural Strategies for Reducing Building Energy Demand

Reya Kundu and Sulata Bhandari

Abstract Building air conditioning being the second largest factor of energy demand, efforts to reduce this even by a small fraction plays an important role in reduction in total global energy demand. Present day concern is to explore for air-conditioning techniques with higher energy efficiency or reduced carbon emissions. As proposed in this paper, another possible solution is to adopt building design strategies with lower need for air-conditioning requirements without compromising on the comfort of its occupants. In this paper, efforts have been made to appreciate and re-interpret the passive housing architectural strategies for temperature control, that were adopted by the vernacular houses of any region. The architectural strategies for thermo-regulation adopted by these vernacular houses were developed along with passage of time, considering the local climate, the available material and of course, the socio-cultural structure of the community. The study was carried out by comparing the architectural strategies adopted by two different vernacular housing forms of south-eastern Turkey namely (i) Domed houses of Harran and (ii) Mardin houses. The aim of this paper is to highlight the fact that though adopting the passive house design techniques of the vernacular house form and including them in modern architecture is a good approach to meet the energy and environment crisis, but the factors like the topography and climate specific to the site should be given due consideration and importance for better results.

Keywords Building air conditioning · Energy demand · Vernacular housing · Mardin houses · Domed houses · Thermo-regulation

---

R. Kundu (✉)
CEPT University, Ahmedabad, India
e-mail: reya.kundu@gmail.com

S. Bhandari
Electrical Engineering Department, Punjab Engineering College, Chandigarh, India
e-mail: sulata.bhandari@gmail.com

D. K. Ashish et al. (eds.), *3rd International Conference on Innovative Technologies for Clean and Sustainable Development*, RILEM Bookseries 29,
https://doi.org/10.1007/978-3-030-51485-3_11

1 Introduction

Air conditioning and fans for cooling amounts to 10% of the total energy consumption of the world today, which in other words is equal to one fifth of the total energy consumption in buildings worldwide. As per analysis reports of IEA (the International Energy Agency) of 2018, the electricity required for air-conditioning of buildings amounts to 20% of total energy consumption in buildings of the world and is predicted to increase by three times by the year 2050.

After the industrial sector energy demand, the maximum share for building energy requirement will be that required for building air-conditioning [1, 2]. With increasing number of residential buildings in urban area, the need of the hour is to explore methods to reduce this cooling and heating energy requirement without compromising on quality of life of the residents [3]. The residential requirement of the increasing volume of urban population is accommodated through vertical expansion. Heat from sun has become a matter of concern for these high-rise concrete structures with thin filler walls.

With easy to implement modern technology and dying or almost lost knowledge about vernacular passive architecture, mechanical/electrical air conditioning techniques seem to be the easiest option for creating comfortable atmosphere [4, 5]. In the present day concern about global warming and ozone depletion because of these modern air conditioning techniques, the purpose of this paper is to review the passive cooling design approaches that were used in the vernacular buildings and appraise their prospects in modern design concept.

Passive houses are buildings where the energy required for heating and cooling is reduce drastically compared to its modern counterparts. They are not limited to residential houses but are equally applicable to other buildings like schools, commercial buildings and swimming pools [6]. Proper assessment of the sources and factors involving heat exchange like the roof, walls, windows etc. is done.

In this paper, an attempt has been made to explore and appreciate the ingenious architecture of vernacular buildings that are time tested over centuries and were in harmony with nature. As the needs, the lifestyle, the aesthetic requirement of people has changed; the techniques cannot be adopted as such but adapted to satisfy the requirements of the present and of the future times to come. This vernacular architecture takes advantage of the site, climate of the region and available material to minimize the use of energy [7].

2 Literature Survey

Till date lots of study has been undertaken to explore, analyze and evaluate the building architectural strategies from the point of view of energy requirements for temperature control. From time immemorial to satisfy the basic needs of shelter and protection from harsh environmental conditions, man has learned to build houses to create a comfortable and safe environment for himself. With the modern air-conditioning techniques not available in ancient times, man has very cleverly and intelligently made use of the strategies learned through experience to counter the adverse climatic conditions, still maintaining harmony with nature. Recent trend is to explore newer techniques of temperature control with lesser or no amount of energy usage which is harmful to environment.

Authors like Subramanian and others presents a detail review of various solar passive strategies studied and presented by various authors for attaining a thermally comfortable indoor microclimate adopting various passive techniques like thermal insulation provided by walls, effect of orientation w.r.t. wind and sun direction, effect of shading on buildings and so on for residential buildings [8]. He also presented comparative studies of thermal performance of these techniques in both traditional as well as the modern buildings done by various authors. Kamal on the other hand presents a detail overview of various strategies which can be followed for thermoregulation and critically analyses their performance and application [9], with detail analysis of techniques like shading of buildings [10] and passive draught evaporative cooling [11]. Urge-Vorsatz and others did a study on drivers of energy use in buildings based on Kaya identity approach and studies its global development trends in present and future based on past data [1]. Role of advanced and Eco-efficient materials to reduce the cooling energy needs is well discussed by authors like Pacheco-Torgal and others [2].

Recent trend of 'Bio mimicry' is copying and following the natural ways and techniques used by the living beings for coping with their surrounding environment without harming or damaging it, rather preserving it for catering to the needs of future generation too. Okeke reviews the literature and discusses the various methodologies and practices followed for adopting such natural techniques in architecture [12]. Amaralraof and others on the other hand presents a passive cooling mechanism for buildings inspired by the nasal turbinate of camels [13]. Natasha and Nazila in their work proposes an investigative tool based on Bayesian Networks to help implement the biomimetic energy efficient strategies in architecture [14].

Many works have been carried out by various authors to analyze the thermoregulatory properties of traditional houses of various regions so that they can be incorporated in modern building for its temperature control. Turkey has number of different traditional housing forms, each unique in its characteristics. Thermal analysis of the domed Harran houses by Basaran [15] and three traditional Turkish houses by Akan and Cakici [16] reflected the efficient thermoregulatory properties of these buildings. Experiments on thermoregulation were carried out by Yilmaz on two similar modern buildings, one in Istanbul and other in Mardin, both in turkey. Though both the buildings were built following the same energy conservation standards, they did not respond in similar way to their surrounding environment. A comparison was carried out with the traditional Mardin houses too. The traditional house showed a better thermal regulatory response [17]. All these studies show that a strategy suitable for a site might not respond in similar way for the other site.

Communities settle in any region based on the local conditions and the size of the community. This settlement then gradually attains its own type of form and concept that evolves over a period and is well accepted by the community, thereby becoming the traditional habitat of the people. In this paper, efforts have been made to appreciate and re-interpret the importance of site specific passive housing architectural strategies for temperature control, that were adopted by the traditional houses of any region.

This habitation is influenced by factors like climate, topography, locally available material, and available technology of the time. Different areas are bound to have different types of buildings because of the differences in the above factors apart from culture and lifestyle of the people. Aim of this study is also to highlight the fact that how the thermoregulatory properties of these building forms are site specific and should not be adopted in other region without proper review and evaluation.

3 Methodology Adopted

As an example this study offers a comparative study and analysis of two well researched vernacular building forms of Harran and Mardin vis-à-vis the needs of the residents, their economic status and above all the climate of the region. Though both of these cities, are located in southeastern Turkey and separated by a distance of 160 km only, they have two totally different and distinct forms of traditional building forms to combat their local site specific weather extremities. Attempt has been to find out how the climate acts as a major guiding factor in the building architecture and how the designing allows tackling the weather extremities resulting in thermal comfort in these dwellings. There are many such examples all over the world which have been time tested through the centuries and are still in use.

4 Domed Houses of Harran

To the southeast of Turkey, situated between Euphrates and Tigris rivers, Harran is a town with dry and clear summers (May–October) and winter is cold and partly cloudy. Annual temperature variation is from 37 to 103 °F. It is a plain with annual rainfall barely 400 mm. The geographical coordinates of Harran are 36.860° latitude, 39.031° longitude. It has an average elevation of 1189 ft. above sea level, with a maximum variation of elevation of only 732 ft. within 10 miles of Harran [18].

Harran was a temporary colony of nomadic tribal societies for the last five centuries following a patriarchal family format and was the cultural and religious epicenter of these tribes. Being nomadic, the houses were flexible type using "reusable building materials".

Typically, these houses could be assembled and dismantled in a very short time. Accordingly, a small house meant less tax to be paid to feudal lord. After the tax collector's visit, these houses were quickly rebuilt back, as small houses meant less tax.

4.1 Architectural Aspects

The photographs of houses of nineteenth and twentieth century depict their evolution with time, while earlier; the conical domes were erected directly on the ground resembling a tent, later in early part of twentieth century they were erected on square cubical bases. Further, in the later half, planned rectangular rooms with earth roofed, either inclined on all sides or flat were added.

Layout of the streets of Harran was organic vis-à-vis the forms of the houses. These houses were built close to one another as the adage goes "birds of the same feather flock together". People were in a phase of transition from nomadic life to permanent one, made their living by agriculture and stockbreeding. Enough space was thus added to their houses for their livestock and agricultural products.

Apart from the living room, the rooms meant as barn, store and the kitchen were independent building units, each covered separately with single dome. Arches and internal doors connected these units to each other. Few rooms had a small door opening to the courtyard. Number of rooms depended on the size of the family and their necessities, besides the economic condition of the owner. There was provision of adding more rooms depending on future requirement. "Tandrilik", the kitchen was converted into a living room during winters. Some of the courtyards had a well at the center and toilet was in one corner away from the rooms.

4.2 Passive Design Elements

The rooms were arranged in one to three rows along the east–west axes on one side of the courtyard, as shown in Fig. 1. This orientation helped obtain cool spaces by shading the radiation of sun from the west by the domes.

A single hole at the top of the dome served as chimney and as ventilator for hot air during summers.

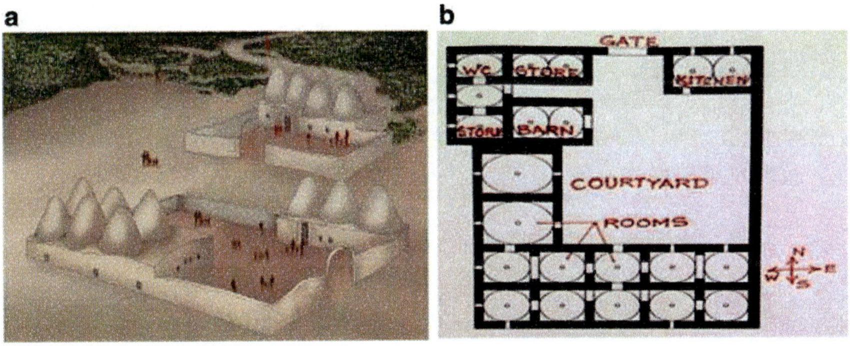

Fig. 1 **a** A typical Harran house. *Source* Ref. [19]. **b** Spatial organization. *Source* Ref. [20]

Though the side holes for ventilation did not have any preferred axis of placement, they faced each other to generate cross breezing. The windows on the side walls facing the courtyard as well as street were kept open in summer for natural ventilation but were kept closed during winters.

Houses were normally built using material like adobe, brick and stones as wood was a rare commodity because of the arid climate [21]. The domes built of sun dried handmade bricks or those collected from old ruins and burned, were constructed by sliding flat bricks inward in each row. The stones and bricks were joined using mud mortar, which being handmade were of different size. Mud mixed with sliced straw for reinforcement was used to render the domes externally each year, but could be rendered only up to human height internally. Stone projections were left on the dome sides for being used as scaffolding during rendering. 60–70 cm thick adobe walls provided the desired temperature control.

5 Mardin Houses

Mardin, a city in south eastern Turkey, is situated on a steep slope terrain facing the Mesopotamian Plain, making the vernacular houses terrace type (Fig. 2a). The region lies in mild temperate zone. It is a town with hot, arid and clear summers (June to mid-September) and winter is cold and partly cloudy. Annual temperature variation is from 29 to 99 °F. January is very cold and at early mornings it is freezing. Mardin experiences the highest rainfall in January, equal to 106 mm. The geographical coordinates of Harran are 37.313° latitude, 40.744° longitude, and average elevation of 2825 ft. above sea level, with notable variation of elevation of 2421 ft. within 10 miles of Mardin [22].

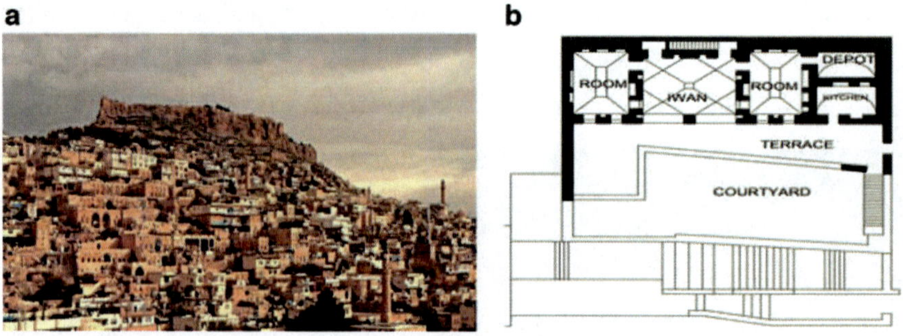

Fig. 2 **a** Mardin houses. *Source* Ref. [24]. **b** Spatial organization. *Source* Ref. [25]

5.1 Architectural Aspects

Hence, mostly each house has separate space suitable for use in summer and in winter, with semi open spaces called eyvans. Stone which acts as an insulator is used as building material [23]. The size of the room is influenced by the constraint of masonry building system.

Houses generally have two stories, with spaces affected greatly by social structure. The upper level (extrovert spaces with windows) is used in summers and the lower (introvert spaces with garden walls and services) in winters. Roof of the lower level rooms or houses serve as terraces to the upper levels.

The plan of a typical multi storeyed Mardin house is U-shaped, with a terrace in the middle oriented to the south. Courtyards on the street level are bound by high walls to ensure privacy. Houses of high income families sometimes have colonnaded courtyards becoming terraces at the south ends. Terraces veiled with white cloth all around are used for sleeping in summer nights. Due to the formation of such cloistered terraces, there is not much space for trees and plants. Further, due to shortage of water, selsebils and pools are found only in public buildings.

Next in the series are the rooms or the closed spaces which are the kitchen, living space, workshop and bedrooms with openings on the south side overlooking the valley.

The Mardin house thus consists of three kinds of spaces: open, semi-open (eyvan) and closed spaces (rooms). Thus, expansion took place by addition of an eyvan and a room which is directly connected to it. The scenery of Mesopotamian Plain had major influence in the shape and placement of these units.

5.2 Passive Design Elements

Tough courtyards and terraces are widely used, they aren't very comfortable during summer middays due to their south orientation and surrounding reflective surfaces. Hence, eyvans, which are semi open spaces that serve as buffer zones, become the liveliest part of the house. It remains shaded during summers and receives a lot of sunlight during winters. These are protected against rain but not closed to air movements. Its arches opening to the courts and paving of colorful stones makes it the living rooms of the houses. The porous flooring retains water in its holes, making evaporation easy for a longer period of time.

Thick exposed stone walls (60–110 cm limestone), high ceilings (reaching 5 m in some houses) that are barrel or cross-vaulted with a thickness of 50–60 cm at the apex, double glazed windows with about a 50 cm space in between (sometimes a shutter replaces the outer glazing), double entrance doors with a space in between to provide insulation, portions of rooms on the ground level buried into the sloping hill, making use of the earth temperature that is lower in summers and higher in winters than the mean air temperature, rooms and cellars created by excavation of the hills are all direct climatic responses in Mardin houses [26]. Due to topography and necessity for protection from the cold winter winds, the north sides of the houses have a very few openings.

At times, the inner exposed wall is whitewashed. Its light colour and smooth texture reflects solar radiation and provides a cool atmosphere inside. In some houses, earthen jars are installed into the ceilings giving extra insulation and reducing the dead load on the structure.

On a larger scale, the overlapping houses with overlapping boundaries and ease of communication in the vertical direction give way to a socially interactive community. The streets formed after the houses have been built and influenced by climatic reasons were narrow, but the houses did not block each other's sun, wind or view.

6 Inference

Though both the building types were in Turkey but the form of building in both the cases are very different and distinct from each other. The primary factor guiding the type of building in these cases is not only the climate but topology of the place too.

As is evident from the above observation the temperature regulation of the buildings was given due consideration in designing the vernacular houses like these.

Table 1 given summaries the various building design elements adopted for thermal conditioning of the two house forms guided by the climatic conditions prevailing there.

Both the houses were in harmony with the topography of their site, which has an important impact on the built form too. Since Harran is a plain area the shading provided by the domed roof structure was well utilized for cooling effect, whereas Mardin houses enjoys the evyans. Using locally available material as building materials also makes the difference in built structure and thus different ways to mitigate the climate extremities. In case of Harran houses the natural ventilation is supported by the hole in top of the dome, whereas the small window on top of the window serves the same purpose in case of Mardin houses. Orientation of the houses is also guided by the topography of the site.

7 Vernacular Strategies in Modern Context

The worldwide concern for degradation of environment due to ozone depletion and global warming is forcing us to explore for alternate sources of energy with reduced carbon emissions, these passive house design techniques can be a potential for thermal design criterion of the modern buildings. At times, adopting these vernacular methods is not free of problems in the present context, but they can be modified and adapted to suit the present and future generations need and requirement and of course the technological availability. For example, vernacular building using

Table 1 List of various building design elements adopted for thermoregulation of the buildings

Comparative assessment of different thermoregulatory design factors					
		1. Domed houses of Harran		2. Mardin houses	
Geographical location/environmental factors		A town southeast of Turkey, situated between Euphrates and Tigris rivers • Hot and arid climate • It is a plain Annual rainfall barely 400 mm		Located in south eastern Turkey and situated on steep sloping region facing the Mesopotamian Plain makes these vernacular houses terrace type • Region lies in mild temperate zone • Cold season is freezing at times • Elevation variation can be as large as 3852 ft. (within 50 miles)	
S. no.	Design factor	Strategy adopted	Impact	Strategy adopted	Impact
1	Shading from sun	1. The rooms arranged in one to three rows along the east–west axes on one side of the courtyard	1. This orientation helps obtain cool spaces by shading the radiation of sun from the west by the domes  Fig. 1.1 Domed shaped houses of Harran. *Source* Wikipedia	1. Public areas like passages are covered, and are known as 'Kabalti' or 'Abbara' 2. Streets are narrow and maintains harmony with the topography and climate	 Fig. 2.1 Covered passages. *Source* Ref. [27]  Fig. 2.2 Streets of Mardin. *Source* Ref. [28]

(continued)

Table 1 (continued)

S. no.	Design factor	Strategy adopted	Impact	Strategy adopted	Impact
2	Ventilation	2. A single holeat the top of the dome acting as solar chimney (of about 20 cm diameter, with a small roof on top for protection from rain) and the side holes facing the courtyard as well as street	2. The top hole act like a suction cowl helps accelerate the ventilation process and the side holes were for cross breezing and ventilation 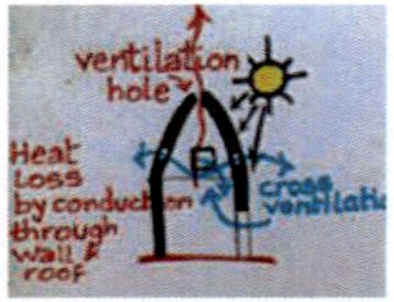 Fig. 1.2 **a** Figure showing air circulation. *Source* Ref. [20] 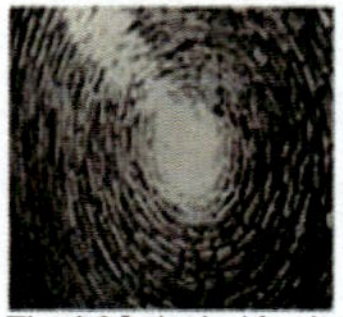 Fig. 1.2 **b** An inside view of dome showing ventilation hole. *Source* Ref. [21]	3. Eyvans (semi open spaces), were the liveliest part of the house	3. Remains shaded during summers and receives lot of sunlight during winters and though protected against rain were not closed to air movements  Fig. 2.3 *Source* Ref. [29]
				4. North sides of the houses had a very few openings	4. Protection from the cold winter winds

(continued)

Table 1 (continued)

S. no.	Design factor	Strategy adopted	Impact	Strategy adopted	Impact
				5. Small window on top of normal window	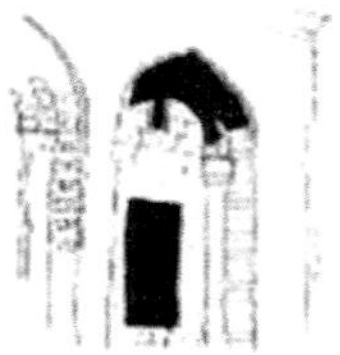 Fig. 2.5 **a** Window 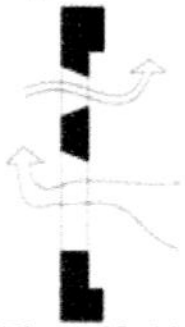 Fig. 2.5 **b** Air circulation. *Source* Ref. [25]

(continued)

Table 1 (continued)

S. no.	Design factor	Strategy adopted	Impact	Strategy adopted	Impact
3	Thermoregulation by walls	3. 60–70 cm thick adobe walls	3. Provided the desired temperature control. Experiments conducted showed that even with temperature difference of 42 °C in the afternoon and 22 °C at night, and because of the thermoregulatory properties of construction, the temperature inside varied by only ±3 °C and remain maintained around 30 °C [15] Even the humidity level was unaffected by the outside humidity level and was maintained within comfortable level	6. Stone used as building material, high ceilings, double glazed windows, double entrance doors Portions of rooms on the ground level buried into the sloping hill	6. Thick stone walls and the other features together act as good thermal insulators and temperature regulators  Fig. 2.6 *Source* Ref. [25]
				7. Light colored, smooth textured whitewashed inner wall and earthen jars installed into the ceilings	7. Provides a cool atmosphere inside Thick walls perform both as heat reservoirs and insulators
Remarks		The strategies adopted in both the buildings is well suited to the site specific characters i.e. its climate and other geographical factors and influenced by the culture and is answer to the specific needs of the society it belongs to. The structures, thus cannot be interchanged, to meet the requirement of the other site			

thick stone walls providing thermal insulation can be replaced by walls with thinner coating of Polystyrene, rock wool, and sheep's wool [30]. The thermal insulating property of a 500 mm thick stone wall is same as 15 mm thick rock wool layer.

The Wallasey School building built in 1961, in Cheshire, is a classic example of modern day passively solar heated building [31]. Even the oil fired system originally fixed for heating was removed, as it was found to be redundant. During the 1990s, in Germany the use of super insulation formed the part of PassivHaus standard, which has been since used extensively worldwide to not only improve the thermal performance of new buildings but also the already existing buildings. It also drastically reduces the requirement of conventional heating systems requirements.

Besides the factors like client requirements, limitations imposed by building codes and regulations, aesthetics etc. influencing the design of a building, the factors influencing the building's energy performance too if given due importance during the initial design phases can results in minimization of the energy loads thus facilitating the sustainability of building industry.

Sensitivity analysis of various influencing factors like the plan shape, Plan Depth, Plan Orientation, Window-to-Wall Ratio and Window Orientation can be carried on to evaluate their impact on total energy use by a building using different softwares available in the market today. One such study was carried out by Babak Raji [32] and others using simulations in Energy Plus platform of Design Builder version 4.7. Results shows how different design respond differently to different climatic conditions, thus emphasizing the fact how and why it is important to consider and include the site specific factors in the design process to achieve for an energy efficient building design.

8 Conclusion

Analysis of the house forms reflects that it is the occupation i.e. the economy and the purpose of building which are the main factors influencing the structure of the given house form. But the factors like site, climate and locally available building materials and the construction techniques developed through generations too are important parameters influencing the thermal behavior and design needs of any built form. These comparison shows that the vernacular housing forms have well established features which can be adopted in modern architecture resulting in reduced air conditioning energy requirements, but simply adopting them does not serve the purpose. The factors like the topography and climate specific to the site if given due consideration and proper evaluation done before adopting them yields better results.

As the energy consumed for air conditioning of buildings forms a substantial portion of the total global energy demand, thus any saving on that front is a welcome step to meet the deficiency in energy scenario. In this context, it was an attempt to re-establish the forgotten passive design strategies of the vernacular buildings, which are the established architectural descriptions of the locality for centuries, and to see if these methods could be reinterpreted, integrated and adapted in the modern architectural context.

The modern passive heating techniques integrated with the vernacular architectural concepts can be adopted for reducing building heating requirements and the influence of the site specific factors and requirements can be taken care of, in achieving more efficient thermal performance of the buildings.

References

1. Urge-Vorsatz, D., Cabeza, L., Serrano, S., Barreneche, C., Petrichenko, K.: Heating and cooling energy trends and drivers in buildings. Renew. Sustain. Energy Rev. **41**, 85–98 (2015)
2. Pacheco-Torgal, F., Labrincha, J., Cabeza, L., Granqvist, C.G.: Eco-Efficient Materials for Mitigating Building Cooling Needs: Design, Properties and Applications. Elsevier Science and Technology, Cambridge (2015)
3. Practical Guidebook for Implementing Smart and Clean Energy Projects in Existing High-Rise Residential Apartments. A Meghraj Capital Advisors Private Limited Publication (2017)
4. Kamath, L.G., Daketi, S.: Jaalis: a study on aesthetics and functional aspects in built environment. Int. J. Sci. Eng. Appl. Sci. (IJSEAS) **2**(2) (2016)
5. Srivastav, S., Jones, P.J.: Use of traditional passive strategies to reduce the energy use and carbon emissions in modern dwellings. Int. J. Low-Carbon Technol. **4**, 141–149 (2009)
6. Collins, T., Michler, A.: Passive houses can reduce energy use by 90%: keep occupants comfortable and healthy. https://smartenergyliving.org. Accessed 15 Apr 2019
7. Passive Solar Home Design: https://www.energy.gov/energysaver. Accessed 20 Apr 2019
8. Subramanian, C.V., Ramachandran, N., Kumar, S.S.: A review of passive cooling architectural design interventions for thermal comfort in residential buildings. Indian J. Sci. Res **14**(1), 163–172 (2017)
9. Kamal, M.A.: An overview of passive cooling techniques in buildings: design concepts and architectural interventions. Acta Tech. Napocensis Civ. Eng. Archit. **55**(1) (2012)
10. Kamal, M.A.: A study on shading of buildings as a preventive measure for passive cooling and energy conservation in buildings. Civ. Environ. Eng. **10**(6) (2010)
11. Kamal, M.A.: Reinventing traditional system for sustainable built environment: an overview of passive downdraught evaporative cooling (PDEC) technique for energy conservation. J. Res. Archit. Plann. **11** (2011)
12. Okeke, F.O., Okeke, O.C.J., Adibe, F.A.: Biomimicry and sustainable architecture: a review of existing literature. J. Environ. Manag. Saf. **8**(1), 11–24 (2017)
13. Abdullah, A., Said, I.B., Ossen, D.R.: A sustainable bio-inspired cooling unit for hot arid regions: integrated evaporative cooling system in wind tower. Appl. Therm. Eng. **161** (2019)
14. Chayaamor-Heil, N., Hannachi-Belkadi, N.: Towards a platform of investigative tools for biomimicry as a new approach for energy-efficient building design. Buildings **7**(19) (2017)
15. Başaran, T.: Thermal analysis of the domed vernacular houses of Harran, Turkey. Indoor Built Environ. **20**(5), 543–554 (2011)
16. Akan, A.E., Cakici, F.Z.: The influence of climate in the formation of traditional Turkish houses. In: Conference of Postgraduate Research in the Built and Human Environment, University of Salford, 14–15 Apr 2005
17. Yilmaz, Z.: Evaluation of energy efficient design strategies for different climatic zones: comparison of thermal performance of buildings in temperature-humid and hot-dry climate. Energy Build. **39**, 306–316 (2007)
18. Weatherspark Homepage: https://weatherspark.com/y/101186/Average-Weather-in-Harran-Turkey-Year-Round#Sections-Topography. Accessed 24 Nov 2019

19. Folkrooms Studio: https://indayear3studio-1718s1.blogspot.com/2017/09/chonnikanjeans-vernacular-architecture.html. Accessed 24 Nov 2019
20. Baran, M., Yılmaz, A.: A study of local environment of Harran historical domed houses in terms of environmental sustainability. J. Asian Sci. Res. **8**(6), 211–220 (2018)
21. Ozdeniz, M.B., Bekleyen, A., Gonuil, I.A., Gonul, H., Sarigul, H., Ilter, T., Dalkilic, N., Yildirim, M.: Vernacular domed houses of Harran, Turkey. Habitat Int. **22**(4), 477–485 (1998)
22. Weatherspark Homepage: https://weatherspark.com/y/101596/Average-Weather-in-Mardin-Turkey-Year-Round. Accessed 24 Nov 2019
23. Torus, B.: Learning from vernacular Turkish house: designing mass customized houses in Mardin. Intercult. Understand. **1**, 105–112 (2011)
24. Travelatelier Homepage: https://travelatelier.com/blog/mardin-city-woven-poetry-mystery. Accessed 20 Apr 2019
25. Çaglayan, M.: The model of vernacular countryside from Turkey: Mardin pavilions. In: 9th International Conference on Structural Analysis of Historical Constructions, Mexico (2014)
26. Gideon, G.: Housing in Arid Lands—Design and Planning. Architectural Press, The University of Michigan (1980)
27. Mardin Travel Homepage: https://www.mardintravel.com/abbara. Accessed 24 Nov 2019
28. Varolgüneş, F.K., Canan, F.: Disidentification of historical city centers: a comparative study of the old and new settlements of Mardin, Turkey. Int. J. Civ. Environ. Struct. Const. Architect. Eng. **11**(5) (2017)
29. Wikimedia Commons: https://commons.wikimedia.org/wiki/File:Mardin_stone_houses_02148.jpg. Accessed 24 Nov 2019
30. Cole, C.: The myth of stone walls as insulation. https://carbonlimited.co.uk. Accessed 20 Apr 2019
31. Everett, B.: Solar thermal energy. In: Boyle, G. (ed.) Renewable Energy: Power for a Sustainable Future, 3rd edn., pp. 21–74. Oxford University Press (1996)
32. Raji, B., Tenpierik, M.J., Dobbelsteen, A.V.: Early-stage design considerations for the energy-efficiency of high-rise office buildings. Sustainability **9** (2017)

A Comparative Study on the Sustainability of Public and Private Road Transportation Systems in an Urban Area: Current and Future Scenarios

Sandeep Singh and Bishnu Kant Shukla

Abstract Transportation system is the fastest growing consumer of fossil fuels and the fastest growing source of fuel emissions. The negative impacts of road transport-related emissions on urban sustainability are huge. Hence there is an immediate need to reduce vehicular emissions by implementing mitigation measures. This study is intended to formulate vehicular consumption and emissions-related mitigation measures by conducting an extensive study on the comparison of both public and private transportation systems. System Dynamic (SD) models are built to forecast the vehicle population, fuel consumption, and fuel emission levels by the vehicles in the urban environment for current and future scenarios. In addition to the current scenario model (do-minimum scenario model), two other scenario models (partial-efforts scenario model and desirable scenario model) are built using the SD method to forecast the future levels of vehicle population, fuel consumption, and fuel emissions. These SD models are built by augmenting the growth rate of the public transportation system (public buses) and simultaneously restricting the growth rate of the private transportation system (two-wheelers and cars). When compared to the do-minimum scenario model, the partial-efforts scenario model and desirable scenario model results indicated a significant reduction in fuel consumption and fuel emission levels. Finally, policies are formulated for mitigating the vehicular consumptions and emissions in line with the context of achieving Sustainable Development Goals (SDG).

Keywords Public transport · Private transport · System dynamics (SD) · Scenario analysis · Sustainable development goals (SDG)

S. Singh (✉)
Department of Civil Engineering, National Institute of Technology Tiruchirappalli, Tiruchirappalli, India
e-mail: sandeepsingh.nitt@gmail.com

B. K. Shukla
School of Civil Engineering, Lovely Professional University, Phagwara, India
e-mail: bishnukantshukla@gmail.com

D. K. Ashish et al. (eds.), *3rd International Conference on Innovative Technologies for Clean and Sustainable Development*, RILEM Bookseries 29,
https://doi.org/10.1007/978-3-030-51485-3_12

1 Introduction

Rapid urbanisation in developing countries is increasingly rising energy use and fuel emissions from urban regions. A big cause of air pollution and greenhouse gasses (GHG) emissions is road transportation system. A few decades ago, sustainability was not noteworthy worldwide. However, due to the alarming issues of global warming and climate change all around the world, sustainability became an inevitable choice. All nations recognized the value of sustainability and agreed to change their policies on the environment. Though, policies were framed based on the environmental conservation grounds, yet these were not supported with adequate inter-disciplinary research.

Emissions from vehicles are important sources of air pollution, especially in urban areas, which contribute significantly to GHG emissions. Many different factors coalesce to promote an urgent need to reduce pollution, transport fuel consumption, and tackle global climate change, energy protection and fuel prices. The relationship between these segments must be carefully considered to resolve these possible risks [1]. Rapid economic and population growth have helped to boost energy demand [2]. It is necessary to develop travel demand models to forecast the trip purpose by transport mode, time of day, travel duration, and location of travel [3]. The critical relationship between transportation, energy, and emissions comes into the picture as soon as one takes into account the exact importance of transportation for economic growth. Collective action and coercive policy measures are the key criteria in order to face up to these challenges in the new paradigm and to reduce these losses so that efficient transport systems can be used more efficiently and effectively. Hence the primary aim of this research work is first to study the energy consumption and emission factors that influence the road transportation sector. Second, to build various alternative scenarios based on System Dynamics (SD) simulation models for forecasting future demands involving the sectors of transportation, energy consumption, and fuel emissions. Thirdly to critically investigate and analyze the SD models which would potentially address and reduce the losses in transportation, energy, and emission sectors and finally, to recommend appropriate transport policies that ensure prosperity in building up a sustainable transportation system.

2 Review of Literature

Various research works have been carried in different countries focusing on understanding the design and analysis of sustainable transportation systems. Al-Osaimi et al. [1] studied the transport-related demand and supply, on fuel and emissions to predict fuel consumption requirements and emissions by developing a System Dynamic (SD) model. The scenario analysis of the SD model revealed that due to the confinement of the private sector vehicles and enhancement of the mass-mobility sector with a model-split value of 30:70 resulted in a reduction of 70% of energy consumption and 50% of emission levels. Giannakis et al. [2] developed an environmentally-extended input–output model to conduct an economy-wide assessment of CO2 emissions in Cyprus. The empirical analysis identified that nearly twenty-two economic activity sectors contribute to the generation of CO2 emissions. James et al. [3] proposed a method for quantifying the energy consumption in the transport sector. The authors reported that in understanding energy consumption vis-à-vis transport, it is crucial to consider not only the decisions which are short-term based, such as mode choice and point of destination, but also to consider the medium to long-term decisions made by individuals and firms concerning residential and workplace locations, auto-ownership, and labor force participation.

Kan et al. [4] adopted the space–time path for constructing an analytical method to more accurately estimate vehicle energy consumption and emissions based on vehicles' stationary and mobile activities. Lee and Tong [5] used grey forecasting model, based on grey theory, to estimate the energy consumption in China. Maheshwari et al. [6] proposed a paradigm model that allows us to identify the equilibrium points, perform stability analysis, and analyze vectorfield diagrams from a macroscopic perspective. The proposed study focused on building dynamic models of performance indices that would help to understand the behavior of interdependent systems. Nesamani et al. [7] proposed a structural equation modeling-based approach. They illustrated a methodology to predict the time spend by the vehicles in different operating conditions and concluded that the speed limit has the most substantial influence on vehicle operations, which eventually has an impact on the emissions. Parikh et al. [8] have studied the energy options for India in the near future for an interdependent world. They have analyzed the implications of the available domestic energy potential for the energy supply and demand situation in the transportation sector. A quantitative model had been built, taking into account the possible forms of energy consumption for transportation of the people and freight. Simulation has been carried out from the base year 2000 up to the year 2050. Demand projection has been carried out based on crucial factors like economic growth and demographics and their corresponding dynamics. Paul [9] has carried out a study on the transportation and energy sector to build a forecasting model for the future number of different types of vehicles and their energy consumption using a System Dynamics (SD) approach. Three scenarios were analyzed to predict the future demand in transportation and energy sector estimation by considering vehicle ownership, growth in income level, and CO_2 emissions rate per annum. These three scenarios were studied by changing their growth rates. Finally, the best scenario has been suggested for policy implementation. Ramachandra and Shwetmala [10] prepared an inventory for emissions from transport sector in India. The study mainly focused on measurement and quantifications of GHG's from transport sector. The Indian Government proposed a viable and sustainable pricing policy for the four major oil products, namely, petrol, diesel, kerosene and liquified petroleum gas (LPG) to limit the fiscal burden [11]. Shepherd and Ortolano [12] had studied the effect of transport emissions on economic development. This research work deals with the pattern of the spatial development of the transportation sector in India and its effect on the environment. The authors make use of a general equilibrium framework for estimating the contributions of transportation emissions to the spatial development of a particular region.

Shuwei [13] introduced a strategy for motor vehicle emission reduction and management using the System Dynamic and Grey Model (SD-GM) approach. The author has compared different policies, which indicated that the motor vehicle trips and the quantity of air pollutants decreased because of the motor vehicle pollution charging management model. Wang et al. [14] suggest that the SD approach is useful in clearly explaining the interactions between the multiple parameters and variables. For modeling the dynamics involved in the transportation system, researchers have used the SD approach based on cause-and-effect analysis and feedback loop structures. There is no convincing indication that market fluctuations caused by shifts in oil markets will continue over the medium term in an increasingly transparent and competitive environment. Furthermore, it is desirable to promote competition on a level basis between public and private transport sectors [15]. Yevdokimov and Mao [16] assessed the impacts of the transport system on the economy, environment, and general social well-being; and measured by systems' effectiveness and efficiency.

As it is evident from the literature mentioned above, several studies have been done in the past by researchers. However, still, an interrelationship based study is not critically analyzed in the sectors involving transportation, energy, and environment, which instigates the need for this research work.

3 Study Methodology

A study methodology is proposed to investigate and analyze the interrelationship between the transportation, energy, and emission sectors, which is shown in Fig. 1.

The data concerning the transport, energy, and emission sector were collected from online sources. The annual vehicle population, trip length based on fuel consumption and fuel emission for the study area is obtained to determine the annual fuel consumption and fuel emission. The SD simulation models with respect to the do-minimum scenario, partial efforts scenario and desirable scenario are developed for determining the fuel consumption and fuel emission levels over the horizon year 2030. From the obtained results the best scenario SD model is adopted and recommendations have been made for implementation of policy measures.

4 Study Area

The Chennai city, which is one of the four metropolitan areas in India with a population of around 9 million, with an area of 174 km^2 was chosen as the study area. The traffic in the city is steadily increasing, and chronic traffic congestion has become a serious problem. Additionally, the Chennai city has the highest vehicle density in India.

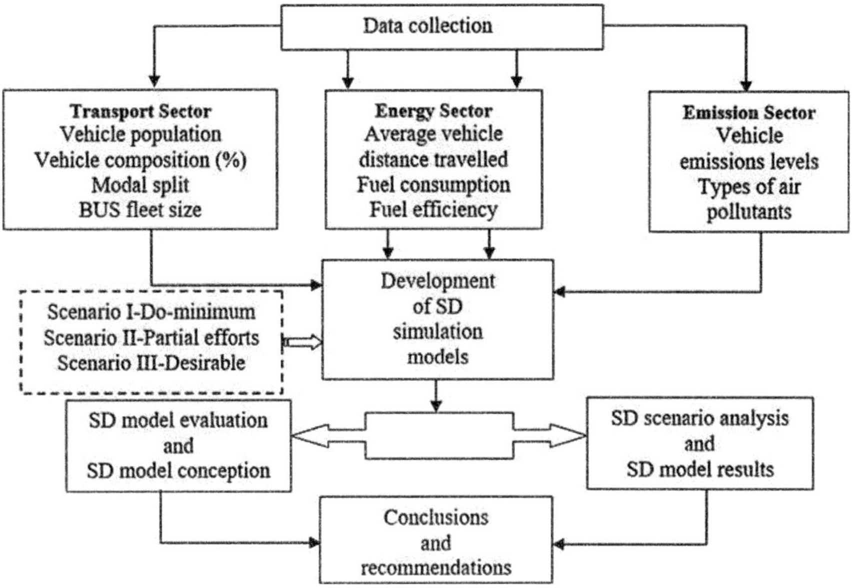

Fig. 1 Flow chart showing the study methodology

The rapid increase in the growth rate of vehicular population has increased fuel demand and emission levels. This has resulted in increased fuel prices due to fuel imports from other nations. Hence it is considered alarming and imperative to study the effect of the increase in vehicles on fuel consumption and fuel emissions through a System Dynamics (SD) approach.

5 Data Collection

5.1 Transportation Sector

The dominant modes of transport in the study area are the private transport vehicles such as Two Wheelers (TW) and Cars (CR) and the public transport vehicles like Metropolitan Transport Corporation Buses (BUS). Other types of vehicles such as Auto-Rickshaws (AR), Taxies (TX), Private Buses (PB), Mini Buses (MB), Light Commercial Vehicles (LCV), and Heavy Commercial Vehicles (HCV) are also found to be operational on the city roads. It is important to know the traffic volume and different class of vehicle in any roadway for strategic planning, improvements and management. The classified vehicle population data on a historical basis from the year 2013 to 2018 was collected from the Statistics of the Transportation Department, Chennai. This classified vehicle population data is used for three different scenario

Table 1 Historical values of vehicle population for Chennai city

Class of vehicle	Year					
	2013	2014	2015	2016	2017	2018
BUS	5132	5196	5291	5481	5697	6386
AR	73,593	95,010	100,011	102,899	106,400	110,781
TX	1889	1902	2031	2213	2276	2442
LCV	17,754	19,104	28,685	32,204	49,704	61,872
HCV	19,269	26,892	33,248	50,357	61,250	73,686
PB	4053	4359	4443	4487	4523	4565
MB	3143	3326	3533	3609	3737	3863
TW	3,302,375	3,872,517	4,337,778	4,580,691	4,919,184	5,263,740
CR	724,455	872,069	898,143	923,630	983,034	1,030,902
Total	4,151,661	4,900,374	5,413,161	5,705,568	6,135,803	6,558,236

Source Statistics of the Transportation Department, Chennai (www.tn.gov.in)

analyses, such as the do-minimum scenario analysis, partial scenario analysis, and desirable scenario analysis. The different vehicle types and their vehicle population for the Chennai city is shown in Table 1.

5.2 *Energy Sector*

The energy sector plays an essential role in the driving mechanism of the transportation system. However, due to the limited availability of energy, there are concerns about it now and in the near future. The consumption of the fuel is measured based on the average trip distance traveled and the efficiency of the fuel. The average distance traveled by the various classes of vehicles in km/day, the efficiency of fuel in km/liter and the consumption of fuel in liter/year for the different vehicle types are taken into account which are shown in Table 2. These data are used in the SD model building and simulation process.

5.3 *Emission Sector*

The vehicles in the Chennai city make the longest trips in daily commute and the city stands second on overall vehicular emissions. It also stands second in fuel consumption and has the second highest number of TW on the roads amongst the metropolitan cities. The data regarding the type of pollutants emitted from the class of each vehicle are given in Table 3.

Table 2 Fuel efficiency and fuel consumption of the different vehicle types

Class of vehicle	Average distance travelled (km/day)	Fuel efficiency (km/l)	Fuel consumption (l/year)
BUS	151	4.1	13,415
AR	96	21	1669
TX	21	13	534
LCV	51	14	1330
HCV	55	4.33	4637
PB	111	05	11,863
MB	22	8.7	897
TW	18	53	124
CR	24	12.9 (Petrol) 15.6 (Diesel)	684 652

Source Report of the Expert Group, Government of India Report

Table 3 Total emissions from the different vehicle types in g/km

Class of vehicle									
Pollutant type	BUS	AR	TX	LCV	HCV	PB	MB	TW	CR
CO_2	515.2	60.3	208.3	423.84	423.84	515.2	515.2	26.6	223.6
CO	3.60	5.10	0.90	1.61	1.61	3.60	3.60	2.20	1.98
NO_x	12.00	1.28	0.50	10.96	10.96	12.00	12.00	0.19	0.20
CH_4	0.09	0.18	0.01	0.05	0.05	0.09	0.09	0.18	0.17
SO_2	1.42	0.02	10.3	1.39	1.39	1.42	1.42	0.01	0.05
PM	0.56	0.20	0.07	0.33	0.33	0.56	0.56	0.05	0.03
HC	0.87	0.14	0.13	0.50	0.50	0.87	0.87	1.42	0.25

Source Ramachandra and Shwetmala [10]

6 Scenario Analysis and System Dynamics (SD) Model Results

Based on the System Dynamics (SD) simulation analysis, the vehicle population, fuel consumption, and fuel emission were forecasted and are presented in this section. The developed SD simulation models for the scenario I (Do-minimum scenario), scenario II (Partial scenario), and scenario III (Desirable scenario) are discussed in the following subsections.

6.1 Scenario I—Do-Minimum Scenario (When the Existing Trend is Allowed to Continue Without Any Change)

The Do-minimum scenario model built on the principles of SD works based on simple assumptions. In this scenario, it is assumed that the current growth rates of the different classes of vehicles like BUS, AR, TX, PB, MB, LCV, HCV, TW, and CR continue up to the horizon year 2030. No change in the growth rate of the vehicles reflects the conditions of this scenario. Based on this, the values of fuel consumption and fuel emissions by each class of vehicles have been simulated and forecasted. The results of the developed SD simulation model for the Do-minimum scenario are shown in Table 4.

Table4 shows that if the existing trend is continued considering the present growth rates for the vehicle population till the horizon year 2030, the number of TW and CR will reach 116 lakhs and 25 lakhs (private transportation), respectively. The BUS mode of public transport has increased to a fleet size of around 8944 in the horizon year 2030. It is very well known that if the vehicle population increases, consequently, the fuel consumption and fuel emissions for different vehicle classes also increase, which are shown in Tables 5 and 6, respectively.

The forecasted fuel consumption for the horizon year 2030 by TW is found out to be 24.59 lakh liters per day, CR is 34.56 lakh liters per day, and BUS is just 2.84 lakh liters per day. Finally, the total fuel consumption by all the vehicles in the horizon year 2030 would be 96.74 crore liters per day. Hence the annual fuel consumption would reach 353.08 crore liters per annum. The increasing consumption of fuel is

Table 4 Predicted vehicle population from the scenario I

Year	BUS	AR	TX	LCV	HCV	PB	MB	TW (lakhs)	CR (lakhs)
2018	6386	11,0781	2442	61,872	73,686	4565	3863	53	10
2019	5545	97,764	2424	61,408	62,571	4303	3573	48	9
2020	5778	103,533	2886	73,138	63,760	4865	3966	51	10
2021	6020	109,641	3438	87,106	64,973	5503	4403	56	11
2022	6274	116,110	4095	103,744	66,206	6224	4886	60	12
2023	6536	122,960	4876	123,559	67,465	7039	5424	65	13
2024	6811	130,215	5808	147,158	68,746	7961	6020	71	15
2025	7098	137,898	6918	175,265	70,053	9004	6683	77	16
2026	7395	146,034	8239	208,741	71,384	10,184	7418	83	18
2027	7706	154,649	9813	248,610	72,740	11,518	8234	90	19
2028	8030	163,774	11,686	296,095	74,121	13,026	9140	98	21
2029	8366	173,436	13,919	352,649	75,530	14,734	10,145	106	23
2030	8944	185,790	15,321	374,355	77,593	15,945	11,808	116	25

Table 5 Predicted fuel consumption from the scenario I

Year	BUS	AR	TX	LCV	HCV	PB	MB	TW (lakhs)	CR (lakhs)
(In lakh liters/day)									
2018	1.69	3.84	0.07	1.71	7.10	1.21	0.07	9.86	12.52
2019	1.76	4.07	0.08	2.03	7.24	1.37	0.08	10.68	13.69
2020	1.84	4.31	0.09	2.43	7.38	1.54	0.09	11.57	14.98
2021	1.90	4.56	0.10	2.88	7.52	1.74	0.10	12.54	16.39
2022	2.00	4.83	0.11	3.44	7.66	1.97	0.10	13.58	17.93
2023	2.07	5.12	0.12	4.10	7.81	2.23	0.11	14.71	19.62
2024	2.17	5.43	0.13	4.88	7.96	2.53	0.14	15.94	21.47
2025	2.26	5.75	0.14	5.81	8.11	2.86	0.15	17.26	23.48
2026	2.35	6.09	0.15	6.93	8.27	3.24	0.17	18.71	25.68
2027	2.44	6.44	0.16	8.25	8.42	3.66	0.18	20.26	28.10
2028	2.52	6.82	0.17	9.83	8.58	4.14	0.21	21.95	30.75
2029	2.66	7.23	0.18	11.71	8.74	4.69	0.23	23.78	33.64
2030	2.84	7.96	0.19	12.39	9.06	5.04	0.24	24.59	34.56

Table 6 Predicted fuel emissions from the scenario I

Year	BUS	AR	TX	LCV	HCV	PB	MB	TW	CR
(In Gg/day)									
2018	0.343	0.477	0.008	0.923	1.185	0.180	0.030	1.936	3.732
2019	0.360	0.502	0.008	0.956	1.253	0.188	0.031	2.174	4.147
2020	0.378	0.530	0.008	0.990	1.324	0.196	0.032	2.442	4.607
2021	0.397	0.558	0.009	1.026	1.400	0.205	0.033	2.742	5.118
2022	0.417	0.588	0.009	1.063	1.479	0.214	0.033	3.079	5.686
2023	0.438	0.620	0.009	1.101	1.564	0.224	0.034	3.458	6.318
2024	0.460	0.653	0.010	1.141	1.653	0.233	0.035	3.883	7.019
2025	0.483	0.689	0.010	1.182	1.747	0.244	0.036	4.361	7.798
2026	0.507	0.726	0.011	1.225	1.847	0.254	0.037	4.897	8.664
2027	0.532	0.765	0.011	1.269	1.952	0.266	0.038	5.500	9.625
2028	0.559	0.807	0.012	1.314	2.063	0.277	0.039	6.176	10.694
2029	0.587	0.850	0.012	1.362	2.181	0.290	0.040	6.936	11.881
2030	0.616	0.896	0.013	1.411	2.305	0.302	0.041	7.789	13.199

seen in this scenario with the increase in the number of vehicles in public and private transport.

It can be seen from Table 6 that if the existing trend is allowed to continue till the horizon year 2030, the fuel emission of TW, CR, and BUS will reach to 7.789 Gg/day, 13.199 Gg/day and 0.616 Gg/day, respectively. The total annual fuel emission by all the vehicles in the horizon year 2030 is 26.57 Gg/day. So the total annual fuel emission would be reaching 9699 Gg/annum. The higher fuel levels pose a major challenge to the city's sustainability. This must be minimized by effective policy adoptation and mitigation steps, for which a comprehensive development and evaluation of other two major scenario-based SD simulation models are carried out in study. These are discussed in the subsequent sections.

6.2 Scenario II—Partial-Efforts Scenario

In the scenario of the partial effort, the growth rate values of BUS (public): TW and CR (private) have been altered to be 50:50 modal split for the horizon year 2030. To achieve the modal split of 50:50 between public and private transport modes, the growth rate of the former one has been incremented to reach 15.34%, and on the other hand, at the same time, the growth rate of the later is restrained to being 8.64% and 9.15% respectively which is nearly half of the already existing growth rate values for TW and CR. Meanwhile, the growth rates of the other modes of transport like the AR, TX, LCV, HCV, PB, and MB is assumed to be the same as that in the do-minimum scenario since the model split process is carried out between the public and private transport modes only. The forecasting process of the SD simulation model is carried out for each class of vehicles based on their incremental growth rate values, which is shown in Table 7.

An important observation from the analysis in scenario II is shown in Table 7. It is observed that if minimal efforts are taken in the partial scenario by altering the

Table 7 Predicted vehicle population from the scenario II

Year	BUS	AR	TX	LCV	HCV	PB	MB	TW (lakhs)	CR (lakhs)
2018	6386	110,781	2442	61,872	73,686	4565	3863	53	10
2019	5786	96,200	2385	60,425	61,570	4234	3515	44	9
2020	6394	101,876	2840	71,967	62,740	4787	3903	45	9
2021	7065	107,887	3383	85,713	63,933	5414	4332	46	9
2022	7807	114,252	4029	102,084	65,147	6124	4808	47	10
2023	8626	120,993	4798	121,582	66,386	6926	5337	49	10
2024	9533	128,132	5715	144,803	67,646	7834	5924	50	11
2025	10,532	135,691	6807	172,461	68,932	8860	6576	51	11
2026	11,638	143,697	8107	205,401	70,242	10,021	7299	52	11
2027	12,861	152,174	9656	244,632	71,576	11,333	8102	53	12
2028	14,211	161,153	11,499	291,357	72,935	12,818	8994	55	12
2029	15,703	170,661	13,696	347,006	74,322	14,498	9983	56	13
2030	16,257	179,249	15,568	361,846	75,281	14,943	11,162	57	14

Table 8 Predicted fuel consumption from the scenario II

Year	BUS	AR	TX	LCV	HCV	PB	MB	TW (lakhs)	CR (lakhs)
(In lakh liters/day)									
2018	1.72	3.91	0.06	1.74	7.23	1.23	0.08	10.03	12.74
2019	1.90	4.14	0.07	2.06	7.37	1.39	0.08	10.27	13.24
2020	2.10	4.38	0.08	2.47	7.51	1.57	0.09	10.51	13.75
2021	2.32	4.64	0.09	2.93	7.66	1.77	0.10	10.76	14.28
2022	2.56	4.92	0.11	3.50	7.80	2.01	0.11	11.02	14.84
2023	2.83	5.21	0.14	4.18	7.95	2.27	0.12	11.28	15.40
2024	3.13	5.52	0.15	4.96	8.10	2.58	0.14	11.54	16.01
2025	3.46	5.85	0.18	5.92	8.25	2.91	0.16	11.82	16.65
2026	3.93	6.19	0.19	7.05	8.41	3.29	0.18	12.10	17.27
2027	4.22	6.55	0.21	8.40	8.57	3.72	0.19	12.39	17.95
2028	4.66	6.94	0.22	10.00	8.73	4.21	0.21	12.68	18.64
2029	5.16	7.35	0.25	11.91	8.90	4.77	0.24	12.98	19.37
2030	5.73	8.10	0.26	12.61	9.22	5.13	0.25	14.45	20.71

growth rate of the existing vehicular trend, the number of TW and CR will reach 57 lakh and 14 lakh, respectively from 116 lakhs and 25 lakhs (scenario I values) in 2030, whereas the public transport vehicles constituting the BUS have increased only to a fleet size of around 16,257 from 8944 (scenario I value) in the horizon year 2030. This happens so because of the consideration of the hypothesis that 40 TW and 10 CR can be replaced by one single BUS of the maximum capacity of 50.

The decrease in the vehicle population concerning the number of TW, CR, and BUS happens due to a modal split value of 50:50 among public and private transport modes. This has led to a decrease in fuel consumption and fuel emission, which is shown in Tables 8 and 9, respectively.

From Table 9 it can be observed that the fuel consumed by TW decreases to 14.15 lakh liters per day from the scenario I's 24.59 lakh liters per day, and the fuel consumed by CR decreases to 20.71 lakh liters per day from the scenario I's 34.56 lakh liters per day and the fuel consumed by BUS increases to 5.73 lakh liters per day from the scenario I's 2.84 lakh liters per day for the horizon year 2030. The change in the growth rate of the number of public and private transport vehicles indicates that the fuel consumption by each vehicle type often varies accordingly.

From the above results, it can be said that the fuel consumption of BUS has seen to increase, but when we take the values of fuel consumption per person traveling in BUS into account with respect to that of persons using TW and CR combinedly, it can be said that the fuel consumption of the former mode of transport is seen to be much lesser than that from the latter one. When these resultant values are compared with the Scenario I results, it can be seen that the total fuel consumed by all the vehicles in scenario II decreases to 278.30 Crore liters per annum from the scenario I is 353.08 Crore liters per annum in 2030, which shows a reduction of 74.77 Crore

Table 9 Predicted fuel emissions from the scenario II

Year	BUS	AR	TX	LCV	HCV	PB	MB	TW	CR
(in Gg/day)									
2018	0.343	0.477	0.008	0.923	1.185	0.180	0.030	1.936	3.732
2019	0.381	0.502	0.008	0.956	1.253	0.188	0.031	2.093	4.024
2020	0.423	0.530	0.008	0.990	1.324	0.196	0.032	2.263	4.338
2021	0.470	0.558	0.009	1.026	1.400	0.205	0.033	2.447	4.677
2022	0.522	0.588	0.009	1.063	1.479	0.214	0.033	2.646	5.042
2023	0.579	0.620	0.009	1.101	1.564	0.224	0.034	2.861	5.436
2024	0.643	0.653	0.010	1.141	1.653	0.233	0.035	3.093	5.861
2025	0.715	0.689	0.010	1.182	1.747	0.244	0.036	3.345	6.319
2026	0.794	0.726	0.011	1.225	1.847	0.254	0.037	3.616	6.812
2027	0.881	0.765	0.011	1.269	1.952	0.266	0.038	3.910	7.344
2028	0.979	0.807	0.012	1.314	2.063	0.277	0.039	4.228	7.918
2029	1.087	0.850	0.012	1.362	2.181	0.290	0.040	4.571	8.537
2030	1.207	0.896	0.013	1.411	2.305	0.302	0.041	4.942	9.203

liters per annum of fuel consumption by all the vehicles in the scenario II. Hence the annual saving in the fuel consumption results in 21.17%.

In the partial efforts scenario, from Table 9 it can be seen that the fuel emissions by TW decrease to 4.92 Gg/day from the scenario I's 7.789 Gg/day and the fuel emissions by CR decreases to 9.203 Gg/day from scenario I's 13.199 Gg/day while the fuel emissions by BUS increases to 1.207 Gg/day from the scenario I's 0.616 Gg/day for the horizon year 2030. The total fuel emission incurred by private transportation is 14.145 Gg/day, whereas that of public transportation is just 1.207 Gg/day.

From the above findings, it is clear that while there was an increase in BUS fuel emissions, but given the per capita fuel emission from the BUS of those who use TW and CR, it is safe to conclude that the former fuel transport mode has much lower emissions than the former. When these resultant values are compared with the Scenario I results, it is found that the total fuel emissions incurred by all the vehicles in scenario II decrease to 7417 Gg/annum from scenario I's 9699 Gg/annum in the horizon year 2030. Hence the annual reduction in fuel emissions is 2282 Gg/annum, which results in a 30.76% reduction in fuel emissions. Hence, it can be affirmed that the increase in growth rate in the number of vehicles and fuel consumption by public and private transport has a corresponding direct effect on the emissions of fuel.

6.3 Scenario III—Desirable Scenario

The growth rate values of BUS, TW, and CR have been altered in the desirable scenario in which 75:25 modal split between public and private transport modes is assumed. In order to achieve a 75:25 modal split between public and private transport modes, the growth rate of the BUS has been increased to reach 19.78%, whereas the growth rate of TW and CR is restrained from being 3.38% and 3.12% respectively. Meanwhile, the growth rates of the other modes of transport like the AR, TX, LCV, HCV, PB, and MB is assumed to be the same as that in the do-minimum scenario since the model split process is carried out between the public and private transport system only. The forecasted vehicle population for each class of vehicle-based on the growth rates mentioned above is shown in Table 10.

Table 10 shows that if the desirable efforts are applied by altering the growth rate of the existing vehicular trend, the number of TW and CR will reach to 49 lakh and 10 lakh in scenario III from 57 and 14 lakh, respectively (scenario II) and from 116 lakhs and 25 lakhs respectively (scenario I) in the horizon year 2030. While the BUS transport system would increase only to a fleet size of around 21,452 from 16,257 (scenario II) and from 8944 (scenario I) in the horizon year 2030, this happens so because of the consideration of the hypothesis that 40 TW and 10 CR can be replaced by one single BUS which has a capacity of 50. This has led to a decrease in fuel consumption and fuel emissions also, which is shown in Tables 11 and 12, respectively.

The alteration of the growth rate in the number of vehicles between the public and private transportation has resulted in the consumption of fuel by each class of vehicles in the transportation system to vary correspondingly. The fuel consumed by TW decreases to 12.78 lakh liters per day in scenario III from scenario II's 14.15 lakh liters per day, and from the scenario I's 24.59 lakh liters per day. Similarly, the fuel consumed by CR decreases to 17.24 lakh liters per day from scenario II's 20.71 lakh liters per day, and from the scenario, I's 34.56 Lakh liters per day. The fuel

Table 10 Predicted vehicle population from the scenario III

Year	BUS	AR	TX	LCV	HCV	PB	MB	TW (lakhs)	CR (lakhs)
2018	6386	110,781	2442	61,872	73,686	4565	3863	53	10
2019	5871	95,417	2366	59,934	61,070	4199	3487	43	9
2020	6637	101,048	2817	71,382	62,230	4748	3871	44	9
2021	7502	107,010	3355	85,016	63,413	5370	4297	44	9
2022	8480	113,323	3997	101,254	64,617	6074	4769	45	9
2023	9586	120,009	4759	120,593	65,846	6870	5294	45	9
2024	10,836	127,090	5668	143,626	67,096	7770	5876	46	9
2025	12,249	134,588	6751	171,059	68,371	8788	6522	46	9
2026	13,846	142,529	8041	203,731	69,671	9939	7239	47	10
2027	15,651	150,937	9577	242,643	70,994	11,241	8036	47	10
2028	17,692	159,843	11,406	288,989	72,342	12,714	8921	48	10
2029	19,999	169,274	13,585	344,185	73,717	14,380	9902	48	10
2030	21,452	177,792	15,442	358,904	74,669	14,822	11,072	49	10

Table 11 Predicted fuel consumption from the scenario III

Year	BUS	AR	TX	LCV	HCV	PB	MB	TW (lakhs)	CR (lakhs)
(in lakh liters/day)									
2018	1.79	4.08	0.07	1.82	7.54	1.28	0.08	10.47	13.29
2019	2.02	4.32	0.08	2.15	7.68	1.45	0.09	10.58	13.50
2020	2.29	4.57	0.10	2.58	7.83	1.63	0.10	10.70	13.72
2021	2.59	4.84	0.12	3.06	7.99	1.85	0.11	10.82	13.94
2022	2.93	5.13	0.15	3.65	8.13	2.09	0.11	10.94	14.21
2023	3.32	5.43	0.16	4.36	8.29	2.37	0.13	11.06	14.40
2024	3.74	5.76	0.19	5.18	8.45	2.69	0.15	11.19	14.64
2025	4.24	6.10	0.22	6.17	8.60	3.04	0.17	11.31	14.88
2026	4.78	6.46	0.23	7.36	8.77	3.44	0.19	11.43	15.13
2027	5.41	6.84	0.26	8.76	8.94	3.88	0.20	11.57	15.37
2028	6.11	7.24	0.27	10.43	9.11	4.39	0.22	11.70	15.62
2029	6.91	7.67	0.29	12.43	9.28	4.97	0.25	11.83	15.89
2030	7.70	8.45	0.31	13.15	9.62	5.35	0.26	12.78	17.24

Table 12 Predicted fuel emissions from the scenario III

Year	BUS	AR	TX	LCV	HCV	PB	MB	TW	CR
(in Gg/day)									
2018	0.343	0.477	0.008	0.923	1.185	0.180	0.030	1.936	3.732
2019	0.395	0.502	0.008	0.956	1.253	0.188	0.031	2.034	3.922
2020	0.455	0.530	0.008	0.990	1.324	0.196	0.032	2.138	4.121
2021	0.523	0.558	0.009	1.026	1.400	0.205	0.033	2.246	4.330
2022	0.602	0.588	0.009	1.063	1.479	0.214	0.033	2.361	4.550
2023	0.693	0.620	0.009	1.101	1.564	0.224	0.034	2.480	4.781
2024	0.798	0.653	0.010	1.141	1.653	0.233	0.035	2.606	5.024
2025	0.918	0.689	0.010	1.182	1.747	0.244	0.036	2.739	5.279
2026	1.057	0.726	0.011	1.225	1.847	0.254	0.037	2.878	5.547
2027	1.217	0.765	0.011	1.269	1.952	0.266	0.038	3.024	5.829
2028	1.401	0.807	0.012	1.314	2.063	0.277	0.039	3.178	6.125
2029	1.612	0.850	0.012	1.362	2.181	0.290	0.040	3.340	6.436
2030	1.856	0.896	0.013	1.411	2.305	0.302	0.041	3.509	6.763

consumed by BUS increases to 7.70 lakh liters per day from scenario II's 5.73 lakh liters per day, and from the scenario, I's 2.84 lakh liters per day.

The total fuel consumption by private transportation is 30.02 lakh liters per day, whereas that of public transportation has been just 7.70 lakh liters per day. From this scenario analysis, it can be observed that the fuel consumption of BUS has seen to increase, but when we take the values of fuel consumption per person traveling in BUS into account concerning that of persons using TW and CR combinedly, it can be said that the fuel consumption of the former mode of transport is seen to be much lesser than that from the latter one. When these resultant values are compared with the scenario II and I results, it can be seen that the total fuel consumed by all the vehicles in scenario III decreases to 272.28 Crore liters per annum from 278.30 Crore liters per annum and from the scenario I's 353.08 Crore liters per annum in the horizon year 2030.

In the scenario III as observed from Table 12, the fuel emission by TW decreases to 3.509 Gg/day from scenario II's 4.942 Gg/day and from the scenario I's 7.789 Gg/day similarly the fuel emission by CR decreases to 6.763 Gg/day from scenario II's 9.203 Gg/day and from the scenario I's 13.199 Gg/day while the fuel emission by BUS increases to 1.856 Gg/day from scenario II's 1.207 Gg/day and from the scenario I's 0.616 Gg/day. The total fuel emissions incurred by private transportation are 10.272 Gg/day, whereas that of public transportation is just 1.856 Gg/day.

The above findings indicate that, although the emissions of fuel from BUS have risen, yet, considering the per capita fuel emissions of the BUS mode of transport and the persons combining both TW and CR, the emission of fuel from the former mode of transport is far less than the latter mode. When these resultant values are compared with the Scenario II & I results, it can be seen that the total fuel emission incurred by all the vehicles in scenario III decreases to 6240 Gg/annum from scenario II's 7417 Gg/annum and from scenario I's 9699 Gg/annum in the horizon year 2030. Hence the annual reduction in the fuel emissions is 1177 Gg/annum and 3459 Gg/annum, which results in 18.86% and 55.43% reduction in the fuel emissions in scenario III for the horizon year 2030.

7 Conclusions and Recommendations

This study has carried out a comprehensive analysis of three different scenarios with respect to the fuel consumption and fuel emissions in the transport sector for a metropolitan city in India. From the scenario analyses, it can be seen that the desirable scenario gives the best results in terms of reduction in fuel consumption and fuel emissions. However, it may be noted that the implementation of the desirable scenario may not be possible immediately. Also, the feasibility of public transport in terms of its accessibility, mobility, and connectivity has to be studied and analyzed in detail. However, at the early stages, the partial scenario measures which suggest the increase in public transport and restriction in private transport may be implemented on a pilot basis for a zone or area, and the impacts may be assessed for future scope.

In order to mitigate vehicular emissions, the number of vehicles plying on the road has to be reduced, providing an alternate public transport facility like the buses, metro-rail, intermediate public transport (IPT) and other shared-vehicles on the city roads. These modes of transport should be accessible, attractive, and affordable. These measures will reduce not only the plying of the private mode of transport vehicles on the city roads but also improves the usage of the public mode of transport vehicles, including sub-urban rails, metro rails, and others. For the smooth functioning of the public transit system, the fleet size of the public transport like the buses has to be increased not only with more excellent reliability but also with promising efficiency to achieve the last mile connectivity.

In addition to these measures, the decrease of vehicle fuel subsidy, with investment in sustainability, can enhance the public transport systems [1]. Other vehicular emission mitigation measures can be the use of public bi-cycles sharing systems, especially during winters, the use of low emissions transport mode, use of electric vehicles, and car-pooling. Such policy initiatives would ultimately reduce the fuel consumption, fuel emissions and fuel cost in the city. This will make the transportation system more efficient, which would eventually, maintain a safer environment that will provide healthier and more sustainable lives for future generations.

The proposed SD modeling approach may be useful for researchers and practitioners to understand and improve such models for robust analysis of sustainable transportation systems. As a result, the SD model can be used as the starting point to understand the behavior of the dynamicity involved in the overall interdisciplinary systems of transportation, energy, and emissions.

References

1. Al-Osaimi, S., Sreekanth, K. J., Al-Foraih, R., Al-Kandari, S.: Trends in road transportation fuel consumption and carbon emissions: a scenario analysis using system dynamic modelling. Int. J. Sustain. Energy, 1–13 (2019). https://doi.org/10.1080/14786451.2019.1696343
2. Giannakis, E., Serghides, D., Dimitriou, S., Zittis, G.: Land transport CO_2 emissions and climate change: evidence from Cyprus. Int. J. Sustain. Energy 1–14 (2020). https://doi.org/10.1080/14786451.2020.1743704
3. James, J., Mark, J., Aruna, S.: A review of urban energy system models: Approaches, challenges and opportunities. Renew. Sustain. Energy Rev. **16**(6), 3847–3866 (2012). https://doi.org/10.1016/j.rser.2012.02.047
4. Kan, Z., Tang, L., Kwan, M.P., Zang, X.: Estimating vehicle fuel consumption and emissions using GPS big data. Int. J. Environ. Res. Pub. Health **15**, 1–23 (2018). https://doi.org/10.3390/ijerph15040566
5. Lee, Y.S., Tong, L.I.: Forecasting energy consumption using a grey model improved by incorporating genetic programming. Energy Consumpt. Manual **52**(1), 147–152 (2011). https://doi.org/10.1016/j.enconman.2010.06.053
6. Maheshwari, P., Khaddar, R., Kachroo, P., Paz, A.: Dynamic modeling of performance indices for the planning of sustainable transportation systems. Netw. Spat. Econ. (2014). https://doi.org/10.1007/s11067-014-9238-6
7. Nesamani, K.S., Saphores, J.D., McNally, M.G., Jayakrishnan, R.: Estimating impacts of emission specific characteristics on vehicle operation for quantifying air pollutant emissions and energy use. J. Traffic Transp. Eng. **4**(3), 215–229 (2017). https://doi.org/10.1016/j.jtte.2017.05.007
8. Parikh, K.S., Karandikar, V., Rana, A., Dani, P.: Projecting India's energy requirements for policy formulation. J. Energy 34(8), 928–941 (2009). https://doi.org/10.1016/j.energy.2008.11.010
9. Paul, H.A.: System dynamics model of transportation energy demand. In: The 1987 International Conference of the System Dynamic Society, China, pp. 383–397 (1987)

10. Ramachandra, T.V., Shwetmala.: Emissions from India's transport sector: Statewise Synthesis. Atmosph. Environ. (2009). https://doi.org/10.1016/j.atmosenv.2009.07.015.
11. Report of the Expert Group on a Viable and Sustainable System of Pricing of Petroleum Products. Government of India Report, New Delhi (2010)
12. Shepherd, A., Ortolano, L.: Strategic environmental assessment for sustainable urban development. Environ. Impact Assess. Rev. **16**, 321–335 (1996)
13. Shuwei, J.: Dynamic analysis of a motor vehicle pollutant emission reduction management model based on the SD-GM approach. Discrete Dyn. Nat. Soc. **2018**, 1–18 (2018). https://doi.org/10.1155/2018/2512350
14. Wang, J., Lu, H., Peng, H.: System dynamics model of urban transportation system and its application. J. Transp. Syst. Eng. Inf. Technol. **8**(3), 83–89 (2008)
15. Statistics of the Transportation Department, Chennai, https://www.tn.gov.in/Traffic and Tranport.pdf, 3(5), pp. 2–26
16. Yevdokimov, Y.V., Mao, H.: A systems approach to measuring sustainability of transportation. In: 38th Annual Conference of Canadian Transportation Research Forum, Ottawa, Canada, pp 136–147 (2004)

Structural Property Assessment of GFRP Reinforced Concrete Beams

Gurbir Kaur, Raju Sharma, and Amol Singh Ramana

Abstract Steel making process exhibits greater environmental impact, including emission of greenhouse gases, wastewater contaminants and solid wastes. Further, steel corrodes rapidly in aggressive environment. Corrosion has a major impact on the life of the structure and thus affects the overall economy in the reinforced concrete project. Glass fibre-reinforced polymer (GFRP) bars provide a satisfactory solution to some of the environmental problems associated with conventional steel. As compared to steel, GFRP rebar has higher strength and good corrosion resistance but exhibits brittle behavior; therefore direct replacement of steel is not always possible with GFRP rebars. The present study examines the flexural behaviour of steel and GFRP reinforced concrete (RC) beams wherein four different reinforcement configurations are considered. Norms in compliance with ACI440.1R-06 have been used in design of specimens carrying GFRP rebars as internal reinforcement. Steel RC beams of M30 and M35 grades are used as control specimens. A total of 32 beams were tested in flexure; the load-deflection response, ductility, energy absorption, cracking patterns and the failure modes were recorded. A comprehensive analytical analysis was conducted to predict the ultimate loads of test beams. The test results evinced that the performance of the proposed configurations was successful in maintaining the flexure strength. ACI440.1R-06 based formulas can fairly be used to predict ultimate deflection values of GFRP RC beams. GFRP RC beams showed much wider cracks before failure but lower ductility than steel RC beams.

Keywords GFRP rebars · Reinforcement ratio · Load-deflection plots · Ductility · Crack patterns

G. Kaur (✉) · R. Sharma · A. S. Ramana
Thapar Institute of Engineering & Technology, Patiala, India
e-mail: gurbir.kaur@thapar.edu

R. Sharma
e-mail: raju.sharma@thapar.edu

A. S. Ramana
e-mail: amolr05@gmail.com

D. K. Ashish et al. (eds.), *3rd International Conference on Innovative Technologies for Clean and Sustainable Development*, RILEM Bookseries 29,
https://doi.org/10.1007/978-3-030-51485-3_13

1 Introduction

Glass fiber reinforced polymer (GFRP) has been confirmed to be the solution as a major development in strengthened concrete technology [1]. These rebars are attractive against steel bars in minimizing the corrosion related problems [2]. The use of GFRP bars as internal reinforcement of concrete structures has been growing, mainly due to the advantages they present over steel reinforcement, namely their low weight, high tensile strength and corrosion resistance [3]. The weight of the FRP bar are considerably are lesser than the steel bar which gives easiness in handling. The characteristics of GFRP rebars are primarily influenced by factors such as type, volume and orientation of fibres, type of resin, quality control during the manufacturing process, etc. [4]. GFRP rebar exhibits linear elastic behaviour and its stress and strain relationship indicates brittle behaviour and lack of ductility comparing to steel rebar [5]. The tensile modulus of elasticity of GFRP rebars is much lower than steel, approximately 25% of that of steel [4]. It is still a workable alternative to conventional steel because literature suggests that GFRP rebars are immune to degradation process under alkaline and corrosive environment. Many experimental studies indicate that by optimizing the amount and the layout, the FRP is suitable for balancing the strength recovery and that it is possible to restore the yield and the ultimate capacity with the same or lower deflection than initially [6]. Zareef and El Madawy [5] proposed a new method for improving the bond between GFRP rebar and concrete using calcium sulfoaluminate. The improved bond behavior between GFRP rebar and concrete due to the proposed method was quantified in terms of failure mode, ultimate pullout capacity and bond-slip response. In another study, better bond properties between concrete and sand coated GFRP bars were observed with smaller cracks and reduced crack widths as compared with helically grooved GFRP reinforced beams [1]. Yang et al. [7] proposed that the use of GFRP bars for compression reinforcement should be controlled when the compression reinforcements are under great stress; strength and corrosion resistance of GFRP bars can be fully exploited by applying these bars for tensile and stirrup-type reinforcement. The analytical procedures developed for the design of reinforced concrete with steel bars are not directly applicable to the design of reinforced concrete with FRP bars because the differences in the mechanical properties of reinforcing material are not taken into account.

There are two schools of thoughts on the modifications proposed in ACI 440.1R-06 [8] on the design of FRP-concrete elements for construction. While some authors recommend the use of these codes for accurate predictions, others believe otherwise because of overestimations in strength values. Also, the hybrid reinforcement (steel and GFRP) to improve the ductility and elasticity of FRP reinforced concrete has not been widely investigated in literature to give conclusive results. The present paper studies the flexural strength at the ultimate moment and cracking moment, deflection-load behaviour, ductility, energy absorption and crack pattern of reinforced concrete (RC) beams with hybrid reinforcement wherein norms in compliance with ACI440.1R-06 [8] have been used in design of specimens carrying GFRP rebars as internal reinforcement. The variable parameters included reinforcement configurations (four different reinforcement ratios) and grade of concrete (M30 and M35).

2 Experimental Work

2.1 *Materials and Specimen Cast*

The concrete mix for two strength grades (M30 and M35) were produced from ordinary Portland cement, natural sand and crushed ballast with maximum size of 20 mm as per IS-10262:2009 [9]. The beams cast using above material were demolded after 24 h from casting, and placed in fresh water tanks and stored under the laboratory conditions for 28 days before proceeding to the testing stage. GFRP bars manufactured by Xian Kig technology Co. Ltd were used in the investigation. GFRP bars of 8, 10 and 16 mm diameter and steel bars of 8 and 12 mm were used as main longitudinal bars in RC beams. Also, 8 mm steel bars Fe 500 were used as shear stirrups. The elastic modulus of GFRP bars was 60,000 MPa.

A total of 32 RC beams were cast and tested under three-point bending until failure. Figure 1 shows the cross section dimensions (150 mm × 150 mm) and length of specimens (700 mm). The RC beams were classified into 2 categories—GFRP series and Steel series on the basis of different reinforcement arrangement (Tables 1 and 2). Figure 2 presents different configurations of reinforcement in RC beams. Two-legged 8 mm diameter steel shear reinforcement was provided throughout the beam at 60 mm from centre to centre (Fig. 3). Minimum concrete cover of 15 mm was provided at every side of beam.

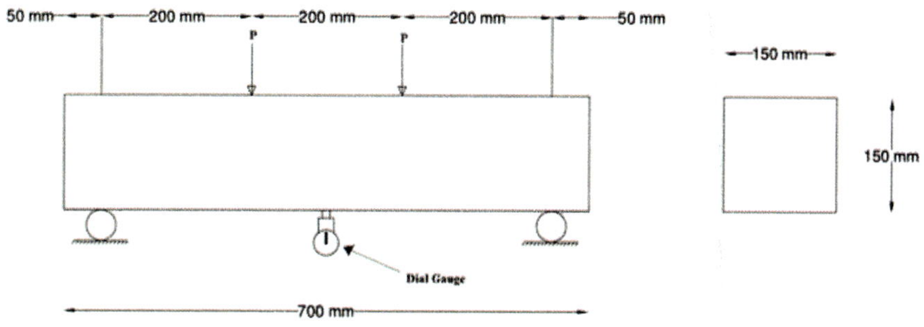

Fig. 1 Dimensions of beam specimens

Table 1 Reinforcement configurations of GFRP RC beams

Beam type	No. of upper bars (steel)/diameter of bars	No. of lower bars (GFRP)/diameter of bars	Reinforcement ratio
G1	2/8	3/8	0.67
G2	2/8	3/10	1.05
G3	2/8	2/16	1.78
G4	2/8	3/16	2.68

Table 2 Reinforcement configurations of steel RC beams

Beam type	No. of upper bars (steel)/diameter of bars	No. of lower bars (steel)/diameter of bars	Reinforcement ratio
S1	2/8	3/8	1.12
S2	2/8	2/12	1.45
S3	3/8	3/12	2.18
S4	3/12	3/12	3.01

2.2 *Test Setup*

The beams were simply supported at two rigid supports as shown in Fig. 4. Two-point load is applied to all the beams and were monotonically increased until failure by universal testing machine of 3000 kN capacity at a loading rate of 0.1 kN/s. The deflection at the mid span was measured by using digital dial gauge with a precision of 0.001 mm. The RC beams were tested as discussed and their ultimate failure moments are calculated from the ultimate force values. To make comparison with the experimental values, the theoretical ultimate moments were calculated in compliance with IS 456:2000 [10] for steel RC beams and ACI440.1R-06 [8] for GFRP-RC beam.

3 Test Results

3.1 *Load Deflection Behaviour*

Figures 5, 6, 7 and 8 presents the load-deflection curves of GFRP series and steel series RC beams. Steel RC beams show steep increase in the load with very less deflection and then at a certain point the rate of load increment starts decreasing and deflection starts to shoot up at a very high rate which will give an indication of the starting of the steel yielding. When the load is increasing steeply the load-deflection curve shows a straight line graph calling this portion of the graph as linearity period of the graph. The load capacity in linearity period increases with

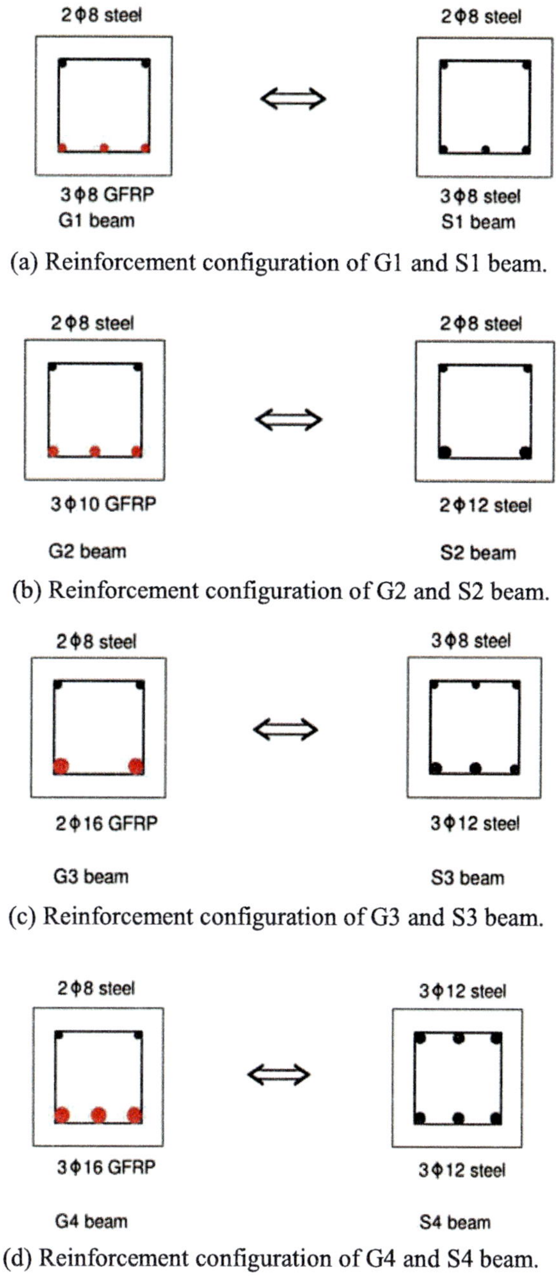

(a) Reinforcement configuration of G1 and S1 beam.

(b) Reinforcement configuration of G2 and S2 beam.

(c) Reinforcement configuration of G3 and S3 beam.

(d) Reinforcement configuration of G4 and S4 beam.

Fig. 2 Reinforcement configurations corresponding for different RC beams. **a** Reinforcement configuration of G1 and S1 beam. **b** Reinforcement configuration of G2 and S2 beam. **c** Reinforcement configuration of G3 and S3 beam. **d** Reinforcement configuration of G4 and S4 beam

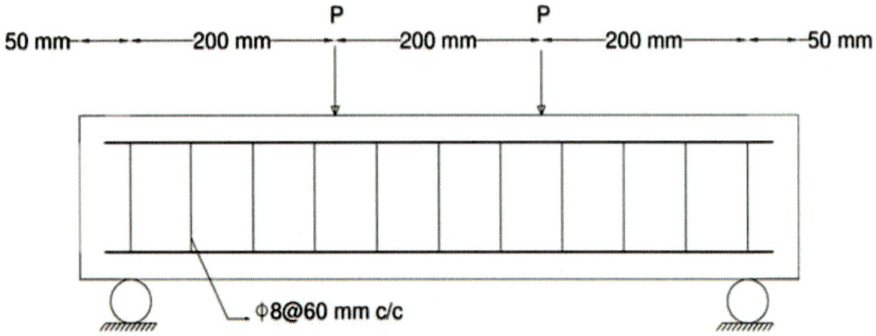

Fig. 3 Longitudinal section of RC beams depicting longitudinal and shear reinforcement

Fig. 4 Schematic of test setup

Fig. 5 Load-deflection curves of S1 and G1 beams of M30 and M35

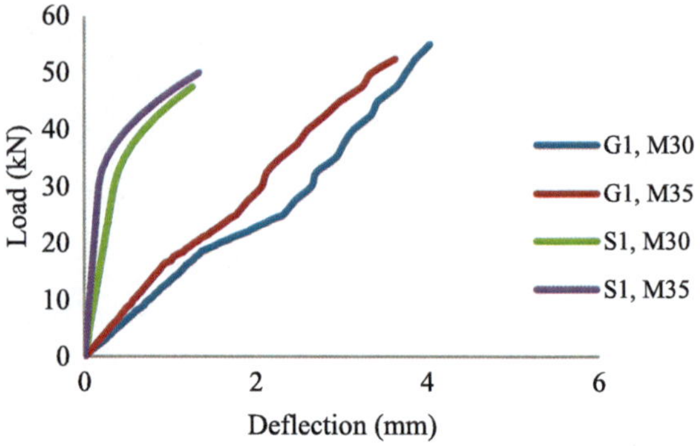

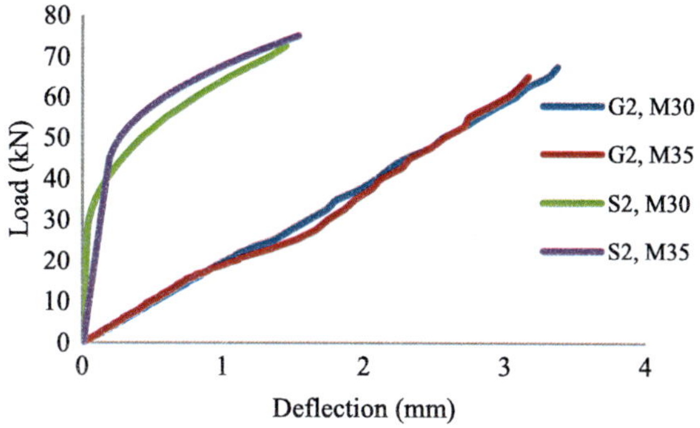

Fig. 6 Load-deflection curves of S2 and G2 beams of M30 and M35

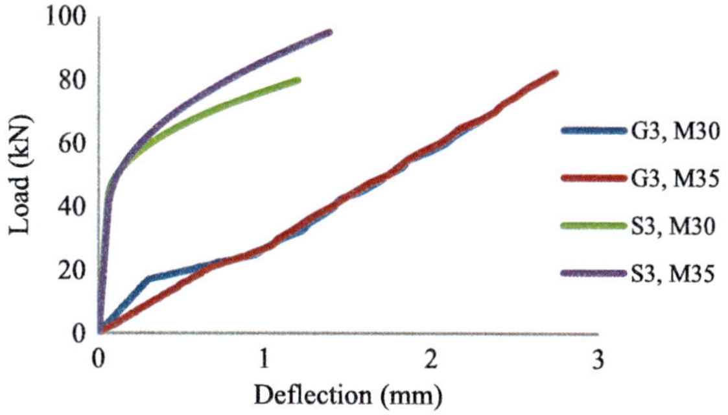

Fig. 7 Load-deflection curves of S3 and G3 beams of M30 and M35

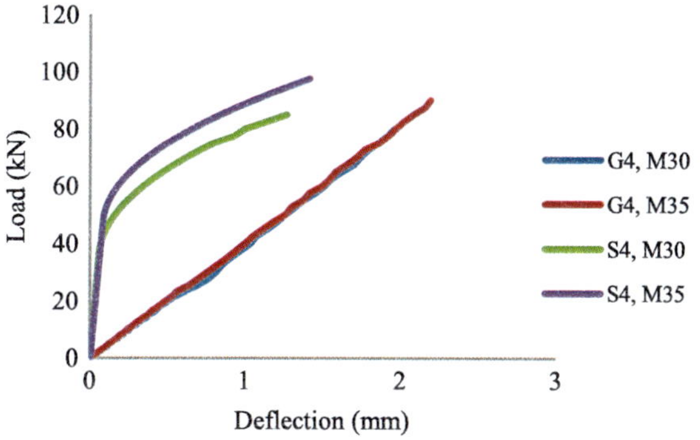

Fig. 8 Load-deflection curves of S4 and G4 beams of M30 and M35

increase in the reinforcement ratio and deflection in the linearity period decreases as the reinforcement ratio increase. GFRP RC beams show nearly a linear behaviour throughout the load taking process without much change in slope. Although some indication of the linearity period is present up to the approximate one-third or one-fourth of peak load in all beams and corresponding deflection at that time is between 0.5 and 1.2 mm. After the linearity period, there are fluctuations in every GFRP RC beam i.e. no exactly straight line is seen. As the reinforcement ratio is increased, the ultimate load also gets increased and the deflection in beams is decreased.

The first difference between both beam types is deflection. The deflection of GFRP RC beams is far more than the steel reinforced RC beams due to less elastic modulus of GFRP rebars than steel rebars. The second difference is the behaviour at failure; steel reinforced RC beams take nearly half load of ultimate load-carrying capacity with very small deflection and afterwards they show large deflections with very less increment in load whereas GFRP RC beams show nearly linear behaviour throughout the graph. Similar behaviour is shown previous authors [7, 11–13].

3.2 *Ultimate Load and Moment Values—Experimental and Theoretical*

Tables 3 and 4 present the experimental and theoretical ultimate loads for GFRP and steel RC beams respectively. The predicted or theoretical ultimate load values were determined using IS-456:2000 [10] for steel RC beams and ACI440.1R-06 [8] for GFRP RC beams.

Tables 5 and 6 present the experimental and theoretical ultimate moments for GFRP and steel reinforced beams respectively. Tables 7 and 8 depict the experimental and theoretical deflection values for GFRP and steel RC beams respectively.

In GFRP and steel RC beams, there is no specific pattern in the trend of % error with reinforcement ratio but GFRP RC has more % error in experimental and theoretical ultimate load or moment than the steel RC beams.

Table 3 Ultimate load of GFRP RC beams

Beam type	Experimental/theoretical ultimate load for M30 (kN)	Percentage error	Experimental/theoretical ultimate load for M35 (kN)	Percentage error
G1	54.72/53.92	1.5	53.12/55.41	4.1
G2	65.46/62.72	4.4	58.09/65.87	11.8
G3	67.20/68.52	1.9	69.92/75.57	7.5
G4	67.98/77.76	12.6	82.28/86.26	4.6

Table 4 Ultimate load of steel RC beams

Beam type	Experimental/theoretical ultimate load for M30 (kN)	Percentage error	Experimental/theoretical ultimate load for M35 (kN)	Percentage error
S1	47.58/48.23	1.3	51.20/48.75	5.0
S2	63.59/57.48	10.6	68.32/70.42	3.0
S3	79.97/81.85	2.3	90.04/84.58	6.4
S4	84/88.27	4.8	93.82/90.81	3.3

Table 5 Ultimate moment of GFRP RC beams

Beam type	Experimental/theoretical ultimate moment for M30 (kNm)	Percentage error	Experimental/theoretical ultimate moment for M35 (kNm)	Percentage error
G1	10.94/10.78	1.5	10.62/11.08	4.1
G2	13.09/12.54	4.4	11.61/13.17	11.8
G3	13.44/13.70	1.9	13.98/15.11	7.5
G4	13.59/15.55	12.6	16.45/17.25	4.6

Table 6 Ultimate moment of steel RC beams

Beam type	Experimental/theoretical ultimate moment for M30 (kNm)	Percentage error	Experimental/theoretical ultimate moment for M35 (kNm)	Percentage error
S1	9.51/9.64	1.3	10.24/9.75	5.0
S2	12.72/11.49	10.6	13.66/14.08	3.0
S3	15.99/16.37	2.3	18.01/16.91	6.4
S4	16.80/17.65	4.8	18.76/18.16	3.3

Table 7 Mid-span deflection of GFRP RC beams

Beam type	Experimental/theoretical ultimate moment for M30 (mm)	Experimental/theoretical ultimate moment for M35 (mm)
G1	3.87/3.85	3.66/3.65
G2	3.25/3.23	2.74/2.73
G3	2.26/2.26	2.29/2.30
G4	1.66/1.66	1.98/198

Table 8 Mid-span deflection of Steel RC beams

Beam type	Experimental/theoretical ultimate moment for M30 (mm)	Experimental/theoretical ultimate moment for M35 (mm)
S1	1.24/1.24	1.31/1.30
S2	1.26/1.26	1.29/1.29
S3	1.19/1.19	1.29/1.29
S4	1.25/1.24	1.35/1.34

The difference in calculated and experimental mid-span deflections is minimal with maximum error of 0.01 mm in S2 grade M35 beam which makes the codal formulas relevant to predict deflection values. In GFRP RC beams the maximum error found was 0.02 mm which was greater than the steel reinforced beams but it is not significant, thus, ACI440.1R-06 [8] based formulas can fairly be used to predict ultimate deflection values. Issa et al. [11] concluded that ACI formulas gives second accurate results after the models given by the Faza and Rao [14].

3.3 Crack Patterns and Failure Modes

In general, the propagation cracks of the element depends on the internal material like type of concrete, type of reinforcements and types of other material which is being used in constructing the material. The mode of failure observed in the RC beams were mainly of three types in which one type of failures (pure flexure failure) is only shown by the steel RC beams, whereas other two types of failures (two sided shear failure and single sided shear failure) were shown by both beam series. The notable difference between the failure of steel and GFRP RC beams was that GFRP RC beams showed much wider cracks before failure than steel RC beams. GFRP RC beams first start showing cracks in the non-shear zone which starts widening and propagating on more depth of the beam, just before beam reaches to ultimate load it starts abruptly showing major shear cracks after which it collapsed due to the shear failure. Thus, it can be concluded that due low elastic modulus of the GFRP bars such type of failure pattern was prominent. Other researchers [1, 11, 12] observed same types of cracks at ultimate failure of the beams. The first noticeable cracks of GFRP RC beams were earlier than the steel RC beams due to less bonding between the concrete and GFRP bars. The different types of crack patterns of beams at time of failure are shown in Fig. 9.

3.4 Ductility

Ductility is the material property that indicates the capacity of the material to undergo substantial plastic deformation before rupture. There are many methods to calculate the ductility like Issa et al. [11] found ductility of the beams with help of deformability factor which was defined as the ratio of energy absorption at ultimate load to the energy absorption at service load. The deformability factor found by Issa et al. [11] was 3.6 for normal grade concrete whereas for high grade concrete it was 5.9. In this study, the method given by Rao et al. [15] was used to determine the ductility factor for calculating ductility. Ductility factor is defined as ratio of deflection at the failure to deflection at yield or at the first crack. More the difference between ultimate deflection and first crack deflection more will be ductility. Table 9 shows the ductility factors for both types of RC beam series. According to Yoo et al. [12], the ductility of GFRP RC beams increased with increase in the reinforcement ratio which is not the case in present study. The ductility factor calculated in the GFRP RC beams, which typically lies between 2.69 and 4.42 does not showing any particular behaviour with increment or decrement of reinforcement ratio, whereas in steel RC beams (with ductility factor between 7.86 and 20.79) roughly indicate increase in

Fig. 9 Crack pattern in RC beams at failure

(a) Complete flexure failure

(b) Two-sided shear failure

(c) One-sided shear failure

Table 9 First crack value and ductility factor

Beams	Concrete grade	First visible crack load (kN)		Ductility factor		Average ductility factor
		First beam	Second beam	First beam	Second beam	
S1	M30	28	20	4.06	11.65	7.86
S2		25	41	11.102	7.44	9.27
S3		37	43	20.591	21.00	20.80
S4		41	37	16.812	12.33	14.53
G1		18	18	3.01	2.87	2.94
G2		19	22	3.46	2.53	2.99
G3		20	22	4.25	3.01	3.63
G4		22	21	3.46	2.54	2.99
S1	M35	30	27	9.49	8.33	8.91
S2		43	32	10.153	9.53	9.84
S3		40	46	12.423	17.07	14.75
S4		49	48	17.025	11.42	14.22
G1		17	16	3.27	3.87	3.57
G2		19	20	2.53	2.85	2.69
G3		17	21	4.90	3.94	4.42
G4		21	24	3.54	3.70	3.62

the ductility factor with increase in the reinforcement ratio. The ductility factor of steel RC beams is nearly 3–5 times of GFRP RC beams. When concrete strength is increased the ductility factor had shown small increment in both the steel and GFRP RC beams which indicates that by increasing the grade of concrete, the ductility factor of beams can be increased. These results are in compliance with Issa et al. [11].

3.5 Energy Absorption

Energy absorption is the area under the load-deflection curves of RC beams. The energy absorption of GFRP RC beams show roughly a decrement on increase in reinforcement ratio whereas in steel RC beam, the energy absorption shows a regular increase with an increase in reinforcement ratio. The details of energy absorption is shown in Table 10. The energy absorption of steel RC beams has increased consequently with increase in the reinforcement ratio but this trend was not visible in the GFRP RC beams. In GFRP beams, first the energy absorption has increased to the optimum level at a certain level after which further increase in the reinforcement ratio gives a decrease in the energy absorption.

Table 10 Energy absorption in both RC beam series

Beam type	Concrete grade	Energy absorption (kN/mm)		Average energy absorption (kN/mm)
		First beam	Second beam	
S1	M30	43.18	40.04	41.61
S2		43.79	79.53	61.67
S3		73.27	78.35	75.81
S4		79.89	83.53	81.71
G1		101.37	87.49	94.43
G2		109.66	91.26	100.46
G3		78.76	69.65	74.20
G4		35.09	75.75	55.42
S1	M35	49.95	54.61	52.28
S2		94.03	47.87	70.95
S3		100.13	79.88	90.00
S4		87.56	109.87	98.71
G1		93.86	99.72	96.79
G2		52.60	94.25	73.66
G3		103.86	56.14	79.99
G4		62.35	96.91	79.63

4 Conclusions

Following conclusions can be drawn from the present investigation:

1. The ultimate load carrying capacity is achievable in both GFRP and steel RC beams with a maximum percentage error of 10.6% in GFRP RC beams and 12.6% in the steel RC beams. The maximum error in the theoretical moment between the steel and GFRP RC beams is nearly 28.8%, whereas in the experimental data, it decreased to 19.5%.
2. On the basis of deflection, both in GFRP and steel RC beams, a maximum difference of 2.63 mm is observed in S1 beam and when the reinforcement ratio is increased, the % error or difference decreased i.e. in S4 beams, % error is 0.4 mm in S4.
3. In load-deflection curve in steel RC beams, first the load increases very rapidly with very less deflection in a linear manner and then achieving half the ultimate load the curve changes from linearity to different degree by which load increases very slowly and deflection increase at a very high rate that change in behaviour tells the starting of yielding of steel. On the other side, GFRP RC beams do not show such behaviour it just shows nearly straight line till ultimate failure. In, GFRP RC concrete beams had no specific yielding point is shown so deflection increases due to de-bonding of the GFRP rebars with concrete with an increase in force. De-bonding occurs due to low elastic modulus of the GFRP rebars.

4. Ductility is calculated by taking the ratio of ultimate deflection to the deflection at first cracking point. The ductility of steel RC beams is greater than the GFRP RC beams. The steel RC beams have 2–10 times more ductility than GFRP RC beams.
5. Energy absorption of steel RC beams has increased with increase in reinforcement ratio whereas in GFRP RC concrete beams, the energy absorption started to decrease with increase in reinforcement ratio. The maximum energy absorption difference is observed in S1 and G1 beam of grade M30 of 52.81 kN/mm whereas the minimum is noticed in S3 and G3 beam of grade M30 of 1.61 kN/mm.
6. At ultimate load, few of the under-reinforced steel RC beams shows pure flexure failure pattern where all other beams show mixed failure trend i.e. first the specimens showed flexure cracks, which are very large in GFRP RC beams and low in the steel RC beams and at end all of them show abrupt shear cracks.

References

1. Jabbar, S.A.A., Farid, S.B.H.: Replacement of steel rebars by GFRP in concrete structures. Karbal Int. J. Mod. Sci. **4**, 216–277 (2018)
2. Hassan, A., Khairallah, F., Mamdouh, H., Kamal, M.: Structural behaviour of self-compacting concrete columns reinforced by steel and glass fibre-reinforced polymer rebars under eccentric loads. Eng. Struct. **188**, 717–728 (2019)
3. El-Hassan, H., El Maaddawy, T.: Microstructure characteristics of GFRP reinforcing bars in harsh environment. Adv. Mater. Sci. Eng., Article ID 8053843, 19 pages (2019)
4. Benmokrane, B., Chaallal, O., Masmoudi, R.: Glass fibre reinforced plastic (GFRP) rebars foe concrete structures. Constr. Build. Mater. **9**, 353–364 (2016)
5. El Zareef, A.M., El Madawy, E.M.: Effect of glass-fiber rods on the ductile behaviour of reinforced concrete beam. Alexandria Eng. J. **57**(4), 4071–4079 (2018)
6. Toutanji, H.A., Saafi, M.: Flexural behaviour of concrete beams reinforced with glass fibre-reinforced polymer (GFRP) bars. ACI Struct. J. **97**, 712–719 (2000)
7. Yang, W., He, X., Dai, L.: Damage behaviour beams reinforced with GFRP bars. Compos. Struct. **161**, 173–186 (2017)
8. ACI Committee 440, 1R-06: Guide for the design and construction of concrete reinforced with FRP bars, Flangminton Hills (MI): American Concrete Institute (2006)
9. Indian Standard Code IS 10262:2009, Guidelines of Concrete mix proportioning—Guidelines (First Revision). Bureau of Indian Standards, Manak Bhavan, 9 Bahadur Shah Zafar Marg, New Delhi 110002
10. Indian Standard Code IS456:2000, Plain and reinforced concrete—code of practice (fourth revision), Bureau of Indian Standards, Manak Bhavan, 9 Bahadur Shah Zafar Marg, New Delhi 110002
11. Issa, M.S., Metwally, I.M., Elzeiny, S.M.: Influence of fibres on flexural behaviour and ductility of beams reinforced with GFRP rebars. Eng. Struct. **33**, 1754–1763 (2011)
12. Yoo, D., Banthia N., Yoon Y.: Predicting service deflection of ultra-high-performance fiber reinforced concrete beams reinforced with GFRP bars. Compos. Part B **99**, 381–397 (2016)
13. Sheikh, S.A., Kharal, Z.: Replacement of steel with GFRP for sustainable reinforced concrete. Constr. Build. Mater. **160**, 767–774 (2018)

14. Faza, S.S., Ganga Rao, V.: Pre-and post-cracking deflection behaviour of concrete beams reinforced with fibre-reinforced plastic rebars. In: Proceedings of the First International Conference on Advanced Composite Materials in Bridges and Structures, pp. 129–137 (1992)
15. Rao, G.A., Vijayanand, I., Eligehausen, R.: Studies on the ductility of RC beams in flexure and size effect. Mater. Struct. **41**(4), 759–771 (2017)

Concrete with Encapsulated Self-healing Agent: A Critical Review

Maulik Mistry and Santosh Shah

Abstract Concrete due to its brittle nature easily cracks. Such cracks in concrete structures become the pathway for the penetration of aggressive agents which in turn affect the durability, functionality, and strength of the structure. To ensure the desired functioning of the structure suffering from concrete cracks, regular maintenance and special types of treatments are necessary which may be sometimes expensive and time-consuming. The application of smart materials, especially self-healing concrete has received significant attention due to its ability to heal cracks automatically, which is a relatively quick and economical alternative to conventional methods. In the present study, a comprehensive review of encapsulated self-healing concrete is discussed. The study of the literature reveals that the selection of suitable healing material and the capsules shell are major challenges reported by various researchers. In this paper, a detailed review of various healing and shell materials is carried out along with a review of the performance of self-healing efficiency through various experimental techniques. Challenges in the selection of suitable healing and shell materials for the development of encapsulated concrete keep the research area active.

Keywords Cracks · Self-healing concrete · Shell material · Healing material · Healing-efficiency

1 Introduction

The use of easily available and low-cost raw materials, as well as high compressive strength, makes concrete the most extensively used man-made material, globally. However, due to its brittleness and low flexural strength, concrete suffers from

M. Mistry (✉)
Applied Mechanics Department, Government Engineering College, Bhuj, India
e-mail: maulik.civil9@gmail.com

S. Shah
Civil Engineering Department, Babaria Institute of Technology, Vadodara, India
e-mail: santoshgshah@gmail.com

D. K. Ashish et al. (eds.), *3rd International Conference on Innovative Technologies for Clean and Sustainable Development*, RILEM Bookseries 29,
https://doi.org/10.1007/978-3-030-51485-3_14

cracking and hence reinforcements are embedded into the concrete matrix to overcome such limitations [1]. Micro or Macro cracks may appear at any stage during the design life of the concrete structure due to factors such as shrinkage, thermal stresses, harsh environmental exposure conditions, flaw of the design and/or at construction stage or a combination of these factors [2]. Cracks become the pathway for aggressive liquids and gases in concrete, which results in degradation in concrete strength and durability as well as that affect the functionality of the concrete structure. Moreover, such cracks lead the reinforcement to be exposed to come in direct contact with aggressive agents causing corrosion and subsequently result in further deterioration of concrete, which may sometimes result in a premature failure of the structure. Through the studies conducted on European countries, it is established that half of the annual budgets of constructions are spent on the repair and maintenance of structures [3]. Hence, it is imperative to carry out continuous monitoring, maintenance, and repairing of the structures. But in large-scale structures such as bridges, dams, storage tanks, etc., continuous inspection, maintenance, and repairs can be difficult. Also, a huge amount of capital investment is required to carry out such activities. Though various conventional crack repair methods are developed to repair the surface cracks by external intervention, these are costly and time-consuming as well as require special expertise to implement these methods [4]. However, it is difficult to identify and repair the cracks which are not propagated to the concrete exterior surface.

Self-healing concrete is considered to be a promising solution for addressing the aforementioned problems [5]. Self-healing with capsules occurs due to the rupture of the capsules and the healing agent is poured into the crack, which in turn prevents the aggressive agents from penetrating into the concrete. As a result, the service life and durability of the concrete structures get extended. Moreover, mechanical properties of concrete structures can be enhanced by the presence of self-healing mechanisms. Self-healing concrete as a novel idea has received considerable attention from numerous researchers in their study over the last few decades. Various researchers in their study published state-of-the-art reviews focusing on various approaches to self-healing mechanisms as well as methods for measuring self-healing performance [6–8]. This paper presents a detailed review of available literature focused on the encapsulated self-healing concrete.

Self-healing concrete is broadly classified into two major approaches, namely autogenous and autonomous, [6]. In autogenous self-healing, concrete cracks can be healed naturally by the reaction between anhydrate cementitious particles with available water and carbon dioxide by forming C–S–H gel. However, the self-healing efficiency of this approach is relatively lower as compared to autonomous as the efficiency of the former is highly affected by various factors such as the width of cracks, source of water, age of concrete, etc. Self-healing efficiency of concrete can be improved autonomously by additives such as mineral admixtures, crystalline admixtures, bacteria, vascular tube network and chemical encapsulation [9]. Among these, the chemical encapsulation based approach is found to be most versatile due to its ability to respond directly to the damaged areas.

In encapsulated self-healing concrete, microcapsules are embedded homogeneously into the concrete matrix. Once the concrete starts cracking, the microcapsules present on the crack path get ruptured and release healing agents to heal the newly developed cracks [10]. Crack-repair effectiveness of the encapsulated self-healing concrete mainly dependents on various factors such as immediate rupture of capsule shell wall, flow-ability of healing agents into the crack, and reaction between a healing agent with cementitious materials and/or catalyst. The effects of various healing agents and shell materials on the performance of the self-healing mechanism of encapsulated self-healing concrete are attempted by researchers. Self-healing efficiency of encapsulated self-healing concrete is assessed in terms of the mechanical and durability properties of self-healing concrete [11, 12].

In the present study, a detailed review of various healing materials and shell materials is carried out along with the review of the performance of self-healing efficiency obtained through various experimental techniques for the assessment of durability and mechanical properties of self-healing concrete. Challenges in the selection of suitable healing and shell materials for the development of encapsulated concrete keep the research area active.

2 Materials for Self-healing Agent

The selection of suitable healing agents plays a vital role in improving the self-healing efficiency of encapsulated self-healing concrete. Single or dual-component healing agents are enclosed in the shell wall of the capsule. Single healing agents start reacting upon once the crack is developed resulting in the healing agent to come in the contact with moisture, air or cementitious material, whereas in dual-component healing agents, the reaction between catalyst and the healing agent takes place before flowing through the crack [10]. A brief discussion of commonly used self-healing agents is carried out in this section.

2.1 Cyanoacrylate Healing Agent

Cyanoacrylates, also known as superglues, having very short shelf-life is more effective in applications showing cracks at the early age of the concrete structures [13]. Joseph et al. [14] have studied using cyanoacrylate as a single healing-agent through assessing properties of self-healing concrete such as viscosity, curing time and the bond strength between the concrete surfaces. It is also reported that the micro-cracks are healed through capillary action due to very low viscosity of cyanoacrylate which results in gaining rapid bond strength during the curing process. Based on the experimental findings, the elasticity module of Passive Smart Self-Healing Engineered Cementitious Composite (PSS-ECC) specimens has been enhanced when these specimens have been subjected to several loading cycles. Le et al. [15] have shown that at laboratory exposure conditions cyanoacrylate proved to be an effective healing agent. Through several numerical and experimental studies to investigate the rate of capillary rise of cyanoacrylate under different crack configurations, Gardner et al. [13] have reported that the viscosity of cyanoacrylate remains constant over a time of 15 min. Owing to a very short shelf-life and limited range of applicability, limited research efforts on the application of cyanoacrylate were attempted in the existing volume of the literature.

2.2 *Sodium Silicate Healing Agent*

Sodium silicate is one of the silica-based self-healing agents, which heals the cracks by producing calcium silicate hydrates while reacting with calcium hydroxide in cementitious materials present in the concrete. Huang and Ye [16] have shown that the self-healing efficiency of ECC specimens, treated with sodium silicate as a self-healing agent applied through encapsulation, improved significantly in pre-cracked specimens. It is also reported that the self-healing recovery is influenced by the concentration of sodium silicate. According to Gilford et al. [17], the size and morphology of encapsulated sodium silicate microcapsules are directly influenced by preparation parameters such as temperature control at the emulsion stage, agitation rate and pH of emulsion. The authors have also reported that the experimental results of cylindrical concrete test specimens shown an 11% enhancement in the value of modulus of elasticity after healing treatment. Giannaros et al. [18] have quantified the effect of the addition of microencapsulating sodium silicate embedded, both in liquid and solid form, into the cement paste on rheological and mechanical properties of cracked specimens and have reported that after just 28 days sorptivity is reduced by 45%. Mostavi et al. [19] have developed double-walled polyurethane/urea–formaldehyde (PU/UF) microcapsules containing sodium silicate as healing agent and studied the effect of different parameters such as agitation rate, pH, and temperature on the performance of prepared microcapsules and to optimize the micro-encapsulation procedure as well as evaluate their healing performance in concrete. Authors have reported that the best microcapsule developed at a pH value of 3.1, and agitation rate of 1000 rpm, and curing temperature 57°C. It is also reported that by the addition of 5% microcapsules, effective self-healing achieved.

2.3 *Methyl Methacrylate (MMA) Based Healing Agent*

Due to the prolonged shelf-life, Methyl Methacrylate (MMA) - acrylic-based healing agent, has received the attention of researchers over the past few decades. Other promising factors, such as extreme temperature efficiency, low viscosity, high bond strength and the ability to fix cracks by avoiding chloride penetration and carbonation, offer widespread use of the same [20].

Based on the experimental results obtained with concrete specimens treated by two distinct approaches, surface treatment and treatment through hollow tubes, Dry and McMillan [20] have shown that the effect of Methyl methacrylate (MMA) applied through the later treatment controls effectively the permeability and compressive strength of the test specimens. A similar trend in permeability enhancement is reported by Yang et al. [21] through the experimental investigation on MMA microcapsules embedded in cement mortar test specimens and they have reported a reduction in permeability by 50.2 and 66.8% at the test specimen age of 3 and 30 days, respectively, compared to untreated mortar test specimens.

2.4 Epoxy Resin-Based Healing Agent

Low viscosity among other supporting parameters responsible for the healing agent's successful efficiency in fixing micro-cracks, more epoxy resin-based healing agents are a solution for widespread applications in self-healing concrete. Earlier applications of epoxy resin-based healing agents suggest the use of high temperature or pressure techniques for hardening as discussed by Xue et al. [22]. ACI recommends viscosity values in the range of 100–500 cps for the epoxy-based healing-agent for the effective performance of self-healing in concrete, ACI 515.2R-13 [23]. Thao et al. [24] suggest a higher lower bound value of viscosity to 250 cps for the smooth flow of healing agents through the cracks for achieving better performance in terms of the self-healing efficiency. According to Mihashi et al. [25], the correct mix stoichiometry significantly depends on the curing method of epoxy resins. Cosco et al. [26] reported that the amount of epoxy resin stored during the synthesis process directly depends on agitation speed and temperature of reacting emulsion. Dong et al. [27] have used micro-encapsulated epoxy resin in the urea–formaldehyde shell and found considerable improvements in mechanical and a reduction of permeability properties of mortar specimens. Li et al. [28] have reported that a higher healing efficiency can be achieved in the cracked specimen as a result of the reaction among liquid resin and hardener provided through dual component epoxy microcapsules. Abdul et al. [29] have shown that the compressive strength of cement mortar gets significantly improved by the addition of 10% epoxy resin as a healing agent.

2.5 Polyurethane (PU) Based Healing Agent

Polyurethane based self-healing agents can expand, which enables it to heal wider cracks in concrete structures [30]. Maes et al. [31] have reported that the crack width up to 100–300 μm can be effectively healed in cement mortar specimens by encapsulated polyurethane. Experimental investigations carried out by Hu et al. [32] on encapsulated diluted polyurethane applied through the glass tube and embedded into the concrete show improvements in the sealing efficiency by about 67%. Tittelboom et al. [33] used tubular capsules filled with polyurethane healing agents in their study and have observed that 50% of the original strength and stiffness are regained under cyclic loading after self-healing along with the improvement in water permeability. Based on the results of four-point bending testing on real-scale concrete beams encapsulated with polyurethane Tittelboom et al. [34] have shown improvements in self-healing efficiency.

2.6 Dicyclopentadiene

Due to excellent healing properties such as low initial viscosity, high toughness after curing and suitability in various environmental conditions, the dicyclopentadiene (DCPD) has been used as a healing agent by researchers [35]. Gilford III et al. [17] have used encapsulated dicyclopentadiene microcapsule and reported that the modulus of elasticity of concrete has increased by 30% after cracking under cyclic loading. Brown et al. [35] have developed a synthesis process to encapsulate dicyclopentadiene in the urea–formaldehyde shell which results in achieving a higher yield (It is the ratio of weight of microcapsules to the theoretical weight of ingredients, expressed in %) in the range of 80–90% and higher fill content in the range of 83–92%.

2.7 Calcium Nitrate

Calcium nitrate has been selected by researchers as an alternative healing agent due to its low cost and its ability to react with anhydrate cement particles to form calcium silicate hydrates. Hassan et al. [36] have developed a synthesis procedure for the production of calcium nitrate microcapsules and studied the effect of production parameters on the shape, size and shell wall thickness. The authors also have evaluated the self-healing efficiency of microcapsules for concrete and have reported that the modulus of elasticity of concrete increases after healing. Arce et al. [37] have encapsulated the calcium nitrate into the urea–formaldehyde microcapsules to evaluate the self-healing efficiency of cement mortar under dry and wet healing conditions and reported that the self-healing efficiency has improved after 14 days of healing. Milla et al. [38] have developed a modified micro-encapsulation procedure and evaluated the effect of calcium nitrate tetra hydrate as a healing agent and have reported significant improvement in the compressive strength as well as in modulus of elasticity, which implies better self-healing efficiency. Al-Ansari et al. [39] modified the calcium nitrate microencapsulation procedure by altering the concentration of surfactant in the synthesis process and have suggested that the concentration of 0.75% microcapsules by weight of cement enhances mechanical properties in terms of compressive strength and modulus of elasticity.

3 Materials for Capsules

The healing agent is stored in a capsule shell; the effective release of healing agents into the cracks depends on the fracture behavior of the shell wall. Minnebo and Hemelrijck [40] proposed that a good encapsulating shell material should possess properties like low porosity, robustness during mixing, no interaction between shell wall and encapsulating material, easy to fracture, a good interfacial bond between shell wall and cementitious materials and resist an aggressive environment. The self-healing efficiency of cementitious materials also depends on the shape, size, and morphology of the capsule. Several studies conducted by researchers for selecting suitable encapsulating shell material are discussed here.

3.1 Gelatin

Mihashi et al. [25] have micro-encapsulated the acrylic resin into gelatin shells and embedded them into a concrete matrix to evaluate the self-healing efficiency of the concrete matrix. The authors have observed a smaller improvement in strength recovery after cracking due to very little amount of healing agents available for reactions. Two different healing agents, tung oil and $Ca(OH)_2$, were encapsulated in gelatin microcapsules and embedded in a concrete mortar to evaluate their robustness in the mixing process as well as their self-healing efficiency [3]. Cailleux and pollet [3] have observed that during the mixing process microcapsules filled with epoxy resin survived, whereas some of the microcapsules filled with $Ca(OH)_2$ ruptured. They also found an improvement in the self-healing efficiency of microcapsules filled with tung oil.

3.2 Silica Gel

Yang et al. [21] have developed silica gel microcapsules with an average diameter of 4.15 μm, consisting of methyl methacrylate as a healing agent and triethyl borane as a catalyst. They dispersed microcapsules into the concrete matrix and observed that when micro-cracks initiated the capsule shell walls ruptured which resulted in crack healing by polymerization between healing agent and catalyst. Improvement in permeability and mechanical strength were also observed after self-healing.

3.3 Polyurethane (PU)

A detailed study was conducted by Beglarigale et al. [41] to optimize, characterize and observe the fracture behavior of polyurethane microcapsules consisting of sodium silicate and it is reported that the developed microcapsule is a free-flowing powder with a spherical shape. Optimum shell-forming monomer to core material ratio obtained was 0.67 assessed by the yield, SEM, and TG analyses. Authors also have observed adequate interfacial bond strength between the shells of microcapsules and cement mortar. Pelletier et al. [42] have optimized polyurethane microcapsules with sodium silicate as core material and obtained diameters of microcapsules between 4 and 800 μm. The developed microcapsules were incorporated into cement mortar and it was observed that the compressive strength of mortar after healing by polyurethane microcapsules was not affected.

3.4 Urea–formaldehyde (UF)

Gilford et al. [17] have reported experimental procedures for the development of two different types of urea–formaldehyde microcapsules containing a healing agent as dicyclopentadiene (DCPD) and sodium silicate. They studied the effect of temperature, agitation rate and pH of emulsion on size and shell wall thickness of the developed microcapsules and also evaluated the self-healing efficiency of microcapsules in a concrete matrix. After executing an experimental study, they reported that uniform and coherent microcapsules were developed at 55°C temperature in both DCPD and sodium silicate microcapsules. However, the shell thickness of sodium silicate microcapsules found to be increased with an increase in pH of emulsion, unlike DCPD microcapsules. Also, the size of microcapsules reduced as the agitation rate increased in the case of DCPD microcapsules, but in the case of sodium silicate microcapsules, the size remained constant as the agitation rate increased from 250 to 550 rpm. Also, they observed that at a lower dosage of DCPD microcapsules are more effective in self-healing efficiency as compared to sodium silicate microcapsules. Blaiszik et al. [43] have developed a robust method for the preparation of urea–formaldehyde microcapsules containing a reactive resin and solvent to observe its morphology and size after the synthesis process. They reported that the size of developed microcapsules was in the range from 300 to 30 μm, with a uniform shell thickness of 160 nm. The surface of developed microcapsules is smooth from inside and found rough outside, which provides a better bond with the host matrix. Hassan et al. [36] have developed a synthesis procedure for micro-encapsulating calcium nitrate with a urea–formaldehyde shell and also studied the effect of preparation parameters on the size and morphology of microcapsules. From their study, they observed that size decreased at a higher agitation rate and the shell wall thickness reduces at a higher temperature. Also, the surface texture of the shell wall is smooth from inside and rough from outside which enhances bond strength with the concrete matrix.

3.5 Polyurethane/Urea–Formaldehyde (PU/UF)

Brown et al. [35] have micro-encapsulated dicyclopentadiene in the poly (urea–formaldehyde) shell and assessed the shape, size, and morphology of microcapsules influenced by the agitation rate and pH of the emulsion. Fill content and yield of microcapsules were analyzed by the CHN test. They observed that the size of microcapsules was in the range of diameters from 10 to 1000 μm by varying the agitation rate from 200 to 2000 rpm, the shell wall thickness obtained was 160–200 nm with a smooth inner surface and a rough exterior surface. Free-flowing spherical shape powder with high yield and fill content observed through CHN analysis. Giannaros et al. [18] have developed poly (urea–formaldehyde) microcapsule containing sodium silicate in solid form as a healing agent and observed the effect of microcapsules on rheological and durability properties of the cement paste. By the addition of 4% microcapsules in cement paste, the viscosity of cement paste was increased by 47%, as well as the time until the setting was shorter. It is also reported that the durability of cement paste is improved due to a reduction in sorptivity by 45%.

3.6 Glass

Qureshi et al. [44] have encapsulated expansive minerals into glass capsules and embedded them into the mortar matrix. They studied the behavior of glass capsules under different curing regime conditions and found that crack sealing is achieved up to 95% after cracking; also the strength is recovered up to 25% after 28 days. Joesph et al. [14] have encapsulated cyanoacrylate adhesive into the glass sphere and dispersed it into a reinforced mortar beam. It was observed that little improvement in self-healing was achieved due to a small amount of self-healing agents available for healing. Restuccia et al. [45] have filled the sodium silicate solution and measured the effectiveness of glass capsules in cement mortar. The authors have observed that a small amount of sodium silicate solution is available for the reaction during the self-healing process and it results in low load recovery after cracking under cyclic loading. Tittelboom et al. [46] have observed that glass capsules did not survive during concrete mixing even though coated with cement mortar.

3.7 Ceramics

Ceramic capsules were used as they were an alternative material for glass capsules, to overcome the limitation of the glass capsules which may react with the host matrix by the alkali-silica reaction. Tittelboom et al. [46] have encapsulated polyurethane in a ceramic capsule. They compared the two different shell materials glass capsules and ceramic capsules with the same healing agent and measured the self-healing efficiency as well as the water permeability. From the observations, they reported that ceramic capsules provide better results in both self-healing efficiency as well as water permeability.

3.8 Polymeric Material

Polymeric capsules are designed to overcome the limitation of glass and ceramic capsules, as they are brittle and maybe not able to survive during the mixing process of concrete. Kanellopoulos et al. [47] have suggested the emulsification procedure for the development of polymeric capsules in oil-in-water-in-oil emulsion. They observed that the developed microcapsules have average diameters ranging from 300 to 700 μm. A shell wall thickness in the range of 5–20 μm provides robustness during the mixing procedure of concrete. The shell material of microcapsules is designed in such a way that it becomes soft and rubbery during the hydration process and becomes stiff and glassy after drying. They also found good thermal resistivity up to 190 °C. Hilloulin et al. [48] have developed three different types of polymeric capsules and heated the developed capsules to change their state from brittle to rubbery before the mixing process of concrete. It is reported that the survival ratio of capsules was considerably increased during the concrete mixing process. Kanellopoulos et al. [49] have encapsulated silica-based healing agents in polymeric capsules and embedded them in different cement-based matrices to evaluate mechanical and durability properties. It is observed that there is a correlation between microcapsule content and compressive strength, by increasing the concentration of microcapsules substantially decreased the compressive strength of concrete. Also, it is evaluated that by increasing the concentration of microcapsules, the durability of concrete is improved by effective self-healing. Lv et al. [50] have micro-encapsulated dicyclopentadiene in a polymeric capsule. The authors have studied the effect of preparation parameters on the size and morphology of microcapsules; they also assessed the stability and fracture behavior of microcapsules during cracking. Microcapsules of spherical shape with a smooth continuous shell wall were obtained and also assessed that the size of microcapsules was affected by the emulsification agitation rate. They observed good bond strength between microcapsules shell and cementitious material, as well as an easy fracture in the shell wall during a cracking event.

A summary of various healing agents and shell materials used for the preparation of microcapsules and cylindrical capsules along with its major findings reported by various researchers are depicted in Table 1.

4 Tests and Methods for Evaluating Self-Healing Efficiency

To design an effective self-healing mechanism and its application in cementitious materials, it is required to evaluate the effectiveness of the proposed self-healing mechanism through assessment of its material property, based on the application area. Self-healing efficiency can be influenced by a number of factors like the width of the crack, curing time, and age of the crack [31, 44].

To evaluate self-healing efficiency at the laboratory level, the concrete specimen needs to be pre-cracked precisely in a controlled manner. It is reported that for plain cement concrete or ordinary fibre reinforced concrete, a controlled crack can be induced either by three-point bending or four-point bending testing, whereas for high-performance fibre reinforced concrete, a direct or indirect tensile testing needs to be carried out [11]. After pre-cracking the self-healing efficiency of the specimen can be evaluated by measuring the healed crack width, and/or by assessing the mechanical and durability properties of the specimen concerning virgin properties of the specimen. Tests and techniques are summarized in Table 2, evaluating the recovery of mechanical and durability properties as well as visualizing the process of self-healing. Several research studies conducted by researchers to assess the recovery of mechanical strength, recovery of durability properties and visualization of self-healing mechanisms are discussed here.

4.1 Recovery of Mechanical Strength

Hu et al. [32] have carried out a three-point bending tests on the cementitious specimen to evaluate the self-healing efficiency of polyurethane healing agents and reported that 75% strength recovery was obtained for crack widths up to 300 μm. Tittelboom et al. [33] have performed three-point bending and splitting tensile tests on concrete mortars to evaluate the mechanical properties of mortar specimens and found that 50% strength and stiffness recovered after self-healing. Qureshi et al. [44] have carried out three-point bending tests on concrete specimens to evaluate the self-healing efficiency and strength recovery. They observed that an optimum self-healing efficiency is up to 95% and the mechanical strength was 25% with conditioning underwater.

Table 1 Summary of shell materials and healing materials

Type of capsule	Capsule material	Healing agent	Dimensions			Reported major findings	References
			External diameter (μm)	Thickness (μm)	Length (μm)		
Micro-capsule	Gelatin	Acrylic resin	125–297	–	–	A lower volume occupied by the healing agent in capsules results in a lower healing efficiency	[25]
	Gelatin	Tung oil	50	–	–	Specimens with tung oil and calcium hydroxide showed a good healing efficiency compared to specimens without microcapsules	[3]
	Gelatin	Calcium Hydroxide	50	–	–		
	Silica Gel	Methylmethacrylate	4.15	–	–	Reduction in permeability results in an improvement in the self-healing efficiency of mortar	[21]
	Silica Gel	Triethylborane	4.15	–	–		
	Polyurethane	Sodium Silicate	40–800	–	–	Improvement in flexural strength as well as significant corrosion inhibition observed by adding microcapsules	[42]

(continued)

Table 1 (continued)

Type of capsule	Capsule material	Healing agent	Dimensions			Reported major findings	References
			External diameter (μm)	Thickness (μm)	Length (μm)		
	Urea–formaldehyde	Epoxy resin	100–250	–	–	Microcapsules with excellent surface texture, suitable size, robust and remarkable thermal stability are produced with relatively high crack healing ratios against chloride penetration	[27]
	Urea–formaldehyde	Calcium nitrate	45.2–91.5	–	–	Microcapsules had good healing efficiency after 7 days	[38]

(continued)

Table 1 (continued)

Type of capsule	Capsule material	Healing agent	Dimensions			Reported major findings	References
			External diameter (μm)	Thickness (μm)	Length (μm)		
	Melamine urea formaldehyde	Epoxy resin	10–1800	–	–	Various factors affect the preparation of microcapsules. i.e. stirring rate, pH, core-wall ratio and temperature	[51]
	Polyurethane/urea formaldehyde	Sodium Silicate	–	–	–	Double-walled PU/UF microcapsules provide better durability compared to single-walled microcapsules	[19]
Cylindrical capsule	Glass	Cyanoacrylate	800	–	75	Immediate release of cyanoacrylate observed after distinct cracking of capillary tubes and proves effective self-healing under cyclic loading	[14]
	Glass	Cynoacrylate	1500	–	75		
	Glass	Cynoacrylate	3000	–	100		

(continued)

Table 1 (continued)

Type of capsule	Capsule material	Healing agent	Dimensions			Reported major findings	References
			External diameter (μm)	Thickness (μm)	Length (μm)		
	Glass	Polyurethane	2200–3350	200–350	100	More than 50% of original strength and stiffness recovered after the healing process	[33]
	Glass	Polyurethane	3350	350	50	Cracks were partially filled with hydration products	[34]
	Glass	Magnesium Oxide	11,400	–	50	Expansive minerals first hydrated and then slowly carbonated over time Optimum healing efficiency with 95% crack sealing and 25% strength recovery in 28 days	[44]
	Glass	Bentonite	11,400	–	50		
	Glass	Calcium oxide	11,400	–	50		
	Glass	Water	6140	–	50		

(continued)

Table 1 (continued)

Type of capsule	Capsule material	Healing agent	Dimensions			Reported major findings	References
			External diameter (µm)	Thickness (µm)	Length (µm)		
	Glass	Polyurethane	2300	600	50	For glass capsules, resistance against the mixing process of cement mortar is very low as compared with ceramic and polymeric capsules and results in poor self-healing efficiency	[52]
	Ceramic	Polyurethane	3000–4000	500	15–50	Better healing efficiency obtained as compared with glass capsules	[33]
	Ceramic	Polyurethane	3000	–	100	More than 80% of original strength and stiffness recovered after the healing process	[53]

(continued)

Table 1 (continued)

Type of capsule	Capsule material	Healing agent	Dimensions			Reported major findings	References
			External diameter (µm)	Thickness (µm)	Length (µm)		
	Ceramic	Polyurethane	3000	600	50	Provided higher resistance against mixing process of cement mortar resulted in higher healing efficiency	[52]

Table 2 Summary of tests and techniques used to assess healing and recovery of mechanical and durability properties after healing

Test	Scale of test	Purpose of test	Limitations	Reference
For recovery of mechanical properties				
Compression and tensile test	Macro	To evaluate recovery in strength in terms of compression and flexure	Low accuracy due to result influenced by moisture content, size and curing time of specimen	[17, 18, 21, 27, 29, 31–33, 36–39]
Three-point and four-point bending test	Macro	To initiate pre-cracking in the specimen and also used to evaluate mechanical properties after healing	Result affected by number and state of crack, also the effect of an empty capsule on strength could not be excluded	[3, 14–16, 24, 32, 34, 37, 44, 45, 47]
Resonance frequency analysis	Macro	To evaluate recovery in toughness property after healing	Results influenced due to the size and geometry of the specimen	[55]
Acoustic emission analysis	Macro	To analyze rupture of the capsule after cracking by signals from the sensor attached at the surface	Very sensitive in measurement, results are affected by noise and signal strength	[53, 54]
For recovery of durability property				
Water permeability	Macro	Measure water-tightness after self-healing in terms of water permeability coefficient	Measurements are affected by size and state of the crack	[3, 18, 33]
Air permeability	Macro	Measures flow rate of air after healing to evaluate the resistance of cracks against moisture	Results may vary based on the composition of the specimen	[21]

(continued)

Table 2 (continued)

Test	Scale of test	Purpose of test	Limitations	Reference
Chloride diffusion	Macro	To determine the resistance against chloride penetration generally applicable for coastal structures	Indirectly realized	[27, 31]
For visualization and determination of self-healing efficiency				
Scanning electron microscopy	Micro	To visualize developed micro-capsule in terms of size and morphology, also it is used for visualization of crystal deposits near the crack after self-healing	Precise instrumentation is needed which may be very costly and image accuracy depends on the resolution of the image	[17, 19, 26–29, 32, 35, 36, 38, 39, 41, 44, 50]
Infrared analysis	Micro	Evaluation of composition and nature of healing products	Not accurate because it is influenced by the presence of moisture in the specimen	[18, 26, 28, 41, 44]
Environmental scanning electron microscopy	Micro	To observe the morphology and chemical nature of healing products	Results are influenced by very low pressure has to maintain at the time of microscopy which alters the microstructure of the specimen	[15, 16, 21, 37]
Optimal microscopy with image analysis	Micro	For visualization of crystal deposits after healing and also used to monitor healing time	Accuracy of results depends on the resolution of the microscope	[50]
Digital image correlation	Micro	To visualize crack patterns	The high-resolution camera needed for accuracy	[34, 47]
X-ray computed tomography	Micro	To observe breakage of a capsule and crack filling	Not accurate due to limitation of resolution	[33, 50]

4.2 Recovery of Durability Property

Tittelboom et al. [33] have carried out water permeability tests to assess the durability properties of cementitious materials and found that the water permeability reduced by 10^2 to 10^4 times compared to the cracked sample. Dong et al. [27] have carried out a rapid chloride migration test for the mortar specimen and evaluated the chloride penetration with healing ratios from 20.71% to 45.59%, which proves that self-healing systems work effectively. They also found that the crack width affects the self-healing efficiency by comparing different crack width from 100 to 300 μm. Maes et al. [31] have observed that chloride penetration in cementitious materials increases as crack width increased from 100 to 300 μm.

4.3 Visualization of Self-healing Performance

Dong et al. [27] have used urea–formaldehyde/epoxy microcapsules for their study and carried out scanning electron microanalysis and observed that microcapsules fracture under pre-cracking conditions. The authors have obtained that good bond strengths between the microcapsules and the mortar matrix and the cracked area have been filled with a healing agent. Hu et al. [32] have carried out scanning electron microscopy analysis and observed that the polyurethane healing agent is released effectively to heal the crack after the damage of the glass capsule. Tittelboom et al. [33] observed crack filling by high-resolution X-ray computed tomography, and found that polyurethane foam effectively filled the crack. Arce et al. [37] have carried out environmental scanning electron microscopy with energy dispersive spectroscopy to investigate the morphology and chemical nature of the healing products. They reported that crystal-like and gel-like healing products are observed near a crack in the form of calcite crystal and the chemical nature of healing products was presumed as CSH. Qureshi et al. [44] have carried out FTIR, SEM and EDX analysis to study the self-healing material formation and the nature of self-healing products. They reported that encapsulated healing agents were released successfully into the crack and most healing products were mostly composed of Ca, Si, Al and Mg hydrated and carbonated products. Tsangouri et al. [54] analyzed the impact of small size capsules on resistance to concrete damage and found that the resilience of the fracture improved by 35% to virgin sample. Authors also analyzed the impact of the capsules through acoustic emission on the fracture mechanism and found that the process zone of fracture expands by 40% through localization of events.

5 Conclusion

Encapsulated self-healing concrete has been adopted as the most effective approach to healing concrete cracks. The selection of suitable healing materials and the capsules shell material are major challenges reported by various researchers. Also, various experimental studies were conducted to assess the self-healing efficiency. Based on this review following conclusions are drawn:

1. Polymeric healing agents are most effective among all other reported self-healing agents. Also, calcium nitrate is a newly emerging self-healing agent at a low cost.
2. Brittle materials like glass and ceramic are most widely used as a shell material for encapsulating healing agents. However, the major challenge is reported as glass and ceramic capsules did not survive during the mixing process of cementitious materials. The polymeric shell material is used to improve the survival ratio during the mixing process. Urea–formaldehyde is the most compatible shell material with different healing agents.
3. Various mechanical and durability tests are conducted to quantify the self-healing efficiency of encapsulated self-healing concrete. To observe the effectiveness of healing agents in the crack filling, micro-scale testing is required.

References

1. Van Breugel, K.: Self-healing material concepts as solution for aging infrastructure. In: 37th Conference on Our World in Concrete and Structures, 1–17 (2012)
2. ACI-Committee-224: Causes, Evaluation, and Repair of Cracks in Concrete Structures, (ACI 224.1R-07) American Concrete Institute (2007)
3. Cailleux, E., Pollet, V.: Investigations on the development of self-healing properties in protective coatings for concrete and repair mortars In: 2nd International Conference on Self Healing Materials, 1–4 (2009)
4. Van Breugel, K.: Is there a market for self-healing cement-based materials?. In: 1st International Conference on Self-Healing Materials, 1–9 (2007)
5. White, S.R., Sottos, N.R., Geubelle, P.H.: Erratum: autonomic healing of polymer composites. Nature **415**(6873), 794–797 (2002)
6. Van Tittelboom, K., De Belie, N.: Self-healing in cementitious materials-a review. Materials **6**(6), 2182–2217 (2013)
7. Gupta. S., and Harn.W.K.: Encapsulation technology and techniques in self-healing concrete. J. Mater. Civil Eng. **28**(12), 1–15 (2016)
8. Aissa, B., Therriault, D., Haddad, E., Jamroz, W.: Self-healing materials systems: Overview of major approaches and recent developed technologies. Adv. Mater. Sci. Eng., 1–17 (2012).
9. Li, V.C., Herbert, E.: Robust self-healing concrete for sustainable infrastructure. J. Adv. Concr. Technol. **10**(6), 207–218 (2012)
10. Hager, M.D., Greil, P., Leyens, C.: Self-healing Materials. Adv. Mater. **22**, 5424–5430 (2010)
11. Ferrara, L., Van Mullem, T.V., Alonso, M.C., Antonaci, P., Borg, R.P., Cuenca, E., De Belie, N.: Experimental characterization of the self-healing capacity of cement based materials and its effects on the material performance: a state of the art report by COST action SARCOS WG2. Constr. Build. Mater. **167**, 115–142 (2018)

12. Muhammad, N.Z., Shafaghat, A., Keyvanfar, A., Majid, M.Z.A., Ghoshal, S.K., Yasouj, S.E.M., McCaffer, R.: Tests and methods of evaluating the self-healing efficiency of concrete: a review. Constr. Build. Mater. **112**, 1123–1132 (2016)
13. Gardner, D., Jefferson, A., Hoffman, A., Lark, R.: Simulation of the capillary flow of an autonomic healing agent in discrete cracks in cementitious materials. Cem. Concr. Res. **58**, 35–44 (2014)
14. Joseph, C., Jefferson, A.D., Isaacs, B., Lark, R., Gardner, D.: Experimental investigation of adhesive-based self-healing of cementitious materials. Mag. Concr. Res. **62**(11), 831–843 (2010)
15. Li, V.C., Lim, Y.M., Chan, Y.W.: Feasibility study of a passive smart self-healing cementitious composite. Compos. B Eng. **29**(6), 819–827 (1998)
16. Huang, H., Ye, G.: Application of sodium silicate solution as self-healing agent in cementitious materials. In: Application of Sodium Silicate Solution as Self-healing Agent in Cementitious Materials, 1–8 (2011)
17. Gilford, J., III., Hassan, M.M., Rupnow, T., Barbato, M., Okeil, A., Asadi, S.: Dicyclopentadiene and sodium silicate microencapsulation for self-healing of concrete. J. Mater. Civ. Eng. **26**(5), 886–896 (2014)
18. Giannaros, P., Kanellopoulos, A., Al-Tabbaa, A.: Sealing of cracks in cement using microencapsulated sodium silicate. Smart Mater. Struct. **25**(8) (2016)
19. Mostavi, E., Asadi, S., Hassan, M.M., Alansari, M.: Evaluation of self-healing mechanisms in concrete with double-walled sodium silicate microcapsules. J. Mater. Civ. Eng. **27**(12), 1–8 (2015)
20. Dry, C., McMillan, W.: Three-part methylmethacrylate adhesive system as an internal delivery system for smart responsive concrete. Smart Mater. Struct. **5**(3), 297–300 (1996)
21. Yang, Z., Hollar, J., He, X., Shi, X.: A self-healing cementitious composite using oil core/silica gel shell microcapsules. Cement Concr. Compos. **33**(4), 506–512 (2011)
22. Xue, C., Li, W., Li, J., Tam, V.W.Y., Ye, G.: A review study on encapsulation-based self-healing for cementitious materials. Struct. Concr. **20**(1), 198–212 (2019)
23. ACI 515.2R.: Guide to Selecting Protective Treatments for Concrete (2013)
24. Thao, T.D.P., Johnson, T.J.S., Tong, Q.S., Dai, P.S.: Implementation of self-healing in concrete–proof of concept. IES J. Part Civil Struct. Eng. **2**(2), 116–125 (2009)
25. Mihashi, H., Kaneko, Y., Nishiwaki, T., Otsuka, K.: Fundamental study on development of intelligent concrete characterized by self-healing capability for strength. Concr. Res. Technol. **11**, 21–28 (2000)
26. Cosco, S., Ambrogi, V., Musto, P.: Properties of Poly(urea-formaldheyde) Microcapsules Containing an Epoxy Resin. J. Appl. Polym. Sci. **105**, 1400–1411 (2007)
27. Dong, B., Fang, G., Ding, W., Liu, Y., Zhang, J., Han, N., Xing, F.: Self-healing features in cementitious material with urea-formaldehyde/epoxy microcapsules. Constr. Build. Mater. **106**, 608–617 (2016)
28. Li, Q., Siddaramaiah, Kim, N.H., Hui, D., Lee, J.H.: Effects of dual component microcapsules of resin and curing agent on the self-healing efficiency of epoxy. Compos. Part B: Eng., **55**, 79–85 (2013)
29. Sam, A.R.M., Ariffin, N.F., Hussin, M.W., Lee, H.S., Ismail, M.A., Majid, M.Z.A.: Performance of epoxy resin as self-healing agent. J. Teknol. **77**(16), 9–13 (2015)
30. Van Tittelboom, K., Van den Heede, P., De Belie, N. D.: Self-healing Concrete with Encapsulated Polyurethane. Eco-efficient Repair and Rehabilitation of Concrete Infrastructures, 429–C66. Elsevier Ltd. (2018)
31. Maes, M., Van Tittelboom, K., De Belie, N.: The efficiency of self-healing cementitious materials by means of encapsulated polyurethane in chloride containing environments. Constr. Build. Mater. **71**, 528–537 (2014)
32. Hu, Z.X., Hu, X.M., Cheng, W.M., Zhao, Y.Y., Wu, M.Y.: Performance optimization of one-component polyurethane healing agent for self-healing concrete. Constr. Build. Mater. **179**, 151–159 (2018)

33. Van Tittelboom, K., De Belie, N., Van Loo, D., Jacobs, P.: Self-healing efficiency of cementitious materials containing tubular capsules filled with healing agent. Cement Concr. Compos. **33**(4), 497–505 (2011)
34. Van Tittelboom, K., Wang, J., Araújo, M., Snoeck, D., Gruyaert, E., Debbaut, B., De Belie, N.: Comparison of different approaches for self-healing concrete in a large-scale lab test. Constr. Build. Mater. **107**, 125–137 (2016)
35. Brown, E.N., Kessler, M.R., Sottos, N.R., White, S.R.: In situ poly(urea-formaldehyde) microencapsulation of dicyclopentadiene. J. Microencapsul. **20**(6), 719–730 (2003)
36. Hassan, M.M., Milla, J., Rupnow, T., Al-Ansari, M., Daly, W.H.: Microencapsulation of calcium nitrate for concrete applications. Transp. Res. Rec. **2577**, 8–16 (2016)
37. Arce, G.A., Hassan, M.M., Mohammad, L.N., Rupnow, T.: Characterization of self-healing processes induced by calcium nitrate microcapsules in cement mortar. J. Mater. Civ. Eng. **29**(1), 1–10 (2017)
38. Milla, J., Hassan, M.M., Rupnow, T.: Evaluation of self-healing concrete with microencapsulated calcium nitrate. J. Mater. Civ. Eng. **29**(12), 1–12 (2017)
39. Al-Ansari, M., Abu-Taqa, A.G., Hassan, M.M., Senouci, A., Milla, J.: Performance of modified self-healing concrete with calcium nitrate microencapsulation. Constr. Build. Mater. **149**, 525–534 (2017)
40. Minnebo, P., Hemelrijck, D. V.: Ideal material properties for capsules or vascular system used in cementitious self-healing materials. In: International Conference on Self-healing Materials, 1–4 (2015)
41. Beglarigale, A., Seki, Y., Demir, N.Y., Yazlcl, H.: Sodium silicate/polyurethane microcapsules used for self-healing in cementitious materials: monomer optimization, characterization, and fracture behavior. Constr. Build. Mater. **162**, 57–64 (2018)
42. Pelletier, M.M., Brown, R., Shukla, A., Bose, A.: Self-healing concrete with a microencapsulated healing agent. University of Rhode Island, Kingston (2010)
43. Blaiszik, B.J., Caruso, M.M., McIlroy, D.A., Moore, J.S., White, S.R., Sottos, N.R.: Microcapsules filled with reactive solutions for self-healing materials. Polymer **50**(4), 990–997 (2009)
44. Qureshi, T.S., Kanellopoulos, A., Al-Tabbaa, A.: Encapsulation of expansive powder minerals within a concentric glass capsule system for self-healing concrete. Constr. Build. Mater. **121**, 629–643 (2016)
45. Restuccia, L., Reggio, A., Ferro, G.A., Tulliani, J.M.: New self-healing techniques for cement-based materials. Procedia Struct. Int. **3**, 253–260 (2017)
46. Van Tittelboom, K., Tsangouri, E., Van Hemelrijck, D., De Belie, N.: The efficiency of self-healing concrete using alternative manufacturing procedures and more realistic crack patterns. Cement Concr. Compos. **57**, 142–152 (2015)
47. Kanellopoulos, A., Litina, C., Giannaros, P., Al-Tabbaa, A.: Effect of different types of polymeric microcapsules on the self-healing efficiency of cement based composites.In: 9th International Conference on Fracture Mechanics of Concrete and Concrete structures, FraMcos-9, 1–7 (2016)
48. Hilloulin, B., Van Tittelboom, K., Gruyaert, E., De Belie, N., Loukili, A.: Design of polymeric capsules for self-healing concrete. Cement Concr. Compos. **55**, 298–307 (2015)
49. Kanellopoulos, A., Giannaros, P., Palmer, D., Kerr, A., Al-Tabbaa, A.: Polymeric microcapsules with switchable mechanical properties for self-healing concrete: Synthesis, characterisation and proof of concept. Smart Mater. Struct. **26**(4) (2017)
50. Lv, L., Yang, Z., Chen, G., Zhu, G., Han, N., Schlangen, E., Xing, F.: Synthesis and characterization of a new polymeric microcapsule and feasibility investigation in self-healing cementitious materials. Constr. Build. Mater. **105**, 487–495 (2016)
51. Li, W., Zhu, X., Zhao, N., Jiang, Z.: Preparation and properties of melamine urea-formaldehyde microcapsules for self-healing of cementitious materials. Materials **9**(3) (2016)
52. Gruyaert, E., Van Tittelboom, K., Sucaet, J., Anrijs, J., Van Vlierberghe, S., Dubruel, P., De Belie, N.: Capsules with evolving brittleness to resist the preparation of self-healing concrete. Mater. de Constr. **66**(323) (2016)

53. Van Tittelboom, K., De Belie, N., Lehmann, F., Grosse, C.U.: Acoustic emission analysis for the quantification of autonomous crack healing in concrete. Constr. Build. Mater. **28**(1), 333–341 (2012)
54. Tsangouri, E., Gilabert, F., De Belie, N., Van Hemelrijck, D., Zhu, X., Aggelis, D.: Concrete fracture toughness increase by embedding self-healing capsules using an integrated experimental approach. Constr. Build. Mater. **218**, 424–433 (2019)
55. Abd-Elmoaty, A.E.M.: Self-healing of polymer modified concrete. Alexandria Eng. J. **50**(2), 171–178 (2011)

Effect of Masonry Infills on Seismic Response of RC Framed Buildings

Surender Kumar Verma, Kuldeep Kumar, and Sameer Dogra

Abstract While performing structural design of multi-storeyed RC framed buildings with masonry infills, it is a common practice to exclude the increase in stiffness of the structure caused by the presence of masonry infills. This practice can lead to incorrect estimation of various seismic responses and hence, the seismic design results obtained for the structure shall be unreliable. In the present study, the effect of variation of structural size and stiffness on base shear and fundamental period has been investigated. Response spectrum analysis has been performed on all the frames under investigation and analysis has been carried out on ETABS 2015 software. Masonry infills are modelled as equivalent diagonal compressive struts to simulate the compressive behaviour of masonry walls under lateral seismic forces. Results have been compared for base shear and fundamental period in lateral direction of earthquake vibration. Building models were prepared with variation in infill panel stiffness and infill opening percentages were varied as 20, 40 and 60%. The incorporation of the stiffness of the infills shows an increase in the base shear but decreases the fundamental period of the building. However, an increase in infill openings decreases the base shear but increases the fundamental period of the building.

Keywords Masonry infills · Fundamental period · Base shear · Response spectrum analysis · Equivalent diagonal strut · Infill panel stiffness

1 Introduction

Structural design of multi-storeyed RC framed buildings primarily involves designing the structure for vertical loads such as dead, live loads, and for lateral loads such as wind, seismic loads etc. These lateral loads produce critical stresses, which primarily

---

S. K. Verma · K. Kumar · S. Dogra (✉)
Punjab Engineering College (Deemed to be University), Chandigarh 160012, India
e-mail: sameer12dogra@gmail.com

S. K. Verma
e-mail: skverma5724@yahoo.com

D. K. Ashish et al. (eds.), *3rd International Conference on Innovative Technologies for Clean and Sustainable Development*, RILEM Bookseries 29,
https://doi.org/10.1007/978-3-030-51485-3_15

govern the design of structure. Hence, estimation of these lateral loads needs precise assessment. Most of the RC framed buildings consists of masonry infill walls, both with and without openings. Although, these infilled masonry walls have a positive influence on the stiffness and strength of the building frame, but still their contribution is not taken into account while performing structural design of buildings [1, 6, 27, 32, 36, 38]; (Ko et al. 2014). This practice is taking place primarily because there is a great degree of uncertainty associated in the estimation of fundamental period of vibration as has been shown in literature and as proposed in different standards worldwide [10, 15, 36].

RC frames infilled with brick masonry show a complex nonlinear inelastic behaviour under seismic action. This complex behaviour primarily arises from the various factors causing material non-linearity, namely, (i) cracking and crushing of the masonry panel, (ii) failure of surrounding RC frame and local bond slip in the surrounding frame, and (iii) degradation of the bond-friction mechanism and variation of the contact length along the panel-frame interface [11]. The above listed nonlinear causes are responsible for making the analytical investigations quite complex and cumbersome and require sophisticated computational techniques to accurately incorporate the nonlinear effects in structural analysis softwares. Because of these complications, the structural designers tend to disregard the structural effect of infill panels while carrying out the analysis. As per the past information for infilled RC frames, when subjected to seismic loading, tend to get detached from masonry infills and the contact between the two remains only at diagonally opposite corners.

However, despite the availability of several infill modelling concepts in the literature, from diagonal strut modelling (single or multiple compression struts and shear struts) to FE modelling, [11, 15, 19, 26, 31, 35] the masonry infills due to a large amount of uncertainty in their behaviour are usually not accounted for. Therefore, the determined dynamic properties in this case are unreliable in seismic performance assessments. Also, the formulas for the determination of the fundamental period are mostly dependent on only a few parameters (overall building height and plan dimensions) and the effect of stiffness of the infills might not be considered. This leads to inaccurate seismic analysis results.

It is established fact that the height of the building and the characteristics of masonry are crucial factors, which affect the performance of RC masonry infilled frames when subjected to seismic action. In the present investigation, the effect of span lengths, number of spans, stiffness of infills and infill openings have also been studied. Structural analysis results of base shear and fundamental period, in the direction of earthquake vibration have been compared for the various analytical models considered for the present study.

2 Description of Structures

2.1 RC Building Frame Parameters

The present investigation consists of buildings with varying number of spans and span lengths in longitudinal X-direction and a single span of length 6 m in transverse Y-direction. Fully infilled models are prepared using equivalent diagonal strut modelling concept for infilled frames without openings, to examine the influence of masonry infills on base shear and fundamental period [19, 26, 31, 35]. Finally, to study the effect of percentage of openings in the infill panel, building models are prepared with variation in percentage of openings in the infill panel. A bare frame, fully infilled frame and three infilled frames with openings varying between 20, 40 and 60% were considered in this part of investigation. Table 1 lists the various building parameters considered for the present analytical investigation.

A total of 34 number of frames have been prepared for the current study, out of which 27 are bare frame models with variation in number of storeys, number of spans and span lengths. The frames have been designed as per the latest Indian Standards. Member Sizes have been kept such that the design longitudinal reinforcement in these members does not exceed 1.5% of gross cross-sectional area of member. 9″ thick masonry infilled walls of frames along X-direction with openings were modelled as equivalent diagonal struts with reduced strut widths calculated using stiffness reduction factor.

2.2 Analysis and Design of Structures

Response spectrum method of seismic analysis has been employed for carrying out the present investigation using ETABS 2015 software. In the response spectrum method, as per the codal provisions, the modal mass must be more than 90% of the total seismic mass and accordingly number of modes have to be decided (Cl. 7.7.5.2) [23].

Column sections of square shape were used in all the frames. A live load of 25% has been considered in seismic weight calculations (Table)10, [23]. Moment of inertia for columns shall be taken as 70% of I_{gross} of columns, and 35% of I_{gross} of beams, in case of beam members (Cl. 6.4.3.1) [23].

Table 2 describes the various member sizes considered for each type of building.

Table 1 Building parameters

Type of structure	Multi-storeyed RC framed buildings
No. of storeys	10, 14 and 18
No. of spans (in X-direction)	4, 6 and 8
Span lengths (in X-direction)	3.5, 4.5 and 5.5 m
Infill panel stiffness (Et)	3.542×10^5 kN/m – 9.98×10^5 kN/m
Typical storey height	3.5 m
Masonry infill opening percentage	20, 40 and 60%
Grade of concrete	M35
Grade of longitudinal reinforcement	Fe500
Grade of transverse reinforcement	Fe500
Slab thickness	150 mm
Plinth beam sizes	230 mm × 325 mm
Zone factor	0.24 (Zone – IV)
Foundation depth from plinth	1.5 m
Importance factor	1.5 (Community building)
Response reduction factor	5 (SMRF)
Soil type	Medium stiff
Damping ratio	0.05
Dead load (floor loads)	(i) 6 kN/m^2 for all internal floors (ii) 6.2 kN//m^2 for roof slab
Dead load (Masonry wall loads)	(i) 14 kN/m for 9″ thick wall (ii) 7 kN/m for 4.5″ thick wall (iii) 4.2 kN/m for parapet loading
Live load	(i) 3 kN/m^2 for all internal storeys (ii) 1.5 kN/m^2 at roof level

3 Modelling

3.1 General Description

All buildings were modelled as space frames using ETABS 2015 software. Modelling of masonry infills is crucial for assessing seismic analysis responses. From various

Table 2 Member specifications for building frames with ten, fourteen and eighteen storeys with varying number of spans and span lengths

No. of spans	Span length (in m)	Roof beam sizes (in mm)	Column sizes, in mm (varies from top to bottom storey)		
			Ten storeyed building	Fourteen storeyed building	Eighteen storeyed building
4	3.5	350 × 600	350–650	350–700	350–750
6	3.5	350 × 650	350–700	350–700	350–750
8	3.5	350 × 700	400–700	400–750	400–800
4	4.5	350 × 700	400–700	400–750	400–800
6	4.5	350 × 750	400–750	400–750	400–800
8	4.5	350 × 800	400–800	400–800	400–850
4	5.5	350 × 800	400–800	400–800	400–800
6	5.5	350 × 850	400–850	400–850	400–850
8	5.5	350 × 900	400–850	400–900	

analytical and experimental investigations [8, 17, 19, 26, 29–31, 35, 37; 18], it is indicated that the diagonal axial compressive strut model with suitable geometrical and mechanical properties could provide a reasonably accurate solution to the complex behaviour of infilled RC framed buildings under seismic action.

3.2 Modelling of Masonry Infills Without Openings

From the review of the research work of previous scholars [1, 6, 27, 32, 36, 38]; (Ko et al. 2014) it is clear that the behaviour of infilled RC frames under seismic excitation is a complex one. Hence, it is important to choose a suitable model for infill walls so as to account for the infill wall stiffness for accurate analysis results. In this study, the equivalent diagonal strut has been employed in structural modelling of infill walls.

Polyakov [31] suggested the use of diagonal bracing concept to incorporate the effects of masonry infills. Holmes [19] adopted this concept to employ the stiffness of infills in the structural analysis and suggested to replace the infill walls by an equivalent pin-jointed diagonal strut. The strut had the same material properties as of the masonry infill, with the same infill thickness. The following Eq. (1) gives the expression for the width of strut, w:

$$\frac{w}{d} = \frac{1}{3} \tag{1}$$

where d is the length of diagonal of the masonry panel.

Smith and Carter (1969) further using additional experimental data proposed to express the width of strut as a function of stiffness parameter relative to infill panel and frame, as:

$$\lambda_h = h\sqrt[4]{\frac{E_w t_w \sin 2\theta}{4EIh_w}} \tag{2}$$

where, E_w = young's modulus of the masonry infill,
EI = flexural rigidity of the columns enclosing the masonry panel,
t_w = thickness of the infill panel and equivalent strut,
h = height of columns between beam centre line,
h_w = clear height of masonry infill panel from the enclosing beams, and.
θ = angle, whose tangent is the height to length ratio of masonry infill.

Figure 1 depicts various components of a masonry infilled frame. In Fig. 1, 'z' refers to the contact length between column and infill.

The equivalent width of strut can be evaluated from the above expression as proposed by Mainstone and Weeks (1970) in relation to the panel to frame stiffness parameter, as per Eq. (3).

$$\frac{w}{d} = 0.175\lambda_h^{-0.4} \tag{3}$$

This equation has been used in the present research work for modelling of masonry infill panels without openings. It is to be noted that the axial compressive action of masonry is activated in a single diagonal of masonry panel, for a particular direction of lateral force. If the direction of lateral force is reversed, the other diagonal of masonry acts as a compressive strut. Hence, in frame modelling, the masonry struts within a beam-column panel shall be modelled similar to a X-bracing (cross bracing)

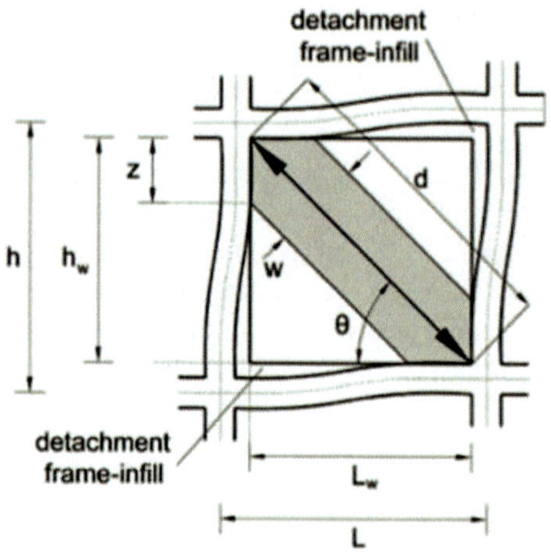

Fig. 1 Components of masonry infill frame

and tension limit is set to zero in both the struts so as to simulate actual compressive behaviour of strut (Fiore et al. 2012; Uva et al. 2012).

3.3 Modelling of Masonry Infills with Openings

In most of the RC framed buildings, normally openings are present in many masonry walls. Therefore, to take into consideration the effect of openings in reducing the stiffness of infill walls, a suitable model needs to be adopted so as to obtain more realistic analysis results. In a study conducted by Asteris (2003), a relation between opening percentage, location of wall opening and stiffness reduction factor has been proposed, for masonry infill walls with openings. For the present investigation, the model with openings lying on the compressed diagonal has been adopted, and the relationship between stiffness reduction factor, λ and wall openings is described in Eq. (4) as:

$$\lambda = 1 - 2\alpha_w^{0.54} + \alpha_w^{1.14} \tag{4}$$

where α_w is calculated by dividing the opening area with the total area of infill panel. To find the equivalent width of the strut, one needs to simply multiply the stiffness reduction factor with the width of strut obtained for the case when infill wall has no openings.

4 Masonry Infill Properties

The selection of brick masonry properties is carried out as per guidelines of various related Indian Standards [21, 24, 25]. Poisson's ratio considered for masonry is 0.25. Thickness of brick masonry walls adopted is 230 mm for all peripheral external walls and 115 mm for all internal walls for all building frames considered in the present investigation. In RC infilled frames, only external peripheral masonry walls along longitudinal X-direction are modelled as equivalent diagonal struts.

The modulus of elasticity and compressive strength of masonry infill has been calculated in accordance with Clause 7.9.2.1 of Indian Standard IS 1893 (Part-1) [23]. The modulus of Elasticity of masonry infill shall be obtained from the expression:

$$E_m = 550f_m = 550 * 0.433 * f_b^{0.64} * f_{mo}^{0.36} \tag{5}$$

where, f_m = compressive strength of masonry prism, in MPa,

f_b = compressive strength of brick, in MPa; and.

f_{mo} = compressive strength of mortar, in MPa.

Table 3 shows the masonry properties and Table 4,5 shows the equivalent strut sizes chosen for various infilled RC framed building models in the present study.

Table 3 Brick masonry properties considered for ten storeyed infilled RC framed buildings with 4 spans and 3.5 m span length

Type of structure	Class of bricks	Grade of masonry mortars
FIF1 (Panel stiffness - Et = 3.542×10^5 kN/m)	Class 12.5	MM 2
FIF2 (Panel stiffness - Et = 5.53×10^5 kN/m)	Class 15	MM 5
FIF3 (Panel stiffness - Et = 7.696×10^5 kN/m)	Class 20	MM 7.5
FIF4 (Panel stiffness - Et = 9.98×10^5 kN/m)	Class 30	MM 7.5

Table 4 Masonry equivalent strut widths for ten storeyed infilled RC framed buildings with 4 spans, 3.5 m span length and varying infill panel stiffness's

Storey No	Column sizes (in mm)	Strut widths (in mm)			
		Et = 3.542×10^5 kN/m	Et = 5.53×10^5 kN/m	Et = 7.696×10^5 kN/m	Et = 9.98×10^5 kN/m
10	400	485	465	450	435
7–9	450	500	480	465	455
4–6	500	520	500	480	470
2–3	550	535	515	495	485
1	600	565	540	520	510
	650	580	550	535	520

Table 5 Masonry equivalent strut widths for ten storeyed infilled RC framed buildings with 4 spans, 3.5 m span length and varying infill openings (Panel stiffness, Et = 3.542×10^5 kN/m)

Storey No.	Column sizes (in mm)	Strut widths (in mm)		
		20% openings	40% openings	60% openings
10	400	155	64	19.75
7–9	450	160	66	20.365
4–6	500	165	69	21.18
2–3	550	170	71	21.79
1	600	180	75	23.01
	650	185	77	23.62

5 Results and Discussion

5.1 General

In order to investigate the influence of number of storeys, number of spans and span lengths on base shear and fundamental period, ten, fourteen and eighteen storeyed bare reinforced concrete framed buildings are considered for comparing analysis results. Buildings with 4, 6 and 8 number of spans were analyzed for a given number of storeys and span length. Further, for a building with a particular number of storeys and number of spans, span lengths were varied from 3.5, 4.5 and 5.5 m. Span lengths and number of spans varies in the direction of earthquake excitation considered for comparing analysis results.

Furthermore, to investigate the influence of infill wall panel stiffness, the analysis results of a ten storeyed bare frame structure with four number of bays in X-direction and each bay of length 3.5 m were compared with four infilled reinforced concrete framed buildings with varying infill panel stiffness's. Also, to investigate the influence of percentage of openings within the infill panel, the analysis results of ten storeyed bare and fully infilled (FIF1) reinforced concrete framed buildings with four number of bays in X-direction and each bay of length 3.5 m were compared with three buildings with 20, 40 and 60% openings in masonry infill walls. Overall, a total of 34 frames were modelled in the present study.

5.2 *Influence of Number of Storeys on Base Shear and Fundamental Period*

The base shear results for the bare RC framed models considered for the present investigation have been enumerated in Table 6.

From the results, it is observed that as the number of storeys increase from ten to fourteen and eighteen storeys, the base shear increases but at a very slow rate. Average increase in base shear values, from response spectrum analysis, for buildings with fourteen storeys is 4.38% with respect to the base shear values for buildings with ten storeys and for buildings with eighteen storeys, the average increase in base shear from response spectrum analysis is 9.76% with respect to the base shear values for

Table 6 Base shear results for bare RC framed buildings

Number of bays/spans	Span length (in m)	Base shear, in kN (Response spectrum method)		
		Ten storeyed buildings	Fourteen storeyed buildings	Eighteen storeyed buildings
4	3.5	354.44	377.15	398.26
6	3.5	585.96	601.88	642.68
8	3.5	845.09	894.76	954.82
4	4.5	452.59	476.45	499.67
6	4.5	726.52	758.74	801.88
8	4.5	1083.53	1133.88	1195.47
4	5.5	561.58	589.02	592.8
6	5.5	920.52	933.02	985.2
8	5.5	1330.3	1381.22	1451.7

buildings with ten storeys. Figure 2 depicts the relation between the base shear of the structure and the number of storeys.

Table 7 exhibits the results of fundamental period for all the bare frames considered for the present investigation.

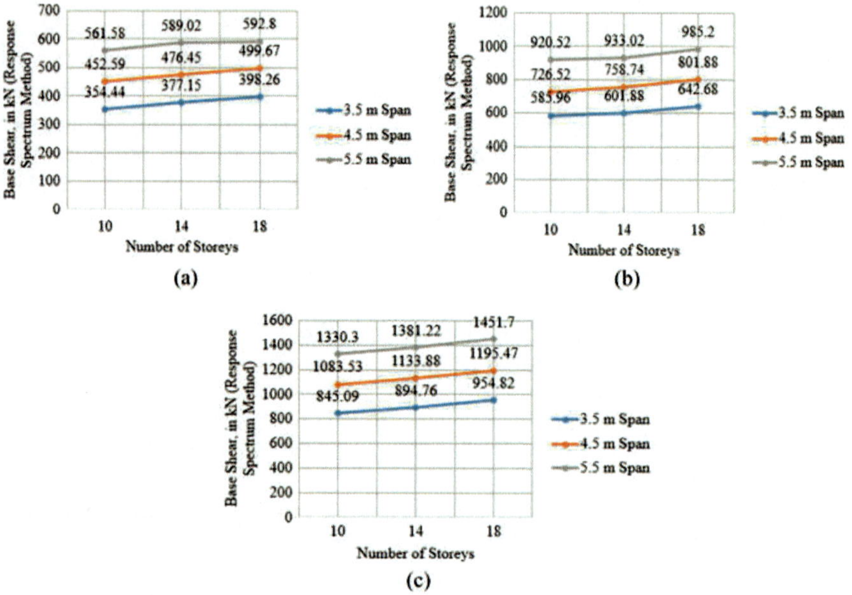

Fig. 2 Base shear distribution for bare RC framed buildings using response spectrum method, with **a** four spans, **b** six spans, and **c** eight spans

Table 7 Fundamental period for bare RC framed buildings

Number of bays/spans	Span length (in m)	Base shear, in kN (Response spectrum method)		
		Ten storeyed buildings	Fourteen storeyed buildings	Eighteen storeyed buildings
4	3.5	354.44	377.15	398.26
6	3.5	585.96	601.88	642.68
8	3.5	845.09	894.76	954.82
4	4.5	452.59	476.45	499.67
6	4.5	726.52	758.74	801.88
8	4.5	1083.53	1133.88	1195.47
4	5.5	561.58	589.02	592.8
6	5.5	920.52	933.02	985.2
8	5.5	1330.3	1381.22	1451.7

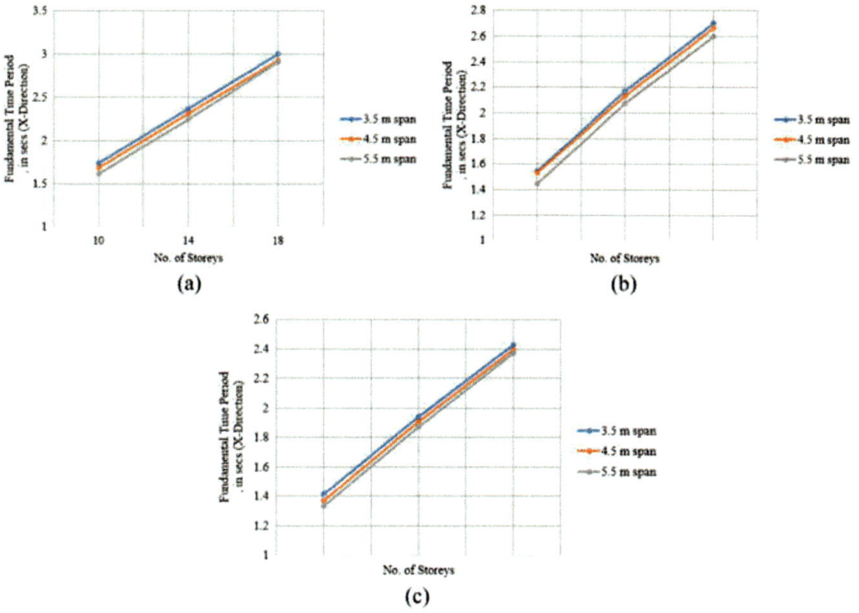

Fig. 3 Distribution of fundamental period for bare RC framed buildings using response spectrum method, with **a** four spans, **b** six spans, and **c** eight spans

Average increase in fundamental period, for buildings with fourteen storeys is 39.12% with respect to the fundamental period for buildings with ten storeys. For buildings with eighteen storeys, the average increase in fundamental period is 75.51%, when compared with respect to the fundamental period of ten storeyed buildings. Figure 3 depicts the relation between the fundamental period and the number of storeys.

5.3 Influence of Number of Spans on Base Shear and Fundamental Period

Figure 4 presents the relation between the base shear of the structure and the number of spans, for a given span length, in the direction of earthquake excitation. It is observed that as the number of spans increase from four to six and eight spans, for a given span length, in the direction of earthquake excitation, the base shear shows a significant increase in its value. Average increase in base shear values, from response spectrum analysis, for buildings with six number of spans is 61.67% with respect to the base shear values for buildings with four number of spans. For buildings with eight number of spans, the average increase in base shear is 138.70% with respect to the base shear values for buildings with four number of spans.

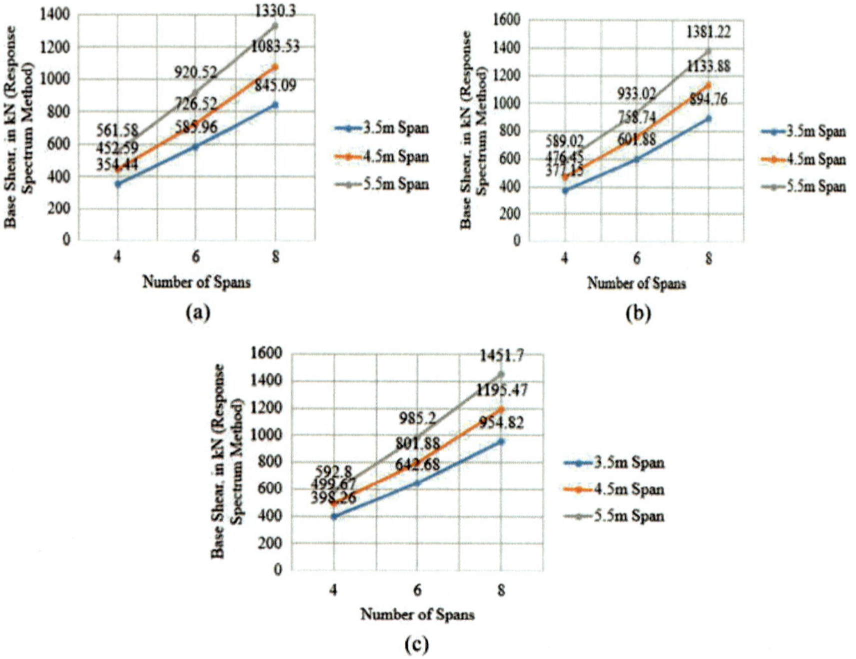

Fig. 4 Base shear distribution for bare RC framed buildings using response spectrum method, with **a** ten storeys, **b** fourteen storeys, and **c** eighteen storeys

Average decrease in fundamental period, for buildings with six number of spans is 9.4% with respect to the fundamental period for buildings with four number of spans. For buildings with eight number of spans, the average decrease in fundamental period is 18.13%, when compared with respect to the fundamental period for buildings with ten storeys. Overall, it is observed that the increase in number of spans in a particular direction of building, for a given span length shall decrease the fundamental period of the building in the same direction. Figure 5 presents the relation between the fundamental period of the structure and the number of spans, for a given span length, in the direction of earthquake vibration.

5.4 *Influence of Span Length on Base Shear and Fundamental Period*

The effect of length of each span, for a given number of spans, on base shear and fundamental period was subsequently investigated. Figure 6 depicts the relation between the base shear of the building and its span length, for specific number of spans in the direction of earthquake vibration.

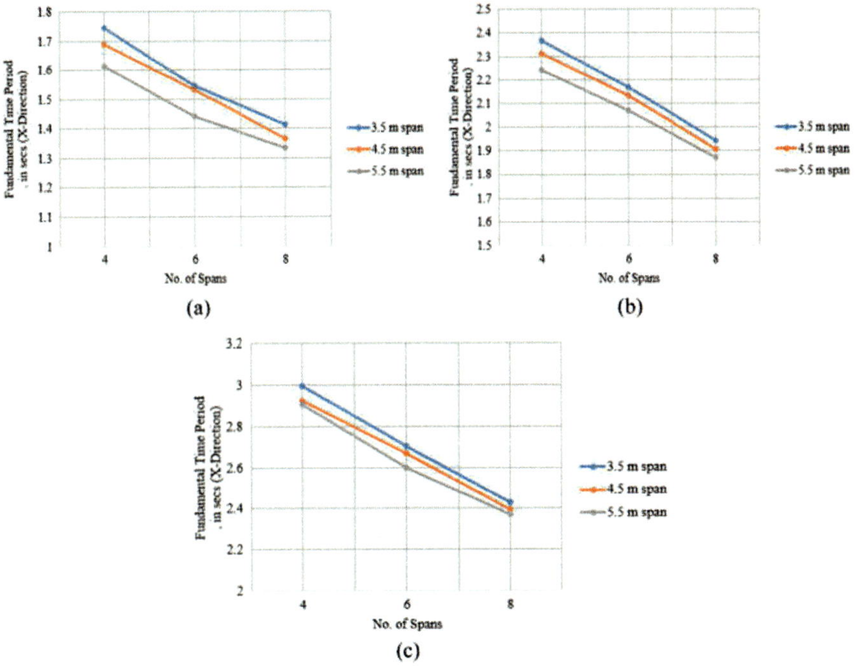

Fig. 5 Distribution of fundamental time period for bare RC framed buildings using response spectrum method, with **a** ten storeys, **b** fourteen storeys, and **c** eighteen storeys

It is observed that as the span length increases from 3.5 to 4.5 and 5.5 m, for specific number of spans in the direction of earthquake excitation, the base shear shows a significant increase in its value. Average increase in base shear values, from response spectrum analysis, for buildings with 4.5 m span length is 26.05% with respect to the base shear values for buildings with 3.5 m span length. For buildings with 5.5 m span length, the average increase in base shear from response spectrum analysis is 54.74% with respect to the base shear values for buildings with 3.5 m span length.

Figure 7 presents the relation between the fundamental period of the structure and span length, for specific number of spans in the direction of earthquake excitation. Average decrease in fundamental period, for buildings with 4.5 m span length is 2.78% with respect to the fundamental period for buildings with 3.5 m span length. For buildings with 5.5 m span length, the average decrease in fundamental period is 5.38%, when compared with respect to the fundamental period for buildings with 3.5 m span length. Overall, it is observed that the increase in span length in a particular direction of building shall slightly decrease the fundamental period of the building in the same direction, for a specific number of spans.

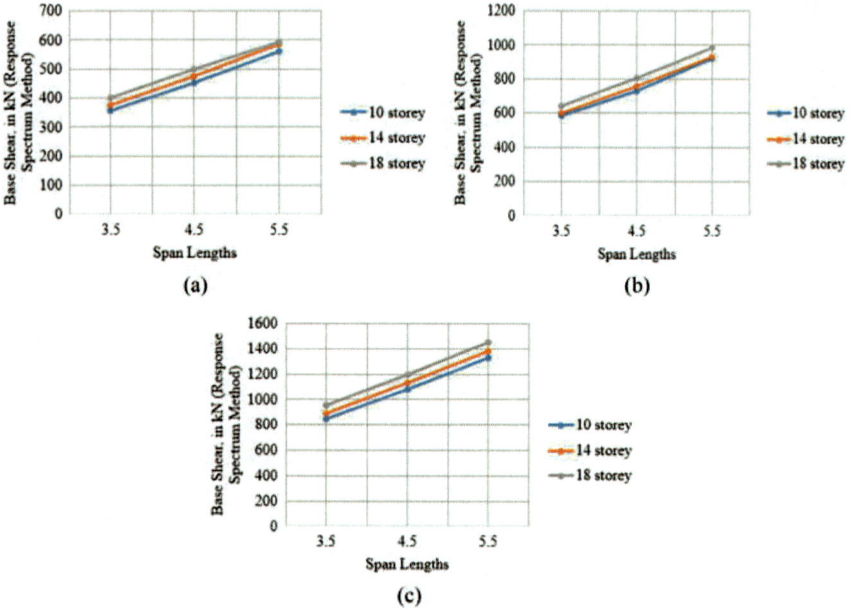

Fig. 6 Base shear distribution for bare RC framed buildings with varying span lengths, and **a** four spans, **b** six spans, and **c** eight spans

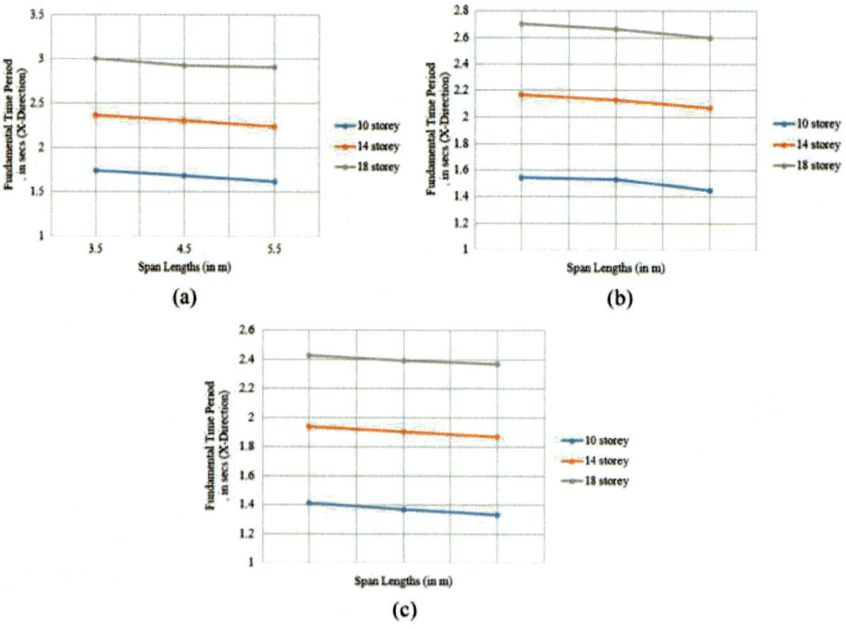

Fig. 7 Distribution of fundamental time period for bare RC framed buildings with varying span lengths, and **a** four spans, **b** six spans, and **c** eight spans

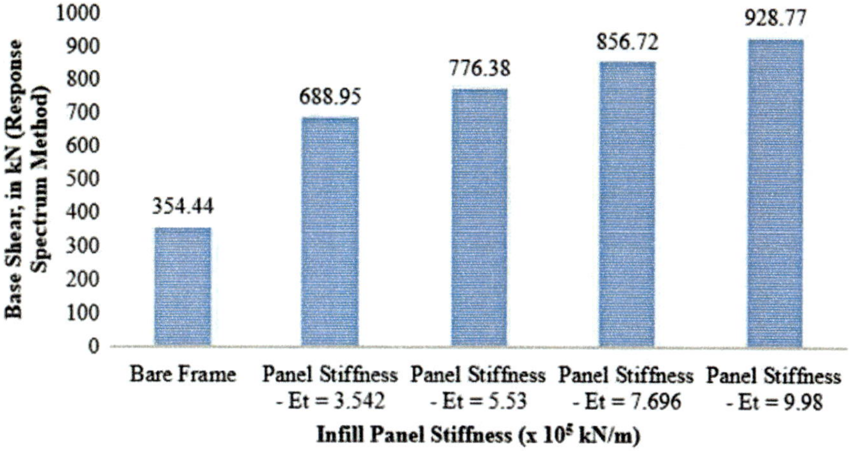

Fig. 8 Base shear distribution for ten storeyed RC infilled buildings with varying infill panel stiffness

5.5 *Influence of Stiffness of Masonry Infills on Base Shear and Fundamental Period*

Figure 8 shows the variation between the base shear and the stiffness of masonry infills of the structure.

From the analysis results, it is observed that the base shear and fundamental period of vibration are quite sensitive to the stiffness of masonry infill. As the stiffness of the structure increases, it attracts more seismic forces to act on it. Hence, the base shear increases with increase in the infill panel stiffness. From model FIF1 to FIF4, the increase in base shear varies from 94.38 to 162.04% in comparison with the reference bare frame model. In addition, the rate of increase in base shear reduces as the infill wall panel stiffness increases.

Fundamental period decreases with increasing stiffness of masonry infills. From model FIF1 to FIF4, the decrease in fundamental period varies from 45.76 to 59.68% in comparison with the reference bare frame model. Figure 9 presents the variation between the fundamental period and stiffness of masonry infill of the structure.

5.6 *Influence of Percentage of Openings in Masonry Infill Panels on Base Shear and Fundamental Period*

The results from the response spectrum analysis show a decrease in base shear values as the infill opening percentage increases. The base shear reduces by 27.37% when infill opening percentage increases from 0 to 20%, and it reduces by 45.07% when the infill opening percentage increases to 60%. Figure 10 displays the variation between

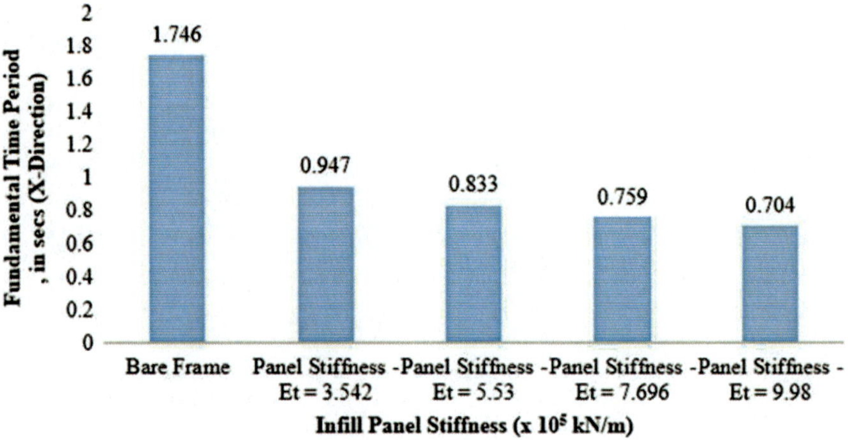

Fig. 9 Distribution of fundamental period for ten storeyed RC infilled buildings with varying infill panel stiffness

the base shear and percentage of openings within the masonry infills of the structure.

Figure 11 depicts the variation in fundamental period and percentage of openings within the masonry infills of the structure. The analysis results show that fundamental period increases as the infill opening percentage increases. The fundamental period increases by 35.59% when the infill openings increase from 0 to 20%, and it increases by 74.66% when the infill openings increase to 60%.

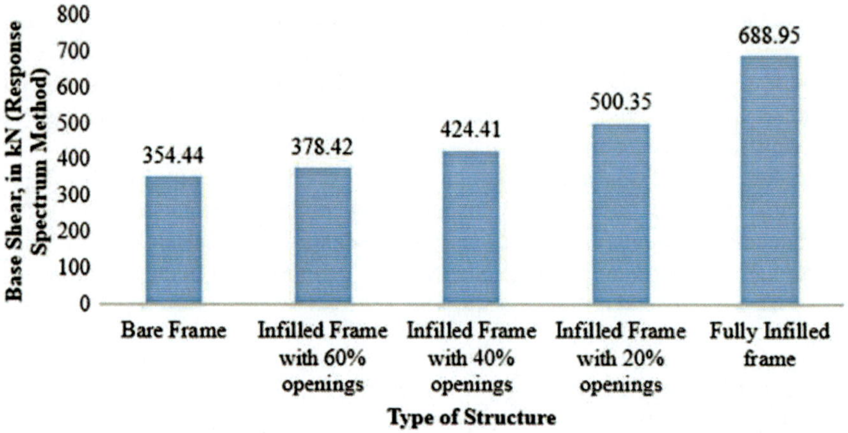

Fig. 10 Base shear distribution for ten storeyed RC infilled buildings with varying infill openings ($Et = 3.542 \times 10^5$ kN/m)

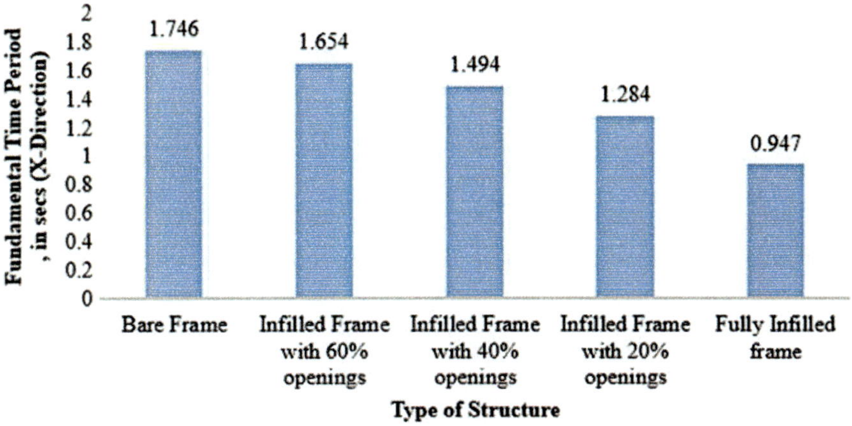

Fig. 11 Distribution of fundamental period for ten storeyed RC infilled buildings with varying infill openings (Et = 3.542×10^5 kN/m)

6 Conclusions

Based on the results obtained, the following conclusions can be drawn out of the present study:

1. As the number of spans, for a specific span length increases, base shear increases substantially whereas only a slight increase in base shear is observed when the number of storeys of the structure are increased. Also, the percentage increase in base shear is maximum when the number of spans of the structure are increased, for a specific span length, whereas the percentage increase in base shear is least when the number of storeys of the structure are increased.
2. The increase in span length of the structure, for a given number of spans, in the direction of earthquake vibration, shall increase the base shear of the structure.
3. With the increase in infill panel stiffness, the base shear also increases but the rate of increase of base shear is marginal.
4. The base shear value decreases as the percentage of openings within infill increases, when compared with respect to fully infilled RC frame. The rate of reduction in base shear is significant when openings increase from 0 to 60%, with the maximum rate of reduction in base shear at 20% openings, and as the openings increase beyond 60%, the base shear value is comparable to the base shear for a bare frame.
5. The increase in number of spans in one direction of a building, for a particular span length, shall decrease the fundamental period of the building in the same direction. Also, when the span length is increased, for a given number of spans, in the direction of earthquake vibration, a slight decrease in fundamental period of building is observed in the same direction.

6. The fundamental period is quite sensitive to the number of storeys of the building and an increase in the number of storeys shall significantly increase the fundamental time period of the building.
7. Fundamental period decreases with an increase in the infill panel stiffness, whereas, it shows an increasing trend as the infill opening percentage increases.

7 Recommendations

As observed from the results of the present investigation, the response of the structure by considering the stiffness of masonry infill walls is quite different than the one obtained for the bare RC frame. Therefore, for obtaining a realistic assessment of infilled RC framed buildings under seismic action, the effect of stiffness of masonry infill walls must be taken into account while carrying structural analysis.

References

1. Arulselvan, S., Subramanian, K., Pillai, E.B.P., Santhakumar, A.R.: RC infilled frame-RC plane frame interactions for seismic resistance. J. Appl. Sci. **7**(7), 942–950 (2007)
2. Arunkumar, S., Nandini Devi, G.: Seismic demand study of soft storey building and it's strengthening for seismic resistance. Int. J. Emerg. Trends Technol. Comput. Sci. **5**(2), 52–57 (2016)
3. Asteris, P.G.: Lateral stiffness of brick masonry infilled plane frames. J. Struct. Eng. **129**(8), 1071–1079 (2003)
4. Asteris, P.G., Repapis, C.C., Tsaris, A.K., Trapani, F.D., Cavaleri, L.: Parameters affecting the fundamental period of infilled RC frame structures. Earthq. Struct. **5**(9), 999–1028 (2015)
5. Bhatt, M.R., Pradhan, P.M., Jha, S.: Study on the effect of soft story on infill RC frames under seismic effect. J. Sci. Eng. Technol. **13**(2), 79–91 (2017)
6. Braga, F., Manfredi, V., Masi, A., Salvatori, A., Vona, M.: Performance of non-structural elements in RC buildings during the L'Aquila, 2009 earthquake. Bull Earthq. Eng. **9**, 307–324 (2011)
7. Celarec, D., Ricci, P., Dolšek, M.: The sensitivity of seismic response parameters to the uncertain modelling variables of masonry infilled reinforced concrete frames. Eng. Struct. **35**, 165–177 (2012)
8. Chidananda, H.R., Raghu, K., Narayana, G.: Analysis of RC framed structures with central and partial openings in masonry infill wall using diagonal strut method. Int. J. Res. Eng. Technol. **4**(4), 640–647 (2015)
9. Chopra, A.K., Goel, R.K.: Building period formulas for estimating seismic displacements. Earthq. Spectra **16**, 533–536 (2000)
10. Choudhury, T., Kaushik, H.B.: Seismic response sensitivity to uncertain variables in RC frames with infill walls. J. Struct. Eng. **144**(10), 04018184 (2018)
11. Crisafulli, F.J., Carr, A.J.: Proposed macro-model for the analysis of infilled frame structures. Bull. N. Z. L. Soc. Earthq. Eng. **40**(2), 69–77 (2007)
12. Duggal, S.K.: Earthquake Resistant Design of Structures. Oxford University Press (2013)
13. Egyptian Seismic Code: Regulations for Earthquake Resistant Design of Buildings in Egypt. Egyptian Society for Earthquake Engineering, Cairo (1988)

14. European Committee for Standardization CEN: Eurocode 8: Design of structures for earthquake resistance—Part 1: General rules, seismic actions and rules for buildings. European Standard EN 1998-1:2004, European Committee for Normalization, Brussels (2004)
15. Fiore, A., Porco, F., Raffaele, D., Uva, G.: About the influence of the infill panels over the collapse mechanisms activated under pushover analyses: two case studies. Soil Dyn. Earthq. Eng. **39**, 11–22 (2012)
16. Fiore, A., Netti, A., Monaco, P.: The influence of masonry infill on the seismic behaviour of RC frame buildings. Eng. Struct. **44**, 133–145 (2012)
17. Genidy, M., Hanna, N., Tahoon, H., Mahmoud, S.: Seismic response evaluation of moment-resisting-frame multi-story buildings with soft story. Int. J. Civil Struct. Eng. Res. **3**(1), 118–127 (2015)
18. Grover, A., Verma, S.K.: Effect of infill walls on RC framed structure. Int. J. Innov. Res. Sci. Eng. Technol. **5**(7), 12710–12720 (2016)
19. Holmes, M.: Steel frames with brickwork and concrete infilling. Proc. Inst. Civ. Eng. **19**(2), 473–478 (1961)
20. Indian Standard IS 456: Plain and Reinforced Concrete - Code of Practice, 4th Revision, Bureau of Indian Standards, New Delhi (2000)
21. Indian Standard IS 1077: Common Burnt Clay Building Bricks—Specification, 5th Revision, Bureau of Indian Standards, New Delhi (1992)
22. Indian Standard IS 1893 (Part 1): Criteria for Earthquake Resistant Design of Structures—part 1: General Provisions and Buildings, 5th Revision, Bureau of Indian Standards, New Delhi (2002)
23. Indian Standard IS 1893 (Part 1): Criteria for Earthquake Resistant Design of Structures—Part 1: General Provisions and Buildings, 6th Revision, Bureau of Indian Standards, New Delhi (2016)
24. Indian Standard IS 2212: Brick Works—Code of Practice, 1st Revision, Bureau of Indian Standards, New Delhi (1991)
25. Indian Standard IS 2250: Code of Practice for Preparation and Use of Masonry Mortars, 1st Revision, Bureau of Indian Standards, New Delhi (1981)
26. Mainstone, R.J., Weeks, G.A.: The influence of bounding frame on the racking stiffness and strength of brick walls. In: Proceedings of the 2nd International Brick Masonry Conference, pp. 165–171. Building Research Establishment, Watford, England (1970)
27. Mehrabi, A.B., Shing, P.B.: Finite element modeling of masonry-infilled RC frames. J. Struct. Eng. **123**(5), 604–613 (1997)
28. New Zealand Society of Earthquake Engineering NZSEE: Recommendations for the Assessment and Improvement of the Structural Performance of Buildings in Earthquakes. NZSEE Study Group on Earthquake Risk Buildings, New Zealand Society of Earthquake Engineering (2006)
29. Patel, Y.P., Jamani, A.: Performance and analysis of masonry infill in RC frame structure. Int. Res. J. Eng. Technol. **4**(4), 2141–2145 (2017)
30. Pokhrel, A., Gautam, D., Chaulagain, H.: Effect of variation on infill masonry walls in the seismic performance of soft story RC building. Austr. J. Struct. Eng. **20**(1), 1–9 (2019)
31. Polyakov, S.V.: On the interaction between masonry filler walls and enclosing frame when loaded in the plane of the wall. Transl. Earthq. Eng. **2**, 36–42 (1960)
32. Qian, K., Li, B.: Effects of masonry infill wall on the performance of RC frames to resist progressive collapse. J. Struct. Eng. **143**(9), 04017118 (2017)
33. Singh, V.P.: Mechanical Vibrations: Mechanical Engineering (A Modern Approach). Dhanpat Rai & Co. (P) Ltd., Delhi (2006)
34. Singhal, V., Rai, D.C.: Role of toothing on in-plane and out-of-plane behavior of confined masonry walls. J. Struct. Eng. **140**(9), 04014053 (2014)
35. Smith, B.S., Carter, C.: A method of analysis for infilled frames. Proc. Inst. Civ. Eng. **44**, 31–48 (1969)
36. Uva, G., Porco, F., Fiore, A.: Appraisal of masonry infill walls effect in the seismic response of RC framed buildings: a case study. Eng. Struct. **34**, 514–526 (2012)

37. Yadunandan, C., Kiran Kuldeep, K.N.: Study on behaviour of RC structures with infill walls due to seismic loads. Int. Res. J. Eng. Technol. **4**(6), 2494–2500 (2017)
38. Zhai, C., Kong, J., Wang, X., Chen, Z.: Experimental and finite element analytical investigation of seismic behavior of full-scale masonry infilled RC frames. J. Earthq. Eng. **20**(7), 1171–1198 (2016)

Role of FRP in Developing Sustainable Infrastructure—A Review

Gurpreet Singh, Maninder Singh, and Babita Saini

Abstract Fibre-reinforced polymer (FRP) sheets are frequently used due to their ability to provide intentive and rational solutions to resolve the developing ageing issue in RC structural elements. FRP composites create a great interest in their application in newly built as well as already existing civil infrastructure for repair/rehabilitation and strengthening purposes due to their superiority over the traditional steel reinforcement. FRP materials were introduced into the construction industry for more than 50 years ago. This paper presents the review of the performance and recent advances of FRP in structural applications. The prime focus of this study is to present types of FRP composite with their properties, features, and applications. The configuration and strengthening technique of FRP has a great impact on the overall strength of the RC element. A beam-column joint is a critical region that undergoes heavy shear stresses when subjected to seismic loads; therefore, these joints should be designed as such, they maintain stability and overall integrity of the structure. The review from the past studies revealed that the lamination of FRP enhanced the durability parameters, the life span of the structure and contributed to making sustainable infrastructure. The usage of FRP provides a better solution to overcome these stresses, but the premature brittle failure due to debonding is a major shortcoming of FRP.

Keywords Fibre-reinforced polymer (FRP) · Beam-column joint · Polymeric · Strengthening · Sustainable infrastructure

---

G. Singh (✉) · M. Singh · B. Saini
Civil Engineering Department, National Institute of Technology, Kurukshetra, Kurukshetra 136119, India
e-mail: gurpreet_61900051@nitkkr.ac.in

M. Singh
e-mail: maninder_6160049@nitkkr.ac.in

B. Saini
e-mail: bsaini@nitkkr.ac.in

D. K. Ashish et al. (eds.), *3rd International Conference on Innovative Technologies for Clean and Sustainable Development*, RILEM Bookseries 29,
https://doi.org/10.1007/978-3-030-51485-3_16

1 Introduction

By definition, Sustainability is the use of resources available in the environment to full-fill the needs of societies, but care should be taken such that the right of future generation not get violated. The objective of sustainable construction is to minimize the environmental effect of building for its entire life span and at the same time by optimizing its economy. FRP material properly fit into the equation of sustainability conforming all the condition required, for being a sustainable material. FRP looks like a fragile polymeric layer that constitutes of different fibres. They are flexible and planar (sheet-like) usually machined from a polymeric material and sometimes from natural materials. From the past few decades, FRP becomes a popular strengthening material due to various advantages like lightweight material with easy installation, low maintenance cost, resistance to corrosion, acoustics, and shaping versatility. Various types of FRP products are available such as glass fibre reinforced polymer (GFRP), carbon fibre reinforced polymer (CFRP), aramid fibre-reinforced polymer (AFRP) and basalt fibre reinforced polymer (BFRP). The traditional practice for the strengthening or rehabilitation of structural element involves concrete covering or anchoring with steel bars, but due to their susceptibility to the environmental effect like emission of carbon-dioxide, corrosion and shortened-life span it is better to choose a material like FRP which provide better sustainability than conventional materials. Therefore, usage of CFRP either sheet or plates in the beam-column joint for strengthening with the proper orientation reduces the chances of failure at defected joint [1]. The moment resisting ability of BC (beam-column) joint can be increased up to 60% by providing the external FRP reinforcement [2]. According to the previous studies special focus should be given to the thickness of wrap and overlay thickness to prevent the failure of wrap [2, 3]. Prota et al. [4] used carbon rod and carbon wrap in the consolidation for the retrofitting of 11 BC joints and exhibited that retrofitting in both interior and exterior side increased the ductility of members. If the initially stressed BC joint is retrofitted with the CFRP jacketing, it recovers its full strength even if it is stressed up to 85% of the ultimate load [5]. By the usage of the U-shaped GFRP jacket, the concrete element act as a single unit and it confines the concrete, which ultimately boosts ductility and load-bearing capacity of joint repaired [5, 6].

2 Background of FRP

Extraction and Production of engineering materials involve the consumption of high energy intensities. Moreover, the consumption of energy totally depends on the type of technology using for the production and extraction of material. The energy intensities involve for various material and their constituents are provided in Table 1.

From the fibres already mentioned in Sect. 1, glass fibre consumes the lowest energy and it is most commonly used FRP material in the construction industry. In general, FRP composite leads to lighter construction, if compared with conventional

Table 1 Energy input for extraction and production of selected engineering materials [7]

Materials	Energy input (MJ/kg)
Polymer	
Polyester	63–78
Epoxy	76–80
Fibers	
Glass	13–32
Carbon	183–286
Metals	
Steel	30–60
Aluminium	196–257

engineering material and also the lightweight ability of FRP increase its versatility [7–12]. Being a lightweight material, the transportation energy input and carbon emission is much low as compared to conventional material like steel and concrete. With the evolution of the wrapping system, FRP sheets because of their stiffness-to-weight and high strength to weight ratios have become the alternative for strengthening and retrofitting of reinforced structural element [13–20]. No doubt composite materials have some limitations like the vulnerability to delamination and brittle failure. Due to the low compressibility of FRPs, it cannot be used in those structures where high compressive strength is required [16]. The modulus of elasticity for the CFRP fibre is 165 GPa and Poisson's ratio of 0.183 [19]; whereas, for the CFRP sheets the modulus of elasticity is 230 GPa and ultimate tensile strength of 3500 MPa [20]. Depending on the strength requirement, cost, availability and the present environmental conditions, the strengthening material is chosen accordingly. The tensile behaviour and shear behaviour of Kevlar fibre reinforced polymer is 55 and 180% greater than the E-glass fibre [2]. However, the carbon fabric has almost 10% greater tensile strength than that of Kevlar fibre; but, the cost of carbon fibre price is approximately double that of Kevlar fibres. BFRP shows the ability to control strain at failure, which makes it more suitable than CFRP [21]. CFRP provides high resistance to an alkaline environment; whereas, the GFRP is highly reactive to all forms of alkali because it comprehends silica [22–24]. The orientation of the fibres used influence the rupture strength, CFRP, if used with horizontal fibres, shows less rupture strength as if provided with the vertical fibres [20]. The ductility and loading capacity of the BC joint improved using unidirectional glass carbon hybrid fibre [6]. For the resistance against the torsional moment, the usage of CFRP is more effective for strengthening than GFRP [25]. The beam strengthened by CFRP fails suddenly when subjected to ultimate load; but, if GFRP is used as strengthening material then beam exhibits some residual strength after peaking. The fibreglass composite reinforcement is more effective and has tremendous potential for enhancing the ductile strength of the beam-column joint in earthquake-prone zones [6]. Moreover, the hybrid combination of different FRPs meets the stringent requirements. The combination of CFRP and GFRP is used to achieve high strength, high ductility and break resistance. The combination of BFRP and CFRP is recommended if the structures have to withstand the high temperature.

3 FRPs Configuration

The configuration for the FRP (wrap and laminates) provided depends on the area to be strengthened and the type of stress to overcome. Some researchers have used unidirectional CFRP, unidirectional GFRP and bidirectional hybrid HFRP [3, 4]. The performance of the FRP strengthened structural elements is positively influenced by the type and properties of the FRP sheet and laminates used on the cementitious structural elements [1, 2, 5, 6]. Moreover, the configuration of the FRP composite plays an important role to withstand the structural failure. The properties of various types of FRP have been given in Table 2.

4 Preparation of Surface

The overall bond strength of the strengthened elements remarkably changes by surface preparation. The procedure involves the cleaning of the surface by removing all the impurities present and removal of the weak external layer [35, 36]. After cleaning, grinding is done in some cases to expose the coarse aggregates for the purpose of better adhesion with members [5, 28]. In externally bonded FRP strengthening, post to the removal of concrete mortar, pores are created on the surface by some innovative techniques [5, 6, 37, 38]. The FRPs are not directly wrapped on the sharp edges of the BC joint; because this reduces the strength [39], the sharp edges are made round by grinder that alternatively lower down the stress concentration and spikes the confinement effect. By modifying the shape of edges, the resistance to impact is increased by 10% [40]. Epoxy is applied on the BC joint surface as well as sheets and properly rolled with the roller until it starts to come out [5, 40]. The literature study reveals that the selection of methods for the preparation depends on the type and orientation of FRP and loading cases.

5 Adhesive Resin Requirement

Fibre-reinforced polymer (FRP) composite primarily originated from a polymer which is produced from fossil fuels, is still an issue that acts as a barrier for sustainability. The development of plant-based binders is the prime focus of future research [41]. The ideal bonding between the surface of concrete and FRP composite to ensure the proper strength can only be attained by applying the high-quality adhesive resins [42–44]. The epoxy and hardener component, in the desired proportion, are mixed

Table 2 Properties of different types of FRP

References	FRP type	Thickness of fibre (mm)	Tensile strength (MPa)	Elastic modulus (GPa)	Adhesive used	Strength of concrete (MPa)	Failure strain (%)
[1]	CFRP	0.13	3500	230	–	25	1.50
[2]	Kevlar (aramid fibre)	–	3150	–	Shell^{TM} chemical epoxy	34	–
[3]	Unidirectional carbon fibre wrap	0.13	3650	238	Epoxy resin (Sika Dur52)	39	1.5
	Unidirectional glass fibre wrap	0.17	2200	76			2.8
	Bi-directional hybrid fibre wrap	–	–	–			–
[5]	CFRP	0.117	3800	240	Blue pigment epoxy glue	20	–
[6]	E-glass fabric	2.30	–	17.9	–	27.6	–
	Kevlar fabric			34.5			
	Carbon fabric			59.3			
[26]	CFRP sheets	0.167	3600	240	–	–	1.5
[27]	CFRP	1.4	2800	210	Resin	–	0.0135
[28]	CFRP	0.36	3500	–	Epoxy putty	30	0.0239
	GFRP	0.11	2250				0.0117
[29]	Unidirectional CFRP sheets	0.165	–	–	Inorganic epoxy	–	
[30]	Carbon sheets	–	–	270	–	23.7	0.016
	Glass sheets			70			0.031
[31]	Glass FRP	–	–	–	–	20.7	0.02
	Carbon FRP						0.012
[32]	CFRP	–	3500	230	Sika-Egypt epoxy	30	0.005
[33]	CFRP laminate	0.400	1050	84.0	–	45	1.25
[34]	CFRP	0.111	4130	242	–	–	1.7
	GFRP	0.273	3400	73			2.7

to form adhesive resin. Singh et al. [5] mixed a ratio of 100 (base):40 (hardener) which form a translucent blue liquid epoxy. Usually, the base and hardener component are assorted at 1:1 to 1:5 ratios. The unit weight of epoxy resin lies between 1.1 and 1.4 kg/m^3 and density 0.5 kg/m^2 [40]. Cracks can be repaired by injecting epoxies into RC members through drilled holes [32]. The tensile strength of epoxy resin lies between 30 and 90 MPa and elastic modulus 1.1–6 GPa [2, 5, 6, 40, 45,

46]. CFRP has zero coefficient of thermal expansion; however, literature indicated variation in thermal expansion has a minimal effect on bonding for given temperature change. Epoxies resin should be able to resist elevated temperature and should have the capacity to bear different loading conditions [5].

6 Strengthening Techniques

Various types of techniques are used to overcome the failure and strengthen the RC structure with FRP material. Conventional techniques involve for the retrofitting of BC joint include bonding or bolting of steel plates in the cracked section or enlargement of section by FRC jacketing [45], these techniques are now replaced by the externally bonded (EB) laminates, NSM (near-surface mounted) bars/strips, by grooving RC element methods with or without adhesives and anchorage system [26, 37, 47, 48]. Various researchers have reported the performance of FRP reinforcement in improving seismic resistance of defected BC joint [6, 26]. Externally bonded FRP techniques for retrofitting can be applied with the required layers with desired orientation like side bonding, full wrapping, and U wrapping. FRP laminates if anchored to the concrete structure close to the BC joint improve the performance and decrease the feasibility of FRP sheets debonding [49]. The main purpose of providing the anchorage system is to increase flexural or shear strengthening for the existing beam with FRP material [27]. Anchoring system controlled the overall strength of structural elements.

7 Strengthening Performance

The beam-column joint is a critical zone in RC moment-resisting frame. During an earthquake, the behaviour of the entire structure is significantly influenced by the beam-columns joint [1]. The RC joint is the most critical and vulnerable element that resist the lateral load. RC joint strengthening is a tough task as it involves various practical difficulties. Conventional methods for strengthening beam-column joint like concrete jacketing and steel jacket (steel cages) need skilled labour and artful detailing. Moreover, this jacketing technique increases weight as well as the dimension of structural element [5]. More than three decades from now, FRP came into existence with numerous desirable characteristics such as resistance to corrosion, ease of installation, availability, etc. [30].

7.1 *Shear Resistance*

From the past decades, the "strong column weak beam", "strong shear weak flexure" and "strong joint weak member" concept have been widely accepted in the seismic design of RC moment resisting building. FRP composite is an applicable solution for enhancing the shear capacity of exterior RC joints. Antonopoulos and Triantafillou [30] reveals if the mechanical anchorages are provided with the effectiveness of FRP sheets in term of overall performance and energy increase by 30 and 40% respectively. FRP strengthened joint provides a positive effect on the shear capacity of BC joint, an increase in strength from 65 to 85% is noted corresponding to 2.5 times higher load. The effect on shear strength due to an increase in axial load is quite positive. Cracks in the concrete lead to the reduction of compressive strength. Issa and Debs [50] from his experimental result found a total reduction in compressive strength due to crack as 40.93% and in case of cracks healed with epoxy injection compressive strength reduction was only 11.25%. The degree of enhancement of beam-column joint at real-life scale 1:1 can be increased by using a 30 mm thick layer of UHPFRC and by adopting high modulus epoxy or chemical anchorage [51].

7.2 *Ultimate Capacities*

From the literature, it is observed that the use of FRP in structural elements not only increase the strength and stiffness parameters; but, also minimize cracks. To achieve the best strengthening for beams, it is recommended to start rehabilitation from the spot where there is minimum damage to RC beam. All beam repaired by using this technique showed an improvement in the ultimate load ranging from 49 to 57% [32]. Delamination due to concentrated load can be prevented by providing transverse straps (U-shaped CFRP sheets) anchored to flexural CFRP sheets [33]. The structural elements exposed to sulfate attack showed tiny cracks in on all sides of beams that cause a reduction in compressive and shear behaviour by 26%, and 24% respectively [52]. The impact resistance of wrapped structural element under shear enhanced by 5–15% [53, 54]. Moreover, the angular wrap system performance configured at 45° along with the FRP strip is better than that of vertical U-type wrapping under impact [55]. The flexural strength of the beam effectively improved by applying the FRP wrapping along the tension side [18, 46]. FRP strengthened beam element with two-layer FRP sheets (U-wrapped) reduce the deflection by 39% compared to the strengthening with one layer of FRP [18]. The flexural response in the previous studies depicted that the strength of EB and NSM FRP strengthened beam enhanced by 74% and 111% respectively [56]. Meier [57] experimentally (four-point bending with 60 specimens) concluded that by using 0.3 mm thick and 200 mm CFRP sheets, flexural capacity increased by 100% over that of unstrengthened beams. Smith et al. [58] reported an increase in flexural capacity ranging between 44 and 216% for RC slab strengthen with FRP.

7.3 Future Trends

FRP composite has a vital role in economic, convenient, fast and sustainable construction. The pre-fabricated structural elements composed of GFRP composite integrated to form a superstructure without any heavy lifting equipment. The FRP panel construction of bridge deck system facilitates the quick building of the structure with relatively low cost and in a short time as compared to conventional materials [41, 59–62]. For the military purpose where they have to do temporary but fast construction, FRP composite panel construction plays an important role and also it is easy to shift the same when needed. Moreover, GFRP less attenuates waves and almost transparent to radar and radio waves and also act as a barrier to protect airport infrastructure against malicious action or natural event [61].

8 Summary and Conclusions

The present review study clearly states that the usage of FRP composites into RC structural elements plays a vital role in continual integration and strengthening. Designers always prefer lightweight material having superior properties like resistance against different loading conditions, potential to withstand varied environmental conditions to achieve desired service life. Results state that FRP composite with adequate anchoring provides an excellent solution for strengthening both in terms of structural behaviour and cost. Though FRPs composite offers a good life span but at a stage when it comes to the end of service life, either the material get discarded or recycled, since dumping or burning is not in favour of environmental conditions, it is better to choose for recycling. The strength of recycled polymer is a major constraint as they are not able to acquire the same strength as in their initial state. The outcome of this review can be concluded in the following points.

- For the design of the modern structure, FRP can be considered as an integral part due to their superior properties compared to traditional material.
- FRP material has a great potential to enhance the ductility and to withstand the shear stresses produced in beam-column joint during the seismic activities.
- Prevention of the brittle failure of fibres can be attained by giving special attention to the ratio of wrap thickness to overlay thickness.
- FRP exhibits high tensile strength, that boost up ultimate strength up to twice that of RC beam cast without FRPs.
- The failure mode such as premature debonding and peeling can be resisted by providing good quality material, proper configuration (according to loading condition) and overall techniques for applying FRP material on the surface.
- Future research can be done on the torsional and axial application for FRP strengthening in the RC beam and RC wall respectively. Moreover, the effect on durability by the wastewater treatment plant, chemical plant, and nuclear plant have not yet been investigated.

References

1. Mahmoud, M.H., Afefy, H.M., Kassem, N.M., Fawzy, T.M.: Strengthening of defected beam-column joints using CFRP. J. Adv. Res. **5**, 67–77 (2014)
2. Granata, P.J., Parvin, A.: An experiment on Kevlar strengthening of beam column connection. Compos. Struct. **53**, 163–171 (2001)
3. Attari, N., Youcef, Y.S.: Seismic performance of reinforced concrete beam-column joint strengthening by FRP sheets. Structure **20**, 353–364 (2019)
4. Prota, A., Nanni, A., Manfredi, G., Cosenza, E.: Selective upgrade of beam-column joints with composites. ACI Struct. J. **101** (2004)
5. Singh, V., Bansal, P.P., Kumar, M.: Experimental study on strength and ductility of CFRP jacketed reinforced beam column joints. Constr. Build. Mater. **55**, 194–201 (2014)
6. Parvin, A., Granata, P.: Investigation on effect of fibre composites at concrete joint. Compos. Part B **31**, 499–509 (2000)
7. Song, Y.S., Youn, J.R., Gutowski, T.G.: Life cycle energy analysis of fiber-reinforced composites. Compos. Part A **40**, 1257–1265 (2009)
8. Karbhari, V.M.: Durability of FRP composites for civil infrastructure myth, mystery or reality. Adv. Struct. Eng. **6**, 243–256 (2003)
9. Keller, T.: Recent all-composite and hybrid fibre-reinforced polymer bridges and buildings. J. Prog. Struct. Eng. Mater. **3**, 132–140 (2001)
10. Hollaway, L.C.: The evolution of and the way forward for advanced polymer composites in the civil infrastructure. Constr. Build. Mater. **17**, 365–378 (2003)
11. Hollaway, L.C.: Case studies, chap. 13. In: Hollaway, L.C., Teng, J.G. (eds.) Strengthening and Rehabilitation of Civil Infrastructures Using Fibre-Reinforced Polymer (FRP) Composites. Woodhead Publishing, Cambridge (2008)
12. Hollaway, L.C.: Fibre-reinforced polymer composite structures and structural components: current applications and durability issues, chap. 10. In: Vistasp, K. (ed.) Durability of Composites for Civil, Structural Applications. Woodhead Publishing, Cambridge, UK (2007)
13. Karbhari, V.M.: Materials considerations in FRP rehabilitation of concrete structures. J. Mater. Civ. Eng. **13** (2001)
14. Hadi Muhammad, N.S., Minh, T.T.: Retrofitting non-seismically detailed exterior beam–column joints using concrete covers together with CFRP jacket. Constr. Build. Mater. **63**, 161–73 (2014)
15. Attari, N.: Efficiency of strengthened beam-column joint by FRP laminates. Adv. Compos. Mater. **19**, 171–183 (2010)
16. ACI 440.2R-02: Guide for the Design and Construction of Externally Bonded FRP Systems for Strengthening Concrete Structures. American Concrete Institute, Farmington Hills, MI, USA (2002)
17. Garcez, M., Meneghetti, L., da Silva Filho, L.C.: Structural performance of RC beams post strengthened with carbon, aramid, and glass FRP systems. J. Compos. Constr. **12**, 522–530 (2008)
18. El-Hacha, R., Soudki, K.: Prestressed near-surface mounted fibre reinforced polymer reinforcement for concrete structures—a review. Can. J. Civ. Eng. **40**, 1127–1139 (2013)
19. Camata, G., Spacone, E., Al-Mahaidi, R., Saouma, V.: Analysis of test specimens for cohesive near-bond failure of fibre-reinforced polymer-plated concrete. J. Compos. Constr. **8**, 528–538 (2004)
20. Hussein, M., Afefy, H.M.E.D., Khalil, A.H.A.K.: Innovative repair technique for RC beams pre damaged in shear. J. Compos. Constr. **17**, 04013005 (2013a)
21. El Refai, A., Ammar, M.A., Masmoudi, R.: Bond performance of basalt fibre-reinforced polymer bars to concrete. J. Compos. Constr. **19**, 04014050 (2015)
22. Petersen, M.R., Yossef, M., Chen, A.: Gap between code requirements and current state of research on safety performance of fibre-reinforced polymer for nonstructural building components. Pract. Period. Struct. Des. Constr. **22**, 04017005 (2017)

23. Benmokrane, B., Elgabbas, F., Ahmed, E.A., Cousin, P.: Characterization and comparative durability study of glass/vinylester, basalt/vinylester, and basalt/epoxy FRP bars. J. Compos. Constr. **19**, 04015008 (2015)
24. Dong, Z., Wu, G., Xu, B., Wang, X., Taerwe, L.: Bond durability of BFRP bars embedded in concrete under seawater conditions and the long-term bond strength prediction. Mater. Des. **92**, 552–562 (2016)
25. Ameli, M., Ronagh, H.R., Dux, P.F.: Behavior of FRP strengthened reinforced concrete beams under torsion. J. Compos. Constr. **11**, 192–200 (2007)
26. Dai, G., Bai, Y.L.: Seismic retrofit of exterior RC beam-column joint with bonded CFRP reinforcement. Compos. Struct. **224**, 111018 (2019)
27. Ascione, L., Berardi, V.P.: Anchorage device for FRP laminates in the strengthening of concrete structure close to beam column joints. Compos. Part B **42**, 1840–1850 (2011)
28. Mukherjee, A., Joshi, M.: FRPC reinforced concrete beam-column joints under cyclic excitation. Compos. Struct. **70**, 185–199 (2005)
29. Toutanji, H., Zhao, L., Zhang, Y.: Flexural behavior of reinforced concrete beams externally strengthened with CFRP sheets bonded with an inorganic matrix. Eng. Struct. **28**, 557–566 (2006)
30. Antonopoulos, C.P., Triantafillou, T.C.: Experimental investigation of FRP strengthened RC beam-column joints. J. Compos. Constr. ASCE **7**, 39–49 (2003)
31. Kachlakev, D., McCurry, D.D.: Behavior of full-scale reinforced concrete beams retrofitted for shear and flexural with FRP laminates. Composites **31**, 445–452 (2000)
32. Hussein, M., Afefy, H.M.E.D., Khalil, A.H.A.K.: Innovative repair technique for RC beams predamaged in shear. J. Compos. Constr. **17**, 04013005 (2013b)
33. Al-Saidy, A.H., Al-Harthy, A.S., Al-Jabri, K.S., Abdul-Halim, M., Al-Shidi, N.M.: Structural performance of corroded RC beams repaired with CFRP sheets. Compos. Struct. **92**, 1931–1938 (2010)
34. Dong, J., Wang, Q., Guan, Z.: Structural behaviour of RC beams with external flexural and flexural–shear strengthening by FRP sheets. Compos. B Eng. **44**, 604–612 (2013)
35. Deniaud, C., Cheng, J.J.R.: Shear behaviour of RC T-beams with externally bonded fiber reinforced polymer sheets. ACI Struct. J. **93**, 386–394 (2001a)
36. Deniaud, C., Cheng, J.J.R.: Review of design methods for reinforced concrete beams strengthened with fiber reinforced polymer sheets. Can. J. Civ. Eng. **28**, 271–281 (2001b)
37. Mostofinejad, D., Tabatabaei Kashani, A.: Experimental study on effect of EBR and EBROG methods on debonding of FRP sheets used for shear strengthening of RC beams. Compos. B Eng. **45**, 1704–1713 (2013)
38. Kang, T.H.K., Howell, J., Kim, S., Lee, D.J.: A state-of-the-art review on debonding failures of FRP laminates externally adhered to concrete. Int. J. Concr. Struct. Mater. **6**, 123–134 (2012)
39. Ghobarah, A., Ghorbel, M.N., Chidiac, S.E.: Upgrading torsional resistance of reinforced concrete beams using fibre-reinforced polymer. J. Compos. Constr. **6**, 257–263 (2002)
40. Pham, T.M., Hao, H.: Behavior of fibre-reinforced polymer-strengthened reinforced concrete beams under static and impact loads. Int. J. Prot. Struct. **8**, 3–24 (2017)
41. Constable, P.A.: Bridge modification approach—a value for money approach paper 3. In: Proceedings of Stronger and Safer Bridges—Bridge Modification, vol. 2, p. 21
42. Basler, M., Mungal, B., Fan, S.: Strengthening of structures with the Sika CarboDur composite strengthening systems. In: Teng, J.G. (ed.) Proceedings of the International Conference on FRP Composites in Civil Engineering—CICE 2001, pp. 253–262. Elsevier Science, Oxford, England (2001)
43. Uomoto, T., Nishimura, T.: Determination of aramid, glass and carbon fibres due to alkali, acid and water in different temperatures. In: Proceedings 4th International Symposium, FRPRCS-4, ACI SP-188, pp. 515–522. American Concrete Institute (1999)
44. Sen, R.: Advances in the application of FRP for repairing corrosion damage. Prog. Struct. Eng. Mater. **5**, 99–113 (2003)
45. Engindeniz, M., Kahn, L.F., Zureick, A.H.: Repair and strengthening of reinforced concrete beam-column joints: state of the art. ACI Struct. J. **102**(2), 187–197 (2005)

46. Chen, W., Pham, T.M., Sichembe, H., Chen, L., Hao, H.: Experimental study of flexural behaviour of RC beams strengthened by longitudinal and U-shaped basalt FRP sheet. Compos. B Eng. **134**, 114–126 (2018)
47. Charalambidi, B.G., Rousakis, T.C., Karabinis, A.I.: Fatigue behavior of large-scale reinforced concrete beams strengthened in flexure with fibre-reinforced polymer laminates. J. Compos. Constr. **20**, 04016035 (2016)
48. Ghobarah, A., El-Amoury, T.: Seismic rehabilitation of deficient exterior concrete frame joints. J. Compos. Constr. **9**, 408–416 (2005)
49. Teng, J.G., Chen, J.F., Smith, S.T., Lam, L.: FRP Strengthening of RC Structures. Wiley, New York, Chichester, England (2002)
50. Issa, C.A., Debs, P.: Experimental study of epoxy repairing of cracks in concrete. Constr. Build. Mater. **21**(1), 157–163 (2007)
51. Irfan, K.M., et al.: Seismic behavior of beam-column joints strengthened with ultra-high performance fibre reinforced concrete. Compos. Struct. **200**, 103–119 (2018)
52. Haddad, R.H., Al-Rousan, R.Z., Al-Sedyiri, B.K.: Repair of shear-deficient and sulfate-damaged reinforced concrete beams using FRP composites. Eng. Struct. **56**, 228–238 (2013)
53. Tang, T., Saadatmanesh, H.: Behavior of concrete beams strengthened with fibre-reinforced polymer laminates under impact loading. J. Compos. Constr. **7**, 209–218 (2003)
54. White, T.W., Soudki, K.A., Erki, M.A.: Response of RC beams strengthened with CFRP laminates and subjected to a high rate of loading. J. Compos. Constr. **5**, 153–162 (2001)
55. Pham, T.M., Hao, H.: Impact behavior of FRP-strengthened RC beams without stirrups. J. Compos. Constr. **20**, 04016011 (2016)
56. Firmo, J.P., Arruda, M.R.T., Correia, J.R., Rosa, I.C.: Three-dimensional finite element modelling of the fire behaviour of insulated RC beams strengthened with EBR and NSM CFRP strips. Compos. Struct. **183**, 124–136 (2018)
57. Meier, U.: Carbon fibre-reinforced polymers: modern materials in bridge engineering. Struct. Eng. Int. **2**, 7–12 (1992)
58. Smith, S.T., Zhang, H., Wang, Z.: Influence of FRP anchors on the strength and ductility of FRP-strengthened RC slabs. Constr. Build. Mater. **49**, 998–1012 (2013)
59. Concrete Society: Strengthening Concrete Structures Using Fibre Composite Materials: Acceptance, Inspection and Monitoring. TR57, Camberley, UK (2003)
60. Daly, A.F., Duckett, W.A.: The design and testing of an FRP highway bridge deck. J. Res. **5**(3) (2002). Transport Research Laboratory, Crowthorne, UK
61. Hollaway, L.C.: A review of the present and future utilisation of FRP composites in the civil infrastructure with reference to their important in-service properties. Constr. Build. Mater. **24**, 2419–2445 (2010)
62. Ceroni, F., Pecce, M.: Evaluation of bond strength in concrete elements externally reinforced with CFRP sheets and anchoring devices. J. Compos. Constr. **14**, 521–530 (2010)

Applications of Fiber Reinforced Polymer Laminates in Strengthening of Structures

Yuvraj Singh and Harvinder Singh

Abstract This paper presents the results of the investigation carried out through numerical simulation on the use of Fiber Reinforced Polymer (FRP) laminates for the strengthening of structural elements. For a structure to be sustainable, strengthening or upgrading a structure may be required because of several reasons including material deterioration and structural distress. Keeping in view the environmental and economic factors, re-construction of the structure seeking attention is not always possible or recommendable. This is because the demolition followed by re-construction will lead to the utilization of the resources besides contributing to construction and demolition waste. The use of FRP laminates for the strengthening of structures is emerging as an alternative to corrosive and bulky conventional strengthening techniques like steel plates. As the sustainability of a structure is of key importance, it is therefore very important to determine the behaviour of such emerging strengthening techniques under loading before their actual application. In the present study, the investigation has been carried out on determining the effectiveness of different types of FRP laminates by simulating the finite element models. It has been found that the use of FRP laminates enhances the performance of the strengthened structural elements by augmenting the load-carrying capacity.

Keywords Finite element modelling · Fiber reinforced polymer (FRP) · Laminates · Numerical simulation · Strengthening · Sustainability

1 Introduction

Rightly said, "Change is the law of the universe". This is also applicable as far as infrastructure is concerned. Repair or upgrading of a structure may be required due

Y. Singh (✉) · H. Singh
Department of Civil Engineering, Guru Nanak Dev Engineering College, Ludhiana, India
e-mail: uvraj_23@yahoo.co.in

Y. Singh
Department of Civil Engineering, IK Gujral Punjab Technical University, Kapurthala, India

D. K. Ashish et al. (eds.), *3rd International Conference on Innovative Technologies for Clean and Sustainable Development*, RILEM Bookseries 29,
https://doi.org/10.1007/978-3-030-51485-3_17

to several reasons. One of the major reasons arising the need to upgrade or repair any infrastructure is the deterioration that may be caused due to time, exposure to contaminants, excessive loading or natural disasters like earthquakes. On the other hand, upgrading of a structure may also be required due to several other reasons including improper structural design, insufficient strength, change of utility of the building, updated/revised codes, etc. In such cases, structures need to be strengthened to fulfill the current serviceability requirements or to satisfy the updated/revised codal guidelines. Therefore, structures may require minor or major repairs or even retrofitting involving a complete technical procedure. One convenient option to fulfill the above-listed requirements can be opting for the complete demolition of a building followed by reconstruction. This practice brings with itself several consequences violating the concept of sustainability. This includes the use of new resources, manpower, capital, construction material whose manufacturing leads to environmental pollution and above all it gives rise to huge construction as well as demolition waste whose disposal itself is a huge challenge in order to achieve sustainability in infrastructure development.

One vital concept of sustainability that demands optimum utilization of resources without degrading environment cannot be achieved by producing more waste whose disposal becomes another challenge. So, as far as possible, for a structure to remain serviceable repairing, retrofitting or strengthening should be opted for. Keeping this vital concern in mind and in an effort to step towards sustainable development, an overview of various conventional and modern strengthening techniques have been presented in this paper. Furthermore, the emerging use of fiber reinforced polymer laminates has been discussed in detail along with the results of numerical simulation and its practical applications.

2 Conventional Strengthening Techniques

Strengthening of structures is not a new concept. In the past many decades, numerous studies have been carried on the investigation of strengthening reinforced concrete and masonry structures as well. Plesu et al. [1] presented an overview of the traditional approaches to strengthen existing masonry structures. Various retrofitting and strengthening techniques including the use of similar materials including concrete, wall buttresses, repointing, covering and injecting the cracks, stitching/sewing large cracks have been employed in case of masonry structural elements [1–8]. Furthermore, several techniques involving external surface treatments have been employed including jacketing, shotcrete, ferrocement, seismic bands in the past decades [3, 9, 10]. Besides these surface treatments, several techniques involving external reinforcement using steel plates, steel bars, post-tensioning steel tie-rods, wood elements and pipes bracing systems have also been practiced in the past [3, 7, 8, 11–13]. Jacketing of structural elements like beams, columns, and slabs using steel plates have been employed in the past many decades. Steel plates apart from being corrosive in nature are very bulky. This contributes to an enormous increase in the dead load and thus defeats the purpose of strengthening or leads to an uneconomical approach. Moreover, the use of steel plates further requires connections which are prone to corrosion or damage. Conventional strengthening techniques evidently involve the inclusion of new structural members to the existing structural system or enlarging the existing elements. In this context, the inclusion of a bracing system or a shear wall is commonly practiced due to its relative ease and effectiveness as compared to jacketing of structural elements like beams and columns which turns out to be economically on the higher side. Achieving the desired structural performance, with economical and reliable strengthening techniques is required for a sustainable approach and hence the conventional techniques involving retrofit design seek optimization.

3 Modern Strengthening Techniques

Keeping in view the limitations of conventional strengthening techniques, researchers have been investigating modern strengthening techniques over the past few decades [14]. This has resulted in several useful advancements. Amongst the modern strengthening techniques, the application of fiber reinforced polymer (FRP) composites emerges as a promising material. Applications of FRP reinforcement for repair, retrofitting or strengthening of a structure have attractive merits including ease of installation, lightweight, non-corrosive nature, less requirement of manpower, and above all greater structural efficiency over other conventional techniques. These amazing merits of FRP make strengthening procedures convenient besides least modification in the geometry or utility of the strengthened structure [15]. Moreover, there may be several situations where the only possible and efficient way of strengthening may be the use of FRP, especially at the places where there is restricted access for heavy machinery or where it is not possible to cease the function of the structure under consideration. It may also be realized that the cost of strengthening using FRP may be as competitive as using traditional techniques and materials, but the greater efficiency of FRP materials leading to lesser consumption compensates and justifies the choice of opting FRPs over the conventional materials like the use of bulky and corrosive steel plates [10]. Fiber Reinforced Polymers may be used externally by plate bonding, confining or jacketing. It may also be used as sprayed FRP composite, near-surface mounted technique or as prestressed composite systems [11].

4 Fiber Reinforced Polymer Laminates

Fiber Reinforced Polymer (FRP) laminates are the emerging and effective material for the external strengthening of a structural element. These are in the form of flexible uni- or bi-directional multilayer sheets. These laminates are bonded to the surface of concrete or steel structural elements using special FRP epoxy resin systems. It proves to be instrumental to reinforce and enhance the strength of masonry, steel, pretensioned, and reinforced concrete structures. The significant enhancement in stiffness and strength results in controlling or preventing cracks which further contributes to reduced permeability leading to increased durability and hence sustainability. These laminates are available in varied materials like glass, carbon, aramid, basalt, etc. and can be used depending upon the requirement [14, 16].

5 Numerical Simulation

In order to obtain a better insight into the effectiveness of Fiber Reinforced Polymer (FRP) laminates in the strengthening of structural elements, it is vital to determine the behaviour of the strengthened structural elements under loading before their actual application. Furthermore, in the case of the investigation involving the use of different types of fibers applied in varying configuration, the numerical investigation becomes instrumental by calibrating/validating the model with the experimental test results. Hence, the study presented below is the result of a numerical investigation carried out by simulating the strengthened structural elements in the form of finite models that were validated with the experimental test results and these models were analyzed using the finite element method package ANSYS [17, 18].

5.1 *Methodology*

The numerical investigation of strengthened RC structural elements (beams and slabs) were performed using FE modelling with the intention to study the behaviour of RC structural elements strengthened with varying configuration of Glass and Carbon FRP laminates. In the preliminary stage of the investigation, finite element models were modelled using ANSYS [17, 18]. In the pre-processing stage, the modelling of the strengthened structural element involved the creation of a volume of concrete followed by discretization. The reinforcement was modelled by connecting node to node by discrete approach, followed by modelling of the FRP laminates. The element type was suitably chosen from the element library of the finite element package [17]. Once the models were calibrated, an analysis of the same was carried out using the solver. And finally, after the successful analysis, the results obtained in the post-processor were compared for investigating the performance of the strengthened structural elements.

5.2 *Finite Element Models and Results of Numerical Analysis*

In the first phase of the investigation, RC beams were strengthened using Glass FRP (GFRP), Carbon FRP (CFRP) and combination of Glass and Carbon FRP (Hybrid, HB) laminates. The laminates were applied in varying configurations. In order to enhance the flexure strength of the beam, laminates were applied at the bottom soffit of the beam. While the laminates were applied on the side faces of the beam in order to enhance the shear strength of the beam. For enhancing both flexural and shear strength as well, the laminates were applied in the form of U-wrap configuration. The finite element models of RC beams are illustrated in Fig. 1. From the finite element analysis, it was observed that in case of strengthening of a beam with Glass FRP laminates, the load-carrying capacity increased to 107.67%, 114.33%, and 147.67% in case of the bottom, side face and U-wrap configuration respectively with respect to that of un-strengthened RC beam. Whereas, in the case of strengthening a beam with Carbon FRP laminates, the load-carrying capacity increased to 116.67%, 174.33%, and 200% in case of the bottom, side face and U-wrap configuration respectively with respect to that of un-strengthened RC beam. Besides increasing the load-carrying capacity of the RC beam, FRP laminates also tend to decrease the deflection when compared to that of the un-strengthened RC beam. In case of strengthening a beam

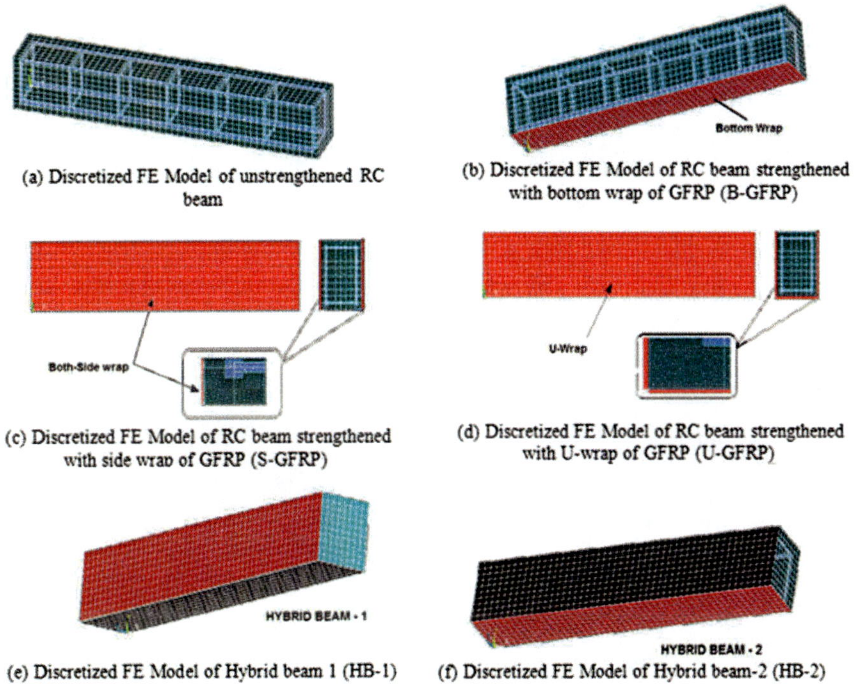

Fig. 1 Finite element models of reinforced concrete (RC) beams strengthened using glass FRP (GFRP), carbon FRP (CFRP) and combination of glass and carbon FRP (hybrid, HB) laminates

using a combination of both Glass and Carbon laminates in two different U-wrap configurations referred as hybrid beams HB-1 and HB-2 as illustrated in Fig. 1e, f, the load-carrying capacity increased to 184.33% and 193.33% respectively with respect to that of un-strengthened RC beam [13, 17].

In continuation, the investigation was extended to RC slabs with circular and square opening as illustrated in Fig. 2. In the case of RC slab with a square opening, the load-carrying capacity was found to be increased to 105.88% and 141.18% on strengthening with Glass and Carbon FRP laminates respectively with respect to that of the un-strengthened slab. Whereas, in the case of RC slab with a circular opening, the load-carrying capacity was found to be increased to 121.43% and 150%

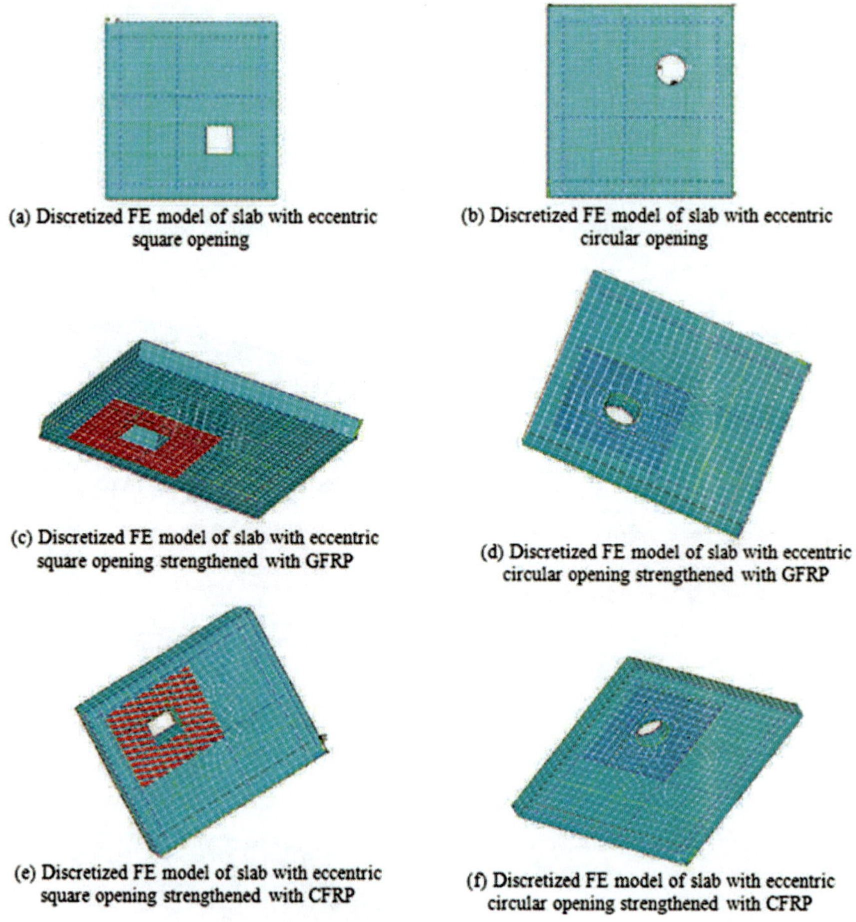

Fig. 2 Finite element models of reinforced concrete (RC) slabs with square and circular opening strengthened using glass FRP (GFRP) and carbon FRP (CFRP) laminates

on strengthening with Glass and Carbon FRP laminates respectively with respect to that of the un-strengthened slab [18].

The load-carrying capacity, delay in crack initiation and propagation can further be enhanced by increasing the layers of FRP laminates, if desirable, thus increasing both the strength and durability as well. The results of numerical simulation justify the effectiveness of FRP laminates in the strengthening of RC structural elements by confinement and the choice of laminate and configuration thus depends upon the extent to which the enhancement in the strength is required.

6 Potential Applications

Being effective in strengthening besides favouring characteristics including ease of handling and installation, non-corrosive nature, lightweight, and availability in any length, fiber reinforced polymer laminates proves to be a potential material for retrofitting or strengthening purpose. It finds its applications in strengthening structural elements including beams, columns, slabs, beam-column joints, walls, bridge deck slab, etc. FRP laminates may be employed locally for strengthening a part of the entire structure or any of its structural elements, or it may even be used for strengthening the entire structure. It also finds application in strengthening or retrofitting of tanks, pipes, chimneys, silos, poles, and tunnels [19]. It helps in providing seismic strengthening, confinement, flexural strengthening, increased ductility, shear strengthening, increased axial load carrying capacity in the case of columns. Besides this, it finds its application in strengthening the walls, in-plane flexure/shear retrofit, out-of-plane flexural retrofit, concrete shear walls, and unreinforced masonry walls [19].

With numerous merits and applications, FRP laminates have limitations too. FRP laminated are prone to failure due to de-lamination, exposure to contaminants, fire, etc. Also, due to wrapping around the structural element and excessive increase of the load-carrying capacity, RC structural elements strengthened with laminates might undergo brittle failure due to a reduction in the deflection corresponding to the loads when compared to un-strengthened beams. But, such limitations can be taken care of by careful considerations like proper design, wet bonding to prevent de-lamination, or use of strong epoxy adhesives, application of fire-resistant coatings, etc. In addition to this, FRP laminates find limited application in strengthening or retrofitting of monuments or structures of historic or architectural significance. Keeping in mind the high cost involved in the re-construction or replacement of any structural or bridge component, or the difficulty that may arise due to its non-functioning, it becomes crucial to consider modern strengthening techniques like the use of FRP laminates that will prove to be economical, effective and hence sustainable.

7 Concluding Remarks

Consideration of the possibility and an effort towards rehabilitation, repair, retrofitting or strengthening over demolition and reconstruction is crucial for achieving sustainability in the construction industry. Keeping in view the limitation or repelling characteristics of conventional strengthening techniques like corrosive nature, heavyweight, difficult handling, and application, it becomes vital to research and switch to better and alternate techniques of strengthening. From the present study, the following conclusions have been made

- Fiber Reinforced Polymer (FRP) laminates augment the load-carrying capacity of the strengthened structural elements and are effective in delaying the initiation and propagation of cracks.
- In the case of reinforced concrete beams, flexural strength and shear strength can be enhanced by the application of laminates at the bottom soffit and side faces, respectively. Furthermore, a U-wrap configuration is most effective in enhancing the overall performance of the beam.
- In the case of reinforced concrete slabs with openings, the load-carrying capacity is augmented on the application of FRP laminates. Carbon fiber is found to be more effective as compared to glass fiber laminates and the effect is more pronounced in case of slabs with circular openings.

Therefore, Fiber Reinforced Polymer (FRP) composites emerge as a potential modern strengthening material over conventional materials due to its effectiveness in enhancing the strength and other numerous merits like non-corrosive nature, lightweight, ease of handling and application. The results of the numerical investigation have substantiated the experimental test results reported in the literature. It is recommended that the type of FRP laminate, its configuration and the number of layers should be decided depending upon the structural element and the enhancement in the strength required. Hence, for the external applications, the use of FRP laminates proves to be an efficient and promising technique provided proper measures are taken to prevent its de-lamination or damage due to fire or other environmental exposures.

References

1. Plesu, I., Taranuu, R., Covatariu, G., Grădinariu, D.: Strengthening and rehabilitation conventional methods for masonry structures. Bull. Polytech. Inst. Jassy Constr. Archit. Sect. **57**(4), 165–176 (2011)
2. Bothara, J.K., Brzev, S.: A Tutorial: Improving the Seismic Performance of Stone Masonry Buildings, pp. 53–71. Earthquake Engineering Research Institute, Oakland, California (2011)
3. El Gawady, M., Lestuzzi, P., Badoux, M.: A review of conventional seismic retrofitting techniques for URM. In: 13th International Brick/Block Masonry Conference, vol. 9, pp. 1–9, Amsterdam, 4–7 July 2004

4. Heiza, K., Nabil, A.: State-of-the-art review: strengthening of reinforced concrete structures—different strengthening techniques. In: Sixth International Conference on Nano-Technology in Construction, NTC (2014)
5. Islam, R.: Inventory of FRP Strengthening Methods in Masonry Structures, pp. 10–33. Polytechnic University of Catalonia, Barcelona (2008)
6. Jeffs, P.A.: Core consolidation of heritage structure masonry walls & foundation using grouting techniques—Canadian case study. In: 9th Canadian Masonry Symposium, vol. 12, pp. 1–12, Canada (2000)
7. Mack, R.C., Speweik, J.P.: Repointing Mortar Joints in Historical Masonry Buildings, pp. 1–6. FAIA, Technical Preservation Service, National Park Service, U.S. Department of the Interior (1998)
8. Trujillo, L.F.R.: Seismic response and rehabilitation of historic masonry buildings. M.Sc. Diss., University of Sheffield, Department of Civil and Structural Engineering, vol. 30, pp. 17–47 (2007)
9. Bothara, J.K., Guragain, R., Dixit, A.: Protection of Educational Buildings Against Earthquakes: A Manual for Designers and Builders, vol. 12, pp. 50–62. National Society for Earthquake Technology, Nepal (NSET) (2002)
10. Singh, Y., Paul, D.K.: Retrofitting of masonry buildings. In: Lecture Notes for National Programme for Capacity Building for Engineers in Earthquake Risk Management, pp. 219–232, India (2006)
11. Mahmoudi, M., Ebadi, F.: Seismic rehabilitation of unreinforced masonry buildings using pipe bracing system. In: Proceedings of the 14th European Conference on Earthquake Engineering 2010: Ohrid, Republic of Macedonia, vol. 5, pp. 1–5 (2010)
12. Taghidi, M.: Seismic retrofit of low-rise masonry and concrete walls by steel strips. Ph.D. Diss., Department of Civil Engineering, University of Ottawa, Canada (2000)
13. Vinzileou, E., Skoura, A.: Seismic behaviour of timber reinforced masonry buildings. Prot. Hist. Build. **5**, 403–408 (2009)
14. Motavalli, M., Czaderski, C., Empa: FRP composites for retrofitting of existing civil structures in Europe: state-of-the-art review. In: Swiss Federal Laboratories for Materials Testing and Research. Composites & Polycon 2007, 17–19 Oct 2007. American Composites Manufacturers Association, Tampa, FL USA
15. Pannirselvam, N., Nagaradjane, V., Chandramouli, K.: Strength behavior of fiber reinforced polymer strengthened beam. ARPN J. Eng. Appl. Sci. **4**(9) (2009)
16. FRP Laminates: https://afzir.com/en/frp-laminate/
17. Singh, Y.: Numerical investigation of RC beams strengthened with FRP laminates. M. Tech. thesis, Guru Nanak Dev Engineering College, Ludhiana (2015)
18. Tanu, Kaur, I., Singh, Y.: Finite element analysis of RC slab with or without opening strengthened with FRP wraps. Int. J. Comput. Appl. **0975**, 8887, pp. 9–12 (2016)
19. SikaWrap: Composite Fabrics for Structural and Seismic Strengthening. https://usa.sika.com/content/dam/dms/us01/a/bro-cpd-SikaWrap-us.pdf

Behaviour of RC Beam-Column Joint Subjected to Opening Moments: Test and Numerical Validation

Ahmad Fayeq Ghowsi, M. Adil Dar, and A. R. Dar

Abstract This paper describes the experimental investigation carried out on a reinforced concrete (RC) beam-column joint subjected to opening moments, using monotonic loading. A half scale beam-column model was designed as per the non-ductile Indian Standard for reinforced concrete members. All the relevant material tests were performed and reported accordingly, mainly for facilitating the numerical study. The behaviour of the joint was primarily studied in terms of the load–displacement response, energy dissipated, ductility ratio and crack pattern. The test strength was also compared with the strength predicted by the Indian Standard, and a good agreement between the two was observed. Lastly, a numerical model was developed for simulating the behaviour of RC beam-column joint using ANSYS. There was a good match between the test result and the numerical one.

Keywords Beam-column joint · Experiment · Strength · Numerical study · RC member

1 Introduction

Unlike structural steel members, the construction quality of reinforced concrete (RC) members depends upon numerous factors that govern the chances of minimizing heterogeneity within the RC members, which substantially influences their performance under different types of loading. However, due to various other favourble qualities, concrete is consumed in large quantities, mainly in the form of RC structural members, and accounts for the second largest material used globally. In framed structures, the beam-column joints are critical members, as they transfer the loads from the beams on to the columns, and are subjected to interaction of different types

A. F. Ghowsi · M. A. Dar
Indian Institute of Technology Delhi, New Delhi 110016, India

A. R. Dar (✉)
National Institute of Technology Srinagar, Srinagar, Jammu & Kashmir 190006, India
e-mail: abdulrashid@nitsri.net

D. K. Ashish et al. (eds.), *3rd International Conference on Innovative Technologies for Clean and Sustainable Development*, RILEM Bookseries 29,
https://doi.org/10.1007/978-3-030-51485-3_18

of forces [1]. Also, from constructional considerations, the detailing of the beam-column joints is complex, especially when the load demands are large, leading to congestion of reinforcement, which affects their heterogeneity [2]. The beam-column joints either open or close under the influence of moment developed due to different loading conditions. The past research has revealed that the opening of the beam-column joints makes them more vulnerable to severe damage, as the diagonal shear cracks develop and widen, which lead to catastrophic failures in the past [3]. The beam-column joints should be designed adequately with the primary focus being to dissipate large seismic energy without substantial degradation in its strength and ductility [4]. Both inadequate design as well as detailing of the beam-column joint puts the entire frame under jeopardy, despite the other structural members being adequate, by conforming to the relevant codal provisions [5, 6]. Different codal provisions recommend different beam-column detailing to ensure that such structural elements experience less damage with the attainment of the desired performance level. However, in India, IS 456 [7] is the widely adopted code that provides design guidelines for reinforced concrete members. But the detailing configurations provided in the same adhere to non-ductile behaviour. Considering that India's 60% area is vulnerable to devastating earthquakes, which has been proven through the losses experienced in the past earthquakes that struck the country numerous times. It is essential to understand the behaviour of a non-ductile RC beam-column joint under opening moments. Although, there are strengthening techniques that have been developed to improve the performance of the capacity deficit structural members [8–10], but that comes at a cost and may not always be a feasible option.

This paper presents an experimental investigation carried out on an RC beam-column joint subjected to opening moments, using monotonic loading. A half scale beam-column model was designed as per the non-ductile Indian Standard for reinforced concrete members. All the relevant material tests were performed and reported accordingly, mainly for facilitating the numerical study. The behaviour of the joint was primarily studied in terms of the load–displacement response, energy dissipated, ductility ratio and crack pattern. The test strength was also compared with the strength predicted by the Indian Standard, and a good agreement between the two was observed. Lastly, a numerical model was developed for simulating the behaviour of RC beam-column joint using ANSYS.

2 Details for the Test Specimen

Full scale tests are always preferred as they bring out more realistic outputs and help in better understanding of the structural engineering problem in to consideration. However, it may not always be possible to adopt full-scale testing, as the available laboratory facilities play a major role in that. Due to that above stated reason, a half scale model was prepared for investigating the behaviour of the non-ductile RC beam-column joint under opening moments. The breadth of the beam was adopted as 200 mm while as the depth was fixed at 250 mm. A square column of size 200 mm was considered. The reinforcement was determined using the basic code of practice for RC members (IS 456: 2000) [7]. The bean-column joint was designed for an arbitrary load of 17 kN. The philosophy of strong column and weak beam was fulfilled. The main reinforcement of the beam comprised of four bars of size 16 mm (diameter) on the tension face and two bars of the similar size on the compression face. Two-legged stirrups made up of 8 mm bars with a 100 mm regular spacing between the centres of the stirrups was considered for shear. A 480 mm development length was considered for the beam, and 600 mm for the column. The reinforcement details of the specimen are shown in Figs. 1 and 2. The concrete used for casting the test specimen belonged to M20 grade, which was prepared in using IS 10262: 2019 [11] under ambient temperature conditions. For binding the various ingredients of the concrete, Ordinary Portland cement (OPC) following grade 53 was adopted.

Aggregates available locally, which were restricted to 12–14 mm in size were considered as coarse aggregates, while as river sand available locally, and passing through 4.75 mm sieve (as per IS 383:2016) [12] was considered as fine aggregate. The ratio of cement to water was adopted as 0.53. Both for the preparation of concrete as well as for curing of the specimen, potable water was used. To ensure a smooth finished surface of the specimen, a CFS sheet of thickness 2 mm was used. This would help in conducting the experimentation smoothly. Furthermore, a 10 mm thick Polytetrafluoroethylene (PTFE Pad) was adopted for providing a frictionless surface to ensure free sliding of the test specimen during the loading process, as shown in Fig. 3.

3 Material Testing

Various mechanical tests were performed to determine the actual properties of the various materials used in the specimen preparation. This would help in better numerical validation, which is one of the research objectives of this study. Tensile coupon tests were performed on reinforced steel bars, conforming to the Indian Standard (IS 1608: 2005) [13], for determining their mechanical properties. The yield strength, ultimate strength percentage elongation and modulus of elasticity of 10 mm reinforcement bars was found to be 503.2 MPa, 567.4 MPa, 21.3% and 196.4 GPa respectively. The same for 8 mm bar was found to be 514.8 MPa, 576.3 MPa, 25.4% and 197.2 GPa, in the same order. Cubes were prepared in accordance with Indian Standards (IS 516: 1959) [14] for the determination of the compressive strength of concrete adopted. For the cement. the percentage standard consistency, fineness, initial setting time, final setting time, specific gravity, compressive strength after 7 days and compressive strength after 28 days were noted as 31.5%, 0.75%, 28 min, 421 min, 24 MPa and 35 MPa respectively. For the fine aggregates, the specific gravity, bulk density, fineness modulus, water absorption and grading were noted as 2.63, 1.31 kg/l, 2.32, 2.41% and grade-II respectively. For the coarse aggregates, the maximum size, specific gravity, water absorption and fineness modulus were note as 14 mm, 2.61, 1.87% and 6.55 respectively. For performing the tensile tests on

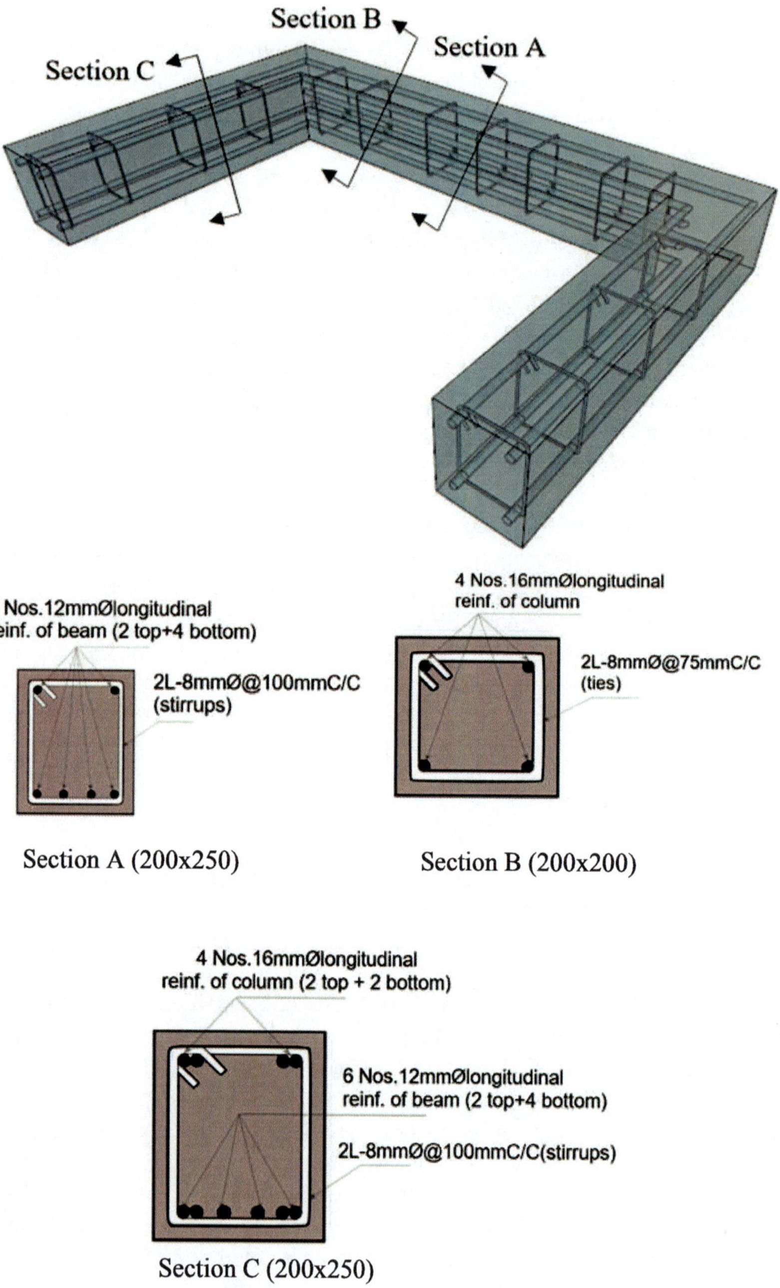

Fig. 1 Reinforcement details of the specimen

Fig. 2 Reinforcement placement in the formwork cage

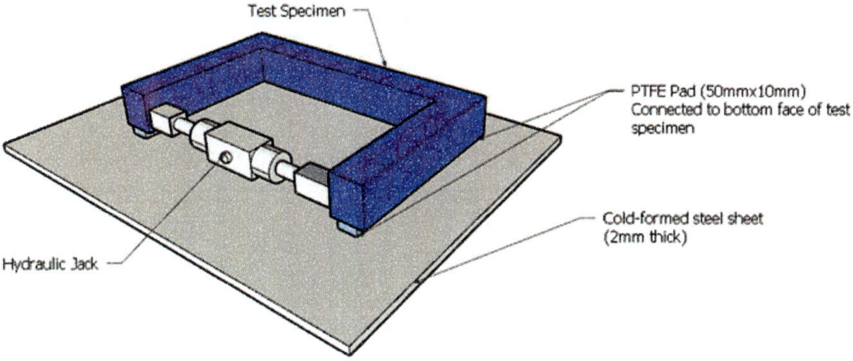

Fig. 3 Position of PTFE pads

the tensile coupons and the compression tests on the concrete cubes, a computerized universal testing machine was employed.

4 Test Set-Up and Loading

As discussed earlier, that the testing facilities play a vital role in quality of experimentation, both in terms of choice of full-scale or reduced scale model specimens as well as the type of tests performed. Accordingly, the specimen was prepared which looked like a portal frame. The specimen was placed on the flat surface horizontally, on the strong floor of the Structural Engineering Laboratory at NIT Srinagar, Jammu & Kashmir. Frictionless supports were used below the specimen at the two far ends

Fig. 4 Test set-up

of the columns and the two joints, to allow better assessment of the specimen while loading. The loading was adopted in the force controlled mode by using a hydraulic jack of 1000 kN capacity, adjusted between the column end of the specimen. This would ensure that the joints would be subjected to opening moments under the loading application, as desired. Dial gauges were used to record the deflections at the critical locations (at the joint near the beam and the column), and the far end of the columns. Since monitoring the cracks in RC members is important from results interpretation point of view, a crack detection microscope (with 0.01 mm least count) was used. As the testing progressed, the loads and their respective displacements were noted at regular loading increments. The monitoring of crack initiation and propagation was also observed. A view of the test set-up is shown in Fig. 4.

5 Test Results and Discussion

After placing the specimen horizontally on the strong floor, and applying the lateral load at the free end of the column in the increments of 2 kN using the hydraulic jack, the deflection at specific locations was measured. As the loading progressed, the initial crack started to originate and were observed at the inner face of the beam-column junction. It was observed so for both the joints simultaneously, at the monotonic loading of 3.32 kN. Further, this crack propagated in vertical direction and as the loading progressed further, numerous other cracks started to form in and around the joint. This ultimately lead to the failure of the joint, at the peak load of 14.94 kN. The overall failure of the joint was observed as a combination of flexural and shear failure. The maximum average deflection at the free end of the columns was noted as 9.77 mm at the time of failure, with the crack width being approximately 1 mm. The yielding of the specimens started at a load of 6.64 kN, beyond which the stiffness of

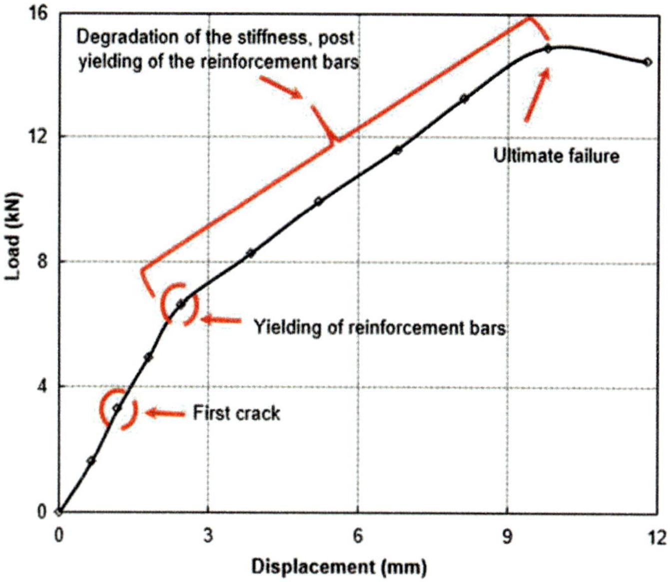

Fig. 5 Load versus displacement plot of the test specimen

the stiffness of the joint dropped. The initial stiffness of the specimen was observed as 2.67 kN/mm. The energy absorbed by the specimen and its displacement ductility were found to be 78 kN mm and 1.45 respectively. The load versus displacement plot and the crack pattern during the loading of the specimen are shown in Figs. 5 and 6 respectively.

6 Numerical Validation

The numerical validation of the test results was carried out using the finite element based commercial package ANSYS. The beam column joint was modelled with Solid 65 and Beam 188 elements. SOLID65 was used for simulating the 3-D modeling of concrete. The element is defined by eight nodes having three degrees of freedom at each node: translations in the nodal x, y, and z directions. BEAM188 element is used to model steel reinforcement. BEAM188 is a linear (2-node) beam element in 3-D with six degrees of freedom at each node. The degrees of freedom at each node include translations in x, y and z directions, and rotations about x, y and z directions. Warping of the cross-section is assumed to be unrestrained. This element is capable of plastic deformations and can take either tension or compression or both. The model was analogous to the tested specimen. Initially key points were created to obtain the desired area and then it was extruded in z-direction using pre-processor modeling commands. Meshing was done using volume sweep command in mesh

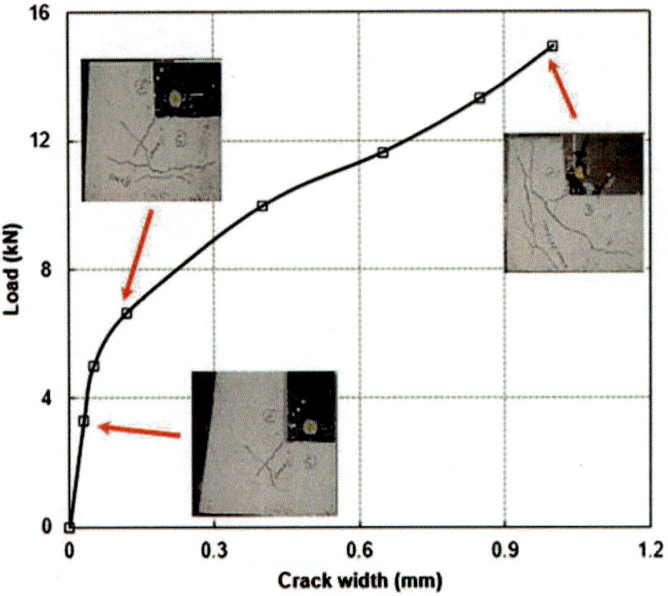

Fig. 6 Crack pattern observed during loading

tool option. A total of 1260 nodes were created to join the assemblage of finite elements. The reinforcement was not meshed as individual elements were created in the modeling through the nodes created by the mesh of the concrete volume. Merging was done carefully in order to avoid the possibility of the occurrence of orphaned nodes which may result into surface load transfer or change in boundary condition thus resulting into early failure of model. Isotropic bilinear model was used to model the reinforcement bars. Figures 7 and 8 show the simulation of the reinforcement bars and the fine meshing of the specimen for the numerical analysis.

Figure 9 shows the deformed shape of the beam column joint. The crack pattern of the specimen is shown in Fig. 10. There was a reasonable match between the test and numerical results, as shown in the comparison of the load versus displacement plot of the specimen in Fig. 11.

7 Conclusions

This paper presented an experimental investigation carried out on an RC beam-column joint subjected to opening moments, using monotonic loading. A half scale beam-column model was designed as per the non-ductile Indian Standard for reinforced concrete members. All the relevant material tests were performed and reported accordingly, mainly for facilitating the numerical study. The behaviour of the joint

Fig. 7 Simulation of the beam-column reinforcement

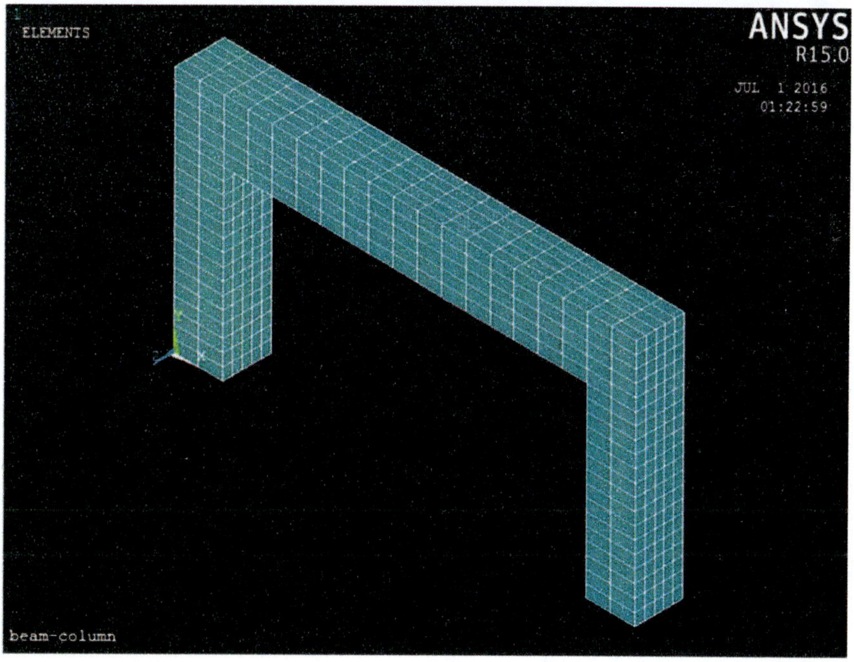

Fig. 8 Fine meshing of the beam-column specimen

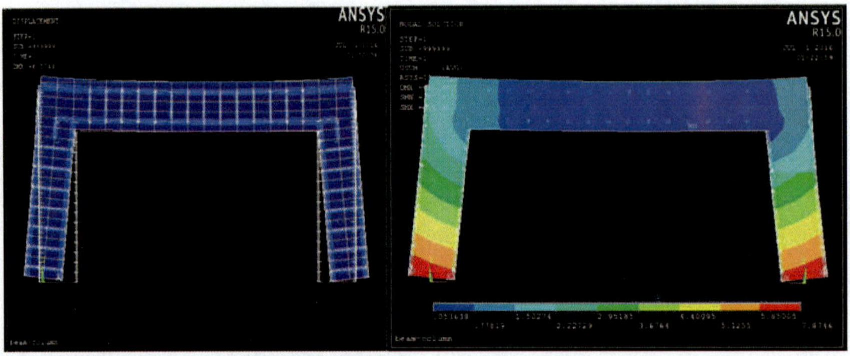

Fig. 9 Deformed shaped of the beam-column specimen

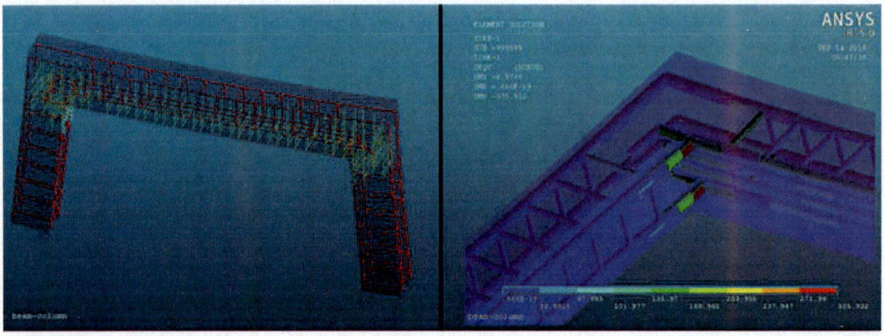

Fig. 10 Crack pattern observed during loading

was primarily studied in terms of the load–displacement response, energy dissipated, ductility ratio and crack pattern. The test strength was also compared with the strength predicted by the Indian Standard, and a good agreement between the two was observed. Lastly, a numerical model was developed for simulating the behaviour of RC beam-column joint using ANSYS. Following are some important conclusions drawn out of this research work:

- Based on the test result obtained from this study, the Indian Standard IS 456, i.e., the conventional standard for the design of RC members under predicts the strength of the non-ductile beam-column within the range of 10–15%. This standard needs to be revised for reliable design strength predictions. However, extensive studies are needed to substantiate this conclusion.
- The first crack was observed at around 20% of the peak load and the first yielding of the reinforcement was identified at around 45% of the peak load.
- The initial stiffness of the beam column joint was observed as 2.67 kN/mm. Right after the first crack initiation took place, the stiffness of the joint dropped to 1.13 kN/mm. This lead to the stiffness degradation of around 60%.

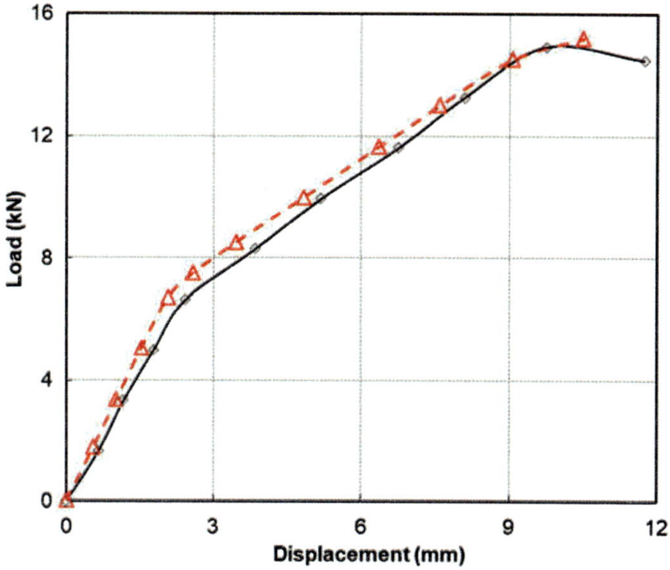

Fig. 11 Comparison of the load versus displacement plots

- The energy absorbed by the beam-column joint was noted as 78 kN mm, while as it displacement ductility as 1.45. These parameters are important from seismic energy dissipation point of view, and must be carefully looked into.
- There was a good agreement between the test and numerical results if the beam-column joint.

This study discussed the structural behaviour of RC beam joint under opening moments. To strengthening the outcomes of this study, further research is needed particularly on the cyclic behaviour of these specimens. Also the effect of variation in the cross-sectional dimensions needs to be studied as well.

The authors are planning to evaluate the efficiency of different strengthening techniques like FRP wrapping around the joint region, adopting a steel haunch, etc., on the performance of these joints. The authors are further planning to improvise the numerical model, by adopting advanced modelling techniques for better behaviour prediction of these joints.

References

1. Subramanian, N.: Design of RC Beam Column Joints. The Masterbuilder, pp. 136–148 (2015)
2. Subramanian, N.: Design of Reinforced Concrete Structures, 1st edn. Oxford University Press, New Delhi (2013)
3. Dar, M.A., Subramanian, N., Pande, S., Dar, A.R., Raju, J.: Performance evaluation of different strengthening measures for exterior RC beam-column joints under opening moments. Struct.

Eng. Mech. **74**(2), 243–254 (2020)
4. Elmasry, M.I.S., Abdelkader, A.M., Elkordy, E.A.: An analytical study of improving beam-column joints behavior under earthquakes. In: Rodrigues, H., Elnashai, A., Calvi, G. (eds.) Facing the Challenges in Structural Engineering. GeoMEast 2017. Sustainable Civil Infrastructures. Springer, Cham (2018)
5. Subramanian, N., Prakash Rao, D.S.: Seismic design of joints in RC structures—a review. Indian Concr. J. **77**(2), 883–892 (2003)
6. Uma, S.R., Meher, P.A.: Seismic behaviour of beam column joints in reinforced concrete frame structures. Indian Concr. J. **80**(1), 33–42 (2006)
7. IS 456: 2000 Plain and Reinforced Concrete—Code of Practice, Fourth Revision. Bureau of Indian Standards, New Delhi
8. Dar, M.A., Subramanian, N., Dar, A.R., Raju, J.: Experimental investigations on the structural behaviour of a distressed bridge. Struct. Eng. Mech. **56**(4), 695–705 (2015)
9. Dar, M.A., Subramanian, N., Dar, A.R., Raju, J.: Rehabilitation of a distressed steel roof truss—a study. Struct. Eng. Mech. **62**(5), 567–576 (2017)
10. Dar, M.A., Subramanian, N., Dar, A.R., Rather, A.I., Mir, A., Syed, S.: Strengthening of capacity deficient RC beams—an experimental approach. Struct. Eng. Mech. **70**(3), 303–310 (2019)
11. IS 10262: 2019 Guidelines for Concrete Mix Design Proportioning, Second Revision. Bureau of Indian Standards, New Delhi
12. IS 383: 2016 Coarse and Fine Aggregate for Concrete—Specification, Third Revision. Bureau of Indian Standards, New Delhi
13. IS 1608: 2005 Mechanical Testing of Metals—Tensile Testing, Third Revision. Bureau of Indian Standards, New Delhi
14. IS 516: 1959 Method of Tests for Strength of Concrete. Bureau of Indian Standards, New Delhi

Reduction of Annual Energy Consumption of Multifamily Dwellings Using BIM and Simulation Tools

Subbarao Yarramsetty, M. V. N. Sivakumar, and P. Anand Raj

Abstract Designing and constructing a Sustainable building is an emerging area of interest with AECOO (Architectural, Engineering, and Construction, Owner and Operator) trades, which is evidenced by the increased use of the green building rating systems throughout the world. Building information modeling (BIM) is one of the effective ways of deciding the suitable building orientation and envelope that controls project cost, time and energy. In this study, multifamily residential building is taken as a case study, which is located in Afghanistan. Energy consumption analysis was conducted by using building energy performance tools. The building is modeled in Autodesk-Revit and different orientations, building envelopes and Wall to window ratio analysis were carried out to find the minimal energy consumption scenario. A total of 126 simulations were conducted. Ultimately, the most energy-efficient option in the context of Afghan dwellings was figured out. Locally available building materials were used in the study. The best energy efficient orientation of the building is evaluated by rotating the building in 15° rotation each time. Furthermore, varying the glazing area from 10 to 60% for energy-efficient WWR ratio. 10% glazing consumes minimal energy consumption for the C48 (combination 48) with wall type (W3)—Adobe brick wall, roof type (R4)—Mud Roof with timber as core structure and Floor type (F4)—Mud floor with timber as structure, was the best option which consumes the least amount of energy leading to reduction in annual electricity demand.

Keywords Building information modelling (BIM) · Building energy performance (BEP) tools · Afghanistan residential buildings · Energy conservation

1 Introduction

With the available limited natural resources, reduction of energy consumption is only the best way that leads to sustainability. Minimizing the energy consumption and

S. Yarramsetty (✉) · M. V. N. Sivakumar · P. Anand Raj
Department of Civil Engineering, National Institute of Technology, Warangal, Telangana State, India
e-mail: sraoystp@student.nitw.ac.in

D. K. Ashish et al. (eds.), *3rd International Conference on Innovative Technologies for Clean and Sustainable Development*, RILEM Bookseries 29,
https://doi.org/10.1007/978-3-030-51485-3_19

maximizing the energy efficiency of the building is the need of the hour to support the present growth rate of population and electricity and fuel consumption. The energy consumption in Architecture, Engineering, and Construction (AEC) sector plays a major role in environmental impacts, the life-cycle cost of projects, carbon reduction, global warming and sustainability issues. Thus, energy conservation and clean energy technologic solutions are very much necessary for the future. The building sector is responsible for 40% consumption of the world's energy and 30% carbon emissions [1]. The application of building insulation plays a major role in the reduction of energy consumption. The majority of the afghan buildings don't use the insulations resulted in higher energy consumption of about 71% of the total domestic electricity [2]. In the initial design phase, select the design alternate which consumes the lowest energy from the energy simulation. The outdated energy calculations are lack in their accuracy due to their hidden drawbacks and limitations. Among various BEP tools selecting an appropriate software tool is very much essential, BIM is used as a foremost priority program for this purpose.

The energy consumption of buildings depends on various features of the building, like the peripheral envelopes and the associated components—Plaster, doors, windows etc., some of these features can be altered to attain higher energy efficiency [3–6]. Literature shows that enhanced insulation resulted in improved energy savings, the behavior of the building user also plays a role in the energy estimates for the cooling-heating system [7–11]. Furthermore, the size, shape, and orientation of the building and density of the occupants, influences the energy demand. The building orientation plays a major role in the proper gain of solar heat and lightening for the minimal amount of energy demand for cooling and lighting purposes [12–15].

Climate variation in Afghanistan is complicated. It depends on different zones and different seasons. Some regions of the country are too hot and some areas are very cold. Generally, Afghanistan is known as a dry continental climate country with cold and relatively rainy weather in winter and hot and sunny weather in summer. Even though numerous researches have been conducted on the building energy behavior on energy consumption based on climate variation worldwide, no enough literature has been published on this topic in the Afghan context. Energy consumption will increase in the future all over the world because of the increase in population, enhancement of social and economic conditions, and high demand for services that need energy [16].

According to the International energy agency (IEA), global total primary energy consumptions and CO_2 emissions grew by 49% and 43% with an average annual raise of 2% and 1.8% from 1984 to 2004, while in 2012 energy use and CO_2 discharges increased by 85% and 75% respectively. Total world energy use is anticipated to increase by almost 32% and CO_2 emission is projected to grow by approximately 16% until 2035 [17, 18]. Table 1 represents global energy demand projection of different organizations including British multinational oil and gas Company (BP), European Commission (EC), Energy Information Administration (EIA), International Atomic Energy Agency (IAEA), International Energy Agency, International

Table 1 World energy anticipation (from 2020 to 2100) in EJ [19]

Institutes	Year	2020	2030	2050	2100
BP	2011	565–635	600–760	NA	NA
EC	2006	570–610	650–705	820–935	NA
EIA	2010	600–645	675–780	NA	NA
IAEA	2009	585–650	670–815	NA	NA
IEA	2010	NA	605–705	NA	NA
IIASA	2007	555–630	NA	800–1175	985–1740
Shell International	2008	630–650	690–735	770–880	NA
WEC	2008	615–675	700–845	845–1150	NA
Tellus Institute	2010	504–644	489–793	425–1003	243–1200

Institute for Applied Systems Analysis (IIASA), Shell International, World Energy Council (WEC) and Tellus Institute for the future. According to IIASA's perception, global energy. consumption is anticipated to increase up to 1740 EJ (exajoules) in 2100 [19].

1.1 Building Energy Performance Tool

An accurate building energy assessment can be done by considering the type of equipment, geographic location, number of occupants, type of construction material, size and shape of buildings. For a better BEP assessment, we need to provide more realistic details as possible [20]. It is evidenced that the construction sector is the most complicated and risky industry [21]. Today BIM is used in the USA, Europe, and some other countries for energy simulation and validation purposes [22]. Compared to BIM, other energy simulation tools are very expensive, time-consuming, difficult to run and they are not able to evaluate all characteristics of energy in building simulation trends [23]. BIM allows designers to select the most energy-effective construction materials for the buildings by investigating various alternatives and systems early in the conceptual phase of the project. By utilizing BIM, users don't need to reinter geometry and other data required for energy simulation which leads to the comprehensive saving of time [24].

Building energy performance (BEP) is a vast any of computer-based building analysis software which are used for different purposes such as assessment of loads, complete energy analysis of buildings, inside air quality and its flow analysis, lighting system design, code compliance, retrofit, analysis of renewable energy and economic assessment [25]. BEP tools are commonly divided into two parts, design tools, and simulation tools. Design tools are used to study the worst-case status for determining the optimum size of HVAC systems. Energy analysis tools take dynamic calculations into consideration for whole year studies [26]. Despite numerous building analysis software tools, the real implementation of the tools is mostly confined to the computation of energy load for complete designing of heat, ventilation as well as air-conditions (HVAC) systems [27]. Energy Plus, eQuest, and Ecotect are examples of Building Performance Simulation (BPS) software tools. Engineers, scholars, and architects broadly use BEP tools for a detailed evaluation of energy consumption such as heating and cooling loads as well as other required parameters of both new and existing buildings by preparing energy models to increase the thermal efficiency of the buildings [28].

1.2 Buildings in Afghanistan

The selection of construction materials and techniques depends on the local situations and conditions such as economy affordability, climate changes, material availability, culture, and social living schema, etc. People in Afghanistan build different types of houses. The main types of construction material used in buildings are Sun-dried (Adobe) brick or mud walls with wooden roofs and floors. Fired brick walls (350 mm) with barrel-arch (Taqe-Zarbi) roofs and floors, fired or adobe brick with dome roofs. In contemporary buildings, they generally use fired brick (220 mm) and concrete masonry unit blocks (200 mm) walls with Reinforced Cement Concrete (RCC) roofs and floors. The majority of existing houses are constructed with adobe brick walls and wooden slabs as well as mud walls and dome roofs all over the country. The main local construction material is Mud which contributed to 95% of construction material in Afghanistan. Contemporary materials are used only in urban areas [29]. Family size in Afghanistan is relatively big that varies between 8 and 14 persons. Traditionally married and unmarried brothers used to live with parents together in the same building. For this reason, people prefer to build multifamily houses. The area of the buildings starts from 100 m^2 onwards. In urban areas, people extend their houses vertically while in rural areas extension of buildings is horizontal.

Today energy is one of the most important factors which helps countries to enhance their economy and bring numerous facilities and services in societies. Afghanistan is a rich country from an energy resource viewpoint. According to the Ministry of Energy and Water 2016, Afghanistan has the capacity of generating 2300 MW of electricity from hydropower plants, 158,500 MW from wind power, 4000 MW from biomass, and 3500 MW from sun radiation as a clean and environmentally friendly energy source. But due to political and security issues, no significant investment has been done in this sector yet. Thus, the major needs of the country are supplied from abroad. On the other side, residential houses in Afghanistan contributed to the consumption of 71% of overall electricity use in the country, industrial building 11%, public—11% and commercial buildings—7% [2]. Nowadays most of the householders consume wood, coal, charcoal, gas, diesel, kerosene, and even animal dung, which leads to various environmental problems. Houses in Afghanistan, need the highest amount of energy in the winter season compared to the summer [30–32].

2 Methodology

The main purpose of this research was to minimize annual energy consumptions of multifamily residential without compromising comfort, health and productivity levels to reduce lifetime cost and life cycle adverse environmental impacts of the buildings by utilizing BIM-based modeling and Energy simulation software tools. Present research Objectives are presented below

- Gathering information regarding available materials used in buildings in Afghanistan.
- Creating 2D Plans of the benchmark building based on-site measurement data in AutoCAD.
- Creating a 3D BIM model by importing 2D plans of the model to Autodesk Revit Architecture.
- Utilizing Revit energy plug-in and GBS for energy simulation.
- Evaluating energy analysis results of the different combinations of the building's components to achieve the best combination from an energy use point of view.
- Investigate energy performance by varying the orientation of the building.
- Check Impacts of increasing glazing area on energy consumption of the building.

3 Case Study Building

The case study building is an existing three-story residential typical dwelling, located in Kabul city with latitude and longitude as 34°28′42.3″ N 69°07′41.2″ E shown in Fig. 1. For research purposes, we are assuming that the building is still in the design phase. The building plan area is 125 m^2 (Fig. 2).

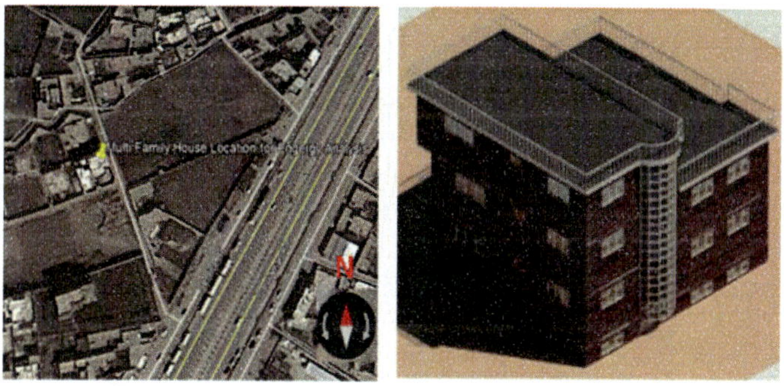

(a) (b)

Fig. 1 **a** Location—satellite image, **b** 3D front view of the building

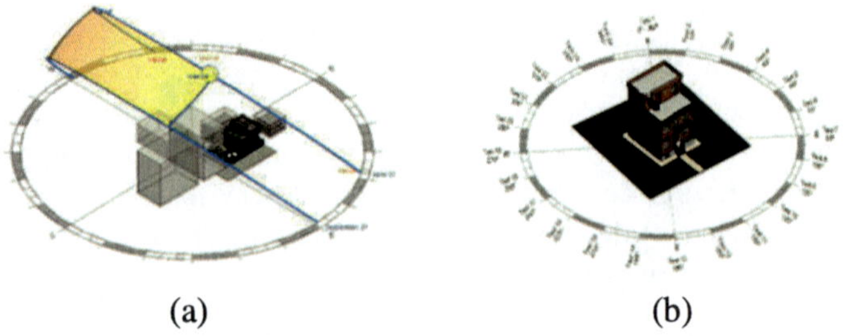

Fig. 2 **a** Existing surrounding buildings, **b** test scenarios for orientation analysis

Basement walls of the building consist of 600 mm thick stone masonry, which conventionally stone is used in basement walls and foundations in construction of buildings in Afghanistan. This wall is remaining unchanged with all combinations across the energy performance assessment process. The nearest weather station for the building is station No. 699799 which was assigned to the software for energy simulation. The building has a box shape and its roof is flat. South-East façade of the building has its entrance on the Ground floor. The North-West and South-West facades of the building have small windows while North-East and South-East façades have larger windows.

4 BIM Modeling and Analysis

BIM modeling is the most essential step involved in the thermal analysis process. Modeling for energy performance analysis slightly different than that of architecture designs. Any likely mistakes in this process hugely influence the result of the analysis. For this purpose, sufficient and correct information must be set into the model. Energy analysis is one of the important features of the rivet which performs whole building energy analysis. There are two ways in Revit for energy analysis simulation, one is by using conceptual masses where actual physical parts of building competent are not required to have existed in the model. And the other by using building elements when actual models within all components and elements including their properties need to have existed in the model. In the current study, the building elements method was used as the base to perform simulation analysis. Figure 3 shows the model generated using Revit and the process involved in extracting the energy analysis. As soon as the BIM model was created, a few modifications were made to get an accurate result from the energy analysis process.

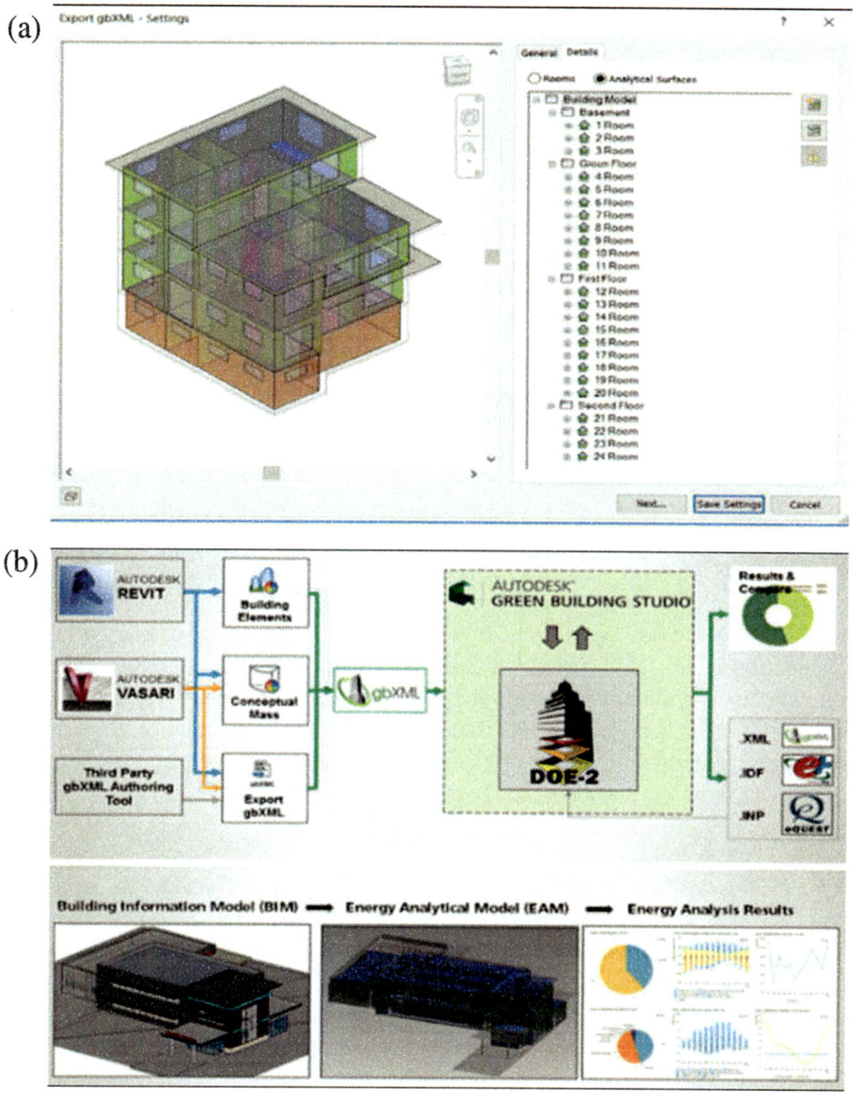

Fig. 3 **a** gbXML file of the model in Revit, **b** energy analysis work flow in Autodesk tools adopted [33]

4.1 Building Envelope Variation Impacts

In the current study, gbXML files of all combinations were imported from Revit to the GBS cloud. Table 2 shows the default simulation parameters used in the GBS studio. Locally available materials that are typically used in the envelope of the residential buildings in Afghanistan are used. A total of 127 models (96 building envelope

Table 2 Default values for multifamily building in GBS

Building type	EPD (W/m^2)	LP (W/m^2)	Outside air flow/person (cfm/person)	Outside air flow/area (cms/m^2)	Infiltration flow (ACH)
Multi family	10.76	7.53	N/A	0.093	0.25

Table 3 Minimum energy demand—combination C48

Wall type	Roof type	Floor type	Electricity demand (kWh)	Electricity annual cost (AF)
W3	R4	F4	29,496.00	213,846.00

+ 24 orientation analysis + 7 analysis for glazing impacts) were analyzed for the energy estimates. For this purpose, 6 types of walls, 4 types of floors, and 4 types of roofs have been taken into consideration. 96 (C1 to C96) different combinations of Afghan local construction materials for the envelope of the residential buildings were studied. As the result of the energy analysis experiments, it was found that a combination C48 with wall type (W3)—Adobe brick wall, roof type (R4)—Mud Roof with timber as core structure and Floor type (F4)—Mud floor with timber as structure, was the best option which consumes the least amount of energy leading to reduction in annual electricity demand. The work carried in identifying the best combination of building material is not presenting here, only its outcome is used for the orientation and glazing analysis. The minimum energy consumption combination is tabulated in Table 3.

4.2 *Orientation Variation Impacts*

The building was rotated clockwise from the North side of the building with a 15° rotational angle interval. A total of 24 tests were performed, test details were marked in Fig. 2b along with the actual orientation. The generated building model is simulated in GBS using the gbXML file or the simulation can be run using Revit plugin-Insight 360. Figure 4 shows the base run in GBS studio, after which the various model parameters can be varied. In this case, we varied the model rotation as shown in Fig. 5 and the typical output of the GBS simulation are shown in Fig. 6. The result shows that test 7 with a 90° rotation angle was the most efficient option from an energy conservation point of view. It indicates 8.22% energy efficiency improvement compared to actual oriented option (test 23 with 330° rotational angle) which is the worst case from the energy conservation viewpoint. Figure 7 shows the complete results of the tests for all rotated angles.

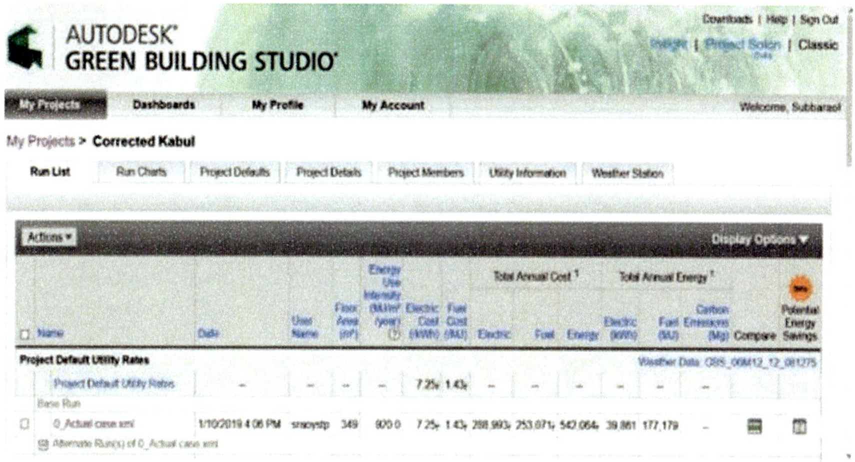

Fig. 4 Base run simulation energy results in GBS

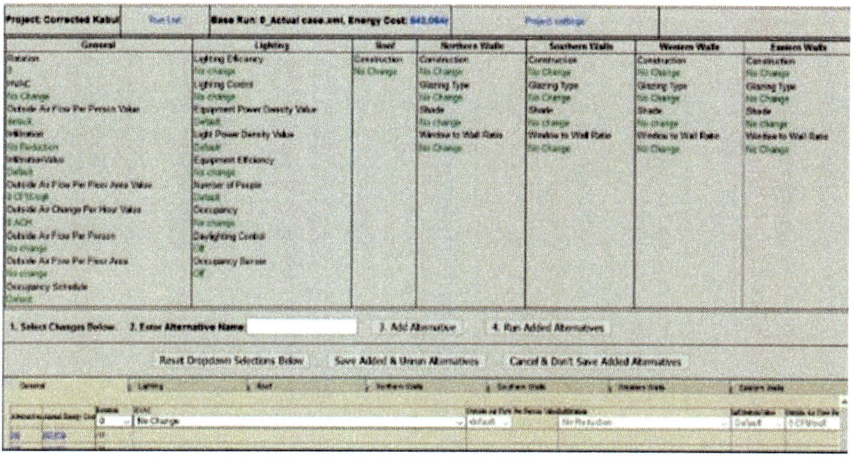

Fig. 5 Alteration of various project parameters

4.3 Glazing Area Variation Results

Glazing area of the building envelope of the most energy-efficient case (C48) was increased from 10 to 60% with 10% WWR increment intervals. For this purpose, the percentage of glazing area of models for 10–60% as well as the actual model were calculated in Microsoft Excel and new windows with new dimensions were assigned to the models accordingly. Then energy simulation was conducted for all new 6 models. Finally, electricity demand results in all models were compared. The detailed result of Energy demand is shown in Table 4 and Fig. 8.

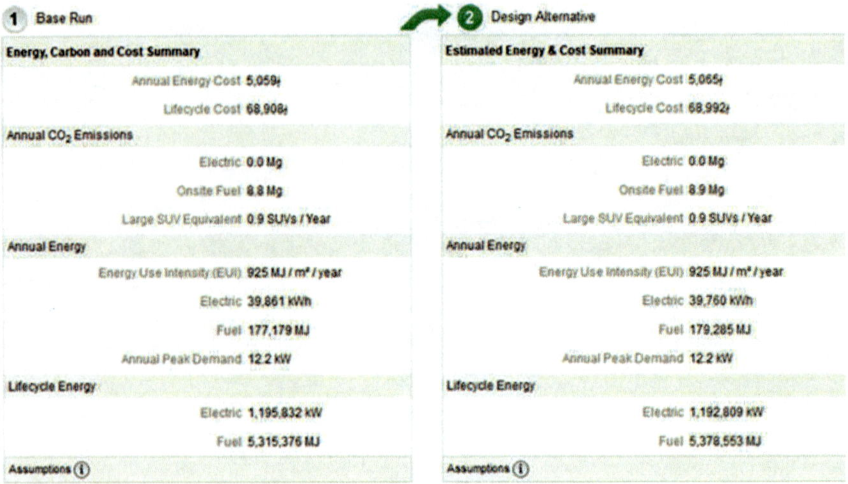

Fig. 6 Typical output result of energy simulation (actual and alternative design)

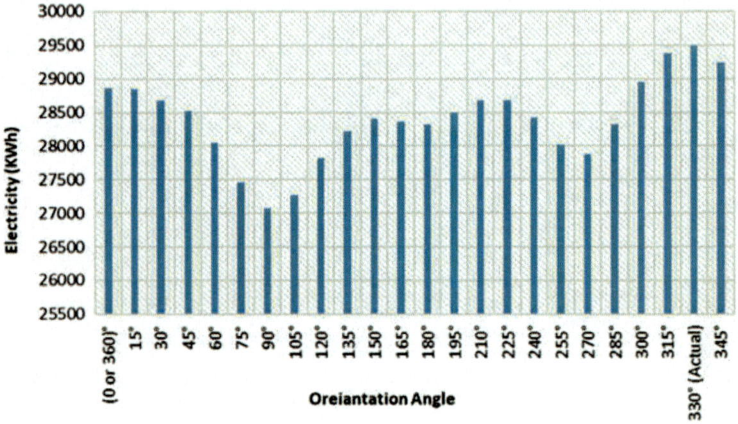

Fig. 7 Electricity demand for all orientation change increment

Table 4 Electricity demand for different glazing area

Com. no.	Windows to wall ratio (WWR) (%)	Electricity demand (kWh)	Electricity cost (AF)
C48_1	10	24,356	176,584
C48_2	20	28,442	206,204
C48_3 (actual)	20.3	29,496	213,846
C48_4	30	33,831	245,276
C48_5	40	39,728	288,025
C48_6	50	44,126	319,911
C48_7	60	48,885	354,417

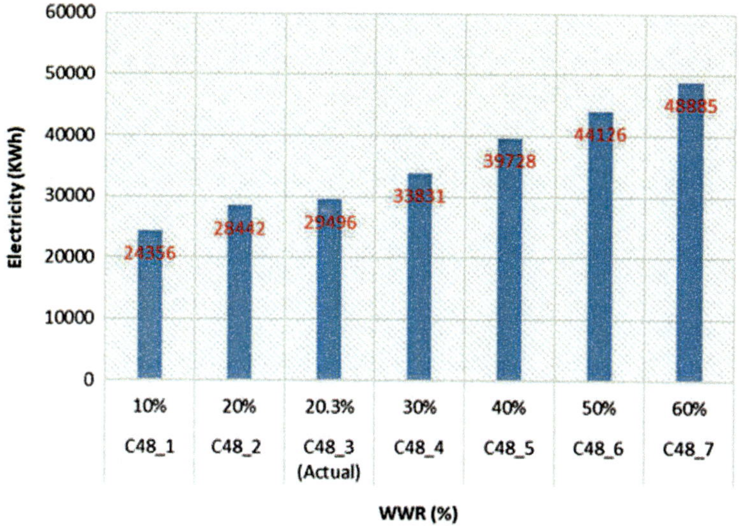

Fig. 8 Variation of electricity demand for variation in the glazing area

5 Conclusion

Due to technological developments, numerous software tools have been developed in energy assessment in the construction industry. Building energy evaluation allows the users to make proper decisions at the different stages of design to enhance sustainability by applying specific solutions.

The main objective of the research was to investigate the local construction material behavior on building energy consumption. For this purpose, the different local construction materials which are conventionally used in buildings were assigned to the main components of the 3D model in Revit. In total 7 types of walls, 4 types of roofs, and 4 types of floors were created from various local materials, and all aforementioned building elements were combined to gather and 96 models were built. gbXML format of all 96 models was imported to the GBS cloud for energy simulation. Finally, it was observed that combination No. 48 which made of adobe brick walls and mud roofs and floors was the best one from the energy conservation point of view as it consumes 29,496 kWh electricity annually which is the minimum electricity consumption option among all combinations.

Furthermore, building orientation variation and glazing area enlargement impacts were investigated to find out the best rotation angle and favorable percentage of WWR from an energy conservation point of view. For this purpose, the most energy-efficient combination model was rotated clockwise with 15° increment angle and 24 new models were built for orientation changing tests and 6 new models were prepared for 10–60% WWR experiments. It was observed that 120° rotation angle for existing building orientation state is the best case. By increasing WWR percentages, electricity demand also increased accordingly. So, to build energy-efficient buildings, smaller WWR is recommended.

References

1. Costa, A., Keane, M.M., Torrens, J.I., Corry, E.: Building operation and energy performance: monitoring, analysis and optimisation toolkit. Appl. Energy (2013). https://doi.org/10.1016/j.apenergy.2011.10.037
2. Mohmmad, O.T.: Energy Efficiency Guidebook for Buildings. Institutional Development for Energy in Afghanistan (IDEA) Program (2017). https://doi.org/10.29171/Azu_Acku_Tk2901_Tay98_1396
3. IEA: Technology Roadmap: Energy Efficient Building Envelopes. IEA Publ. 68 (2013)
4. Khanzadi, M., Kaveh, A., Moghaddam, M.R., Pourbagheri, S.M.: Optimization of building components with sustainability aspects in BIM environment. Period. Polytech. Civ. Eng. **63**, 93–103 (2019). https://doi.org/10.3311/PPci.12551
5. Lapisa, R., Karudin, A., Rizal, F., Nasruddin, K.: Passive cooling strategies in roof design to improve the residential building thermal performance in tropical region. Asian J. Civ. Eng. **20**, 571–580 (2019). https://doi.org/10.1007/s42107-019-00125-1
6. Naji, H.I., Mahmood, M., Mohammad, H.E.: Using BIM to propose building alternatives towards lower consumption of electric power in Iraq. Asian J. Civ. Eng. **20**, 669–679 (2019). https://doi.org/10.1007/s42107-019-00134-0
7. Caldera, M., Corgnati, S.P., Filippi, M.: Energy demand for space heating through a statistical approach: application to residential buildings. Energy Build. **40**, 1972–1983 (2008). https://doi.org/10.1016/j.enbuild.2008.05.005
8. Catalina, T., Virgone, J., Blanco, E.: Development and validation of regression models to predict monthly heating demand for residential buildings. Energy Build. **40**, 1825–1832 (2008). https://doi.org/10.1016/j.enbuild.2008.04.001
9. Abanda, F.H., Cabeza, L.: Investigating occupant's behaviour using emerging building information modelling. In: The Proceedings of the ICSC15—The CSCE International Construction Specialty Conference, pp. 7–10, Vancouver, Canada (2015)
10. Leth-petersen, S., Togeby, M.: Demand for space heating in apartment blocks: measuring effects of policy measures aiming at reducing energy consumption. Energy Econ. 387–403 (2001). https://doi.org/PII:S0140-98830000078-5
11. Santin, O.G.: Occupant behaviour in energy efficient dwellings: evidence of a rebound effect. 311–327 (2013). https://doi.org/10.1007/s10901-012-9297-2
12. Carbonari, A., Rossi, G., Romagnoni, P.: Optimal orientation and automatic control of external shading devices in office buildings. Environ. Manag. Health **13**, 392–404 (2002). https://doi.org/10.1108/09566160210439305
13. Mardookhy, M., Sawhney, R., Ji, S., Zhu, X., Zhou, W.: A study of energy efficiency in residential buildings in Knoxville, Tennessee. J. Clean. Prod. **85**, 241–249 (2014). https://doi.org/10.1016/j.jclepro.2013.09.025
14. Reza Fallahtafti, M.M.: Optimisation of building shape and orientation for better energy efficient architecture. Int. J. Energy Sect. Manag. **9**, 594–618 (2015). https://doi.org/10.1108/IJESM-09-2014-0001
15. Wong, K., Fan, Q.: Building information modelling (BIM) for sustainable building design. Facilities **31**, 138–157 (2013). https://doi.org/10.1108/02632771311299412
16. Grubler, A., Jefferson, M., McDonald, A., Messner, S., Nakicenovic, N.: Global energy perspectives to 2050 and beyond. Int. Inst. Appl. Syst. Anal. Austria. https://pure.iiasa.ac.at/id/eprint/4471/1/WP-95-127.pdf (2004). https://doi.org/10.1016/j.aap.2006.11.009

17. Cao, X., Dai, X., Liu, J.: Building energy-consumption status worldwide and the state-of-the-art technologies for zero-energy buildings during the past decade. Energy Build. **128**, 198–213 (2016). https://doi.org/10.1016/j.enbuild.2016.06.089
18. Pérez-Lombard, L., Ortiz, J., Pout, C.: A review on buildings energy consumption information. Energy Build. https://www.esi2.us.es/~jfc/Descargas/ARTICULOS/PA (2008). https://doi.org/10.1016/j.enbuild.2007.03.007
19. Moriarty, P., Honnery, D.: What is the global potential for renewable energy? Renew. Sustain. Energy Rev. **16**, 244–252 (2012). https://doi.org/10.1016/j.rser.2011.07.151
20. Kim, H., Anderson, K.: Energy modeling system using building information modeling open standards. J. Comput. Civ. Eng. **27**, 203–211 (2013). https://doi.org/10.1061/(ASCE)CP.1943-5487.0000215
21. Clevenger, C.M., Ozbek, M.E., Mahmoud, H., Fanning, B.: Impacts and Benefits of Implementing Building Information Modeling on Bridge Infrastructure Projects (2014)
22. Laine, T., Karola, A.: Benefits of building information models in energy analysis. In: Clima 2007 WellBeing Indoors (2007)
23. Venkataraman, A., Ramesh Kannan, M.: Whole building energy analysis using BIM. ResearchGate (2015). https://www.researchgate.net/publication/281965883
24. Jalaei, F., Jrade, A.: Integrating building information modeling (BIM), energy analysis and simulation tools to conceptually design sustainable buildings. In: Proceedings, Annual Conference, pp. 556–565. Canadian Society for Civil Engineering (2013)
25. Ismet Ugursal, V.: Building performance analysis and simulation: we've come a long way. Buildings **4**. https://www.mdpi.com/2075-5309/4/4/762/ (2014). https://doi.org/10.3390/buildings4040762
26. Senave, M., Boeykens, S.: Link between BIM and energy simulation. In: Building Information Modelling (BIM) in Design, Construction and Operations (2015). https://doi.org/10.2495/bim150291
27. Hopfe, C.J.: Uncertainty and sensitivity analysis in building performance simulation for decision support and design optimization. PhD thesis. https://www.bwk.tue.nl/bps/hensen/team/past/Hopfe.pdf (2009). https://doi.org/10.1016/j.enbuild.2015.09.010
28. Ham, Y., Golparvar-Fard, M.: EPAR: energy performance augmented reality models for identification of building energy performance deviations between actual measurements and simulation results. Energy Build. **63** (2013). https://doi.org/10.1016/j.enbuild.2013.02.054
29. Arya, A.S.: Guidelines for Earthquake Resistant Design, Construction, and Retrofitting of Buildings in Afghanistan (2003)
30. Nawaz, A.: Critical analysis of literature on energy sector of Afghanistan (2016)
31. Ershad, A.M., Brecha, R.J., Hallinan, K.: Analysis of solar photovoltaic and wind power potential in Afghanistan. Renew. Energy **85**, 445–453 (2016). https://doi.org/10.1016/j.renene.2015.06.067
32. Ahmadzai, S., McKinna, A.: Afghanistan electrical energy and trans-boundary water systems analyses: challenges and opportunities. Energy Rep. **4**, 435–469 (2018). https://doi.org/10.1016/j.egyr.2018.06.003
33. Autodesk Press: Autodesk building performance analysis help. https://knowledge.autodesk.com/search-result/caas/CloudHelp/cloudhelp/ENU/BPA-BPAWorkflows/files/GUID-43DAB177-3A4F-496C-BECB-2591FD04FC10-htm.html. Accessed 10 Feb 2020

Evaluating the Effect of Speed Variation on Vehicular Emission Using an Integrated Modelling Approach

Archana Chawla, Mukesh Khare, and Saif Khan

Abstract Globally, vehicular emissions have been attributed as the primary source of air pollution in urban areas. Hence, there is a growing need to accurately estimate the contribution of these vehicular emissions to the overall level of air pollution in order to design and implement pollution-reduction measures effectively. Vehicular emissions are largely influenced by vehicle operating characteristics such as climate, engine temperature, duration of trip, driving speed, acceleration and deceleration, and number of stops. Among all the characteristics, instantaneous speed of the vehicle is the major factor that affects the accuracy of vehicular emissions estimation. For the purposes of accurately predicting the impact of sudden acceleration and deceleration, an integrated modelling approach is proposed to estimate the traffic-related emissions on a micro-scale level. The microscopic traffic simulation model called VISSIM, captures the spatial and temporal variation of vehicles to eventually generate the speed profile for each vehicle. This speed profile will be used as an input for emission estimation model, EnViVer. In the present study, this integrated approach is used to measure the emissions for three different pollutants i.e. Carbon-di-oxide (CO_2), Nitrogen oxides (NO_X) and Particulate matters (PM_{10}) at selected road segments in Delhi City, India.

Keywords Vehicular emissions · Microscopic traffic simulation modelling · Emission modelling · Integrated approach

1 Introduction

The sources of urban air pollution include power plants, domestic heating, industries, vehicles, and biogenic activities among which vehicular emissions are one of the most important contributors of urban air pollution. India has the second-largest road network in the world after the USA [20]. Delhi is the capital of India and the seat of its national government. Apart from being the political centre, it is also a hub

A. Chawla (✉) · M. Khare · S. Khan
Indian Institute of Technology Delhi, Hauz Khas, New Delhi 110016, India
e-mail: archana.chawla222@gmail.com

D. K. Ashish et al. (eds.), *3rd International Conference on Innovative Technologies for Clean and Sustainable Development*, RILEM Bookseries 29,
https://doi.org/10.1007/978-3-030-51485-3_20

for commercial activities, attracting population from different parts of the country which has resulted into an increase in travel demand [28]. Moreover, urbanization, increase in per capita income, city expansion, improved education facilities, rise in employment centres and social activities have led to an increase in passenger vehicle demand. This in turn has resulted in the increase in the number of vehicles outpacing the growth of the road network [2]. A census study by the Government of India has presented that, as of 2015, the number of registered vehicles in India has reached around 210 million. Delhi has the maximum share of 4.21% (8.8 million) of total registered vehicles in India. The exponential growth of vehicles in Delhi in the past two or three decades has also led to an unprecedented rise in the level of vehicular emissions, making them the primary source of the severe air pollution in the city.

The rising number of vehicles has led to an upsurge in traffic congestion and of pollutant emissions to further degrade the air quality especially close to large roadways. Vehicle operational variables such as climate, engine temperature, duration of trip, driving speed of vehicle, acceleration and deceleration, and number of stops greatly influence vehicular emissions. Generally, the speed of a vehicle can be classified into instantaneous speed and average speed. For accurate estimation of emissions, the detailed and actual driving speeds of the vehicle are important [15]. However, in most of the studies, there is a high probability of underestimating the average emissions if they are estimated solely on the basis of average speed because such studies fail to factor in the changes in the driving profiles i.e. idle, acceleration, deceleration [22, 34, 32]. Vehicular emissions directly depend upon the traffic flow conditions of traffic stream. A lot of studies attribute high traffic density and volume in urban areas to more traffic emissions than low traffic density and volume [10, 13, 22]. Vehicular emissions are also affected by road characteristics, such as, road class, road gradient and road altitude. Generally, roads are classified on the basis of speed and accessibility. There is an inverse relation between speed and accessibility of road, if the accessibility of road increases, the speed of the vehicle plying on the road decreases. Such classifications then determine certain traffic flow conditions, such as, free, congested flow and stop-start conditions. The impacts of road infrastructure on vehicular emissions have been investigated by several studies which have found that emissions of congested flows are generally five times those of free flows [5, 23, 35, 27].

2 Literature Review

Motorized vehicle operation is the principal source of urban air pollution and needs immediate attention to prevent its adverse impacts on human health. Many researchers have extensively coupled VISSIM with different emission models to determine the effective concentration of pollutants. Park et al. [26] have combined the VISSIM with Modelo de emisión (MODEM) and concentrations are estimated by using a Gaussian dispersion model. Integrated VISSIM and Comprehensive Modal Emission Model (CMEM) have the ability to determine emissions from a single vehicle and evaluate how traffic control strategy can be helpful in reduction of various pollutant emissions e.g. CO, HC, and NO_x [21, 33]. Ligterink et al. [16] have applied VISSIM with VERSIT + $_{micro}$ and named it as EnViVer. It is a regression-based simplified VERSIT and emission model, which has been devised to be used in conjunction with the output of microscopic dynamic traffic models. This model represents instantaneous speed per kilometer, dynamic driving behavior of a speed profile and acceleration [4]. Huang et al. [12] have studied effects of road maintenance works on emissions by using VISSIM in combination with QUARTET, which is an emission model based on average speed along with MODEM, which is an emission model based on instantaneous speed. Driving speed profiles are the main pre-requisite for instantaneous emission models. Certain studies have utilized microscopic traffic simulation models such as Parallel Microscopic Simulator (PARAMICS), VISSIM, and Advanced Interactive Microscopic Simulation for Urban Network (AIMSUN) for simulating traffic conditions in order to generate vehicle trajectories to eventually estimate emissions [11, 17, 24]. Chauhan et al. [3] have observed the impact of vehicle composition and green signal interval time on emission using an integrated microsimulation approach. The implementation of light rail transport shows 13% decrease in CO_2 emission which impact has been determined by using microscopic simulation software VISSIM along with add on module called EnViVer [7]. As discussed in the literature, there are many techniques including integrated approach which have been implemented in the past to estimate the vehicular pollution. The microscopic models have the ability to precisely simulate the details of the vehicle operating conditions on a second-by-second basis [1]. The VISSIM can analyze traffic operation under limitations such as lane change, traffic composition, traffic signal and time of day. The output generated by VISSIM comprises of second-by-second speed-acceleration profile, link-average speed, link-instantaneous speed, and vehicle trajectory data: speed, acceleration, location [1].

3 Methodology

3.1 Study Area

Delhi, a land-locked metropolitan city and capital of India, attracts population from different parts of the country. As per 2011 census, the population of Delhi was nearly 16.3 million making it the second most populous city in the country. Delhi also accounts for the maximum share of 4.21% (8.8 million) of total registered vehicles in India. This has caused an increase in the passenger vehicle demand which in turn has resulted in the increase in the number of vehicles to outpace the growth of the road network. A mid-block road section of length approximately 800 m on National Highway-19 (Delhi to Agra), near the Central Road Research Institute (CRRI) has been selected for the present study (Fig. 1). The study area includes two links, one is towards Faridabad and other one is towards Ashram Chowk where

Fig. 1 Location map of study site of NH-19 corridor, near CRRI gate

both the links are connected with signalized intersection. The selected road corridor carries almost 0.20 million vehicles per day, which encounters both *inter-* and *intra*-city heterogeneous (mixed) vehicular traffic. Although there is a restriction in entry of commercial vehicles from 07:00 a.m.–10:00 p.m. and data used for the study comprises of morning and evening peak hour. Therefore, the major contributors are cars, two wheelers and three wheelers during study period.

3.2 Data Collection

Collection of traffic flow data such as traffic volume and vehicle composition for peak hour (09:00 a.m.–11:00 a.m. and 05:00 p.m.–07:00 p.m.) of the day has been carried out by installing a portable video camera at suitable height so that it can capture the traffic on both the roadways. Further, the traffic data based on actual traffic fleet moving on road shall be grouped into six vehicle classes which are shown in Table 1. It is clear that the cars, 2-wheelers and 3-wheelers (auto rickshaw) are the predominant vehicle classes which cover almost 96% of the total traffic. The location specific road characteristic data i.e. horizontal and vertical geometry including road alignment, gradient, and lane width; location of parking lots; bus stops; and signal control data including amber, red and green light has been measured

Table 1 Vehicle composition at CRRI gate

S. no.	Vehicle class	Composition (%)	
		Morning	Evening
1	Car	46.24	52.84
2	2-Wheeler	40.96	34.35
3	Auto rickshaw	9.33	8.94
4	Bus	1.71	1.57
5	LCV	1.01	0.94
6	HGV	0.75	1.36

manually. Multiple instantaneous speed of all the categories of vehicles at both the link segment has also been measured by speed gun for morning and evening peak hours individually. This instantaneous speed data will be used as to set the upper and lower bound limit of desired speed distribution in VISSIM simulation.

3.3 Microscopic Traffic Flow Simulation Model: VISSIM

A microscopic model of traffic flow is predicated on the description of motion of each individual vehicle which makes up the traffic stream. These models can simulate traffic operation in great detail but need extensive, input data along with computer execution time for their application. The traffic flow simulation model VISSIM is a discrete, stochastic, time step based microscopic model, which considers driver-vehicle-units as single entities. The model has been formulated after consistent and continued efforts made by Wiedemann [36]. The basic input data needed for the VISSIM can be divided into three categories, first is the network data which contains detail of lane markings at each junction, signal heads, detectors and position of bus stops; second is traffic flow data which is in form of static routing such as turn movements for each connection, input flow in vehicle/h and dynamic assignment routing like Origin-Destination matrix, location of parking lots, traffic composition, road speed limit, travel time and saturation flows; third is signal control data which consist cycle length of each signal group (red, green and amber). Detail of all the input parameter using in the present study are given in Table 2. VISSIM could generate second-by-second trajectory of each vehicle in the entire network but in order to estimate the emission the required extracted output parameters are shown in Table 3. To determine the impact of different speed on the emission generation two different case scenarios have been studied which are illustrated in Table 4. VISSIM network for study site is shown in Fig. 2.

3.4 Calibration and Validation of Model

Calibration is the process used to adjust various parameters in order to accurately generate the modelled data which reflects the field condition. VISSIM has several parameters that can be changed during the simulation. Driving behaviour is the key parameter that affects the output of the model in a significant way. VISSIM is based on the psycho physical car following model which takes into account the movement of vehicles on a single lane without exits. The underlying concept of this model operates on the assumption that a driver can be in any one of the four driving modes. This includes *free mode*, *approaching* mode, *following* mode and *braking* mode [29]. Depending upon speed, relative speed and distance, the driver switches from one mode to another [25]. The calibrated values used in the model have been given in Table 2. Validation is important to measure the accuracy of simulation data

Table 2 Input parameters for VISSIM model

Parameters	Description				
Traffic fleet data	Vehicle type	Morning peak hours		Evening peak hours	
		9:00–10:00	10:00–11:00	5:00–6:00	6:00–7:00
	Car	5275	5443	5658	5884
	2-Wheeler	5295	4200	3434	4070
	Auto rickshaw	1081	1082	993	960
	Bus	204	193	171	173
	LCV	89	144	127	78
	HGV	101	74	162	134
	Total	12,045	11,136	10,545	11,299
Road geometry	Latitude and longitude	28° 37′ 39.99″ N and 77° 14′ 29.04″ E			
	Site terrain	Flat			
	Road length	~800 m			
	Road width	30 m (27 m carriageway and 3 m median)			
	Number of lanes	3 lanes of width 3.75, 3.75 and 6.0 m at each link			
	Land use	Urban area			
	Altitude (MSL)	219 m			

(continued)

Table 2 (continued)

Parameters	Description	
Speed profile	Link 1	Link 2
	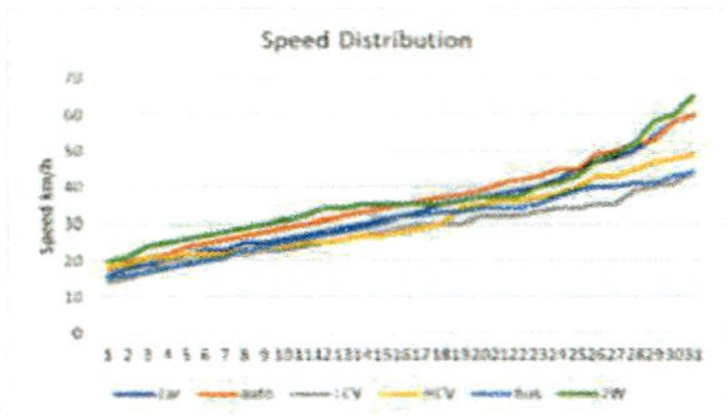	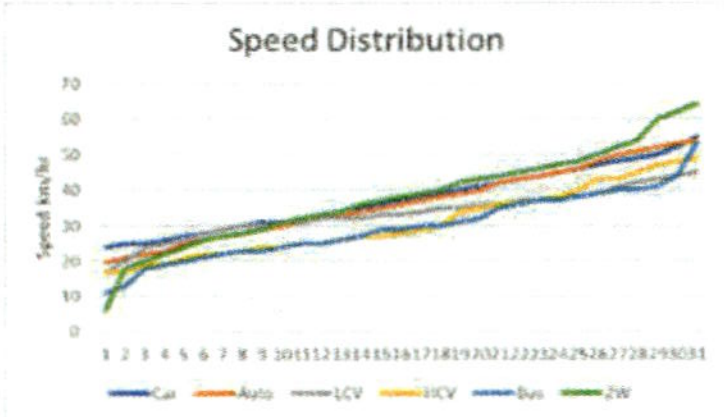
Signal timing	Red	27 s
	Amber	3 s
	Green	120 s
Calibrated values for driving behaviour	Parameters	Value
	Look ahead distance	Max 250 min 30
	Look back distance	Max 150 min 30
	Average standstill distance	0.5 m
	Overtake reduced speed area	Allowed
	Desired position at free flow	Any
	Overtake on same lane	
	On left	Allowed
	On right	Allowed
	Minimum lateral distance	
	Distance standing	0.6 m at 0 km/h
	Distance driving	0.8 m at 50 km/h

Table 3 Output parameters from VISSIM model

S. no.	Attribute	Format
1	Coordinates at the front end of vehicle	Default
2	Vehicle number	Default
3	Simulation time	Seconds
4	Speed at end of time step	Default
5	Number of each vehicle type	Default
6	Name of each vehicle type	Default
7	Gradient on each lane and link	Default
9	Total distance travelled	Meters

Table 4 Baseline and modified speed limit of vehicle

Vehicle class	Baseline scenario (BS)	Scenario 1 (S-1)	Scenario 2 (S-2)
	Desired speed limit in km/h		
Car	70	60	50
2W	60	50	40
Auto rickshaw	50	40	30
Buses	60	50	40
LCV	60	50	40
HGV	60	50	40

Fig. 2 VISSIM network for study site

generated by the model. All the input parameters have been implemented in the software to produce a simulation almost parallel to the real world. Traffic volume is the key parameter used for the validation of the model. Performance measure statisticsthe Geoffrey E. Heaver (GEH) which comprises traffic volume data generated by VISSIM and observed in field has been used [8]. The formula used to determine GEH value are given in Eq. (1). As per guideline given by UK highway agency, 2017 for validation of model, any value of GEH statistics less than 5 indicate good fit.

$$\text{GEH} = \sqrt{\frac{2(v_s - v_o)^2}{(v_s + v_o)}} \quad (1)$$

3.5 *Emission Modelling: EnViVer*

The emission factor is a key parameter in emission estimations and it is defined as the emitted mass of compound per unit driven distance [14]. The average speeds generally determine the emissions factor. But assigning a single average speed to all the vehicles in a road network may cause errors in the prediction of emissions. EnviVer is comprehensive version of VERSIT and micro which is combined with VISSIM to generate real-world emissions based on speed trajectory of various vehicle categories [31]. The traffic fleet file generated by VISSIM which contains all the output data has been installed into the emission model. The input parameters used for emission estimation in EnViVer are given in Table 5. In EnViVer vehicle class is pre-defined for light duty, heavy duty and buses, while fleet properties like; fuel type, age distribution and average regional CO_2 emissions are used based on existing literature [6, 9, 18, 19, 30]. Emission legislation has been defined according to their respective introduction year from Euro 1 to 6 (Emission Standards: India: On-Road Vehicles and Engines). In the present study the EnViVer model has been used to generate the emission data for for three different pollutants i.e. CO_2, NO_X and PM_{10} at selected site.

4 Results and Discussion

4.1 *Validation of VISSIM Model*

The model performance measure GEH has been used for comparison between traffic volume, observed from field and generated by VISSIM. The GEH value obtained for morning (9:00–10:00 and 10:00–11:00) and evening (5:00–6:00 and 6–00-7:00) peak hours are 0.10, 0.16, 0.68 and 0.37 respectively. The GEH value for total volume also obtains as 0.38. All the GEH values are much less than 5 which indicates model is well calibrated and represent a significant match between observed and simulated traffic flow with remarkable accuracy. Table 6 shows the validation of VISSIM model.

Table 5 Input parameters for EnViver

Parameters	2W	Auto rickshaw	Car	LCV	Bus	HGV
Road type	Urban	Urban	Urban	Urban	Urban	Urban
Vehicle type	Light duty	Light duty	Light duty	Light duty	Bus	Heavy duty
Fuel type						
Petrol	100%	2%	63%	0	0	0
Diesel	0	0	27%	36%	29%	92%
LPG	0	1%	1%	0	0	0
CNG	0	97%	9%	64%	71%	8%
Electric	0	0	0	0	0	0
Vehicle age distribution						
Newer than 1 year (%)	9.5	9.2	10	7.5	8.4	7.5
Average vehicle age (year)	5.5	7.5	5.1	8.8	7.5	7.7
Average exit age (year)	10.6	11.0	10	13.4	13.5	13.5
Emission legislation						
Euro I	2000	2000	2000	2000	2000	2000
Euro II	2005	2005	2005	2005	2005	2005
Euro III	2010	2010	2010	2010	2010	2010
Euro IV	2017	2017	2017	2017	2017	2017
Euro V	2019	2019	2019	2019	2019	2019
Euro VI	2020	2020	2020	2020	2020	2020
Average regional CO_2 emission						
Petrol (g/km)	50	79	127	50	–	–
Diesel (g/km)	50	220	150	350	–	–

Table 6 VISSIM model validation

Time duration	Observed volume (Vo)	Simulated volume (Vs)	GEH
9:00 a.m.–10:00 a.m.	12,045	12,056	0.10
10:00 a.m.–11:00 a.m.	11,136	11,153	0.16
5:00 p.m.–6:00 p.m.	10,545	10,475	0.68
6:00 p.m.–7:00 p.m.	11,299	11,260	0.37
Total	45,025	44,944	0.38

4.2 Emission Estimation Due to Reduced Speed Limit

The Government of India has notified the maximum speed of motor vehicles in kilometres per hour for various classes of vehicles on different categories of roads. Selected road sites come under urban category of road and speed limit for different groups of vehicles are given in Table 4, under baseline scenario. The output generated from EnViVer measures the emission in terms of both emission factor and emission rate. The amount of emissions generated for each vehicle category can also be estimated by EnViVer. Figure 3 shows the percentage contribution of all vehicles at different interval of time. For morning peak hour, the extreme amount of emissions are noticed for car; 54% of CO_2, 37% of NO_X and 48% of PM_{10}. The second highest contributor is 2W; 27% of CO_2, 21% of NO_X and 37% of PM_{10}. These two modes of transportation alone can contribute almost 80–90% of total emission generated. To determine the impact of speed reduction on emission, two scenarios have been generated by reducing the speed limit of each vehicle category by 10 km/h. Table 7 shows the quantity of generated CO_2, NO_X and PM_{10} for base line scenario and modified speed limit scenarios. It is clear from the results that by reducing the speed limit the amount of generation of all three pollutants reduced significantly. For S-1 the average reduction is 7.5% for CO_2, 8.5% for NO_X and 3.5% for PM_{10}. Similarly, the values for S-2 is much less than S-1 i.e. 13.75% for CO_2, 8.25% for NO_X and 6.25% for PM_{10}. Emission values mainly depends upon the amount of traffic volume plying on the network. Morning (9:00 a.m.–10:00 a.m.) peak hour has the maximum traffic flow so emission generation at this hour is maximum.

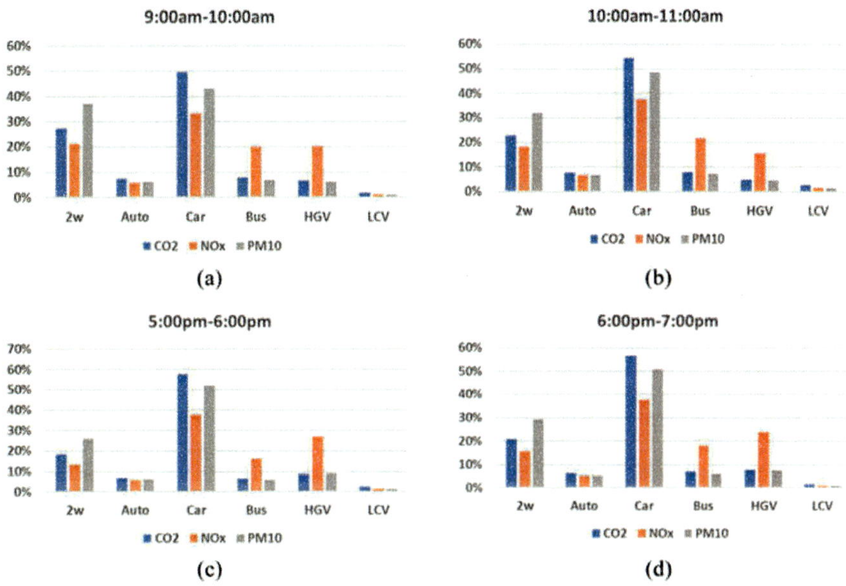

Fig. 3 Percentage contribution by all vehicle class

Table 7 Percentage reduction in emission

Pollutants	Time duration	BS	S-1	S-2	Percentage reduction from BS to S-1 (%)	Percentage reduction from BS to S-2 (%)
CO_2 (kg/h)	9:00–10:00	1574	1433	1290	9	17
	10:00–11:00	1430	1339	1242	6	13
	5:00–6:00	1400	1320	1231	6	12
	6:00–7:00	1496	1400	1299	6	13
NO_X (g/h)	9:00–10:00	6225	5522	5406	11	13
	10:00–11:00	5282	4851	4864	8	8
	5:00–6:00	5475	5118	5228	7	5
	6:00–7:00	5852	5414	5406	7	8
PM_{10} (g/h)	9:00–10:00	532.7	507	471.4	5	12
	10:00–11:00	476	462.7	455.8	3	4
	5:00–6:00	460.5	448.6	441.7	3	4
	6:00–7:00	494.4	480.5	471.4	3	5

Table 8 shows the result of emissions for baseline model and modified speed limit. In case of S-1 it is observed that the amount of pollutant reduced considerably for all the category of vehicle if the speed limit of the road is reduced. While in S-2 some vehicles show increase in emission as compared to S-1. Lowering speed limit to certain amount decrease the emission but may also lead to increase in congestion and consequently increase emission. So, to maintain the condition of free flow the optimum speed limits are required. S-1 shows the substantial amount of emission reduction for all vehicle classes. However, overall reduction in emission is presented by S-2. The emission in terms of g/km of CO_2, NO_x and PM_{10} in VISSIM network is shown in Fig. 4. The red and yellow colour shows the highest amount of emission in network. The emission is more within 200 m range of signal. The tendency of vehicle is to reduce its speed, stop for the moment and then accelerate which increases the amount of emission.

5 Conclusion

Vehicular emissions are the greatest source of air pollution in urban areas. Hence, there is a growing need to accurately estimate the contribution of these vehicular emissions to the overall level of air pollution for designing and implementing effective pollution reduction measures. VISSIM, a traffic simulation model along with it's add on module EnViVer has been used in the present study to model the heterogeneous traffic flow conditions at a Mid-Block section near CRRI in Delhi city. To determine the impact of speed limit reduction on emissions, the VISSIM model

Table 8 Emission estimation for BS, S-1 and S-2

Vehicle class	CO_2 (kg/h)			NO_x (g/h)			PM_{10} (g/h)		
	BS	S-1	S-2	BS	S-1	S-2	BS	S-1	S-2
9:00 a.m.–10:00 a.m.									
2W	427.2	383.6	349.8	1296	1119	1118	196.4	186.2	182.6
Auto	112.3	95.53	94.84	345.1	251.7	223	32.27	29.83	33.98
Car	776.3	711.7	650.5	2041	1663	1572	228.5	217	210.8
Bus	122.7	114	121.2	1233	1210	1351	36.07	35.1	36.87
HGV	103.1	99.06	102.3	1235	1216	1287	32.68	32.28	34.59
LCV	28.36	24.68	22.47	58.27	46.5	38.08	5.342	5.177	5.287
10:00 a.m.–11:00 a.m.									
2W	324.8	301.4	269.5	943.9	874.9	856.2	150.9	146.6	141.6
Auto	109.8	93.62	92.69	340.5	246.8	217.3	31.53	29.52	33.26
Car	774.1	732.2	662.5	1970	1696	1581	229.6	223.1	214.4
Bus	113.2	109.5	113.2	1132	1156	1259	33.84	33.24	34.44
HGV	67.25	65.09	70.44	805	802.3	887.2	21.9	21.97	23.78
LCV	37.14	33.02	29.91	75.74	61.69	50.52	6.981	6.986	7.091
5:00 p.m.–6:00 p.m.									
2W	255.5	236.5	211.7	728.3	672.6	669.2	119.2	115.5	111.1
Auto	96.87	83	81.63	301.5	219.2	190.6	27.8	25.99	29.26
Car	803	755.1	682.5	2034	1694	1601	238.8	230.6	220.7
Bus	87.47	86.18	91.58	873.4	909.4	1014	26.13	25.94	27.19
HGV	121.4	125.9	134.4	1457	1554	1695	41.2	43.19	46.07

(continued)

Table 8 (continued)

Vehicle class	CO_2 (kg/h)			NO_x (g/h)			PM_{10} (g/h)		
	BS	S-1	S-2	BS	S-1	S-2	BS	S-1	S-2
LCV	32.16	29.34	25.87	64.63	55.33	43.84	6.076	6.118	6.095
6:00 p.m.–7:00 p.m.									
2W	310.1	285.9	256	904	822.8	812	143.7	139.1	134.1
Auto	94.9	80.88	79.52	294	213.5	186.3	27	25.31	28.5
Car	845.6	790.8	718.3	2174	1816	1703	249.9	241.1	231.4
Bus	104.1	102.3	105.3	1038	1081	1169	30.61	30.9	31.4
HGV	115.1	115.9	118.1	1381	1426	1490	37.62	38.58	40.48
LCV	22.19	20.75	18.06	45	39.36	30.61	4.164	4.285	4.264

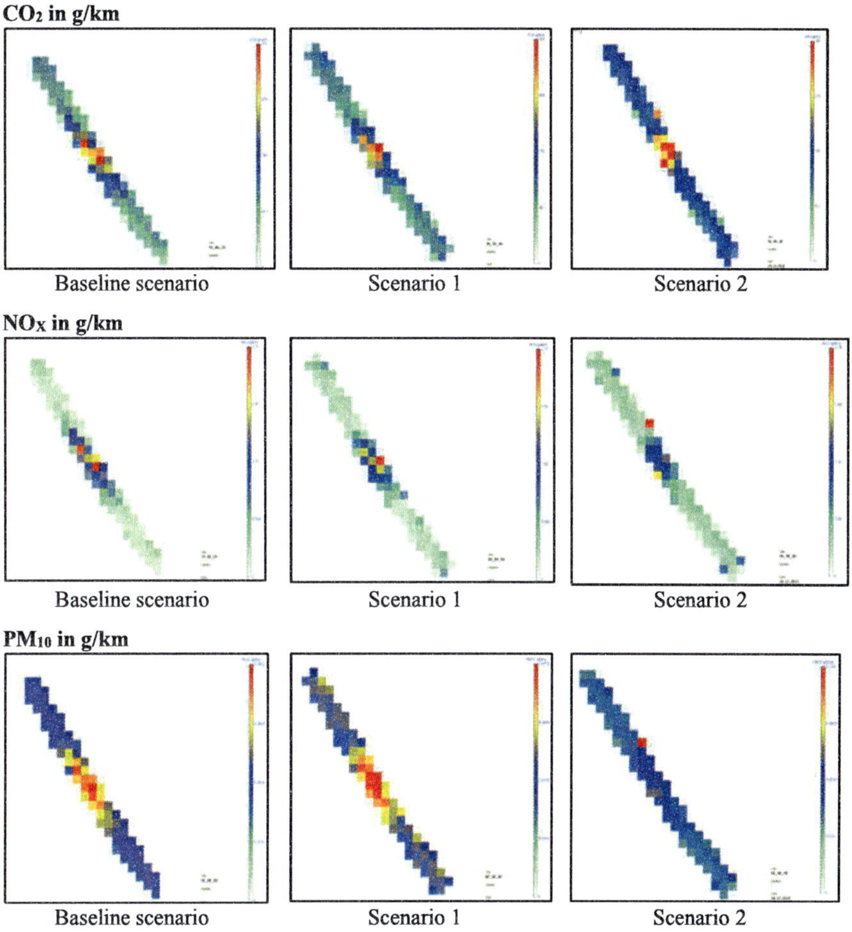

Fig. 4 CO_2, NO_X and PM_{10} emissions for baseline scenario, scenario 1 and scenario 2 in VISSIM network at 9:00 a.m.–10:00 a.m.

has been developed from observed field data. The simulation output carried out by VISSIM comprises of vehicle profile which is then fed into EnViVer for emission estimation. The validation of model is done by GEH statistics which shows significant acceptability of model, since for each case its value is less than 5. The main findings of this study conclude that if the speed limit is reduced by 20 km/h for all the vehicle classes the emission can be reduced significantly as 13.75% for CO_2, 8.25% for NO_X and 6.25% for PM_{10}.

Acknowledgements The authors would like to extend their gratitude to the PTV group for providing the Thesis licence version of VISSIM and EnViVEr. The authors gratefully acknowledge the co-operation and support provided by PTV team throughout the research work.

References

1. Abou-Senna, H., Radwan, E., Westerlund, K., Cooper, C.D.: Using a traffic simulation model (VISSIM) with an emissions model (MOVES) to predict emissions from vehicles on a limited-access highway. J. Air Waste Manag. Assoc. **63**(7), 819–831 (2013)
2. Ahmad, S., Balaban, O., Doll, C.N.H., Dreyfus, M.: Delhi revisited. Cities **31**, 641–653 (2013)
3. Chauhan, B.P., Joshi, G.J., Parida, P.: Car following model for urban signalised intersection to estimate speed based vehicle exhaust emissions. Urban Clim. **29**, 100480 (2019)
4. Dias, H.L.F., Bertoncini, B.V., Oliveira, M.L.M.D., Cavalcante, F.S.Á., Lima, E.P.: Analysis of emission models integrated with traffic models for freight transportation study in urban areas. Int. J. Environ. Technol. Manage. **20**(1–2), 60–77 (2017)
5. El-Fadel, M.U.T.A.S.E.V.I., Najvi, M.A., Seayti, H.: Air quality assessment at a congested urban intersection. In: Statistical Analysis and Modelling of Automotive Emissions, p. 85 (2001)
6. Emission Factor Project/Final Rep. ARAI, Pune
7. Eshghi, S., Saxena, N., Alsultan, A.: Evaluating emission reduction due to a proposed light rail service: a micro-level analysis. Int. J. Transp. Veh. Eng. **13**(2), 159–163 (2019)
8. Fernandes, P., Coelho, M.C., Rouphail, N.M.: Assessing the impact of closely-spaced intersections on traffic operations and pollutant emissions on a corridor level. Transp. Res. Part D Transp. Environ. **54**, 304–320 (2017)
9. Goel, R., Guttikunda, S.K., Mohan, D., Tiwari, G.: Benchmarking vehicle and passenger travel characteristics in Delhi for on-road emissions analysis. Travel Behav. Soc. **2**(2), 88–101 (2015)
10. Gokhale, S.: Traffic flow pattern and meteorology at two distinct urban junctions with impacts on air quality. Atmos. Environ. **45**(10), 1830–1840 (2011)
11. Hirschmann, K., Zallinger, M., Fellendorf, M., Hausberger, S.: A new method to calculate emissions with simulated traffic conditions. In: 2010 13th International IEEE Conference on Intelligent Transportation Systems (ITSC), pp. 33–38, Sept 2010. IEEE
12. Huang, Y., Bird, R., Bell, M.: A comparative study of the emissions by road maintenance works and the disrupted traffic using life cycle assessment and micro-simulation. Transp. Res. Part D Transp. Environ. **14**(3), 197–204 (2009)
13. Kimbrough, S., Vallero, D.A., Shores, R.C., Mitchell, W.: Enhanced, multi criteria-based site selection to measure mobile source toxic air pollutants. Transp. Res. Part D Transp. Environ. **16**(8), 586–590 (2011)
14. Kristensson, A., Johansson, C., Westerholm, R., Swietlicki, E., Gidhagen, L., Wideqvist, U., Vesely, V.: Real-world traffic emission factors of gases and particles measured in a road tunnel in Stockholm, Sweden. Atmos. Environ. **38**(5), 657–673 (2004)
15. Li, M., Yu, L., Zhai, Z., He, W., Song, G.: Development of emission factors for an urban road network based on speed distributions. J. Transp. Eng. **142**(9), 04016036 (2016)
16. Ligterink, N., Van Baalen, J., Eijk, A., Mak, W., Broeders, W., Vortisch, P.: Predicting local vehicle emissions using VERSIT+ and VISSIM. In: Proceedings of the 7th European Congress on ITS, June 2008
17. Madireddy, M., De Coensel, B., Can, A., Degraeuwe, B., Beusen, B., De Vlieger, I., Botteldooren, D.: Assessment of the impact of speed limit reduction and traffic signal coordination on vehicle emissions using an integrated approach. Transp. Res. Part D Transp. Environ. **16**(7), 504–508 (2011)
18. Malik, L., Tiwari, G.: Assessment of interstate freight vehicle characteristics and impact of future emission and fuel economy standards on their emissions in India. Energy Policy **108**, 121–133 (2017)
19. Malik, L., Tiwari, G., Thakur, S., Kumar, A.: Assessment of freight vehicle characteristics and impact of future policy interventions on their emissions in Delhi. Transp. Res. Part D Transp. Environ. **67**, 610–627 (2019)
20. MoRTH: Road Transport Yearbook. Ministry of Road Transport and Highways, Government of India (2016)

21. Nam, E.K., Gierczak, C.A., Butler, J.W.: A comparison of real-world and modeled emissions under conditions of variable driver aggressiveness. In: 82nd Annual Meeting of the Transportation Research Board, Washington, DC, Jan 2003
22. Nesamani, K.S., Chu, L., McNally, M.G., Jayakrishnan, R.: Estimation of vehicular emissions by capturing traffic variations. Atmos. Environ. **41**(14), 2996–3008 (2007)
23. Nicolas, J.P.: Analysing road traffic influences on air pollution: how to achieve sustainable urban development. Transp. Rev. **20**(2), 219–232 (2000)
24. Osorio, C., Nanduri, K.: Energy-efficient urban traffic management: a microscopic simulation-based approach. Transp. Sci. **49**(3), 637–651 (2015)
25. PTV VISSIM 11 User Manual, Germany (2018)
26. Park, J., Noland, R., Polak, J.: Microscopic model of air pollutant concentrations: comparison of simulated results with measured and macroscopic estimates. Transp. Res. Rec. J. Transp. Res. Board **1750**, 64–73 (2001)
27. Robertson, D., Wilson, N., KEMP, S.: The effects of co-ordinated and isolated signal control on journey times and exhaust emissions along the A12 in London. Traffic Eng. Control **37**(1), 4–9 (1996)
28. Sahai, S.N., Bishop, S.: Multi Modal Transport in a Low Carbon Future. India Infrastructure Report (2010)
29. Sajjadi, S., Kondyli, A.: Macroscopic and microscopic analyses of managed lanes on freeway facilities in South Florida. J. Traffic Transp. Eng. (Engl. Ed.) **4**(1), 61–70 (2017)
30. Sharma, N., Kumar, P.P., Dhyani, R., Ravisekhar, C., Ravinder, K.: Idling fuel consumption and emissions of air pollutants at selected signalized intersections in Delhi. J. Clean. Prod. **212**, 8–21 (2019)
31. Smit, R., Poelman, M., Schrijver, J.: Improved road traffic emission inventories by adding mean speed distributions. Atmos. Environ. **42**(5), 916–926 (2008)
32. Smit, R., Smokers, R., Rabe, E.: A new modeling approach for road traffic emissions: VERSIT+. Transp. Res. Part D: Trans. Environ. **12**(6), 414–422 (2007)
33. Stevanovic, A., Stevanovic, J., Zhang, K., Batterman, S.: Optimizing traffic control to reduce fuel consumption and vehicular emissions: integrated approach with VISSIM, CMEM, and VISGAOST. Transp. Res. Rec. J. Transp. Res. Board **2128**, 105–113 (2009)
34. Sun, Z., Hao, P., Ban, X., Yang, D.: Trajectory-based vehicle energy/emissions estimation for signalized arterials using mobile sensing data. Transp. Res. Part D **34**(34), 27–40 (2015)
35. Várhelyi, A.: The effects of small roundabouts on emissions and fuel consumption: a case study. Transp. Res. Part D Transp. Environ. **7**(1), 65–71 (2002)
36. Wiedemann, R.: Simulation des Straßenverkehrsflusses. In: Schriftenreihe des Instituts für Verkehrswesen der Universität Karlsruhe, Heft, vol. 8, p. 5 (1974)

Cold-Formed Steel Concrete Composite Slab: Structural Performance Evaluation Through Experimental Study

M. Adil Dar, Ahmad Fayeq Ghowsi, and A. R. Dar

Abstract Composite slabs with corrugated steel decking are efficient structural systems that have numerous desired key features like simple faster and lighter as well as cost effective structural constructions, particularly in the framed steel buildings. All these advantages attribute to the composite action developed between the steel and concrete component of the composite system, which is mainly governed by the shear transfer between the two. This paper presents the test results on an experimental study performed on a full scale one way cold-formed steel (CFS) concrete slab. The composite slab was subjected to monotonic flexural loading with simply supported end conditions. The primary aim of this investigation was to access the role of shear connector in developing the composite action between the CFS corrugated sheet and the concrete component of the composite slab. M25 grade of concrete was used in the preparation of the CFS concrete composite slab. The structural behaviour assessment was mainly carried out the ultimate load resisted by the composite slab, apart from the mid-span deflection. The mode of failure and the crack pattern was also studied.

Keywords Slabs · Experiment · Strength · Cracks · RC member · Shear connector · Corrugate sheet

1 Introduction

The utilization of cold-formed steel (CFS) members in the construction sector has risen, due to its various key features that suit the constructional requirements in

M. A. Dar · A. F. Ghowsi
Indian Institute of Technology Delhi, New Delhi 110016, India

A. R. Dar (✉)
National Institute of Technology Srinagar, Jammu and Kashmir, Srinagar 190006, India
e-mail: abdulrashid@nitsri.net

D. K. Ashish et al. (eds.), *3rd International Conference on Innovative Technologies for Clean and Sustainable Development*, RILEM Bookseries 29,
https://doi.org/10.1007/978-3-030-51485-3_21

an easy and sustainable manner. In the recent years, the basic weakness of premature localized buckling in the thin walled steel sections has been effectively countered by the continuous research and developmental work on the same [1–12]. One the easy and convenient ways of achieving this objective is by resorting to CFS composite sections. Slabs are important structural members that are directly subjected to different types of loading conditions, mostly vertical in direction. Unlike the conventional reinforced concrete (RC) slab, a cold-formed steel (CFS) composite slab constitutes of conventional concrete and a profiled (corrugated) CFS sheet used beneath the concrete plate. This type of flooring system is extensively adopted in the structural steel framed structures. This serves two purposes, firstly it acts as a part of the tensile steel and secondly it also acts as a formwork for the concrete placement during the constructional phase. Here, the CFS is located at the bottom most fiber, where the tensile stresses are the maximum. Hence, improves the efficiency of this structural system. Ease in construction, lighter sectional weight, higher structural efficiency and development of adequate strength for the intended use are some of the important features of this type of CFS composite construction. The performance of CFS composite members primarily depends on the structural integrity between the two materials used for forming the composite section. The past has witnessed a wide range of research work on the same. Generally, the interaction between the CFS and concrete, which also known as shear bond (horizontal dominates the behaviour of the composite system, particularly in terms of strength parameter [13, 14]. Inadequate shear bonding will make the composite system susceptible to slippage between the two materials, under flexural loading. At this stage, the composite system will experience partial shear. Numerous mechanisms have been developed to improve the friction between the two materials. Some of the common ways of attaining the same is due to embossment on the CFS sheeting which interlocks the CFS and concrete, anchorage assemblies like shear studs welded to the CFS sheet, and crippled CFS sheet towards span end. These minor changes have been identified to affect the shear bond between the two materials in the composite system. The extent of the interaction between steel and concrete influences the flow of shear stresses as well as the distribution of strain in these members [15]. As a result of these changes, the structural behaviour in terms of strength/stiffness as well as the failure mode is seen to alter. De Andrade et al. [16] through an experimental study where the composite action was improved through the installation of self-tapping screws on the top face of the CFS sheets. The implementation of these changes drastically improved shear resistance within the composite system. The results indicated that both the strength and stiffness had improved. This confirmed the importance of effective shear connecting system within the composite slabs.

This paper presents the test results on an experimental study performed on a full scale one way CFS concrete slab. The composite slab was subjected to monotonic flexural loading with simply supported end conditions. The primary aim of this investigation was to access the role of shear connector in developing the composite action between the CFS corrugated sheet and the concrete component of the composite slab. M25 grade of concrete was used in the preparation of the CFS concrete composite slab. The structural behaviour assessment was mainly carried out the ultimate load resisted by the composite slab, apart from the mid-span deflection. The mode of failure and the crack pattern was also studied.

2 Test Specimens

Two composite slab specimens with the sectional details shown in Fig. 1 were prepared. In the first specimen, the slab was directly casted on to the CFS sheet, with no shear connector being used. This was done to set a benchmark specimen, with no external shear connectors, and to get to understand the behaviour of the plain CFS composite slab. This specimen was labelled as benchmark specimen. In the other specimen, circular holes of size 10 mm were punched on the inner side of the CFS sheet, at a regular spacing of 100 mm longitudinally. These holes were made such that the sheet material protruded towards inside (as shown in Fig. 2), which will result in some improvement in the shear bonding between the two materials. The arrangement of the holes was carried out such that a zig-zag pattern was achieved. The holes will further let the concrete infiltrate, which once hardens will serve as

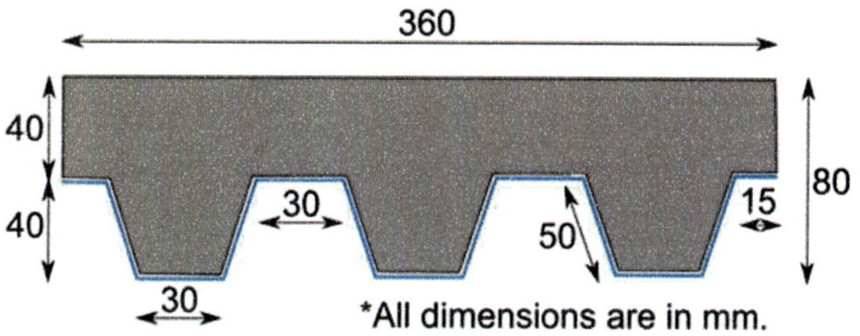

Fig. 1 Section of the composite slab specimen

Fig. 2 A view of the hole punched in the hole punched specimen

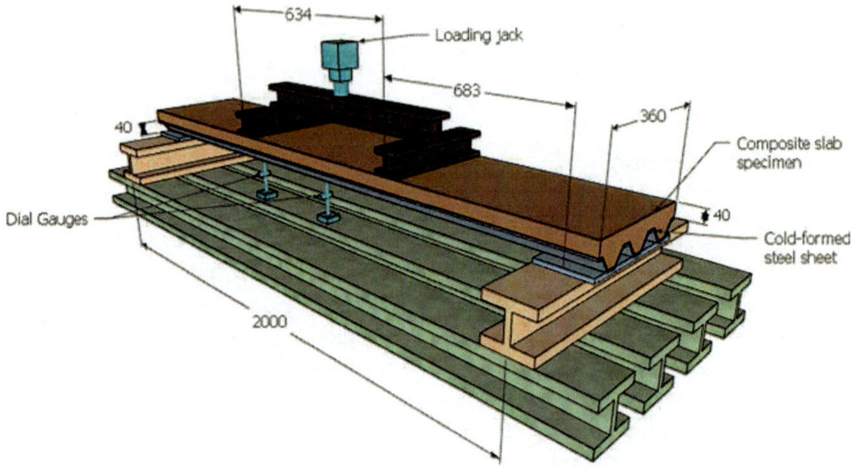

Fig. 3 Slab size and spreader beam arrangement (all dimensions are in mm)

an anchor and will further improve the bond between the two. This specimen was labelled as Hole punched specimen. The effective length of the specimen was fixed at 2 m with its width and thickness being fixed at 80 mm and 360 mm respectively as shown in Fig. 3. To maintain a clear cover of 25 mm throughout, a wire mesh was positioned at the top. The CFS sheet had a wall thickness of 0.8 mm. The CFS sheets were cut in the required dimensions, and the casting of the concrete was carried out, by fully supporting the sheet completely from beneath. M25 grade of concrete was prepared by using IS 10262: 2019 [17].

3 Material Testing

Various relevant tests were performed for the actual material property determination of the CFS sheet and the concrete. The yield strength of the CFS sheet was found to be 379 MPa against the nominal value of 350 MPa, while as the ultimate strength of the same was 433 MPa against the nominal value of 415 MPa. The compressive strength of the concrete was 27 MPa against the design value of 25 MPa.

4 Test Set-Up and Loading

A 1000 kN loading frame was used for carrying out the full-scale tests on the CFS composite slab specimens under monotonic flexural loading. Since the slab was intended to be tested under flexural loading, two spreader beams were provided at a distance of 683 mm from the slab ends. The loading was applied using a manual

500 kN hydraulic jack. The displacements at the mid span and the loading points were recorded using dial gauges. A view of the test set-up is shown in Fig. 4.

5 Results and Discussion

The load versus displacement plot of the benchmark specimen at the mid-span is shown in Fig. 5. The plot was linear up to a load of around 3.5 kN with the corresponding displacement of around 3.5 mm. The slab carried a total load of 5 kN, beyond which the specimen failed to carry further loading. The mid-span displacement observed at the peak load was noted as 10.7 mm. The specimen failed by complete delamination of the concrete part of the specimen with respect to the CFS sheet, as shown in Fig. 6. There was a total separation between the two. This indicated weakness in the bonding between the two. Mechanical locking between the embossments on the CFS sheets and the concrete through friction was the only source of their bonding. The ductility ratio was note as 2.4. The load versus displacement plot of the hole punched specimen at the mid-span is shown in Fig. 7. The specimen started yielding at a loading of around 7 kN and seized to develop any further resistance once the load reached its peak at 8.75 kN. The mid span displacement at the same load was recorded as 23 mm. Separation of the CFS sheet and the concrete component of the composite was observed vertically, as shown in Fig. 8. The ductility ratio was note as 2.5.

6 Conclusions

This paper presented the test results on an experimental study performed on a full scale one way CFS concrete slab. The composite slab was subjected to monotonic flexural loading with simply supported end conditions. The primary aim of this investigation was to access the role of shear connector in developing the composite action between the CFS corrugated sheet and the concrete component of the composite slab. M25 grade of concrete was used in the preparation of the CFS concrete composite slab. The structural behaviour assessment was mainly carried out the ultimate load resisted by the composite slab, apart from the mid-span deflection. The mode of failure and the crack pattern was also studied. The incorporation of the punched holes to offer shear resistance did improve the flexural performance of the composite slabs. The improvement was seen in the order of 75%. The friction between the embossments on the CFS sheets and the concrete were ineffective in developing the necessary shear connection, which governs the performance of composite members. This inadequacy lead to the total separation between the CFS sheet and the concrete part. The use of punching holes did not offer any improvement in the displacement ductility parameter. This study presented the structural performance of shear connector in developing the composite action. More research work needs to be carried

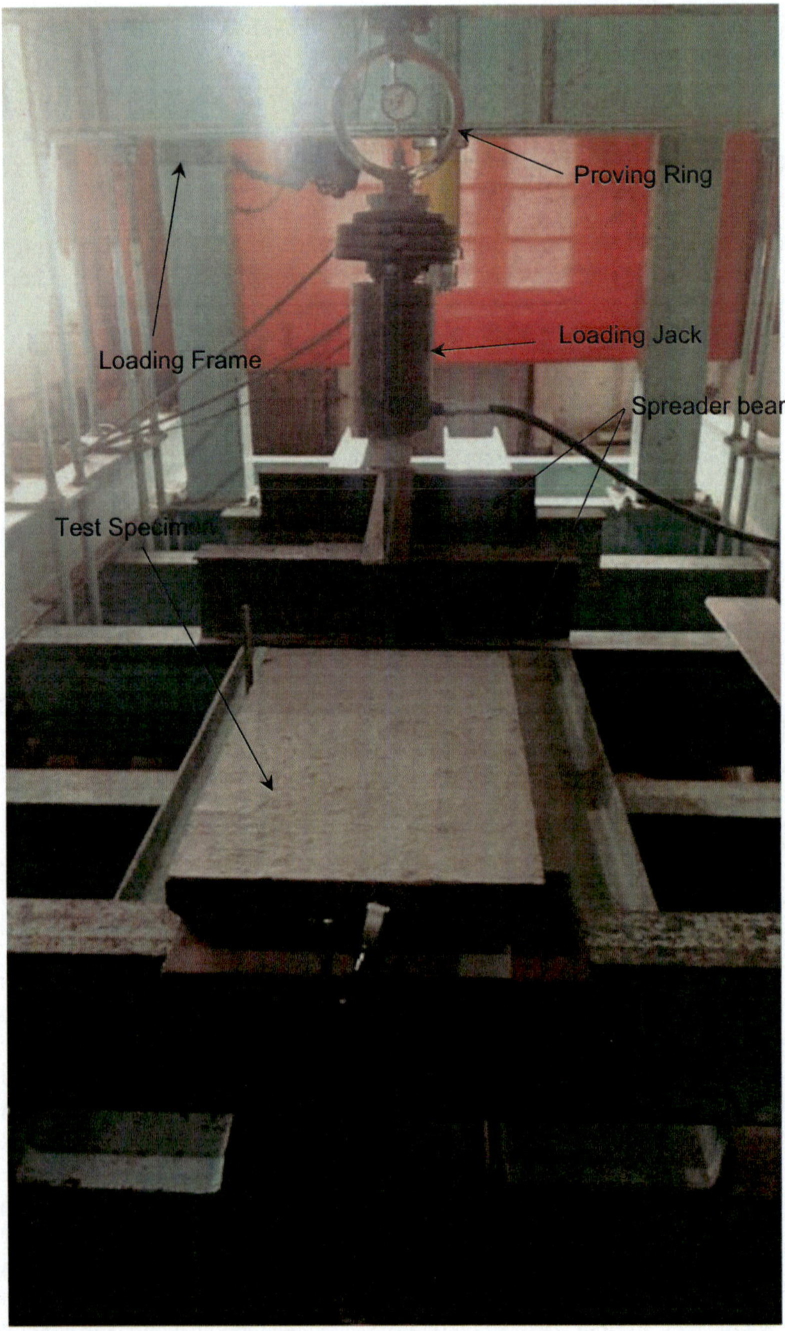

Fig. 4 A view of the test set-up

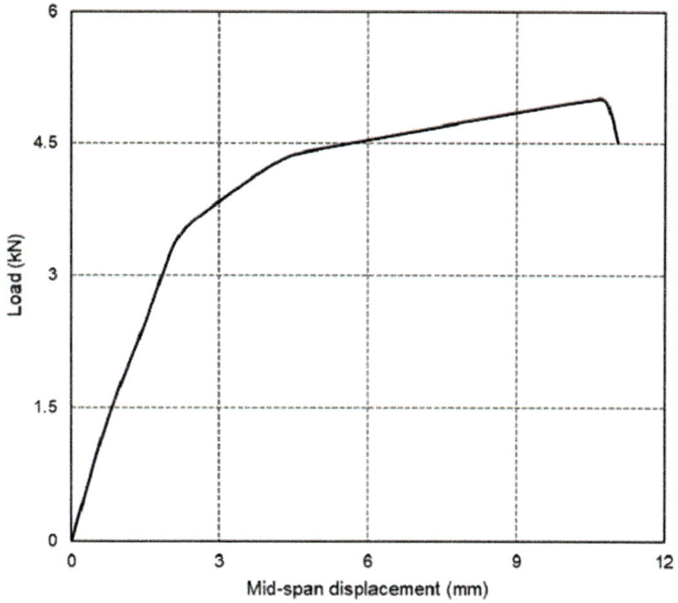

Fig. 5 Load versus displacement plot of the benchmark specimen

Fig. 6 Failure in hole punched specimen

out using other types of shear connectors and bring out the merits and demerits of each one of them. The role of different profiles needs to be investigated as well.

Limited study on the effect of shear connectors was carried out in the current research work. However, to develop better shear connectivity between the steel deck sheet and the concrete slab part, more types of shear connectors must be studied. Further, the test results need to be validated using finite element modelling, for carrying out an extensive parametric study, to bring out reliable design guidelines for the same.

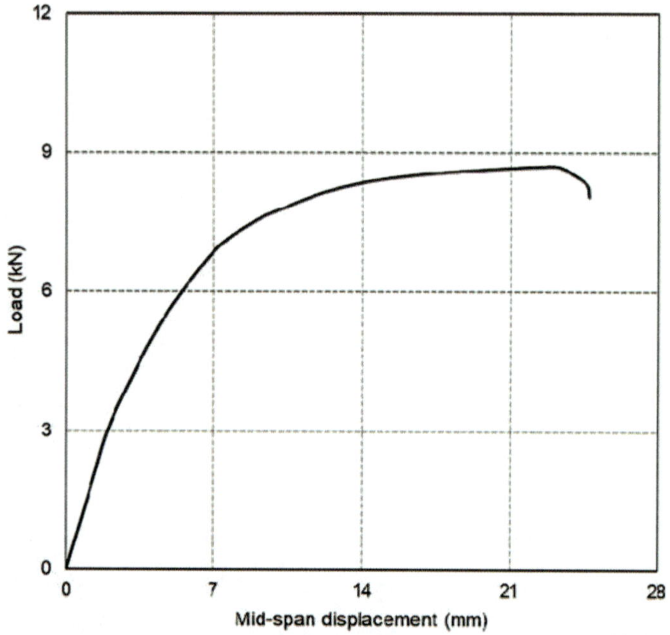

Fig. 7 Load versus displacement plot of the hole punched specimen

Fig. 8 Failure in hole punched specimen

This research work aimed at developing an effective CFS composite slab, with lower depth when compared to conventional RC slabs. This would reduce the consumption of concrete and would contribute towards sustainability.

References

1. Dar, M.A., Subramanian, N., Dar, D.A., Dar, A.R., Anbarasu, M., Lim, J.B.P., Mahjoubi, S.: Flexural strength of cold-formed steel built-up composite beams with rectangular compression flanges. Steel Compos. Struct. **34**(2), 171–188 (2020)
2. Dar, M.A., Subramanian, N., Rather, A.I., Dar, A., Lim, J.B.P., Anbarasu, M., Roy, K.: Effect of angle stiffeners on the flexural strength and stiffness of cold-formed steel beams. Steel Compos. Struct. **33**(2), 225–243 (2019)
3. Anbarasu, M., Dar, M.A.: Improved design procedure for battened cold-formed steel built-up columns composed of lipped angles. J. Constr. Steel Res. **164**, 105781 (2020)
4. Dar, M.A., Sahoo, D.R., Jain, A.K.: Axial compression behavior of laced cold-formed steel built-up columns with unstiffened angle sections. J. Constr. Steel Res. 162 (2019).
5. Dar, M.A., Subramanian, N., Dar, A.R., Muheeb, M., Haseeb, M., Mugees, T.: Structural efficiency of various strengthening schemes for cold-formed steel beams: effect of global imperfections. Steel Compos. Struct. **30**(4), 393–403 (2019)
6. Dar, M.A., Subramanian, N., Dar, A.R., Anbarasu, M., Lim, J.B.P., Mir, A.: Performance of partly stiffened cold-formed steel beams: tests and numerical simulations. Adv. Struct. Eng. **22**(1), 172–186 (2019)
7. Dar, M.A., Sahoo, D.R., Pulikkal, S., Jain, A.K.: Behaviour of laced built-up cold-formed steel columns: experimental investigation and numerical validation. Thin Wall. Struct. **132**(11), 398–409 (2018)
8. Dar, M.A., Subramanian, N., Anbarasu, M., Dar, A.R., Lim, J.B.P.: Structural performance of cold-formed steel composite beams. Steel Compos. Struct. **27**(5), 545–554 (2018)
9. Dar, M.A., Dar, A.R., Yusuf, M., Raju, J.: Experimental study on innovative sections for cold formed steel beams. Steel Compos. Struct. **19**(6), 1599–1610 (2015)
10. Dar, M.A., Sahoo, D.R., Jain, A.K.: Numerical study on the structural integrity of built-up cold-formed steel battened columns. In: Lecture Notes in Mechanical Engineering, pp. 815–823. Springer, Singapore (2020)
11. Anbarasu, M., Dar, M.A.: Axial capacity of CFS built-up columns comprising of lipped channels with spacers: nonlinear response and design. Eng. Struct. **213**, 110559 (2020)
12. Dar, M.A., Sahoo, D.R., Jain, A.K.: Compression capacity of short cold-formed steel built-up columns with double lacing configuration and low sectional compactness. In: Proceedings of the Structural Stability Research Council Annual Stability Conference 2019, vol. 1, pp. 472–482. St. Louis, Missouri, USA (2019).
13. Schuster, R.M.: Composite steel-deck reinforced concrete systems failing in shear-bond. In: 9th Congress. International Association for Bridge and Structural Engineering. Amsterdam, Holland (1973)
14. Johnson, R.P.: Composite Structures of Steel and Concrete. Vol 1: Beams, Slabs, Columns, and Frames for Building. Blackwell Scientific Publication, Oxford (1994)
15. Veljkovic, M.: Behaviour and resistance of composite slabs. Ph.D. thesis. Lulea University of Technology, Lulea, Sweden (1996)
16. de Andrade, S.A.L., da S. Vellasco, P.C.G., da Silva, J.G.S., Takey T.H.: Standardized composite slab systems for building constructions. J. Constr. Steel Res. 493–524 (2004)
17. IS 10262: 2019 Guidelines for Concrete Mix Design Proportioning, 2nd rev. Bureau of Indian Standards, New Delhi

Investigation of Critical Factors Influencing Cost Overrun in Highway Construction Projects

Varun Kumar Sharma, Pardeep Kumar Gupta, and R. K. Khitoliya

Abstract One of the major problems which are becoming very significant and is booming worldwide in the construction industry is the problem of cost overrun. With the increase in demand to meet the need for infrastructure, the problem of cost overrun is also spreading at an alarming rate. The construction companies of India engaged in highway construction are also no longer exempted from this serious issue. The major objective of this research paper is to recognize the major causes of cost overrun in highway projects in the regions of Northern India. A broader review of literature from different sources and interactions with the experts working in highway projects was done to list down the feasible causes of cost overrun. In the next stage, a questionnaire was formulated which consisted of 35 feasible causes of cost overrun. Respondents having vast experience in the field of highway projects, participated in the survey to recognize the critical factors responsible for cost overrun. The analysis of the data was done using IBM SPSS 23 Package. The top five most critical factors causing cost overrun have been identified and discussed in this paper along with some recommendations which are useful to control cost overrun in the highway projects. The top five critical factors identified in this study are escalation in price of raw material, changes in design, conflict/lack of co-ordination between construction parties, poor financial control and constructability issues (construction under limited area, construction under traffic, right of way).

Keywords Cost overrun · Highway construction project · Raw materials · Project planning · Relative importance index value

V. K. Sharma (✉) · P. K. Gupta · R. K. Khitoliya
Punjab Engineering College (Deemed To Be University), Chandigarh 160012, India
e-mail: varunsharmapec@gmail.com

P. K. Gupta
e-mail: p_gupta_2000@yahoo.com

R. K. Khitoliya
e-mail: r_khit_2000@yahoo.com

D. K. Ashish et al. (eds.), *3rd International Conference on Innovative Technologies for Clean and Sustainable Development*, RILEM Bookseries 29,
https://doi.org/10.1007/978-3-030-51485-3_22

1 Introduction

The construction industry has transformed into a basic player and is crucial for the fiscal improvement of any country. In developing countries the scenario is such that, this industry encounters different issues that impact the project success in terms of cost, time and quality. Successful completion of highway construction projects well within the predetermined expense has transformed into an extreme assignment. It is considered very astonishing if an endeavor is completed well within its stipulated period, its specified quality and contracted budget. It is very essential to manage the funds involved in the completion of the project in a properly planned manner. In most of the cases, better planning and management is lacking as a consequence, the project suffers from a remarkable amount of overrun in terms of time as well as cost. To overcome this situation it is very essential to have effective planning from the initial stages of the project. It is very important to understand and prioritize the causes of cost overrun to tackle down these factors and minimize cost overrun in highway projects.

Cost overrun can likewise be comprehended as the difference between real cost which might be characterized as the cost decided toward the completion of the project and introductory cost which is the estimated expense at initial stages [1]. In relative terms cost overrun might be presented in terms of percentage and is computed by utilizing the equation as appeared in "Eq. (1)".

$$\text{Cost Overrun \%} = \frac{[\text{Actual cost at completion of project} - \text{Original estimated cost}] \times 100}{\text{Orignal estimated cost}} \quad (1)$$

2 Literature Survey

The circumstance which creates the situation of cost overrun may or may not differ from country to country. But the problem of cost overrun in highway projects has become a worldwide phenomenon. Cost overrun variation is influenced by various variables depending upon the prevailing environment in the country such as the economic conditions, social conditions, political conditions and also geographical conditions. Some of these factors are due to the involvement of more numbers of stakeholders, different construction materials and equipments, climatic conditions, human resources and so forth.

One of the main considerations for successfully accomplishing a project is completing the project under-considered budget [2]. Project cost overrun is the contrast between the estimated cost during the planning stage and the genuine cost occurred on completion of the project. The construction industries in India are commonly unable to finish the projects under the evaluated expenses [3]. In a study by

Ibrahim et al. [4] it was reported that the cost of the project is the budgeted expenditure, which the client agrees to commit in creating or acquiring the desired construction facility. Cost overrun is the difference between the actual cost of construction incurred after completion of the project and the estimated amount, concurred by the stakeholders during the signing of the contract agreement. Cost overrun is also called cost increase, cost escalation or budget overrun. Besides, cost execution strategies are the key estimation of organizations efficiency and benefit [5].

Several examinations have been performed which led to deal with the issues and concerned factors that are responsible for cost overrun in highway projects. An examination by Flyvbjerg et al. [2] revealed, 9 ventures from the total of 10 construction ventures, faced cost overrun with an average cost overrun of about 28%. Another examination led by Cantarelli et al. [6] demonstrated that cost overrun is a typical issue in construction ventures, which scrutinized 87 projects and established that there were more than 8 projects which confronted cost overruns.

An investigation in the Gaza Strip on cost overruns, revealed the critical causes being lack of construction material, fluctuation in the price of materials, late procurement and availability of materials, differentiation of currency prices, poor control of cost, additional works, poor executives at site, poor correspondence and coordination between stakeholders, delay in payment as per the progress, inadequate planning and scheduling, project complexities, mistakes in configuration of records and delays in obtaining approval for structure endorsement [7].

Another investigation conducted by Le-Hoai et al. [8] considering construction projects of Vietnam found that the vital components that raised cost overrun were, troubles in financing the ventures by the project owner and contractor, change in configuration of designs, poor project management, poor supervision on-site and increment in cost of labor wages. An exploration was completed by Kaming et al. [9] recognizing the huge cost overrun factors. In Saudi Arabia, an examination dependent on construction industries demonstrated that the major factors that provoked cost overrun were the financial dependability, effects of climatic conditions, location of the venture, absence of efficiency norms, socio-cultural impacts, control of the suppliers, deficiency in availability of raw materials, lastly absence of construction cost information [10]. An investigation conducted in Ghana, recognized that the five primary cost overrun factors are poor contractor management, material procurement, difficulty in payment on monthly basis from agencies, poor technical performances, and escalation of material prices [11]. In Nigeria, the key explanations behind cost overwhelms in infrastructure projects are: poor executives, material deficiency, value variation, frequent change in design, financing, late issuing of installments and climate conditions [12]. In India, an examination driven by Iyer and Jha [3] represented ten basic elements prompting cost overrun. To be specific, the clash among members of the venture, ignorance of past experience, poor project attributes, non-existence of cooperation, hostile socio-economic situations, climatic conditions, hesitance hampering timely decision, forceful challenges at tendering stage and short time offered for planning. The outcomes showed that the key factors are the degree of project complexity, material cost increments because of inflation and inaccurate materials estimation.

3 Methodology

To achieve the objective, quantitative methodology was adopted in this study. The study was conducted in two stages. The first stage was emphasized over considering widespread literature survey work and highway project reports to gather the maximum number of factors which might cause cost overrun. It was followed by conducting personal interactions with the expert in the field. As a result of this stage, 35 factors of cost overrun were identified and were further utilized in preparing the questionnaire. The second stage comprised of conducting a questionnaire survey. The questionnaire comprises of two parts: Part A and Part B. Part A comprises of the general particulars of the respondents (name, email-id, contact info, educational qualification, company name, position and working experience), and Part B focused upon factors responsible for cost overrun. In this section the respondents were asked to rank the factors based upon the severity of each factor, depending upon their own judgment and local working experience. The responses received were later used for analysis to understand the severity of each factor. This was done to identify the critical cost overrun causing factors in the regions of Northern India.

4 Data Collection and Analysis

The questionnaire was designed on a five-point Likert scale. Initially draft questionnaire was prepared and discussed with the local experts in highway construction to evaluate the content of the questionnaire. Modifications and changes were incorporated accordingly. For each factor causing cost overrun, used in the questionnaire, respondents had to choose any one option out of the five options to mark the severity of each factor. The severity was categorized on a five-point scale: very low, low, moderate, high, and very high on a scale of 1–5 respectively. To design the sample size for the study, Cochran's (1963) formulas as shown in "Eq. (2)" was used. For the large population, "Eq. (2)" is used to yield a representative sample size whereas for the limited population "Eq. (3)" is used [13].

$$n_o = \left[\left(z^2 * \sigma^2\right)/d^2\right] \tag{2}$$

$$n_1 = [n_o/1 + (n_o/N)] \tag{3}$$

In the above equations, n_0 and n_1 is the sample size. If the computed value of n_0 from "Eq. (2)" is greater than 5% of population size (N), then we have to apply Cochran's correction formula as shown below in "Eq. (3)". In the above equation Z is a numerical measurement used in statistics for any value's relationship to the mean value of the group of values, measured in terms of standard deviations from the mean. It is calculated from the Z table. For a 95% confidence interval the value of Z is taken as 1.96. Since we have used 5 point Likert Scale, so the value of σ will be 1.25. For the desired level of precision (e) of 0.05 and for 5 point Likert scale the value of d is 0.15, corresponding to a 95% confidence interval. N is the population size corresponding to 100 considered highway projects [13].

The value of n_0 is calculated by substituting the respective values of Z, σ, and d in "Eq. (2)" and it comes out to be 278. Now this value is more than 5% of the population size considered in this case. Substituting the value of n_0 and N in "Eq. (3)", we obtain the value of n_1 which is equal to 74. Rounding of the obtained value to 70 we consider the value of sample space as 70 and proceeded further in the study.

Designed questionnaires were distributed using Google Forms and also hand to hand among 70 respondents having considerable experience in the construction of highway projects. A total of 60 responses were received back and considered for evaluation representing an 85% response rate. Ranking of the factors was adopted based on R.I.I (Relative Importance Index) value calculated from the response of each respondent. Higher the value of R.I.I for any cost overrun factors it indicates it is more critical in comparison to other factors with lower value of R.I.I. To find the critical factors of cost overrun, R.I.I value of each factor is calculated using "Eq. (4)".

$$\text{R.I.I value} = \sum \text{W}/(\text{H} \times \text{N}) \tag{4}$$

In the above mentioned "Eq. (4)", R.I.I value is relative importance index value, H is the highest-ranking available as per the Likert scale used in the questionnaire, N is the total number of respondents who answered the questionnaire survey. $\sum W$ is the total weightage given to each factor by the respondents which range from 1 to 5.

Considering the responses for the factor "escalation in the price of raw material" we have, 31 responses which considered it has a very high level of severity corresponding to the value of 5 on Likert scale, similarly 25 responses considered it to be highly critical corresponding to value 4, whereas 4 responses believed it had a moderate level of severity corresponding to value 3, whereas 0 responses for low and 0 responses for very low. From this the total weightage is given to this factor by the respondents ($\sum W$) is calculated and it comes out to be 267.

The calculation is: $\sum W = \{(31 \times 5) + (25 \times 4) + (4 \times 3) + (0 \times 2) + (0 \times 1)\} = 267$. The total number of respondents who answered the questionnaire survey (N) is 60. The highest-ranking available (H) is 5 as per the Likert scale used. So the R.I.I. value for this factor will be computed by substituting the values, $\sum W = 267$, $H = 5$ and $N = 60$ in "Eq. (4)". Therefore the R.I.I. value for the factor "escalation in the price of raw material" comes out to be 0.89. The R.I.I value, based upon "Eq. (4)" was computed for each of these 35 factors responsible for cost overrun. After obtaining the R.I.I value the ranking of the factors was done. The factor having a higher value was ranked top in comparison to other factors having a lower value. The R.I.I. value and the respective ranking for all the factors are shown in Table 1.

Table 1 Ranking of factors causing cost overrun in highway projects based upon R.I.I. value

S. no	Factor code	Cost overrun causing factors	R.I.I. value	Ranking
1	COF1	Escalation in the price of raw material	0.89	1
2	COF2	Changes in design	0.85	2
3	COF3	Conflict/Lack of co-ordination between construction parties	0.84	3
4	COF4	Poor financial control	0.84	4
5	COF5	Constructability issues (construction under limited area, construction under traffic, right of way)	0.80	5
6	COF6	Land acquisition	0.79	6
7	COF7	Delays in shifting existing utilities	0.76	7
8	COF8	On site wastage	0.72	8
9	COF9	Ambiguous or incomplete tender document	0.72	9
10	COF10	Poor workmanship and unskilled labour force	0.72	10
11	COF11	Inaccurate Feasibility report	0.71	11
12	COF12	Construction mistakes	0.69	12
13	COF13	Change is rules and regulations or policies by the Government	0.65	13
14	COF14	Project delivery system	0.64	14
15	COF15	Political issues	0.62	15
16	COF16	Social issues and public agitations	0.60	16
17	COF17	Additional works	0.59	17
18	COF18	Inefficient/Inaccurate estimation during tendering	0.59	18
19	COF19	Dispute on settlement and clearance of bills	0.57	19
20	COF20	Nonavailability of raw material	0.57	20
21	COF21	Increase or unstable interest rates	0.54	21
22	COF22	High quality expectation from owner	0.53	22
23	COF23	Site management and labour relation	0.52	23
24	COF24	Shorting of the contract period	0.52	24
25	COF25	Shifting in the existing schedule	0.52	25
26	COF26	Improper interpretation of Contract and specifications	0.52	26
27	COF27	Dependency on imported materials	0.51	27
28	COF28	Organizational or Institutional Failures	0.51	28
29	COF29	Unrealistic work schedule	0.51	29
30	COF30	Inefficient utilization of resources	0.50	30
31	COF31	Labour strikes	0.48	31
32	COF32	Act of god	0.48	32

(continued)

Table 1 (continued)

S. no	Factor code	Cost overrun causing factors	R.I.I. value	Ranking
33	COF33	Late issuing of approved documents	0.47	33
34	COF34	Negligence of site visit	0.46	34
35	COF35	Fraudulent practices and kickbacks	0.45	35

5 Results and Discussions

The top five, most critical factors have been analysed and discussed below. The R.I.I Value for these five factors were found to be 80% and above. These five factors must be given maximum concern to tackle the problem of cost overrun.

From, Table 1, it is clear that, the factor escalation in the price of raw materials ranks first among all the considered responsible factors. It is caused due to a change in policies by the Government. It might also be due to an increase in wages of labours working in the extraction of raw materials, machinery costs. Inflation is mainly considered responsible for such type of changes in price. Efficient arrangements at the procurement stage will trim down the escalation in price up to a certain extent. Provisions such as contingencies should be considered while making a contract, this will also be useful in avoiding cost overrun.

Change in design, ranks the second most critical factor. Misinterpretation of the data by designer, lack of technical clarity with respect to original site conditions, less experience and carelessness leads to change in design and finally leads to cost overrun. High accuracy of work with sufficient investigation during initial stages is very essential because it will add to the cost to set right the mistakes.

The third-ranked factor is Conflict/Lack of co-ordination between construction parties. Various problems are evolved due to poor communication between construction parties. It is one of the dominant factors and the issues contributing to these factors might be, communication gap prevailing among the stakeholders of the project. Constituting an advisory body which deals with issues such as resolving a disagreement between stakeholders, early sanction of payments and taking initiatives for early decision making for the progress of the project would also play a key role.

The problem of poor financial control ranks fourth. This could occur by the negligence of both the client and contractor. Withholding of payment by the client and unauthorized claim by contractors leads to the cost overrun of the project. To overcome this issue it is necessary to have proper monitoring, analysis and adjustment of the project's cash flow.

Constructability issues have been ranked fifth in this study. This issue arises from various reasons such as constructability under traffic area or under restricted area. Due to this special care and special arrangements are to be made, which directly contributes to cost overrun. These special arrangements when not included might cause severe situations, accidental or fatal injuries to others on site which might lead to the hindrance in the progress of the project by the law and order authorities, causing excessive delays and expenditure on the settlement of the issue. This somehow shows the adverse effect on the cost of the project and leads to cost overrun. To overcome this it is essential to develop policies and teams specifically to deal with the problem of constructability.

6 Conclusions and Recommendations

6.1 *Conclusions*

The issue of cost overrun has become the most widespread problem in the construction of highway projects in India. This study attempts to discover the most critical cost overrun causing factors that affect the highway projects in the regions of Northern India. A questionnaire survey was executed among various Government and private organizations. The major cost overrun causing factors, based on the perspective of respondents include, escalation in the price of raw material, changes in design, conflict/lack of co-ordination between construction parties, poor financial control, constructability issues (construction under the limited area, construction under traffic, right of way), land acquisition, delays in shifting existing utilities, on-site wastage, ambiguous or incomplete tender document, poor workmanship and unskilled labour force, inaccurate feasibility report, construction mistakes. Out of 35 considered cost overrun causing factors, the top most critical factors were selected based upon their R.I.I value. Total of five factors having R.I.I value above 80 were considered the most critical factors.

The analysis of the judgment which was provided by respondents, it may be inferred that the top five most critical factors affecting cost overrun are escalation in price of raw material, changes in design, conflict/lack of co-ordination between construction parties, poor financial control and constructability issues (construction under limited area, construction under traffic, right of way).

6.2 *Recommendations*

To avoid cost overrun occurring due to escalation in the price of raw material it is recommended to conduct a detailed work out on material requirements and its procurement strategies in the initial stages. Preplanned purchase or better procurement strategies would be beneficial to avoid the issue of raw materials which are rarely available and will also resolve the issue of nonavailability of construction materials. Apart from procurement strategies good storing facilities at the site are also recommended. Consideration of realistic overrun percentage in project estimate is very essential to avoid cost overrun in later stages.

Appointment of experienced person/team and providing optimum time for the task of designing would reduce the problem of overruns occurring due to errors in design.

A panel of experienced designers and professionals should be made to verify the designs made at initial stages before the execution stage.

It is highly recommended to have an additional body on board that connects clients, contractors and consultants on frequently occurring problems and resolving issues on an early basis. This will prevent cost overrun occurring due to the issue of conflict and communication gaps in the project stakeholders.

Proper monitoring of cash flow in the project is very essential. The analysis of cash flow in the project will provide better financial control in the project. This will resolve issues related to funding of the projects.

One of the severe issues is of land acquisition. It is recommended to have a separate team that operates and resolve the issue of land acquisition. The major role of this team would be the identification of land to be acquired for the project at an early stage and possessing the land before the execution work of the project starts.

To avoid the cost overrun occurring due to the delays in the removal of utilities from the project site, an effective utilities removal plan should be adopted in the preconstruction stage.

Certain unavoidable situations might occur and it is very difficult to predetermine the extent of overrun caused by it. A situation such as the Act of God is the best example of this. There is no full-proof solution to prevent this but its effect on the project could be minimized up to a certain extent by considering a provision in the contract that will provide a certain allowance in terms of time and cost.

Cost overruns due to unstable interest rates can also be reduced by utilizing a floating rate of interest. Providing sufficient time to prepare tenders and estimation of quantities is a key factor that prevents overrun which is supposed to occur in the construction phase.

References

1. Chitkara, K.K.: Construction Project Management, 3rd edn. Tata McGraw-Hill Publishing Company Limited, New Delhi (1998)
2. Flyvbjerg, B., Skamris Holm, M.K., Buhl, S.L.: What causes cost overrun in transport infrastructure projects? Transp. Rev. **24**(1), 3–18 (2004)
3. Iyer, K.C., Jha, K.N.: Factors affecting cost performance: evidence from Indian construction projects. Int. J. Project Manage. **23**(4), 283–295 (2005)
4. Ibrahim, S., Shakantu, W.M.: A conceptual framework and a mathematical equation for managing construction material waste and cost overruns. Int. J. Soc. Behav. Educ. Econ. Bus. Ind. Eng. **10**(2), 587–593 (2016)
5. Olawale, Y.A., Sun, M.: Cost and time control of construction projects: inhibiting factors and mitigating measures in practice. Constr. Manag. Econ. **28**(5), 509–526 (2010)
6. Cantarelli, C.C., Flyvbjerg, B., van Wee, B., Molin, E.J.: Lock-in and its influence on the project performance of large-scale transportation infrastructure projects: investigating the way

in which lock-in can emerge and affect cost overruns. Environ. Plann. B: Plann. Des. **37**(5), 792–807 (2010)
7. Enshassi, A., Mohamed, S., Abushaban, S.: Factors affecting the performance of construction projects in the Gaza strip. J. Civil Eng. Manag. **15**(3), 269–280 (2009)
8. Le-Hoai, L., Dai Lee, Y., Lee, J.Y.: Delay and cost overruns in Vietnam large construction projects: a comparison with other selected countries. KSCE J. Civil Eng. **12**(6), 367–377 (2008)
9. Kaming, P.F., Olomolaiye, P.O., Holt, G.D., Harris, F.C.: Factors influencing construction time and cost overruns on high-rise projects in Indonesia. Constr. Manag. Econ. **15**(1), 83–94 (1997)
10. Alhomidan, A.: Factors affecting cost overrun in road construction projects in Saudi Arabia. Int. J. Civil Environ. Eng. **13**(3), 1–4 (2013)
11. Frimpong, Y., Oluwoye, J., Crawford, L.: Causes of delay and cost overruns in construction of groundwater projects in a developing countries; Ghana as a case study. Int. J. Project Manage. **21**(5), 321–326 (2003)
12. Elinwa, A.U., Joshua, M.: Time overrun factors in Nigerian construction industry. J. Constr. Eng. Manag. **127**(5), 419–425 (2001)
13. Naing, L., Winn, T., Rusli, B.N.: Practical issues in calculating the sample size for prevalence studies. Arch. Orofacial Sci. **1**, 9–14 (2006)

Use of Char Derived from Waste Plastic Pyrolysis for Asphalt Binder Modification

Abhinay Kumar, Rajan Choudhary, and Ankush Kumar

Abstract Plastic waste management is posing a pressing challenge worldwide. Plastics take thousands of years to degrade naturally, and their disposal in landfills raises grave environmental concerns. Pyrolysis is gaining significant interest as a reliable method for management and energy valorization for plastic wastes. In addition to pyrolytic oil having good calorific value, waste plastic pyrolysis also yields solid char, as a by-product, which doesn't find wide applications and is generally discarded. This study aims to evaluate the waste plastic pyrolytic char as an asphalt binder additive/modifier through different rheological and mechanical investigations. The modification of asphalt binder has been attempted at five char contents (0, 5, 10, 15 and 20% by binder weight). The rheological tests included flow behavior, frequency, and temperature sweeps, multiple stress creep recovery (MSCR). Zero shear viscosity was derived using the Cross model. Burger's model was used to describe the MSCR response. Modified binders showed improved stiffness than the control binder indicating an improvement in their resistance against permanent deformation. The plastic char used in the study showed encouraging results and shows the potential to be used as an additive/modifier for asphalt binders.

Keywords Modified asphalt · Plastic pyrolysis · Pyrolytic char · Rheology · Plastic waste

---

A. Kumar · R. Choudhary (✉) · A. Kumar
Department of Civil Engineering, Indian Institute of Technology Guwahati, Guwahati, India
e-mail: rajandce@iitg.ac.in

A. Kumar
e-mail: abhinay.kumar@iitg.ac.in

A. Kumar
e-mail: ankus174104035@iitg.ac.in

D. K. Ashish et al. (eds.), *3rd International Conference on Innovative Technologies for Clean and Sustainable Development*, RILEM Bookseries 29,
https://doi.org/10.1007/978-3-030-51485-3_23

1 Introduction

Plastics are long-chain hydrocarbons synthesized from petroleum crude and its derivatives. Owing to their durability, light-weight and low cost, plastics remain in high demand in sectors such as packaging, toys, electronics, automobiles, and agriculture. Since 1990, the use of plastics has increased by 5% annually [1]. The increased consumption of plastics has correspondingly increased the quantum of waste plastic to the extent that its management and disposal has now become a pervasive environmental, health and economic issue globally. Plastic wastes are biologically imperishable and may remain in the surroundings for hundreds of years. Their accumulation in landfills and oceans is a major threat to aquatic and terrestrial animal life [2, 3].

Two of the most widely adopted plastic waste management strategies are incineration and recycling. Whereas the incineration of plastics contributes to toxic and harmful emissions, recycling of plastics is also challenging because of the requirement of segregating different plastics, which is a highly labor-intensive activity. Segregation is needed as plastics are comprised of several resins that result in differences in transparency, strength, and color. Plastics that are pigmented have much lower value after recycling [3]. The energy valorization techniques such as pyrolysis, gasification and plasma arc gasification have now garnered wide interest as alternative methods of management of plastic wastes. Pyrolysis involves thermochemical decomposition of the long-chain molecules into simple smaller molecules. The process occurs in a low oxygen environment at higher temperatures. It carries the potential to reduce the dependence on fossil fuel-based energy sources since oil (pyrolytic oil) obtained from the pyrolysis process can be used in energy generation [2, 4].

Asphalt mixes are extensively used for pavement construction all over the world. The performance and mechanical characteristics of asphalt mixes are highly dependent on the asphalt binder properties. Higher temperatures and heavier axle loads have resulted in the early onset of distresses in asphalt pavements with neat (unmodified) binders. Therefore, asphalt binders are nowadays frequently modified to improve the properties of asphalt mixes against pavement distresses (like rutting, cracking, fatigue, and moisture damage). Diverse modifiers in the form of thermoplastic elastomers, thermoplastic plastomers, carbonaceous materials (e.g. carbon black, carbon fiber, coke dust), etc. are used/tried as asphalt binder modifiers/additives. In addition to pyrolytic oil with good calorific value, waste plastic pyrolysis also generates solid carbonaceous char as a by-product. It is needed to find sustainable application channels where the use of this waste plastic pyrolytic char can have a greater impact.

In previous studies, char from pyrolysis of other waste materials [such as scrap tires, biomass wastes (bark residues, seed cover, wood waste)] has been used for asphalt binder modification [5–8]. Lesueur et al. [5] studied char from pyrolyzed tires as a modifier to asphalt binders. The char was compared with other aggregate fillers and asphalt mastics were prepared at similar filler-to-asphalt (weight) ratios of 20–40%. The addition of char enhanced the rutting resistance properties of the mastics and the stiffness increased caused by 20% tire char was equivalent to that caused by 30% mineral fillers. Also, asphalt mixtures fabricated with the tire char exhibited better stripping resistance compared to mixes with other fillers. Chebil et al. [6] used char obtained from vacuum pyrolysis of softwood bark residues as an additive to asphalt binders in the concentrations of 5–25%. A commercial carbon black (CCB) was also used as a reference carbonaceous additive. Results of binder testing indicated that the thermal susceptibility of the char modified binder was lower than that of CCB modified binder. Based on the Superpave rutting parameter, wood-derived char was found to be as suitable as CCB in improving the rutting resistance of the binders. Zhang et al. [7] used char derived from the pyrolysis of waste wood as an asphalt binder modifier at dosages of 0, 2, 4 and 8% by weight. Two sizes of char were used: (1) between 75 and 150 μm, and (2) finer than 75 μm. Scanning electron microscopy (SEM) images revealed that waste wood pyrolytic char had hollow fiber-like particles with a porous and rough structure. The authors stated that fibrous, rough and porous structure caused the char to have a higher specific surface area and hence a better adhesion interaction with the asphalt binder. The addition of char contributed to an increase in rotational viscosities measured at elevated temperatures (110, 135, 150, and 165 °C). Smaller char size resulted in a higher increase in viscosity compared to the larger size. Kumar et al. [8] used pyrolytic char obtained from pyrolysis of seed cover waste of *Mesua ferrea* (rose chestnut tree). Rough, irregular and porous surface features of the char were evident in the SEM images. Modifier dosages for preparation of the blends were 5, 10, 15 and 20% by binder weight. $G^*/\sin\delta$ (G^*: complex modulus; δ: phase angle) of modified binders improved with increasing char dosages, implying increased stiffness of the binder with the addition of char. The modified binders also showed improvement towards resistance to rutting as the multiple stress creep and recovery (MSCR) results showed lower non-recoverable compliance (J_{nr}) and accumulated strain. Review of the literature indicates that pyrolytic char derived from plastic waste pyrolysis has not yet been utilized for asphalt binder modification. Therefore, this study explores the possibility of the use of plastic waste pyrolytic char as a modifier in asphalt binders.

The objectives of the present study are to evaluate the effect of incorporation of different dosages of plastic waste pyrolytic char on the rheological properties of asphalt binders. The binder is modified with five dosages of the char (0, 5, 10, 15, and 20% by weight of the binder). In particular, the present study includes the following rheological investigations on the control and plastic pyrolytic char modified binders: flow behavior (viscosity at varying shear rates), temperature sweep and frequency sweep. Rutting resistance of the binders is then quantified through four different parameters, namely Superpave rutting parameter, Shenoy rutting parameter, zero shear viscosity (ZSV), and multiple stress creep and recovery (MSCR). MSCR results are also submitted for Burger's four-element modeling to derive useful characteristics related to creep and recovery response.

2 Materials and Methodology

2.1 Materials and Preparation of Modified Binder Blends

An unmodified/neat VG-30 grade asphalt binder was used as the base/control binder. Table 1 shows the basic physical properties of the base binder. The plastic pyrolytic char used was obtained from the pyrolysis of polypropylene plastic waste, and was supplied by Innova Engineering & Fabrication (Mumbai). Figure 1 presents the SEM image of the pyrolytic char. The SEM images show a rough surface texture

Table 1 Properties of base asphalt binder (VG-30)

Property	Requirements[a]	Results	Standards
Absolute viscosity at 60 °C (poise)	2400–3600	3410	IS: 1206 [12]
Kinematic viscosity at 135 °C (cSt)	Min 350	525	IS: 1206 [12]
Penetration at 25 °C, 100 g, 5 s, 0.1 mm	Min 45	51.6	IS: 1203 [13]
Softening point (R&B) (°C)	Min 47	52.7	IS: 1205 [14]
Flash point (COC) (°C)	Min 220	280	IS: 1448 [15]
Solubility in trichloroethylene (%)	Min 99	>99	IS: 1216 [16]
Properties for rolling thin film oven (RTFO) residue			
Viscosity ratio, 60 °C	Max 4	1.15	IS: 1206 [12]
Ductility, 25 °C (cm)	Min 40	>100	IS: 1208 [17]

[a]Requirements as per IS: 73 [11], Indian Standard on 'Paving Bitumen Specification'

Fig. 1 SEM image of plastic pyrolysis char

and porous microstructure of the char particles. Such a microstructure is attributed to rapid devolatilization during thermal degradation in the pyrolysis process leading to escape of the volatiles [9, 10]. The pyrolysis process broadly consisted of cleaning and shredding the plastic to about 15–25 mm size and feeding it to the pyrolysis reactor with a proprietary catalyst. The reactor was purged with nitrogen gas to remove all oxygen inside the system. At a temperature of around 250 °C, the plastic cracked and the vapors released were condensed to pyrolysis oil and the uncondensed gases were reverted to the reactor. The products obtained were 70–80% pyrolysis oil, 15–25% carbonaceous char and 10–15% gases. The total process time was about 15 h for an 8-ton batch.

Modified binder blends were prepared on a high shear mixer operating at 12,000 rpm at 160 °C for 30 min. Sulfur was added to all blends (except control/neat binder) at a content of 0.3% by weight of binder for the desired storage stability. Additional high shear mixing for 15 min was conducted after the addition of sulfur. All the blends with sulfur thus prepared were storage stable and met the storage stability requirements of IS 15462 [18]. Immediately after the blending, the blends were stored at -15 °C for 24 h to preserve the modifier dispersion. For rheological tests, an RTFO was used to obtain short-term aged binders in accordance with ASTM D2872 [19]. The binders were placed in the RTFO at 163 °C under an airflow of 4000 mL/min for 85 min to obtain short-term aged binders for further rheological characterization.

2.2 *Rheological Characterization*

Rheology of asphalt binders has a significant influence on the resistance to deformation of the resulting asphalt mixes [20, 21]. The rheological tests were conducted on an Anton Paar MCR 102 dynamic shear rheometer (DSR) with a 25 mm diameter parallel plate geometry with 1 mm gap setting. The viscoelastic rheological parameters complex shear modulus (G^*) and phase angle (δ) were measured on the DSR according to AASHTO T315 [22]. Viscosity was determined at 60 °C at varying shear rates to investigate the flow behavior of the different asphalt binder blends. Frequency sweep was performed at 60 °C where the frequency varied in three decades (0.1–1, 1–10, 10–100 rad/s). Temperature sweep was conducted between 25 and 80 °C at 1.59 Hz (10 rad/s) frequency on all binders.

Four rheological approaches were used in this study to characterize the rutting performance of the binders modified with plastic pyrolytic char: Superpave rutting parameter ($G^*/\sin\delta$), Shenoy rutting parameter ($G^*/(1 - 1/(\tan\delta \sin\delta))$), zero shear viscosity (ZSV), and multiple stress creep and recovery (MSCR). The former three parameters are oscillation-based and are measured at low strain levels, whereas MSCR subjects the binder to multiple stress levels and helps to assess the non-linear response and permanent deformation resistance at higher strain levels. Each of the test approaches is described in the sections that follow.

Superpave Rutting Parameter (G*/sin δ)

The philosophy behind this parameter is that the rutting resistance can be increased by either increasing the total resistance to deformation of the binder (higher G* values) or by raising the relative elastic property (lower δ values). Therefore, binders with higher values of G*/sin δ indicate better rutting performance [23]. The rutting parameter reflects the energy dissipated in each loading cycle when rutting is assumed as a cyclic loading and stress-controlled phenomenon and is mainly attributed to the deformation in the surface asphalt layer [23, 24]. The parameter was measured for short-term (RTFO) aged control and plastic pyrolytic char modified binders at 60 °C temperature and 10% strain (as per the specifications of AASHTO T315 [21]). There are some critical drawbacks with the G*/sin δ parameter that have motivated researchers to suggest new parameters to characterize the rutting performance of asphalt binders [23–26]. First, the strain level used to determine G*/sin δ is quite lesser than the actual binder strain levels in the pavement. Second, it is also realized that the complete reversal of cyclic loading in the stress (or strain) does not represent the contribution of the binder in the rutting resistance of the mix. This is because the rutting occurs due to accumulated plastic (irreversible) strains. Third, the overall effect of the phase angle (δ) on the parameter G*/sin δ is relatively small. Therefore, other rutting parameters were also considered in the study and are discussed below.

Shenoy Rutting Parameter

Based on theoretical rheological concepts, Shenoy [27] derived the expression for unrecovered strain ($\%\gamma_{\text{unr}}$) in a binder when subjected to a stress σ_0 as:

$$\%\gamma_{unr} = \frac{100\sigma_0}{|G*|}\left(1 - \frac{1}{\tan\delta\sin\delta}\right) \tag{1}$$

For a better rutting resistance, the unrecovered strain should be minimized, which can be done by maximizing the following term:

$$\frac{|G*|}{\left(1 - \frac{1}{\tan\delta\sin\delta}\right)} \tag{2}$$

The term in Eq. (2) is the Shenoy rutting parameter. G* and δ at 60 °C and 10 rad/s frequency on short-term aged binders were measured to derive the parameter. The Shenoy rutting parameter maximizes the impact of phase angle (δ) on the binder's rutting resistance [23, 27].

Zero Shear Viscosity (ZSV)

Modified asphalt binders exhibit pseudoplastic (shear-thinning) behavior but show Newtonian behavior at low shear rates with a viscosity independent of the shear rate. This viscosity is termed as ZSV and is measured at very low (near to zero) shear.

Frequency sweep was used to determine ZSV of asphalt binders using the Cross model as per Eq. (3):

$$\eta* = \frac{\eta_0 - \eta_\infty}{1 + (k\omega)^m} + \eta_\infty \tag{3}$$

where, $\eta* =$ complex viscosity (Pa.s); $\eta_0 =$ ZSV (Pa s); $\eta_\infty =$ limiting viscosity; $\omega =$ angular frequency (rad/s), k and m are constant parameters. k has the units of time and its reciprocal is termed as the critical shear rate (γ_c). m is a dimensionless constant. Cross model is a generalized Newtonian model that is able to capture binders' shear-thinning behavior. The frequency sweep was performed between the frequencies of 0.1 and 100 rad/s at 60 °C. Cross model parameters were calculated using non-linear regression on OriginPro software.

Multiple Stress Creep and Recovery (MSCR)
MSCR was performed to evaluate the rutting characteristics of the binders at multiple stress levels (0.1 and 3.2 kPa) as per the specifications of ASTM D7405 [28] at 60 °C. The test consisted of 20 creep-recovery cycles at 0.1 kPa (data was acquired for the second 10 cycles only) and 10 creep-recovery cycles at 3.2 kPa. The loading time period was 1 s and the unloading time period was 9 s that allowed the binder to recover the strain. Let ε_0 and ε_c denote the strain at the commencement and end of the creep (loading) portion for each creep-recovery cycle. The strain at the termination of the recovery portion is denoted by ε_r. Percent recovery (R) is then calculated for the binder as per Eq. (4) (averaged for the 10 cycles at each stress level):

$$R = \frac{(\varepsilon_c - \varepsilon_0) - (\varepsilon_r - \varepsilon_0)}{(\varepsilon_c - \varepsilon_0)} \times 100 \tag{4}$$

Non-recovered strain (ε_{nr}) in each creep-recovery cycle can be calculated as per Eq. (5):

$$\varepsilon_{nr} = \varepsilon_r - \varepsilon_0 \tag{5}$$

For each creep-recovery cycle, non-recoverable creep compliance (J_{nr}) at a stress level (σ) is calculated as:

$$J_{nr} = \frac{\varepsilon_{nr}}{\sigma} \tag{6}$$

A lower J_{nr} indicates lower rutting susceptibility of asphalt binders. The difference in average J_{nr} obtained at 0.1 and 3.2 kPa stress levels are used to calculate percent difference in J_{nr}, denoted as $J_{nr,diff}$ as per Eq. (7):

$$J_{nr,diff} = \frac{J_{nr3.2} - J_{nr0.1}}{J_{nr0.1}} \times 100 \tag{7}$$

The parameter $J_{nr,diff}$ represents the stress sensitivity of the asphalt binder and has a specified upper limit of 75% as per AASHTO MP19 [29].

Rheological Modeling of MSCR Response

Burger's model is frequently used to describe the creep-recovery characteristics of asphalt binders and asphalt-filler mastics [30, 31]. Burger model is able to simulate the three basic elements of viscoelastic behavior (instantaneous elasticity, delayed elasticity, and viscoelasticity) [32]. The schematics of Burger's four-element model are shown in Fig. 2 where a Maxwell element (a spring and dashpot in series connection) and a Kelvin element (a spring and a dashpot in parallel connection) are connected in series. Equations (8) and (9) mathematically describe Burger's model where the strain response under constant stress (σ_0) is presented for creep and recovery phases assuming that the stress is removed at time $t = \tau$:

For creep phase:

$$\varepsilon(t) = \frac{\sigma_0}{E_M} + \frac{\sigma_0 t}{\eta_M} + \frac{\sigma_0}{E_K}\left(1 - e^{\frac{-E_K t}{\eta_K}}\right) \tag{8}$$

For recovery phase:

$$\varepsilon(t) = \frac{\sigma_0 \tau}{\eta_M} + \frac{\sigma_0}{E_K}\left(1 - e^{\frac{-E_K \tau}{\eta_K}}\right)\left(e^{\frac{-E_K t}{\eta_K}}\right) \tag{9}$$

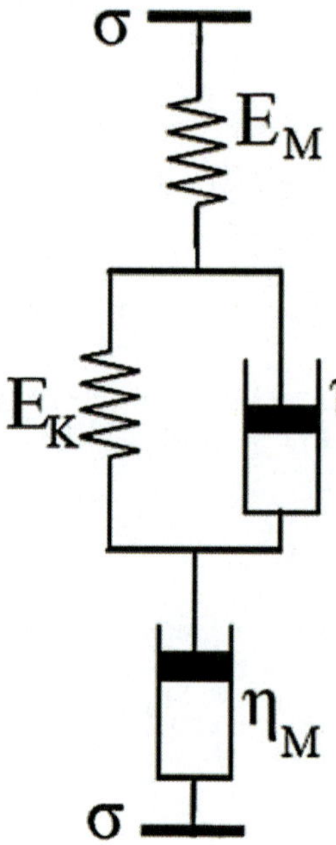

Fig. 2 Schematics of Burger's four-element model

The model is defined by two springs with moduli E_M and E_K and two dashpots with viscosities η_M and η_K. The spring element E_M represents the instantaneous elastic response, while the dashpot element η_M represents the viscous response. The two Kelvin elements (E_K and η_K) relate to delayed elastic response [30]. The parameters of Burger's model were estimated from MSCR data following the procedure reported by Liu and You [33].

3 Results and Discussion

3.1 Flow Behavior

The flow behavior of binders was assessed with the measurement of viscosity at varying shear rates at 60 °C (Fig. 3). Initially, at low shear rates, all binders show Newtonian flow behavior as the viscosity does not depend on the applied shear rate. At higher shear rates, the onset of pseudoplastic (shear-thinning) behavior can be clearly seen for all binders. The addition of pyrolytic char increases the viscosity and with higher char contents, there is a further increase in binder viscosity. Further, the char also influences the flow behavior in that the onset of shear-thinning happens at lower shear rates with an increase in the char dosage. The shear-thinning phenomenon is attributed to changes occurring in the internal structure of the binders [34]. The original internal structure of the binder is initially stable but gets disturbed with an

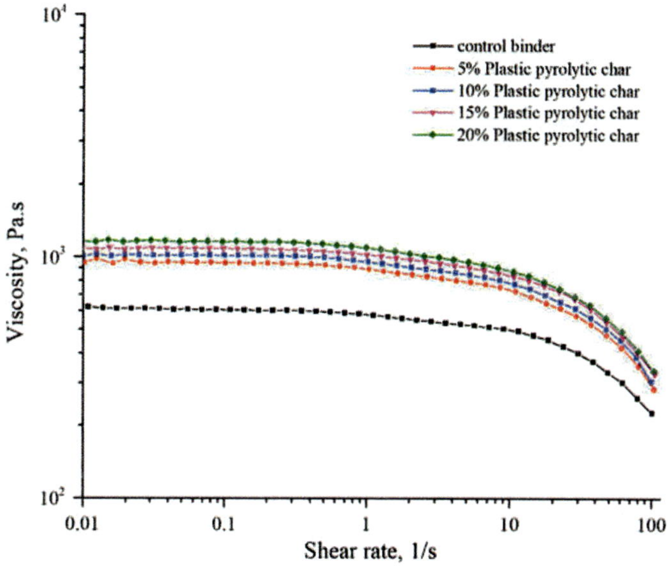

Fig. 3 Flow curve results at 60 °C

increase in shear rates. The resistance to flow is the largest at low shear rates, resulting in high viscosity values. As the shear rate rises, the internal structure rearranges itself to a configuration that is favorable to the deformation, leading to a decline in the viscosity.

3.2 *Frequency Sweep*

Test frequencies were varied in three decades (0.1–1, 1–10, and 10–100 rad/s) in the frequency sweep test. The test provides the rheological response of asphalt binders under low-strain oscillations over a range of frequencies with higher frequencies simulating faster traffic speeds. Storage (G') and loss (G'') moduli of control and modified binders are shown in Fig. 4. The behavior of binder is stiffer at higher frequencies and therefore both moduli increase with an increase in frequency. At all frequencies, the addition of plastic pyrolytic char enhances the storage and loss moduli of modified binders in comparison to the control binder. This result shows that the deformation resistance of modified binders is improved due to additional stiffness derived from char particles, and the improvements become more marked with an increase in char dosage. Both sets of plots indicate that there is a significant increase in storage and loss moduli when the char is added at a dosage of 5%. Thereafter, there is a consistent increase in moduli. The results indicate that the improvement in stiffness of binders achieved with char is manifested in both elastic as well as viscous regimes.

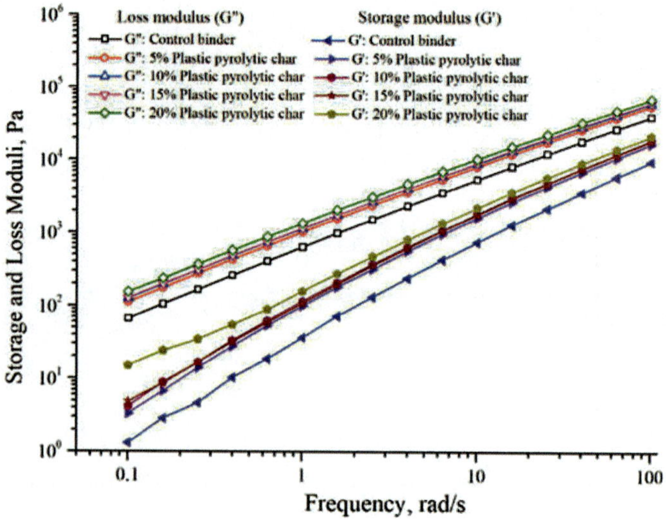

Fig. 4 Frequency sweep results

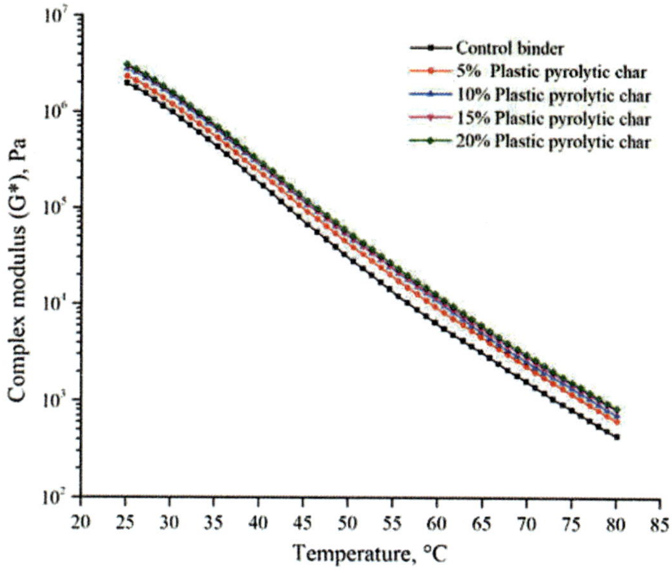

Fig. 5 Temperature sweep results: G*

3.3 *Temperature Sweep*

In the temperature sweep test, G* and δ of the neat and plastic pyrolytic char modified binders were measured over temperatures varying from 25 to 80 °C at 1.59 Hz (10 rad/s) frequency at low strain. The results are presented in Figs. 5 and 6, which are isochronal plots representing the variation of G* and δ, respectively. Figure 5 shows that all modified binders have higher G* than the control binder due to the enhancement in stiffness with the incorporation of waste plastic pyrolytic char. An increase in G* values shows higher stiffness and therefore higher resistance to deformation. Increased char contents further increase the G* values indicating that there is a monotonic improvement in the binders' total resistance to deformation with an increase in the char content. From Fig. 6, at all temperatures, δ values of plastic pyrolytic char modified binders are found lower than that of the control binder, indicating a relatively higher elastic nature of char modified binders than the neat binder.

3.4 *Superpave Rutting Parameter*

The Superpave rutting resistance parameter G*/sin δ was measured at standard test parameters (10 rad/s frequency and 10% strain) at 60 °C. The results are presented in Fig. 7. It is clear that plastic pyrolytic char modified binders have better high-temperature rutting resistance than the control binder. Increases of 41%, 54%, 64%

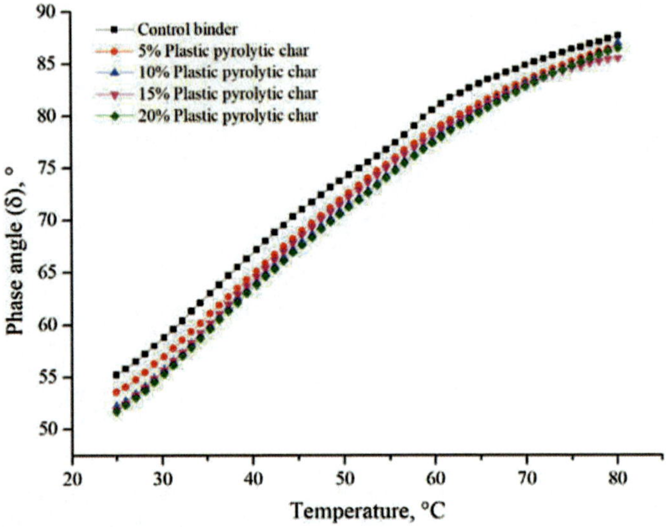

Fig. 6 Temperature sweep results: δ

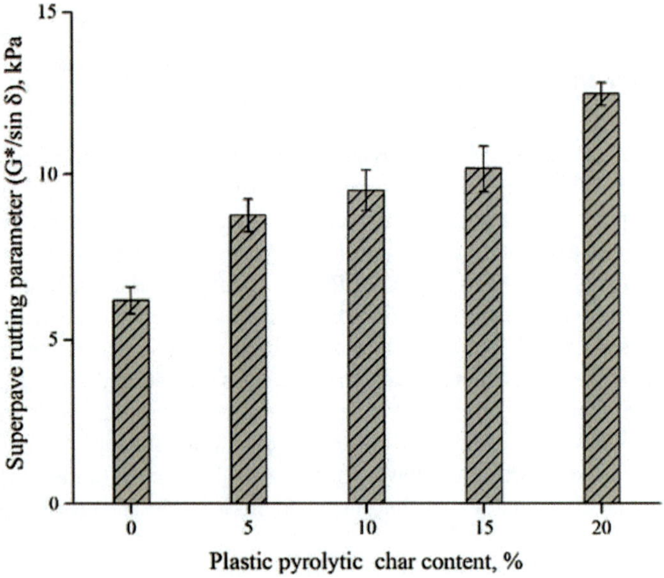

Fig. 7 Superpave rutting parameter results at 60 °C

and 101% in G*/sin δ at char contents of 5%, 10%, 15%, and 20%, respectively are observed. According to the Superpave performance grade specifications, the rutting parameter must be greater than 2.2 kPa for short-term (RTFO) aged binders. All binders meet this requirement at the test temperature. The porous microstructure and rough morphology of biochar results in larger interaction with the asphalt and thus helps to provides better deformation resistance to the binder [7].

3.5 Shenoy Rutting Parameter

The Shenoy's rutting parameter (G*/(1 – 1/(tan δ sin δ))) is an improvement over the Superpave rutting parameter (G*/sin δ) where the effect of elasticity (δ) is maximized. Figure 8 shows the plot of the Shenoy rutting parameter at 60 °C for binders with and without plastic pyrolytic char. The addition of plastic pyrolytic char imparted a considerable rise in the Shenoy's rutting parameter. The modified binders showed 62, 81, 112 and 177% increase in Shenoy rutting parameter for subsequent plastic pyrolytic char contents of 5, 10, 15 and 20%, respectively, compared to the control (unmodified) binder. Similar results were also seen for the Superpave rutting parameter G*/sin δ as discussed in the previous section. Shenoy's parameter is derived by calculating the unrecovered strain in the binder as the difference between the total strain and the elastic and the delayed elastic strain components in the material.

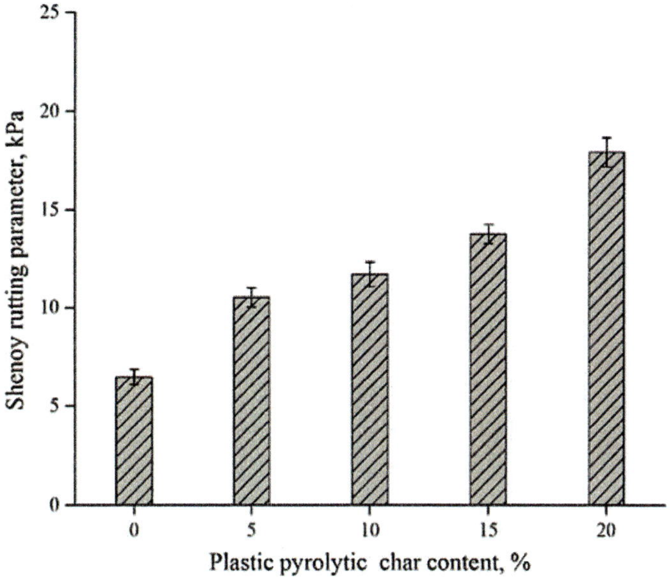

Fig. 8 Shenoy rutting parameter results at 60 °C

Thus, a higher value of this parameter points toward a lower unrecovered (or permanent) strain [27]. The observed results indicate that incorporating plastic pyrolytic char minimized the permanent strain and improved the binder's stiffness resulting in higher resistance against rutting by the plastic pyrolytic char modified asphalt binders.

3.6 Zero Shear Viscosity (ZSV)

Figure 9 presents the ZSV results for asphalt binders with and without plastic pyrolytic char at 60 °C. The addition of plastic pyrolytic char increases the ZSV indicating improved stiffness characteristics of the char modified asphalt binders. A significant increase of 56, 76, 127 and 179% in ZSV values can be noted for subsequent plastic pyrolytic char contents of 5, 10, 15 and 20%, respectively, as compared to the control binder. Similar results were also observed for the Superpave and the Shenoy parameters. Figure 10 shows the critical shear rate (γ_c), which is an indicator of the genesis of the shear-thinning behavior of asphalt binders (the point marking the onset of the decrease in viscosity). It is calculated as the inverse of k obtained from the Cross model as per Eq. (3). The critical shear rate decreases with an increase in plastic pyrolytic char content. This phenomenon indicates earlier commencement of the shear-thinning behavior of modified asphalt binders on the addition of the plastic pyrolytic char. Sybilski [35] attributed this parameter as the limit of the first Newtonian range and defined it as a shear rate at which viscosity differs by 10% from viscosity at zero shear rate. It is reported that the internal structure of the modified

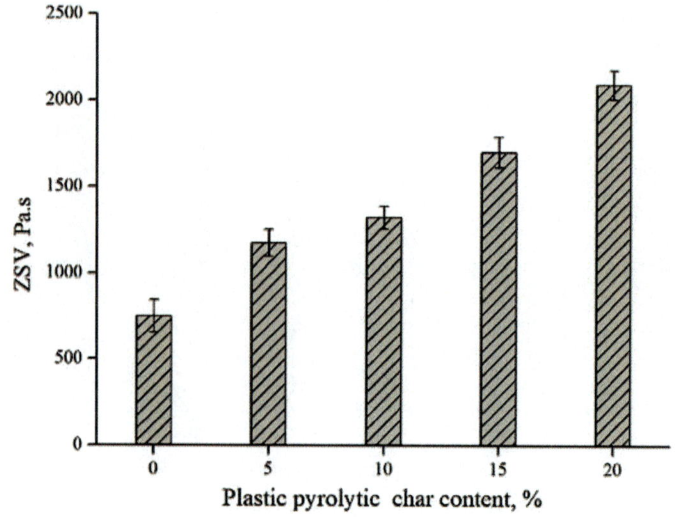

Fig. 9 ZSV results at 60 °C

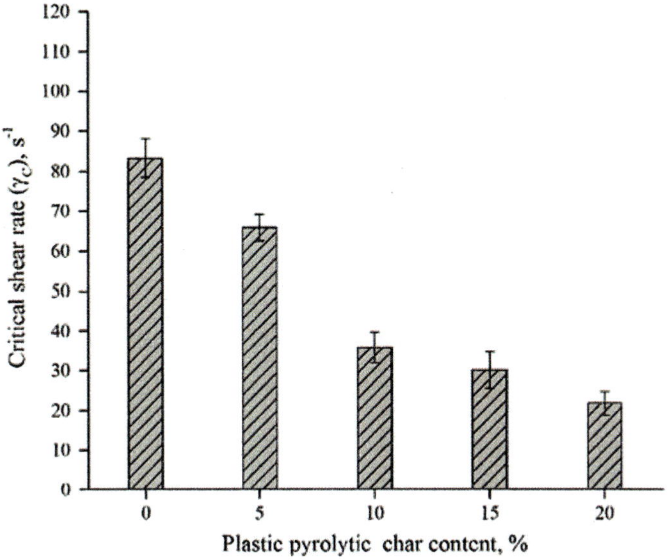

Fig. 10 Critical shear rate results at 60 °C

binder influences the critical shear rate and also the ZSV [35]. Further, it has also been reported that the asphaltenes are mainly accountable for the shear-thinning behavior, while the maltenes (comprising of resins, saturates and aromatics), are responsible for the Newtonian flow [36]. The earlier commencement of non-Newtonian behavior in the modified binders may be explained by the decrease in the effective maltenes or an increase in the effective asphaltenes [37]. However, a more detailed study of the chemical properties of the plastic pyrolytic char modified binders is warranted.

3.7 *MSCR and Burger's Model*

Figure 11 presents the J_{nr} results at the two stress levels (0.1 and 3.2 kPa) obtained from the MSCR test performed at 60 °C. A lesser value of J_{nr} is desired for a better performance against rutting. The J_{nr} decreases with an increase in the pyrolytic char dosage at both stress levels, which suggests higher rutting resistance of the pyrolytic char modified asphalt binders. The addition of pyrolytic char reduces the J_{nr} values (at 3.2 kPa) by 34%, 46%, 55%, and 62%, respectively, at the contents of 5%, 10%, 15%, and 20%. Similar findings have been reported for other pyrolytic char modified binders [38]. The MSCR recovery results are presented in Fig. 12. It is observed that the pyrolytic char also improves the percent recovery values at both stress levels. Further, $J_{nr,diff}$ parameter was found to be lower than 20% for all binders, thus satisfying the maximum limit of 75% stipulated by AASHTO MP19 [29].

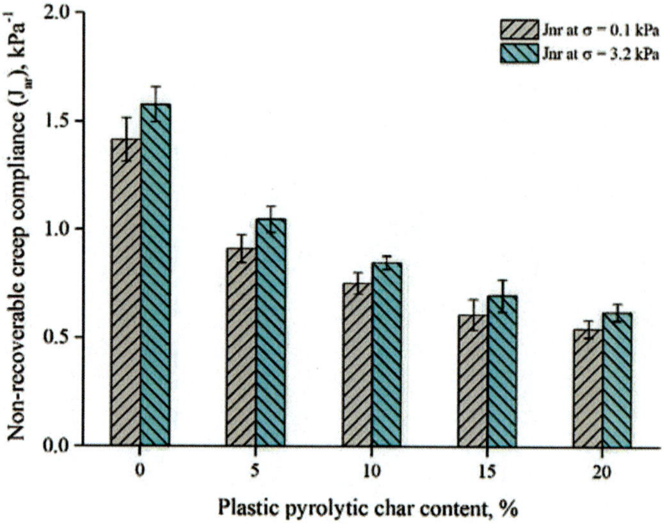

Fig. 11 MSCR J_{nr} results

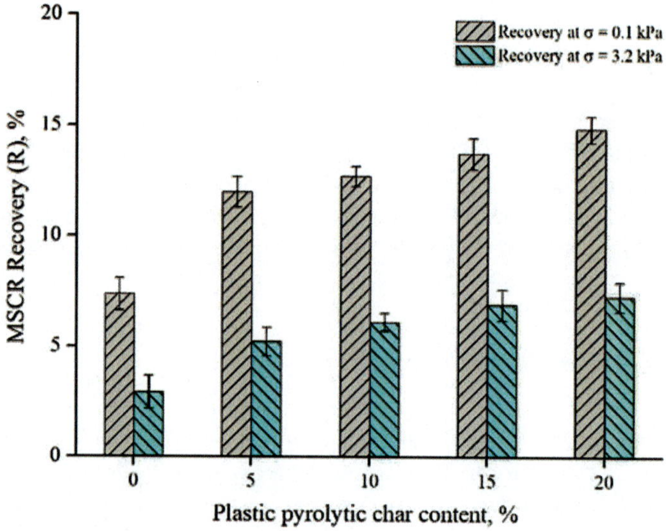

Fig. 12 MSCR recovery results

The Burger's four-element model parameter η_M typifies the viscous flow behavior of the asphalt binder and is an important parameter describing the rutting performance since it represents accumulated (unrecoverable) strain. A higher value of the η_M parameter indicates lower accumulated strain and hence a better resistance to rutting [39, 40]. Figure 13 presents the η_M values obtained for different binders. The η_M

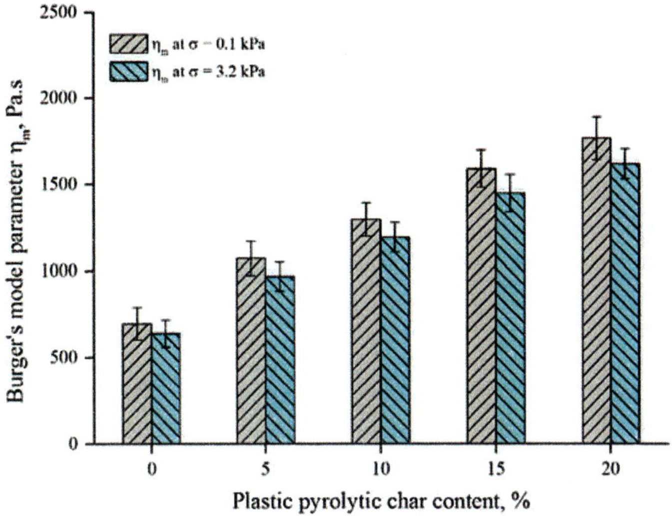

Fig. 13 Burger's model parameter $\eta_M$results

values increase with an increase in pyrolytic char content, at both stress levels and indicate lower unrecovered strains for the plastic pyrolytic char modified binders compared to the control binder. These results are also in agreement with other rutting parameters and portray the enhancement in permanent deformation resistance of the char modified asphalt binders.

4 Conclusions

In this study, the solid carbonaceous char generated as a by-product in plastic waste pyrolysis was used for asphalt binder modification. The rheological tests conducted on the binders were frequency sweep, temperature sweep, and flow behavior. This was followed by an evaluation of rutting resistance of the binders using four rheological approaches/parameters: Superpave rutting parameter, Shenoy rutting parameter, ZSV, and MSCR. Burger's model was used to model the creep and recovery response of the binders recorded from MSCR tests. The main conclusions from the study are drawn as follows:

- The addition of plastic pyrolytic char increased the G* values and decreased the δ values, suggesting an increase in stiffness and elasticity of the asphalt binders. The addition of char also enhanced the storage modulus and the viscous modulus at all test frequencies.
- Plastic pyrolytic char modified binders showed early onset of shear-thinning behavior than the control binder. This was found in the flow curve and also from the Cross-modelling results on frequency sweep data.

- All the four rutting parameters investigated in this study showed similar trends and indicated improvement in the rutting resistance with the incorporation of the pyrolytic char and a further increase with an increase in the char content.
- Results of MSCR tests and Burger's model parameter η_M (signifying accumulated unrecoverable strain) showed enhancement in the rutting resistance of the binders with lower unrecovered strain with an increase in the char content.

Plastic waste pyrolysis is receiving growing attention as a reliable method to manage and recover valuable products from plastic wastes. Pyrolysis generates solid carbonaceous char as a by-product. The findings of the study indicate that plastic pyrolysis char can be a potential modifier for asphalt binders. The use of plastic pyrolytic char in asphalt modification will be highly helpful to delineate an avenue for its bulk utilization. More studies are, however, warranted to understand the influence of plastic pyrolytic char on the properties of long-term aged binders as well as asphalt mixes under different loading and environmental conditions.

Acknowledgements The authors thank Innova Engineering & Fabrication (Mumbai) for providing the plastic pyrolytic char used in this study.

References

1. Kunwar, B., Cheng, H.N., Chandrashekaran, S.R., Sharma, B.K.: Plastics to fuel: a review. Renew. Sustain. Energy Rev. **54**, 421–428 (2016)
2. Miandad, R., Barakat, M.A., Aburiazaiza, A.S., Rehan, M., Nizami, A.S.: Catalytic pyrolysis of plastic waste: a review. Process Saf. Environ. Prot. **102**, 822–838 (2016)
3. Vijayakumar, A., Sebastian, J.: Pyrolysis process to produce fuel from different types of plastic–a review. IOP Conf. Ser. Mater. Sci. Eng. **396**(1), 012062 (2018)
4. Sharuddin, S.D.A., Abnisa, F., Daud, W.M.A.W., Aroua, M.K.: A review on pyrolysis of plastic wastes. Energy Convers. Manage. **115**, 308–326 (2016)
5. Lesueur, D., Dekker, D.L., Planche, J.P.: Comparison of carbon black from pyrolized tires to other fillers as asphalt rheology modifiers. Transp. Res. Rec. **1515**, 47–55 (1995)
6. Chebil, S., Chaala, A., Roy, C.: Use of softwood bark charcoal as a modifier for road bitumen. Fuel **79**(6), 671–683 (2000)
7. Zhang, R., Dai, Q., You, Z., Wang, H., Peng, C.: Rheological performance of bio-char modified asphalt with different particle sizes. Appl. Sci. **8**(9), 1665 (2018)
8. Kumar, A., Choudhary, R., Narzari, R., Kataki, R., Shukla, S.K.: Evaluation of bio-asphalt binders modified with biochar: a pyrolysis by-product of *Mesua ferrea* seed cover waste. Cogent Eng. **5**(1), 1548534 (2018)
9. Brewer, C.E., Schmidt-Rohr, K., Satrio, J.A., Brown, R.C.: Characterization of biochar from fast pyrolysis and gasification systems. Environ. Prog. Sustain. Energy **28**(3), 386–396 (2009)
10. Jamradloedluk, J., Lertsatitthanakorn, C.: Characterization and utilization of char derived from fast pyrolysis of plastic wastes. Adv. Mater. Res. **931**, 849–853 (2014)
11. IS 73.: Paving Bitumen—Specification. Bureau of Indian Standards, New Delhi (2013)
12. IS 1206.: Methods for Testing Tar and Bituminous Materials: Determination of Viscosity. Bureau of Indian Standards, New Delhi (1978)
13. IS 1203.: Methods for Testing Tar and Bituminous Materials: Determination of Penetration. Bureau of Indian Standards, New Delhi (1978)

14. IS 1205.: Methods for Testing Tar and Bituminous Materials: Determination of Softening Point. Bureau of Indian Standards, New Delhi (1978)
15. IS 1448.: Methods of Test for Petroleum and its Products. Bureau of Indian Standards, New Delhi (1969)
16. IS 1216.: Methods for Testing Tar and Bituminous Materials: Determination of Solubility in Carbon Disulphide or Trichloroethylene. Bureau of Indian Standards, New Delhi (1978)
17. IS 1208.: Methods for Testing Tar and Bituminous Materials: Determination of Ductility. Bureau of Indian Standards, New Delhi (1978)
18. IS 15462.: Polymer and Rubber Modified Bitumen—Specification. Bureau of Indian Standards, New Delhi (2004)
19. ASTM D2872.: Standard Test Method for Effect of Heat and Air on a Moving Film of Asphalt (Rolling Thin-Film Oven Test). ASTM International, West Conshohocken (2012)
20. Radziszewski, P.: Modified asphalt mixtures resistance to permanent deformations. J. Civil Eng. Manage. **13**(4), 307–315 (2007)
21. Kumar, A.S., Veeraragavan, A.: Rheological and rutting characterization of asphalt mixes with modified binders. J. Test. Eval. **40**(1), 1 (2012)
22. AASHTO T315.: Standard Method of Test for Determining the Rheological Properties of Asphalt Binder Using a Dynamic Shear Rheometer (DSR). American Association of State Highway and Transportation Officials (AASHTO) (2019)
23. Domingos, M.D.I., Faxina, A.L.: Susceptibility of asphalt binders to rutting: literature review. J. Mater. Civ. Eng. **28**(2), 04015134 (2016)
24. Bahia, H.U., Anderson, D.A.: Strategic Highway Research Program Binder Rheological Parameters: Background and Comparison with Conventional Properties. Transportation Research Record 1488, Transportation Research Board, pp. 32–39. Washington (1995)
25. Bahia, H.U., Hanson, D.I., Zeng, M., Zhai, H., Khatri, M.A., Anderson, R.M.: Characterization of Modified Asphalt Binders in Superpave Mix Design. NCHRP Report 459, National Academy Press, Washington (2001)
26. Anderson, M., D'Angelo, J., Walker, D.: MSCR: A Better Tool for Characterizing High Temperature Performance Properties. Asphalt Magazine of the Asphalt Institute, Lexington (2010)
27. Shenoy, A.: Refinement of the Superpave specification parameter for performance grading of asphalt. J. Transp. Eng. **127**(5), 357–362 (2001)
28. ASTM D7405.: Standard Test Method for Multiple Stress Creep and Recovery (MSCR) of Asphalt Binder Using a Dynamic Shear Rheometer. ASTM International, West Conshohocken (2015)
29. AASHTO MP19.: Standard Specification for Performance-Graded Asphalt Binder Using Multiple Stress Creep Recovery (MSCR) Test. American Association of State Highway and Transportation Officials (AASHTO) (2010)
30. Domingos, M.D.I., Faxina, A.L.: Rheological behaviour of bitumens modified with PE and PPA at different MSCR creep–recovery times. Int. J. Pavement Eng. **16**(9), 771–783 (2015)
31. Yi-qiu, T., Li, Z.H., Zhang, X.Y., Dong, Z.J.: Research on high-and low-temperature properties of asphalt-mineral filler mastic. J. Mater. Civ. Eng. **22**(8), 811–819 (2010)
32. Vale, C.: Influence of vertical load models on flexible pavement response—an investigation. Int. J. Pavement Eng. **9**(4), 247–255 (2008)
33. Liu, Y., You, Z.: Determining Burger's model parameters of asphalt materials using creep-recovery testing data. In: You, Z., Abbas, A.R., Wang, L. (eds.) Pavements and Materials: Modeling, Testing, and Performance, pp. 26–36. American Society of Civil Engineers, Reston (2009)
34. Jiang, X., Li, P., Ding, Z., Yang, L., Zhao, J.: Investigations on viscosity and flow behavior of polyphosphoric acid (PPA) modified asphalt at high temperatures. Constr. Build. Mater. **228**, 116610 (2019)
35. Sybilski, D.: Zero-shear viscosity of bituminous binder and its relation to bituminous mixture's rutting resistance. Transp. Res. Rec. **1535**(1), 15–21 (1996)

36. Shenoy, A.V., Saini, D.R., Nadkarni, V.M.: Rheograms for asphalt from single viscosity measurement. Rheol. Acta **21**(3), 333–339 (1982)
37. Saboo, N., Singh, B., Kumar, P., Vikram, D.: Study on viscosity of conventional and polymer modified asphalt binders in steady and dynamic shear domain. Mech. Time-Dependent Mater. **22**(1), 67–78 (2018)
38. Kumar, A., Choudhary, R., Nirmal, S.K., Pandey, I.K., Kataki, R.: Towards sustainable asphalt binders: Evaluation of bio-asphalt binders and mixes with biochar. J. Indian Roads Congr. **80**(3), 5–15 (2019)
39. Kumar, R., Saboo, N., Kumar, P., Chandra, S.: Effect of warm mix additives on creep and recovery response of conventional and polymer modified asphalt binders. Constr. Build. Mater. **138**, 352–362 (2017)
40. Saboo, N., Kumar, P.A.: Study on creep and recovery behavior of asphalt binders. Constr. Build. Mater. **96**, 632–640 (2015)

Development of Sustainable Masonry Blocks Using Industrial Rejects and Alkali Activation

Nikhil Rathod, Ravijanya Chippagiri, Hindavi R. Gavali, and Rahul V. Ralegaonkar

Abstract An increase in the need for sustainable construction is leading towards the use of alternative construction materials. Inappropriate disposal of industrial rejects has increased the scope for alternate construction materials. Moreover, cementitious materials, which adversely affect the environment, are replaced by alkali activated materials. In the present study, bio-briquette ash (BBA), an agro-industrial reject, was used to determine its potential as an alternative construction material. BBA was subjected to physical, chemical and mineralogical characterization. This identified raw material was mixed with fly ash (FA) in three proportions of 1:1, 1:2 and 1:3. The alkali activator solution was made of liquid Na_2SiO_3 and NaOH (6, 8 and 10 M) with a varying ratio of 1, 1.5 and 2. The alkali activator solution to total dry material ratio was kept constant at 0.3. Developed blocks were tested for physico-mechanical, durability and thermal conductivity properties. The density, compressive strength and water absorption of the blocks were found to be in the range of 1600–1850 kg/m^3, 16–31 MPa and 1.5–3.9% respectively. The criteria for durability and load bearing masonry structures were satisfied. The study concludes that identified bio-briquette ash, an industrial reject, has significant potential in developing sustainable masonry construction.

Keywords Bio briquette ash · Industrial rejects · Alkali activation · Sustainable masonry

1 Introduction

In accordance with rapid industrialization, there have been issues regarding the generation and disposal of solid wastes. These wastes create a burden to the industry and also a threat to the environment [1]. An increase in the demand for construction materials was observed due to the fast growing housing sector. To meet this increasing demand, industrial by-products can be reused as the alternative construction materials

N. Rathod · R. Chippagiri (✉) · H. R. Gavali · R. V. Ralegaonkar
Civil Engineering Department, Visvesvaraya National Institute of Technology, Nagpur, India
e-mail: ravijanya991@gmail.com

D. K. Ashish et al. (eds.), *3rd International Conference on Innovative Technologies for Clean and Sustainable Development*, RILEM Bookseries 29,
https://doi.org/10.1007/978-3-030-51485-3_24

and to save virgin natural resources from getting exploited [2]. Various agricultural and industrial by-products such as recycled paper mill waste (RPMW), sugarcane bagasse ash (SBA), construction and demolition waste (C&DW) [1, 3], bio-briquette ash (BBA) [4], fly ash [5–7], crumb rubber [6], GGBFS [7], cement kiln dust, mine tailings [8], rice husk ash [9] and bottom ash [10] were used to develop sustainable building masonry products.

Alkali activated polymers were synthesized from fly ash and concrete made with it serves the comprehensive need in restricting carbon emissions and energy consumption due to manufacture of cement [5]. These polymers were also synthesized from agro-industrial wastes and can be used as a substitute for cement and other engineered materials from a sustainable perspective [11]. Various techniques were proposed to compute mix proportions which are reliant on the experimental approach and these mix proportions are proved to be dissimilar to plain cement concrete [12–15].

Alkali activated polymer samples were characterized by X-Ray fluorescence (XRF), X-Ray diffraction (XRD), scanning electron microscope (SEM) and compressive strengths [16]. The compressive strength of these alkali activated products depend upon the molarity of NaOH [17, 18], water to solid proportion, alkali activator to solids proportion, sodium silicate to hydroxide proportion [18–20], alkali oxides [21] and dry/wet state of ash [22]. The structural behavior of these kinds of bricks was studied with regard to strength, initial rate of absorption, water absorption, dimensionality and modulus of elasticity [23]. Bricks were made from various industrial by-products and alkali activation. Researchers have maintained the alkali to solid ratio in between 0.35 and 0.8, silicate to hydroxide as 2–2.5 and molarity in between 8 and 18 M of NaOH [7, 8, 20, 24, 25] which resulted in mixed outcomes.

The presented literature shows that BBA has been used in the development of bricks in conventional techniques but its application hasn't been in the alkali activated products. Hence, this paper investigates the applicability of BBA as a substitute material for fine aggregates in the production of alkali activated blocks through experimental works which involve mechanical and chemical properties. The impact of different design parameters of alkali activated blocks was studied and the required optimum combination of materials to obtain maximum compressive strength was obtained. This research work attempts to improve the performance of masonry structure by applying sustainability aspects.

2 Materials and Methodology

2.1 Materials

The materials used for making alkali activated blocks involve fly ash, BBA, alkaline solution consisting of sodium hydroxide solution and sodium silicate solution as binders and water as workability measure.

The fly ash was collected from the Khaperkheda thermal power plant located at Nagpur, India. The BBA (Fig. 1) was collected from the Baidyanath Factory, Nagpur, India. The sodium hydroxide flakes and sodium silicate solution used in the alkaline solution are as shown in Figs. 2 and 3 respectively.

BBA was used as a part of fine aggregate in the product development. After sieving BBA through 2.36 mm sieve, its fineness modulus was determined to be 3.55 and the specific gravity was found to be 2.5. Further, the structure of BBA was examined through tests like SEM, XRD and XRF. The chemical characterization of the BBA (as shown in Table 1) was determined through XRF. The study conducted on the raw material has resulted in its application in the cementitious products [26].

The XRD pattern of BBA shown in Fig. 4 implies that the material predominantly consists of Quartz (SiO_2), Ferric Oxide (Fe_2O_3) and Calcite ($CaCO_3$). Most of the crystalline part is in the state of quartz. The morphological characteristics of BBA

Fig. 1 Bio briquette ash

Fig. 2 Sodium hydroxide flakes

Fig. 3 Sodium silicate solution

Table 1 Chemical characterization of BBA as per XRF results

Compound	SiO_2	CaO	Fe_2O_3	K_2O	Al_2O_3	MgO	P_2O_5	SO_3
% Content	37.54	19.98	8.62	8.56	6.84	5.81	2.72	2.1

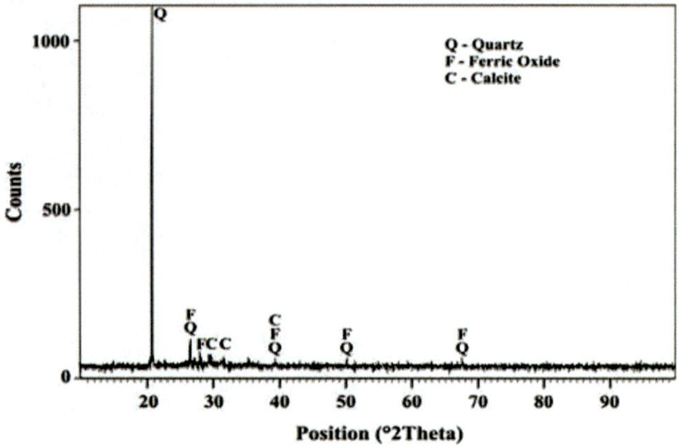

Fig. 4 X-ray diffraction pattern of BBA

particles were captured from SEM (Fig. 5). The XRD and SEM of the fly ash are shown in Figs. 6 and 7 respectively.

The alkali solution is a mixture of sodium hydroxide (or potassium hydroxide) and sodium silicate (or potassium silicate) to different ratio. Due to an increase in the concentration of sodium hydroxide solution, in terms of molarity (M), the mix becomes more brittle in nature with increased compressive strength. Generally, the sodium hydroxides are available in solid state by means of pellets and flakes whereas sodium silicate is also known as water glass or liquid glass, available in liquid (gel) form.

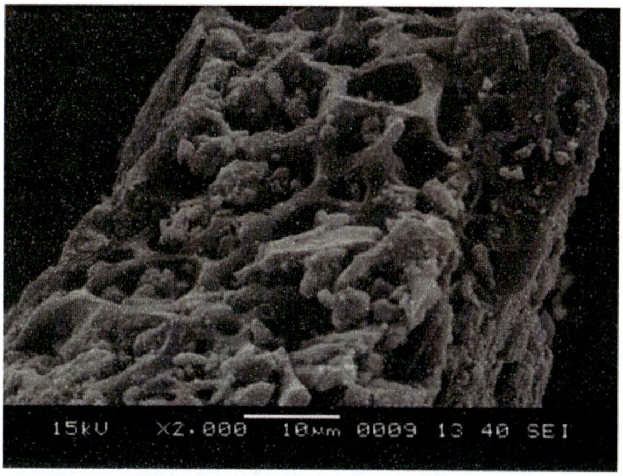

Fig. 5 Scanning electron microscope image of BBA

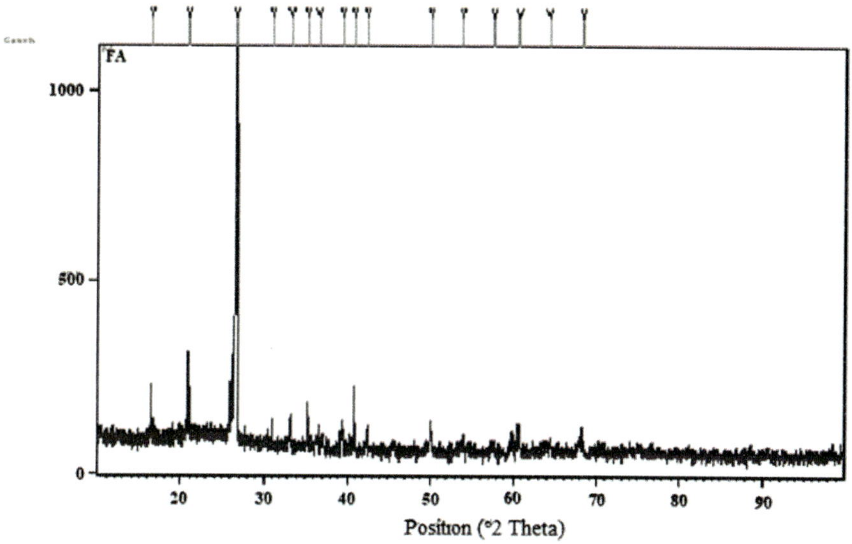

Fig. 6 X-ray diffraction pattern of fly ash

2.2 *Methodology*

The various design parameters with particular variations in materials that have been given an emphasis on are indicated in Table 2.

The effect of these parameters was studied by designing 27 mix combinations as specified in Table 3. Cubes of 70 mm × 70 mm × 70 mm were cast and tested for various physico-mechanical properties as per Indian standards. These mix compositions underwent lab tests for evaluation of compressive strength, density and water

Fig. 7 Scanning electron microscope image of fly ash

Table 2 Design parameters with respective variations

S. no.	Parameters	Variations		
1	Alkali activator solution to dry mix ratio	0.3		
2	Fly ash to BBA ratio (FA:BBA)	1:1	2:1	3:1
3	Molarity of sodium hydroxide solution (M)	6	8	10
4	Sodium silicate to sodium hydroxide ratio	1	1.5	2

Table 3 Nomenclature of the mix designs as per the design parameters

FA:BBA	Molarity (M)	Silicate : Hydroxide		
		1	1.5	2
1	6	M161	M161.5	M162
	8	M181	M181.5	M182
	10	M1101	M1101.5	M1102
2	6	M261	M261.5	M262
	8	M281	M281.5	M282
	10	M2101	M2101.5	M2102
3	6	M361	M361.5	M362
	8	M381	M381.5	M382
	10	M3101	M3101.5	M3102

absorption. The variation in performance of the cubes made with different mix combinations was studied. The masonry blocks of size 300 mm × 150 mm × 120 mm (as shown in Fig. 8) were made of the higher strength mixtures and tested for compressive strength, density, thermal conductivity and durability (chloride and sulphate tests).

Fig. 8 Alkali activated masonry block

The cubes were given codes according to their respective design parameters as shown in the Table 3. Say, M281.5 is the mix prepared of FA: BBA = 2, Molarity = 8 M and Silicate: Hydroxide = 1.5.

3 Results and Discussion

3.1 Results of Test Specimen

The cubes were tested for the compressive strengths at 7, 14 and 28 days respectively and the results are shown in Fig. 9. The cubes were also tested for their densities and water absorption and their results are shown in Figs. 10 and 11. The standard error bars have been indicated in the graphs as the results obtained were the average of

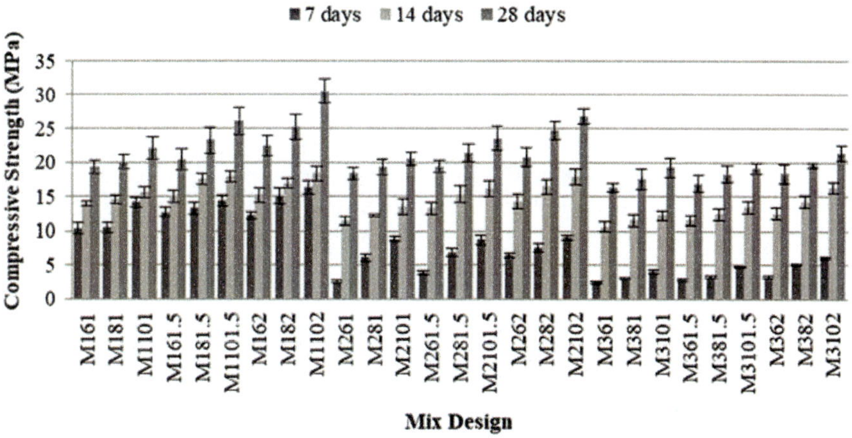

Fig. 9 Compressive strength of mix designs at 7, 14 and 28 days

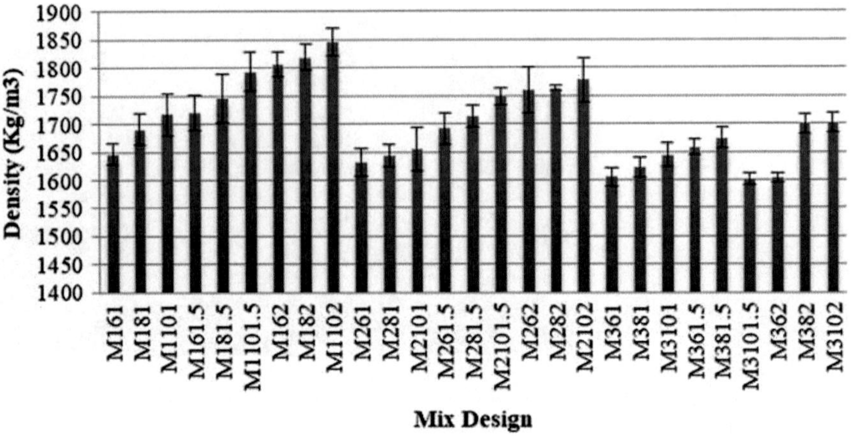

Fig. 10 Density of mix designs

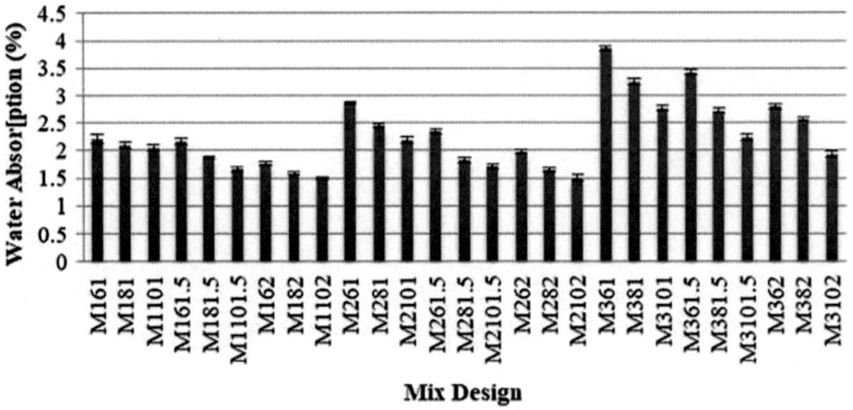

Fig. 11 Water absorption of mix designs

three samples from each mix proportion. The minimum and maximum values of the laboratory tests are presented in the Table 4. From the graphs and table, mix M361 was found to have less compressive strength, less density and more water absorption.

Table 4 Range of the physical/mechanical test results obtained

S. no.	Property	Value (Min.–Max.)	Code of the specimen	
			Min.	Max.
1	Compressive strength (MPa)	16.32–30.61	M361	M1102
2	Density (kg/m^3)	1600–1846	M3101.5	M1102
3	Water absorption (%)	1.49–3.85	M1102	M361

Additionally, M1102 was found to have high compressive strength, high density and low water absorption.

3.2 *Effect of Design Parameters on Compressive Strength*

The effect of design parameters on the 7, 14 and 28 day compressive strengths (Figs. 12, 13 and 14) were observed. Increase in the FA:BBA ratio has decreased the compressive strength of the mix. BBA was added to the mix as a fine aggregate and its contribution towards the strength was found to be more from the drawn results.

Increase in the molarity of NaOH and silicate to hydroxide ratio has increased the compressive strength of the mix proportions and this was observed by other cited researchers too.

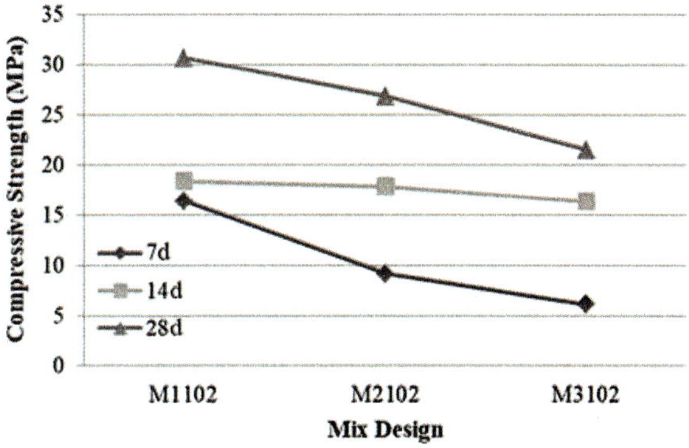

Fig. 12 Effect of FA:BBA ratio on the compressive strength

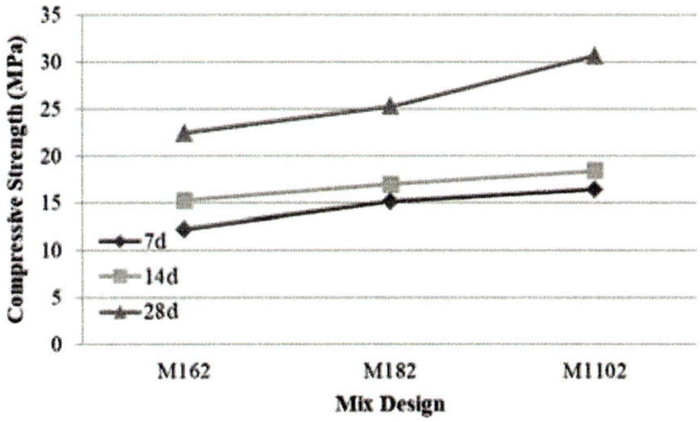

Fig. 13 Effect of molarity on compressive strength

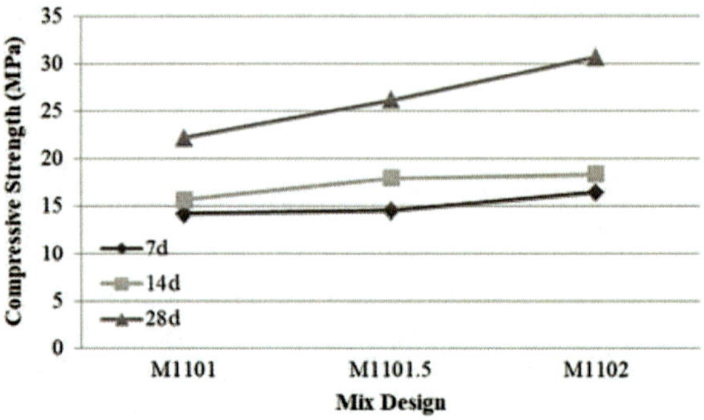

Fig. 14 Effect of silicate: hydroxide ratio on compressive strength

3.3 Test Results of Masonry Block

The cube of the mix M1102 has a high compressive strength, high density and low water absorption. Masonry blocks (Fig. 8) were developed and tested for the same. The compressive strength of the block turned out to be 22 MPa whereas the density was about 1850 kg/m^3. Tests to assess durability and the thermal conductivity were conducted. The chloride content was found to be 0.21 kg/m^3 which was within the permissible limit of 3.0 kg/m^3 [as per IS 456:2000]. The sulphate content was 112.24 mg/l which was found to be compliant with IS 456:2000 due to the absence of cement in the mix. From the Lee's disc apparatus, thermal conductivity of the air-dried cylindrical sample of 100 mm diameter and 50 mm thickness was found to be 1.14 W/m-K when tested after 28 days of casting. These results show that the designed mix is durable and thermally efficient.

3.4 Comparison with Fly Ash Brick

The developed block was further compared to the commercially available fly ash brick (as presented in Table 5) and the properties of this fly ash brick were taken from a study [1]. The physical and mechanical properties such as water absorption and compressive strength were found to be better than the commercial product. Whereas density and thermal conductivity of fly ash brick was found to be a step ahead of the developed product.

Table 5 Comparative analysis fly ash brick and alkali activated BBA (AA-BBA) block

Type of brick	Density (kg/m^3)	Water absorption by weight (%)	Compressive strength (MPa)	Thermal conductivity (W/mK)
Fly ash brick	1800	12	6.5	1.05
AA-BBA block	1850	1.49	22	1.14

4 Conclusion

The use of fly ash and BBA in the development of alkali activated blocks has given satisfactory results. The silicate to hydroxide ratio of alkali activator was highly affecting the strength of the mix. The highest strength was achieved using silicate/hydroxide = 2.0 of about 31 MPa. The increase in fly ash content highly degrades the compressive strength of blocks. The early 7 day strength was affected due to a change in the percentage of fly ash and was decreased by about 63–78% when 1:1 and 3:1 combinations were compared. On the other hand, the latter strength of 28 days was not affected to such an extent.

The densities of blocks of all combinations varied from 1600 to 1850 kg/m^3 and have adequate strength. The block consisting of 1:1 Fly ash to BBA, 10 M of NaOH in alkali activator solution and having silicate to hydroxide ratio of 2 is the optimum mix in terms of compressive strength (31 MPa) and water absorption of about 1.5%. The average density obtained was 1841 kg/m^3 which fulfilled the criteria of load bearing blocks [as per IS SP 62 (1997)]. Thus, these BBA based alkali activated masonry blocks can be used for load bearing walls. Even durability and thermal conductivity showed encouraging results indicating that the blocks are durable and thermally efficient.

5 Future Work

Further studies are to be conducted regarding the assessment of structural properties of the load bearing masonry unit and the feasibility studies of the product for its application in the industry.

References

1. Madurwar, M.V., Mandavgane, S.A., Ralegaonkar, R.V.: Development and feasibility analysis of bagasse ash bricks. J. Energy Eng. **141**, 1–9 (2015). https://doi.org/10.1061/(ASCE)EY.1943-7897.0000200
2. Turgut, P., Murat Algin, H.: Limestone dust and wood sawdust as brick material. Build. Environ. **42**, 3399–3403 (2007). https://doi.org/10.1016/j.buildenv.2006.08.012
3. Madurwar, M.V., Ralegaonkar, R.V., Mandavgane, S.A.: Application of agro-waste for sustainable construction materials: a review. Constr. Build. Mater. **38**, 872–878 (2013). https://doi.org/10.1016/J.CONBUILDMAT.2012.09.011
4. Sakhare, V.V., Ralegaonkar, R.V.: Use of bio-briquette ash for the development of bricks. J. Clean. Prod. **112**, 684–689 (2016). https://doi.org/10.1016/J.JCLEPRO.2015.07.088
5. Davidovits, J.: High-Alkali Cements for 21st Century Concretes. ACI Symposium Publications 144. https://doi.org/10.14359/4523
6. Mohammed, B.S., Liew, M.S., Alaloul, W.S., et al.: Development of rubberized geopolymer interlocking bricks. Case Stud. Constr. Mater. **8**, 401–408 (2018). https://doi.org/10.1016/j.cscm.2018.03.007

7. Radhakrishna, V.K., Sasalatti, V., Venumadhav, T.: Study on geopolymer masonry as sustainable building material. J. Env. Res. Dev. **9**, 925–932 (2015)
8. Ahmari, S., Zhang, L.: Utilization of cement kiln dust (CKD) to enhance mine tailings-based geopolymer bricks. Constr. Build. Mater. **40**, 1002–1011 (2013). https://doi.org/10.1016/j.conbuildmat.2012.11.069
9. Maulana, A.I., Wardani, N.K., Syamsidar, D., et al.: Development of hybrid composite Rice Husk Ash (RHA)-geopolymer for bricks bearing buildings. MATEC Web Conf. **97**, 1–9 (2017). https://doi.org/10.1051/matecconf/20179701009
10. Antunes Boca Santa, R.A., Bernardin, A.M., Riella, H.G., Kuhnen, N.C.: Geopolymer synthetized from bottom coal ash and calcined paper sludge. J. Clean. Prod. **57**, 302–307 (2013). https://doi.org/10.1016/j.jclepro.2013.05.017
11. Asim, N., Alghoul, M., Mohammad, M., et al.: Emerging sustainable solutions for depollution: geopolymers. Constr. Build. Mater. **199**, 540–548 (2019). https://doi.org/10.1016/j.conbuildmat.2018.12.043
12. Montes, C., Gomez, S.A., Khadka, N., Allouche, E.N.: Statistical software to improve the accuracy of geopolymer concrete mix design and proportioning. In: 2013 World Coal Ash Conference, 22–25 Apr 2013. Lexington, KY (2013)
13. Ferdous, W., Manalo, A., Khennane, A., Kayali, O.: Geopolymer concrete-filled pultruded composite beams—concrete mix design and application. Cem. Concr. Compos. **58**, 1–13 (2015). https://doi.org/10.1016/j.cemconcomp.2014.12.012
14. Lloyd, N.A., Rangan, B.V.: Geopolymer concrete with fly ash. In: 2nd International Conference Sustainable Construction Materials and Technologies, vol. 7, pp. 1493–1504 (2010)
15. Anuradha, R., Sreevidya, V., Venkatasubramani, R., Rangan, B.V.: Modified Guidelines for Geopolymer Concrete Mix Design Using Indian Standard (2012)
16. Noor-Ul-Amin, F.M., Muhammad, K., Gul, S.: Synthesis and characterization of geopolymer from bagasse bottom ash, waste of sugar industries and naturally available China clay. J. Clean. Prod. **129**, 491–495 (2016). https://doi.org/10.1016/j.jclepro.2016.04.024
17. Mustafa, A.M., Bakri, A., Kamarudin, H., et al.: Microstructure of different NaOH molarity of fly ash-based green polymeric cement. J. Eng. Technol. Res. **3**, 44–49 (2011)
18. Lokuge, W., Wilson, A., Gunasekara, C., et al.: Design of fly ash geopolymer concrete mix proportions using multivariate adaptive regression spline model. Constr. Build. Mater. **166**, 472–481 (2018). https://doi.org/10.1016/j.conbuildmat.2018.01.175
19. Samantasinghar, S., Singh, S.P.: Effect of synthesis parameters on compressive strength of fly ash-slag blended geopolymer. Constr. Build. Mater. **170**, 225–234 (2018). https://doi.org/10.1016/j.conbuildmat.2018.03.026
20. Morsy, M.S., Alsayed, S.H., Al-Salloum, Y., Almusallam, T.: Effect of sodium silicate to sodium hydroxide ratios on strength and microstructure of fly ash geopolymer binder. Arab. J. Sci. Eng. **39**, 4333–4339 (2014). https://doi.org/10.1007/s13369-014-1093-8
21. Topçu, İ.B., Toprak, M.U., Uygunoğlu, T.: Durability and microstructure characteristics of alkali activated coal bottom ash geopolymer cement. J. Clean. Prod. **81**, 211–217 (2014). https://doi.org/10.1016/J.JCLEPRO.2014.06.037
22. Ferone, C., Colangelo, F., Cioffi, R., et al.: Mechanical performances of weathered coal fly ash based geopolymer bricks. Proc. Eng. **21**, 745–752 (2011). https://doi.org/10.1016/j.proeng.2011.11.2073
23. Venugopal, K., Radhakrishna, K.: Structural behavior of geopolymer masonry. Indian J. Sci. Technol. **9**:1–7 (2016). https://doi.org/10.17485/ijst/2016/v9i25/91586
24. Hadi, M.N.S., Farhan, N.A., Sheikh, M.N.: Design of geopolymer concrete with GGBFS at ambient curing condition using Taguchi method. Constr. Build. Mater. **140**, 424–431 (2017). https://doi.org/10.1016/j.conbuildmat.2017.02.131
25. Chindaprasirt, P., Rattanasak, U.: Fire-resistant geopolymer bricks synthesized from high-calcium fly ash with outdoor heat exposure. Clean Technol. Environ. Policy **20**, 1097–1103 (2018). https://doi.org/10.1007/s10098-018-1532-4
26. Sakhare, V.V., Ralegaonkar, R.V.: Development and investigation of cellular light weight bio-briquette ash bricks. Clean Technol. Environ. Policy **19**, 235–242 (2017). https://doi.org/10.1007/s10098-016-1200-5

Self-compacting Concrete—Optimization of Mix Design Procedure by the Modifications of Rational Method

Peerzada Danish and G. Mohan Ganesh

Abstract Self-Compacting Concrete (SCC) is a special type of concrete recognized for placing in congested reinforced structures without any application of external vibration. Self compactibility can be determined by the properties of material constituents and the design of mix proportions. The absence of an approved code in India on mix design proportions and characteristics of material constituents to achieve compactibility in SCC has necessitated to determine a method for mix design of SCC. Though some researchers have carried out investigations to determine a proper mix design for producing SCC. In this investigation, the optimization of mix proportion design has been determined by adopting the proposed rational mix design method or Japanese method with necessary modifications in consonance with the guidelines of EFNARC. A suitable mix using the marginal aggregates was selected and numerous trial mixes (sixty-five) were carried out with the varying mix parameters like binder content, water-binder ratio, fine aggregate-coarse aggregate ratio, percentage of superplasticizer and viscosity modifying admixtures. The test results of this study are presented in this paper and a successful attempt has been made to determine the suitable mix design for producing SCC.

Keywords Self-compacting concrete · Rational method · Workability · Filling ability · Passing ability and segregation resistance

1 Introduction

The phenomenon of SCC is believed to be the most significant contribution in the history of concrete technology. SCC was acquainted in the construction industry by Japanese researchers in late 1980s [1]. SCC is highly workable concrete with ability to flow or move through the restricted sections under its own weight maintaining stability [2–5]. It does not need external vibration for consolidation or compaction and as such reduces the noise pollution during casting, besides improving the working

P. Danish · G. M. Ganesh (✉)
School of Civil Engineering, Vellore Institute of Technology, Tamil Nadu, Vellore 632014, India
e-mail: gmohanganesh@vit.ac.in

D. K. Ashish et al. (eds.), *3rd International Conference on Innovative Technologies for Clean and Sustainable Development*, RILEM Bookseries 29,
https://doi.org/10.1007/978-3-030-51485-3_25

conditions at the site.[6, 7]. The excellent properties of the SCC such as better workability, good mechanics and durability of the hardened concrete has attracted the focus of various researchers. At present SCC has become a developing direction in the modern concrete technology. The production of an economical SCC having excellent fresh and hardened properties will surely enhance its use in the construction industry [8]. Its use has witnessed a dramatic increase in the last two decades [9]. An important criterion in the efficient production of SCC is a good balance between stability and deformability. The specific properties of SCC have attracted the focus of the researchers for its further development [10]. However, in spite of the interest of the researchers in SCC, its application is still limited in the construction industry. Extensive studies were executed on the mixtures proportioning of the SCC and the researchers were able to set some guidelines on this behalf which include (i) increase in paste volume and decrease in water-binder (w/b) ratio, (ii) limiting volume ratio of aggregates to cememtitious materials (iii) cautiously checking the maximal size of coarse aggregate (CA) and the total volume, (iv) using SP and VMA [1]. The SP was added to enhance the workability and cohesiveness of SCC, while VMA minimizes the bleeding and segregation.

In order to achieve self compatibility in SCC, it is very important to develop a procedure for mixture design of SCC. Mix design forms a vital and core component for the production of successful SCC and thus assumes a significant posture in concrete casting. The mix design means combing optimum proportions of the constituent's materials to satisfy fresh and hardened properties of concrete for a given application [11]. The properties and the type of the constituent materials normally determines the properties of SCC [12]. The review of the literature reveals that there is no approved or accepted procedure for the mixture design of SCC in India. It was only in 1995, Okamura and Ozawa [2] proposed the first well known mix design method (Japanese method) for SCC which was referred to a rational method of mix design of SCC. In this method of mix design, the proportion of CA and FA are generally kept constant to achieve the self-compactibility. The water powder ratio and the percentage of SP are adjusted as per the need. So this mixture design is based on the following fundamental steps: (a) CA of the concrete is kept at 50% of the solid volume of concrete, (b) FA is kept at 40% of the mortar volume, (c) w/p ratio is normally assumed as 0.9–1.0 by the volume depending on the powder properties; and (d) the dosage of SP and the final required w/p ratio are determined by conducting numerous trials in order to ensure the self-compactibility. The empirical mixture design methods are based on the previous experiences and are simplest in execution. But to achieve the desired properties, a lot of trial mixes are required, thus making these methods time consuming [13].

The present investigation which forms a part of the wide research project is aimed to develop a method or procedure for mixture proportion design to achieve self-compacting concrete. The procedure has been developed by modifying the proposed rational mix design method in consonance with the guidelines of EFNARC 2005 [14].

The present investigation under report is innovative because of the fact that the contents of coarse aggregate and fine aggregate were varied instead of them after keeping constant in a fixed range as in rational method of mix design. A suitable mix using the marginal aggregates was selected and numerous trial mixes (sixty-five) were laid with the varying mix parameters like water-powder ratio, fine aggregate-coarse aggregate ratio, percentage of superplasticizer (SP) and viscosity modifying admixtures (VMA). This paper also presents the test results of the fresh characteristics of SCC such as slump flow, V-funnel, J-ring and L-box. Further, the results pertaining to the strength property in terms of compressive strength and concrete density are also presented.

2 Materials Used

2.1 Raw Materials

2.1.1 Cement

Ordinary Portland cement (grade 53) with a specific gravity of 3.15 was used in this study conforming to IS 12269–2013 [15]. The initial and final setting time of cement were recorded as "32 min and 325 min respectively". The chemical characteristics and physical properties of OPC are listed in Table 1.

Table 1 Chemical composition and physical properties of cement

Composition	Cement (%)
Chemical composition	
SiO_2	22.62
Al_2O_3	5.80
Fe_2O_3	3.70
CaO	64.50
SO_3	2.10
Loss on ignition	1.28
Physical Properties	
Bulk density (kg/m^3)	1580
Specific gravity	3.15
Blaine's specific surface area (m^2/kg)	298
Color and physical form	Grey powder

Table 2 Physical characteristics of fine and coarse aggregates

Properties	FA	CA
Specific gravity	2.58	2.65
Water absorption (%)	1.00	0.65
Size (mm)	<4.75	10–12.50
Fineness modulus	2.66	5.80

2.1.2 Coarse Aggregate

Locally available gravel size of 10–12.5 mm was used as coarse aggregate in this study conforming to IS: 383–2016 [16]. The physical characteristics of coarse aggregate are listed in Table 2.

2.1.3 Fine Aggregate

Locally available river sand grading corresponding to zone II from Table 4 of IS: 2386 (part 3) was used as fine aggregate conforming to IS: 383–2016 [16]. The physical characteristics of fine aggregate are listed in Table 2.

2.1.4 Chemical Admixture

Polycarboxylic ether (PCE) based high performance superplasticizer (Master Glenium Sky 8233) (SP) free from chloride was used as water reducing admixture (0.24–1%) conforming to IS: 9013, 2014 [17]. Viscosity Modifying Admixture (VMA) was used to obtain bleed free mixes and these chemical admixtures were procured from BASF India Limited, Kancheepuram, Tamil Nadu, India. The properties of both chemical admixtures are shown in Table 3.

Table 3 Properties of SP and VMA

Properties	SP	VMA
Specific gravity	1.08	–
Aspect	Reddish brown liquid	Colorless free flowing liquid
pH	$\geq$6	$\geq$6 at 25 °C
Relative density	1.08 $\pm$ 0.02 at 25 °C	1.01 $\pm$ 0.01 at 25 °C
Chloride ion content	<0.2%	<0.2%

* The informations are collected the properties of superplasticizer and VMA from the manufacturer's manual

2.1.5 Water

The normal portable tap water available in concrete laboratory conforming to IS 456–2002 [18] was utilized in this study.

2.1.6 Mix Proportion

The blend/mixture was composed of OPC, coarse aggregates, fine aggregates, chemical admixtures and tap water.

2.2 *Composition of Raw Materials*

See Tables 1, 2 and 3.

3 Test Methods

3.1 *Determination of Fresh Properties of SCC*

The different tests were performed to assess the filling ability, passing ability, flowability and segregation resistance as per the EFNARC guidelines. Slump flow test was performed to determine flowability and segregation resistance (Range = 650–800 mm), V-funnel test (Range = 6–12 s) and T_{50} cm flow time test (Range = 2–5 s) were performed to determine the viscosity and segregation resistance. Blocking ratio test (L-box) was performed to determine the passing ability of SCC (Range = 0.8–1.0). The applied test methods are shown by Fig. 1; the test results are summarized in Table 4.

3.2 *Compressive Strength Test*

The concrete mixture was put into the cube mould size of 100 × 100 × 100 mm for performing the hardened concrete tests such as compressive strength test after performing the fresh concrete tests without any external vibration conforming to IS 516-1959 (reaffirmed 1999) [19]. The compressive strength test values of all the mixes are depicted in Table 4 and shown in Fig. 1d.

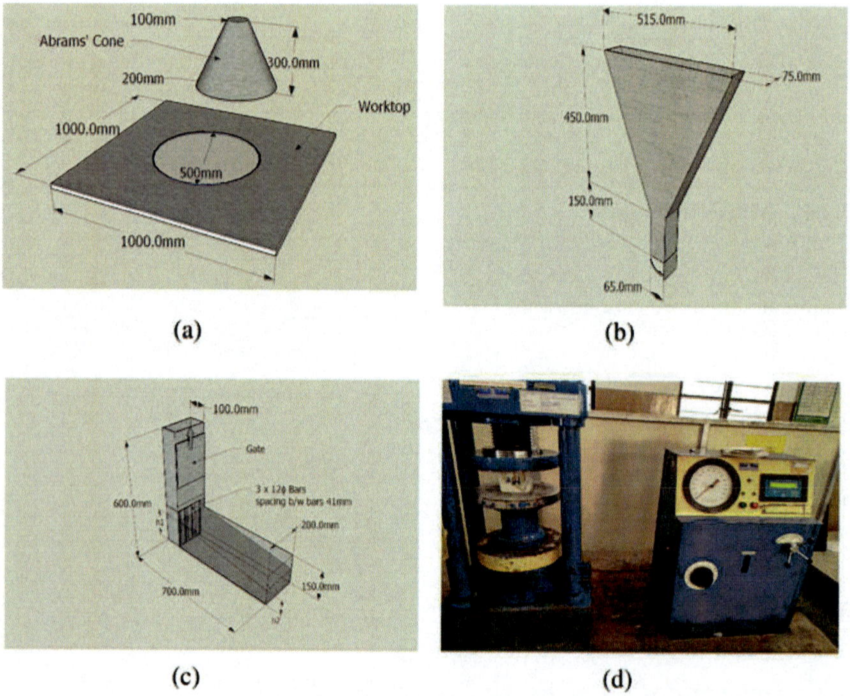

Fig. 1 Workability tests (**a**–**c**) and compressive strength test (**d**)

4 Mixture Design (Modification of Rational Method)

Self compactibility can be determined by the properties of materials constituents and the design of mix proportions. So, it is essential to have an appropriate mix design for SCC using necessary material constituents. A very simple mix proportion order (Japanese Method) has been proposed by Okamura and Owaza in 1995. This mix proportioning proposes that the self-compactibility can be achieved by keeping the fine and coarse aggregates content fixed, with adjustment of water powder ratio and doses of superplasticizer only.

In India there is no approved code on mix design proportions and the optimization of the material constituents in achieving the SCC. In the present investigation the mix proportioning has been carried out by adopting the proposed rational mix design method (Japanese method) in consonance with guidelines of EFNARC 2005 and the following points were considered for suitable and compatible mix design for SCC:

(i) "Generally, the air content is kept at 2%", but in conditions of cold weather concreting or freeze–thaw conditions the air content should be set at more than 2%. The present investigation has been carried out assuming the air content at 2%.

Table 4 Trial mixes with different adjustments

Mix ID	Cement (kg/m^3)	W/B	Fine aggregate (kg/m^3)	Coarse aggregate (kg/m^3)	SP (%)	VMA (%)	Slump flow (mm)	T_{50} cm (s)	V-funnel (s)	L-box	f_{ck} (7 days) MPa	Density (kg/m^3)
TM1	455	0.42	940	700	1.1	0.2	521	14.5	26.2	0.51	23.76	2291.6
TM2	455	0.42	940	700	1.7	0.2	561	11.2	19.2	0.60	24.57	2293.8
TM3	455	0.42	940	730	1.1	0.2	504	15.2	27.3	0.48	24.93	2321.6
TM4	455	0.42	940	730	1.7	0.2	540	13.8	26.1	0.55	25.58	2323.8
TM5	455	0.42	960	700	1.1	0.2	543	13.5	25.2	0.56	23.16	2311.6
TM6	455	0.42	960	700	1.7	0.2	579	9.9	18.1	0.65	24.01	2313.8
TM7	455	0.42	960	730	1.1	0.2	533	14.9	27.2	0.53	24.90	2341.6
TM8	455	0.42	960	730	1.7	0.2	565	11.2	19.2	0.60	25.60	2343.8
TM9	455	0.45	940	700	1.1	0.2	551	11.9	19.9	0.52	18.87	2305.2
TM10	455	0.45	940	700	1.7	0.2	589	8.6	16.5	0.70	19.65	2307.8
TM11	455	0.45	940	730	1.1	0.2	535	14.1	27.0	0.54	20.97	2335.2
TM12	455	0.45	940	730	1.7	0.2	569	11.2	19.2	0.60	21.62	2337.5
TM13	455	0.45	960	700	1.1	0.2	569	11.2	19.2	0.60	19.60	2325.2
TM14	455	0.45	960	700	1.7	0.2	607	7.5	15.1	0.73	22.50	2327.5
TM15	455	0.45	960	730	1.1	0.2	559	11.9	19.9	0.52	21.51	2355.2
TM16	455	0.45	960	730	1.7	0.2	589	8.5	16.1	0.71	22.30	2357.5
TM17	470	0.42	940	700	1.1	0.2	565	10.5	18.9	0.63	26.84	2313.0
TM18	470	0.42	940	700	1.7	0.2	596	8.6	16.0	0.71	27.88	2315.4
TM19	470	0.42	940	730	1.1	0.2	541	13.8	26.1	0.55	27.07	2343.0
TM20	470	0.42	940	730	1.7	0.2	577	11.2	19.2	0.60	28.10	2345.4

(continued)

Table 4 (continued)

Mix ID	Cement (kg/m^3)	W/B	Fine aggregate (kg/m^3)	Coarse aggregate (kg/m^3)	SP (%)	VMA (%)	Slump flow (mm)	T_{50} cm (s)	V-funnel (s)	L-box	f_{ck} (7 days) MPa	Density (kg/m^3)
TM21	470	0.42	960	700	1.1	0.2	581	11.0	18.2	0.60	25.50	2333.0
TM22	470	0.42	960	700	1.7	0.2	618	6.5	13.1	0.75	26.68	2335.4
TM23	470	0.42	960	730	1.1	0.2	565	10.5	18.9	0.63	26.95	2363.0
TM24	470	0.42	960	730	1.7	0.2	602	7.9	15.7	0.72	28.06	2365.4
TM25	470	0.45	940	700	1.1	0.2	592	8.5	16.1	0.71	21.40	2327.1
TM26	470	0.45	940	700	1.7	0.2	635	4.4	8.2	0.83	22.39	2329.5
TM27	470	0.45	940	730	1.1	0.2	571	9.9	18.1	0.65	23.45	2357.1
TM28	470	0.45	940	730	1.7	0.2	613	7.7	15.5	0.73	24.99	2359.5
TM29	470	0.45	960	700	1.1	0.2	622	6.5	13.1	0.75	21.23	2347.1
TM30	470	0.45	960	700	1.7	0.2	664	4.3	7.9	0.85	22.43	2349.5
TM31	470	0.45	960	730	1.1	0.2	605	7.5	15.1	0.73	23.31	2377.1
TM32	470	0.45	960	730	1.7	0.2	641	4.5	8.1	0.84	24.93	2379.5
TM33	485	0.42	940	700	1.1	0.2	622	6.5	13.1	0.75	28.39	2334.5
TM34	485	0.42	940	700	1.7	0.2	664	4.3	7.9	0.85	29.29	2336.9
TM35	485	0.42	940	730	1.1	0.2	605	7.6	15.5	0.72	30.11	2364.5
TM36	485	0.42	940	730	1.7	0.2	639	4.5	8.1	0.84	31.08	2366.9
TM37	485	0.42	960	700	1.1	0.2	645	4.5	8.2	0.84	28.26	2354.5
TM38	485	0.42	960	700	1.7	0.2	684	3.9	6.9	0.88	29.19	2356.9
TM39	485	0.42	960	730	1.1	02	627	6.3	12.5	0.77	30.05	2384.5
TM40	485	0.42	960	730	1.7	0.2	665	4.2	7.7	0.86	31.10	2386.9

(continued)

Table 4 (continued)

Mix ID	Cement (kg/m^3)	W/B	Fine aggregate (kg/m^3)	Coarse aggregate (kg/m^3)	SP (%)	VMA (%)	Slump flow (mm)	T_{50} cm (s)	V-funnel (s)	L-box	f_{ck} (7 days) MPa	Density (kg/m^3)
TM41	485	0.45	940	700	1.1	0.2	645	4.5	8.1	0.84	23.15	2349.1
TM42	485	0.45	940	700	1.7	0.2	680	4.4	7.8	0.86	24.25	2351.5
TM43	485	0.45	940	730	1.1	0.2	620	6.5	13.1	0.75	25.59	2379.1
TM44	485	0.45	940	730	1.7	0.2	659	4.3	7.9	0.84	26.70	2381.5
TM45	485	0.45	960	700	1.1	0.2	679	4.1	7.6	0.87	23.01	2369.1
TM46	485	0.45	960	700	1.7	0.2	708	3.3	6.5	0.90	24.05	2371.5
TM47	485	0.45	960	730	1.1	0.2	655	4.5	8.2	0.84	25.40	2399.1
TM48	485	0.45	960	730	1.7	0.2	689	3.8	7.0	0.88	26.65	2401.5
TM49	500	0.42	940	700	1.1	0.2	665	4.2	7.7	0.86	31.99	2356.0
TM50	500	0.42	940	700	1.7	0.2	691	3.7	7.3	0.88	32.96	2358.5
TM51	500	0.42	940	730	1.1	0.2	650	4.5	8.2	0.84	33.55	2386.0
TM52	500	0.42	940	730	1.7	0.2	673	4.1	7.6	0.87	34.69	2388.5
TM53	500	0.42	960	700	1.1	0.2	692	4.0	7.4	0.87	32.42	2376.0
TM54	500	0.42	960	700	1.7	0.2	719	3.5	7.0	0.90	33.01	2378.5
TM55	500	0.42	960	730	1.1	0.2	668	4.3	7.9	0.85	33.45	2406.0
TM56	500	0.42	960	730	1.7	0.2	697	3.9	7.2	0.88	34.57	2408.5
TM57	500	0.45	940	700	1.1	0.2	687	3.7	6.9	0.89	26.48	2371.0
TM58	500	0.45	940	700	1.7	0.2	721	3.0	6.1	0.92	27.53	2373.5
TM59	500	0.45	940	730	1.1	0.2	660	4.1	7.1	0.87	29.01	2401.0
TM60	500	0.45	940	730	1.7	0.2	692	3.5	6.4	0.91	29.99	2403.5

(continued)

Table 4 (continued)

Mix ID	Cement (kg/m^3)	W/B	Fine aggregate (kg/m^3)	Coarse aggregate (kg/m^3)	SP (%)	VMA (%)	Slump flow (mm)	T_{50} cm (s)	V-funnel (s)	L-box	f_{ck} (7 days) MPa	Density (kg/m^3)
TM61	500	0.45	960	700	1.1	0.2	708	3.3	6.5	0.90	26.38	2391.0
TM62	500	0.45	960	700	1.7	0.2	739	2.6	5.5	0.93	27.42	2393.5
TM63	500	0.45	960	730	1.1	0.2	691	3.5	6.8	0.89	28.94	2421.0
TM64	500	0.45	960	730	1.7	0.2	716	2.9	5.8	0.92	29.82	2423.5
TM65	498	0.42	960	700	1.1	0.25	690	4.0	7.5	0.87	32.40	2373.4

(ii) The volume of the coarse aggregate is defined by bulk density. The range for coarse aggregate is 50–60%. The possibility of collision or contact between coarse aggregate particles rapidly increase in SCC with increasing volume of coarse aggregate. This increases the risk of blockage of SCC at the congested spaces of reinforcement. The optimum coarse aggregate content in SCC is determined by the following parameters:

- Maximum aggregate size: The proportion of the coarse aggregate is higher when the maximum aggregate size is low.
- Crushed/round aggregates: for crushed aggregates a lower content can be used and for rounded aggregates a higher content can be used.

(iii) The optimum value of fine aggregate in mortar is kept between 40–50%.
(iv) The chosen proportions of cement and additions are determined to set the water powder ratio of the paste for zero flow which can be measured by slump flow in pastes at different w/b ratio.
(v) The slump cone and V-funnel tests are used to determine the optimum water powder ratio and optimum dosage of superpalsticizers in the paste.
(vi) The composition of concrete is now determined on trial batches and finally the requisite concrete tests are performed to select the proper dosage of superplasticizers.

Keeping in view the above steps, a suitable mix using the marginal aggregates was selected and numerous trial mixes were laid with the varying mix parameters like water-powder ratio, FA/CA ratio and percentage of SP and VMA. In the present investigation for the optimization of the material constituents, as many as 65 trial mixes have been carried out as shown in Table 4. The cognizance of the previous studies was considered for producing suitable SCC mix design. The modifications were made to the Japanese method of mix design which included changes in volumes of material constituents like binder, coarse aggregate, fine aggregate, SP dosage coupled with changes in w/b ratio. Various tests were performed to evaluate the fresh properties like slump flow, T_{50}, V-funnel, L-box and hardened properties like compressive strength. Initially, the mix design for the present investigation was carried out by using a coarse aggregate content at 30.41% and fine aggregate content at 39.14% by the total volume of concrete. In this study, the CA of size of 10–12.5 mm and the FA of size of less than 4.76 mm were used. The w/b was kept at 0.45 and SP at 1.7% of powder content. The VMA was kept constant for all the mixes at 0.2% of powder content. The binder was kept at 455 kg/m^3 and various trial mixes were prepared. The mixes were tested for fresh properties like slump flow to achieve the workability and hardened property like compressive strength. But the mixes failed to satisfy these properties as per EFNARC guidelines.

5 Results and Discussions

Table 4 represents the trial mixes where in modification to the rational method of mix proportion has been made. The said table also depicts the results of workability tests and compressive strength tests of different trial mixes. In order to achieve SCC mixes, w/c ratio and dosage of superplasticizer were varied and the trials were initiated at 39.16% content of fine aggregate and 30.41% content of coarse aggregate by volume of total concrete. The process of carrying out different trial mixes was continued till all the mix characterizing features were attained. The cement content of TM1 to TM16 mixes were taken as 455 kg/m^3, TM17 to TM32 mixes were taken as 470 kg/m^3, TM33 to TM48 mixes were taken as 485 kg/m^3 and TM49 to TM64 mixes were taken as 500 kg/m^3. The w/b content was varied from 0.42 to 0.45 after every eight mixes. Similarly, the dosage of SP was varied from 1.10 to 1.70% of the trial mixes alternatively. The fine and coarse aggregates dosage was started at 940 kg/m^3 and 700 kg/m^3 which constituted about 39.16% and 30.41% of the total volume of concrete respectively. The trial mixes (TM1 to TM64) were tested for workability such as slump flow, T_{50}, V-funnel, L-box tests and hardened tests like compressive strength.

5.1 Slump Flow

The results in respect of slump flow of the trial mixes from TM1 to TM64 are plotted in Fig. 2 for comparison. From Fig. 2, it was observed that the slump flow values of trial mixes TM1 to TM16 containing a cement content of 455 kg/m^3 were in the range of 504–607 mm which does not fall within the range as per EFNARC guidelines. The slump flow values of trial mixes TM17 to TM32 containing a cement content of 470 kg/m^3 were observed to be in the range of 541–664 and all these mixes except TM30 mix were not within the range in accordance to the EFNARC guidelines. The slump flow values of trial mixes TM33 to TM48 containing a cement content 485 kg/m^3 were in between 605–708 mm and some of mixes were not within the range as per the EFNARC guidelines. The slump flow values of the trial mixes TM49 to TM64 containing a cement content of 500 kg/m^3 were in between 650–739 mm and all the test values were within the specified range as per EFNARC guidelines.

5.2 T_{50} Flow Time

The results in respect of T_{50} flow time of the trial mixes from TM1 to TM64 are plotted in Fig. 3 for comparison. From Fig. 3, it was observed that the T_{50} flow time of trial mixes TM1 to TM16 containing a cement content of 455 kg/m^3 were in the range of 7.5–15.2 s which does not fall within the range as per EFNARC guidelines.

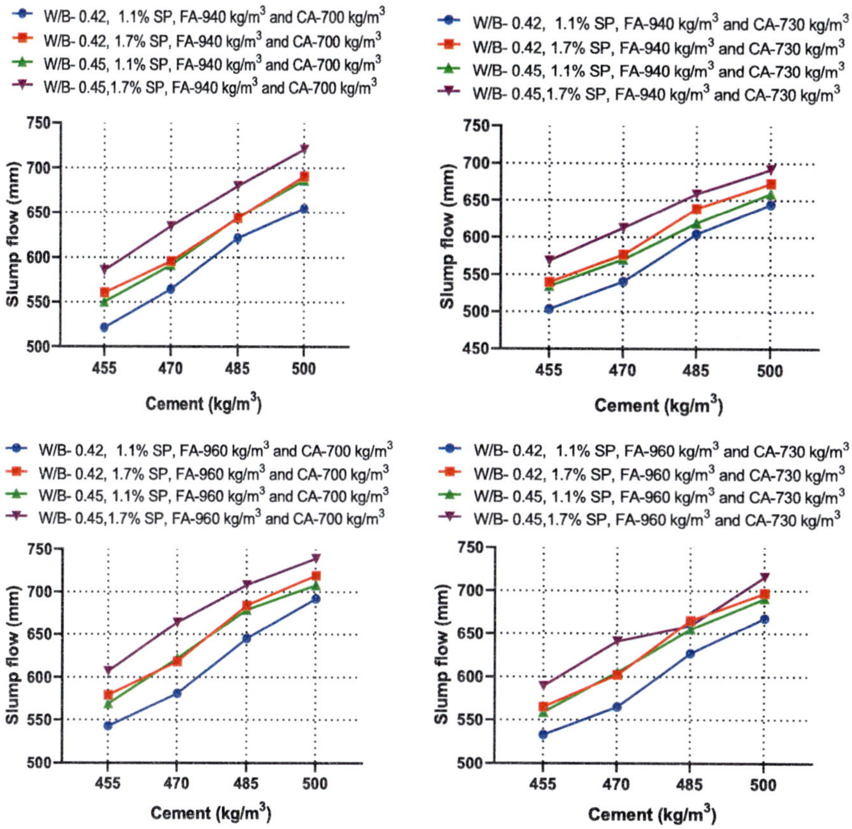

Fig. 2 Variation of slump flow (mm) at different cement contents

The T_{50} flow time of trial mixes TM17 to TM32 containing a cement content of 470 kg/m^3 were observed to be in the range of 4.3–13.8 s and all these mixes except TM26, TM30 and TM32 were not within the range in accordance to the EFNARC guidelines. The T_{50} flow time of trial mixes TM33 to TM48 containing a cement content 485 kg/m^3 were in between 3.3–7.6 s and some of mixes were not within the range as per the EFNARC guidelines. The T_{50} flow time of the trial mixes TM49 to TM64 containing a cement content of 500 kg/m^3 were in between 2.6–4.3 s and all the test values were within the specified range as per EFNARC guidelines.

5.3 *V-funnel*

The results in respect of V-funnel flow time of the trial mixes from TM1 to TM64 are plotted in Fig. 4 for comparison. From the Fig. 4, it was observed that the V-funnel

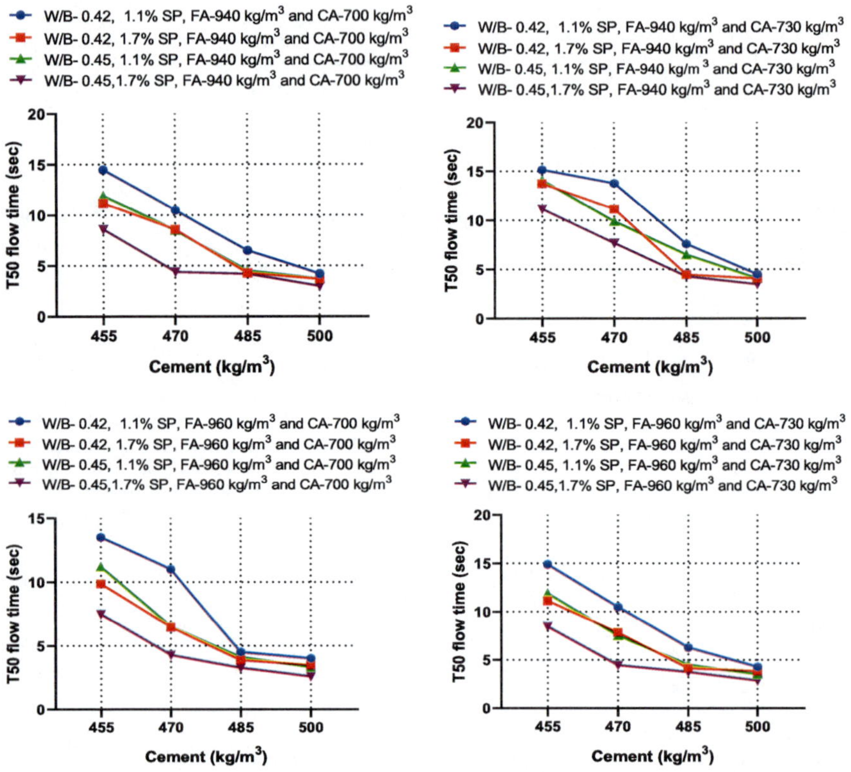

Fig. 3 Variation of T_{50} flow time (s) at different cement contents

flow time of trial mixes TM1 to TM16 containing a cement content of 455 kg/m^3 were in the range of 15.1–27.3 s which does not fall within the range as per EFNARC guidelines. The V-funnel flow time of trial mixes TM17 to TM32 containing a cement content of 470 kg/m^3 were observed to be in the range of 7.9–26.1 s and all these mixes except TM26, TM30 and TM32 mix were not within the range in accordance to the EFNARC guidelines. The V-funnel flow time of trial mixes TM33 to TM48 containing a cement content 485 kg/m^3 were in between 6.5–15.5 s and some of mixes were not within the range as per the EFNARC guidelines. The V-funnel flow time of the trial mixes TM49 to TM64 containing a cement content of 500 kg/m^3 were in between 5.5–8.2 s and all the test values were within the specified range as per EFNARC guidelines.

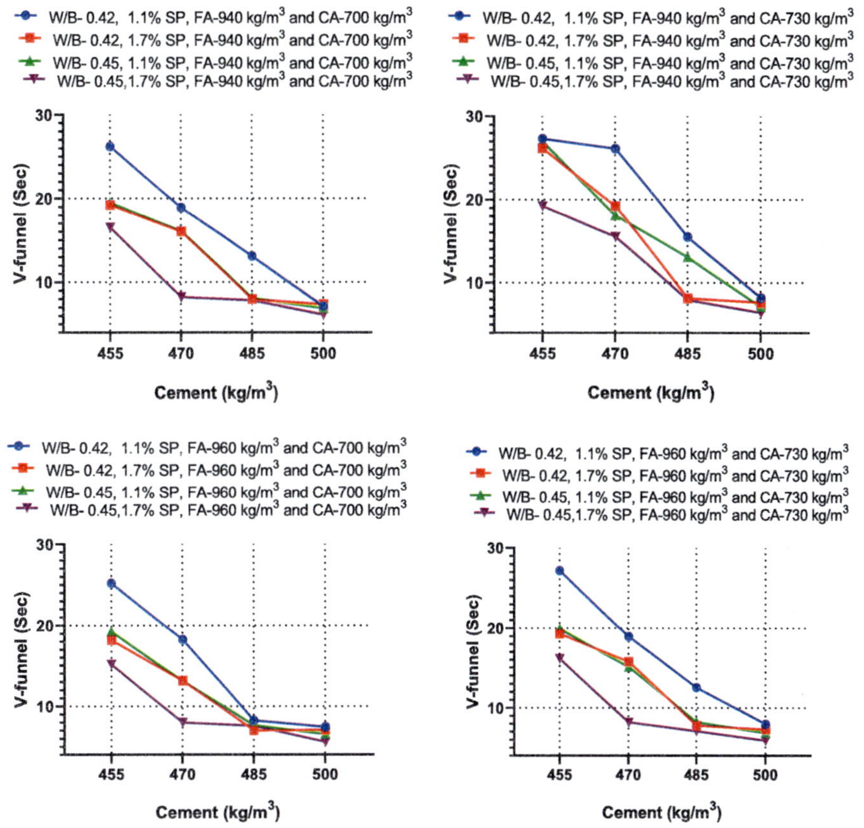

Fig. 4 Variation of V-funnel flow time (s) at different cement contents

5.4 Blocking Ratio (L-Box)

The results in respect of blocking ratio (L-box) of the trial mixes from TM1 to TM64 are plotted in Fig. 5 for comparison. From the Fig. 5 it was observed that the blocking ratio of trial mixes TM1 to TM16 containing a cement content of 455 kg/m^3 were in the range of 0.48–0.73 which does not fall within the range as per EFNARC guidelines. The blocking ratio of trial mixes TM17 to TM32 containing a cement content of 470 kg/m^3 were observed to be in the range of 0.55–0.85 and all these mixes except TM26, TM30 and TM32 mix were not within the range in accordance to the EFNARC guidelines. The blocking ratio of trial mixes TM33 to TM48 containing a cement content of 485 kg/m^3 were in between 0.72–0.90 and some of mixes were not within the range as per the EFNARC guidelines. The blocking ratio of the trial mixes TM49 to TM64 containing a cement content of 500 kg/m^3 were in between 0.84–0.93 and all the test values were within the specified range as per EFNARC guidelines.

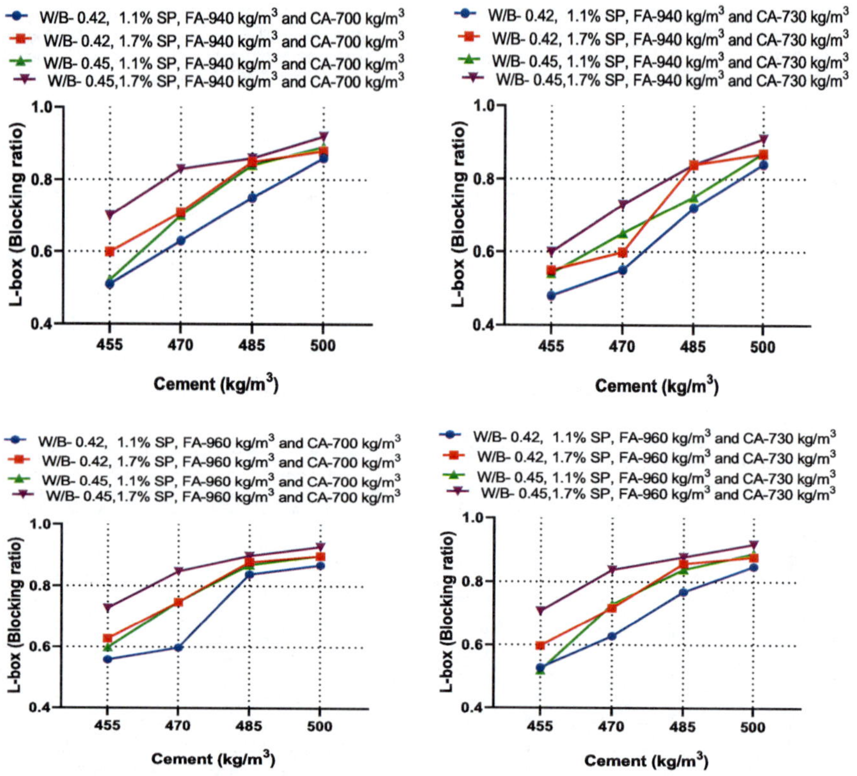

Fig. 5 Variation of blocking ratio (L-box) at different cement contents

5.5 *Compressive Strength*

The compressive strength of all trial mixes (TM1 to TM64) were evaluated at 7 days. The compressive strength results are graphically presented in Fig. 6 for comparison. From the results it can be seen that the compressive strength of the trial mixes TM1 to TM16 were in the range of 18.87–25.60 MPa. The compressive strength of trial mixes TM17 to TM32 was in between 21.23–28.10 MPa. The compressive strength of the trial mixes TM33 to TM48 were in the range of 23.01–31.10 MPa and the compressive strength of the trial mixes TM49 to TM464 were in the range of 26.48–34.69 MPa. The perusal of the results indicates that in the trial mixes where w/b ratio of 0.45 was maintained have exhibited a decrease in compressive strength.

It is pertinent to mention that some trial mixes (TM1 to TM48) which were within the range as per the EFNARC guidelines were rejected owing to the fact that the bleeding and segregation was observed in these mixes due to high w/b ratio. Also, in these mixes the dosage of SP was higher (1.7%) which can escalate the cost of the concrete. The content of fine aggregate was varied from 940 to 960 kg/m^3 and coarse aggregate was varied from 730 to 700 kg/m^3 in different trial mixes to achieve SCC

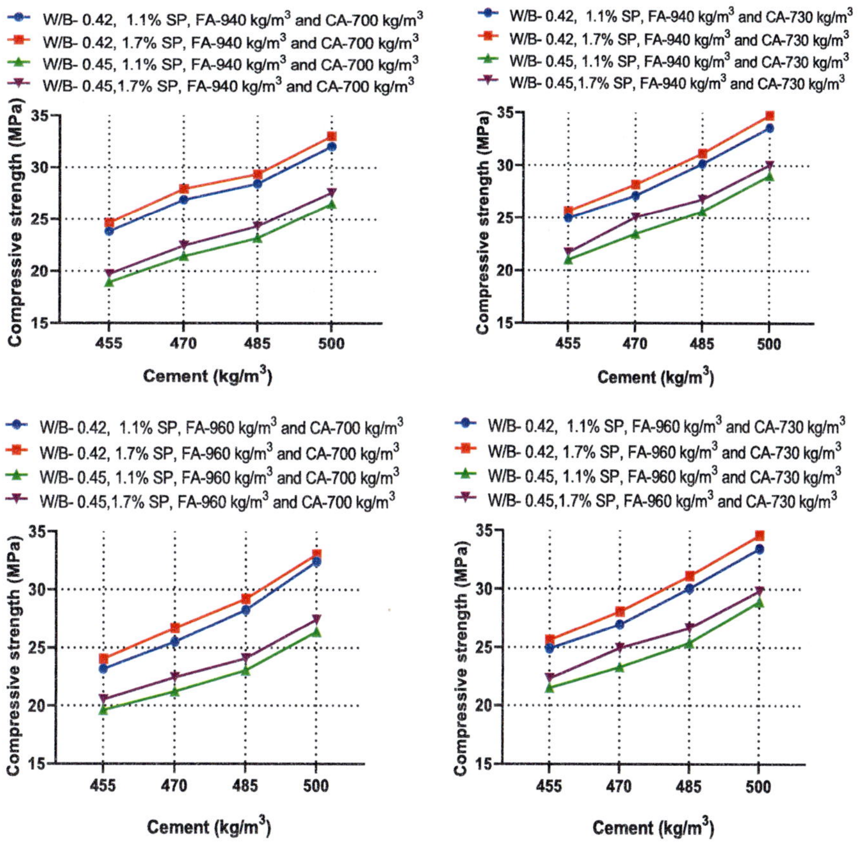

Fig. 6 Variation of Compressive strength (MPa) at different cement contents

characteristics. While testing the trial mixes for workability tests, the trial mixes from TM49 to TM64 achieved almost all the SCC characteristics satisfactorily. The perusal of the literature revealed that a binder content for most of the research studies was taken as 500 kg/m^3 in concrete. So, another trial mix TM65 was designed wherein cement content, fine aggregate content and coarse aggregate content were taken as 498 kg/m^3, 960 kg/m^3 and 700 kg/m^3 respectively. This trial mix TM65 was selected for completion of the ongoing project. The reason for selecting the fine aggregate and coarse aggregate content at 960 kg/m^3 and 700 kg/m^3 respectively was to enhance workability of SCC. In this trial mix the dosage of SP and VMA was taken as 1.10% and 0.25% respectively. The w/b ratio of trial mixes TM49 to TM 64 was kept at 0.42 and for trial mixes TM57 to TM64 was kept at 0.45. But in this research study the w/b ratio was chosen at 0.42 owing to the reason that the lesser water to cement content increases the strength of the concrete as reflected in Table 4. The workability and consistency of trial mix TM65 satisfied all the workability properties such as slump flow, T50 cm, V-funnel and L-box with better compressive strength.

5.6 Standard Deviation and Coefficient of Variation

A statistical measure expressing the arrangement of data points in a data series close to the mean value is termed as coefficient of variation (COV). The COV denotes the ratio of standard deviation (SD) from the mean value. Despite the variations of mean values from each other, the COV is a powerful statistical tool for examining the degree of variation from one data series to other data series [20]. The mean and SD values of different SCC trial mixes were determined and the cognizance of the method proposed by the Naji et al. [21] was considered to determine the COV of different SCC trial mixes for each test. The various SCC trial mixes with different adjustments were ranked taking into the consideration the COV values of each test. The statistical results of trial mixes with different adjustments are reflected in Tables 5, 6, 7 and 8. From the test results, it is obvious that the lower COV values shows the lower scattering of the test results indicating the concrete is more robust in the desired test.

6 Conclusion

This investigation was carried out to develop a procedure for optimization of mixture proportion design to achieve self-compacting concrete. Based on the results, the following conclusions can be drawn:

- Various trial mixes were designed by the modifications of rational method mix design to obtain an optimum mixture design.
- The optimum mixture design so produced have satisfied all workability properties as prescribed by EFNARC guidelines.
- In order to achieve the target slump flow, T_{50} flow time, V-funnel flow time and blocking ratio (L-box) as well as the compressive strength, the coarse aggregate content was decreased to 29.16% by total volume of concrete and fine aggregate was increased to 40% by total volume of concrete. The binder was increased from 455 kg/m^3 to 470 kg/m^3 then to 485 kg/m^3 and finally increased to 500 kg/m^3. The w/b was varied from 0.45 to 0.42 and the dosage of SP was varied from 1.1 to 1.70% of powder content.
- The variation in w/b ratio from 0.45 to 0.42 has enhanced the compressive strength of the concrete. The variation in the dosage of SP from 1.70% to 1.10% has reduced the cost of the concrete making it economically more viable.
- The slump flow, T_{50}, V-funnel and Blocking ratio (L-box) of the optimum mixture design were in the range of 650–739 mm, 2.6–4.3 s, 5.5–8.2 s and 0.84–0.93 respectively.
- The compressive strength of the optimum mixture design was in the range of 26.48 to 34.69 MPa.

The present investigation under report is innovative because of the fact that the contents of coarse aggregate and fine aggregate were varied instead of keeping

Table 5 Standard deviation of trial mixes with FA and CA constant at 940 kg/m^3 and 700 kg/m^3 respectively and variation in binder content, W/B and SP dosage

Test	Parameter	Trial mixes			
		Binder-455 kg/m^3 W/B-0.42, 1.1% SP, FA-940 kg/m^3 and CA-700 kg/m^3	Binder-470 kg/m^3 W/B-0.42, 1.7% SP, FA-940 kg/m^3 and CA-700 kg/m^3	Binder-485 kg/m^3 W/B-0.45, 1.1% SP, FA-940 kg/m^3 and CA-700 kg/m^3	Binder-450 kg/m^3 W/B-0.45, 1.7% SP, FA-940 kg/m^3 and CA-700 kg/m^3
Slump flow	Mean	591	623	618.8	655.5
	Standard deviation	59.14	56.68	59.59	58.14
	Coefficient of variation (%)	10.01	9.098	9.631	8.870
	Rank	4	2	3	1
T_{50}	Mean	8.925	6.950	7.150	5.050
	Standard deviation	4.538	3.576	3.800	2.446
	Coefficient of variation (%)	50.84	51.46	53.14	48.44
	Rank	2	3	4	1
V-funnel	Mean	16.33	12.60	12.65	9.650
	Standard deviation	8.158	5.925	6.126	4.657
	Coefficient of variation (%)	49.97	47.02	48.43	48.25
	Rank	4	1	3	2
L-box	Mean	0.6875	0.7600	0.7375	0.8275
	Standard deviation	0.1511	0.1299	0.1658	0.09287

(continued)

Table 5 (continued)

Test	Parameter	Trial mixes			
		Binder-455 kg/m^3 W/B-0.42, 1.1% SP, FA-940 kg/m^3 and CA-700 kg/m^3	Binder-470 kg/m^3 W/B-0.42, 1.7% SP, FA-940 kg/m^3 and CA-700 kg/m^3	Binder-485 kg/m^3 W/B-0.45, 1.1% SP, FA-940 kg/m^3 and CA-700 kg/m^3	Binder-450 kg/m^3 W/B-0.45, 1.7% SP, FA-940 kg/m^3 and CA-700 kg/m^3
	Coefficient of variation (%)	21.98	17.09	22.48	11.22
	Rank	3	2	4	1
Compressive strength	Mean	27.75	28.68	22.48	23.46
	Standard deviation	3.422	3.475	3.196	3.309
	Coefficient of variation (%)	12.33	12.12	14.22	14.11
	Rank	2	1	4	3

Table 6 Standard deviation of trial mixes with FA and CA constant at 940 kg/m^3 and 730 kg/m^3 respectively and variation in binder content, W/B and SP dosage

Test	Parameter	Trial mixes			
		Binder-455 kg/m^3 W/B-0.42, 1.1% SP, FA-940 kg/m^3 and CA-730 kg/m^3	Binder-470 kg/m^3 W/B-0.42, 1.7% SP, FA-940 kg/m^3 and CA-730 kg/m^3	Binder-485 kg/m^3 W/B-0.45, 1.1% SP, FA-940 kg/m^3 and CA-730 kg/m^3	Binder-500 kg/m^3 W/B- 0.45, 1.7% SP, FA-940 kg/m^3 and CA-730 kg/m^3
Slump flow	Mean	573.8	607.3	596.5	633.3
	Standard deviation	63.22	59.91	54.82	53.71
	Coefficient of variation (%)	11.02	9.866	9.191	8.481
	Rank	4	3	2	1
T_{50}	Mean	10.28	8.400	8.650	6.675
	Standard deviation	5.072	4.855	4.343	3.524
	Coefficient of variation (%)	49.37	57.79	50.21	52.79
	Rank	1	4	2	3
V-funnel	Mean	19.28	15.25	16.33	12.25
	Standard deviation	9.090	8.999	8.418	6.110
	Coefficient of variation (%)	47.16	59.01	51.57	49.88
	Rank	1	4	3	2
L-box	Mean	0.6475	0.7150	0.7025	0.7700
	standard deviation	0.1632	0.1634	0.1408	0.1354

(continued)

Table 6 (continued)

Test	Parameter	Trial mixes			
		Binder-455 kg/m^3 W/B-0.42, 1.1% SP, FA-940 kg/m^3 and CA-730 kg/m^3	Binder-470 kg/m^3 W/B-0.42, 1.7% SP, FA-940 kg/m^3 and CA-730 kg/m^3	Binder-485 kg/m^3 W/B-0.45, 1.1% SP, FA-940 kg/m^3 and CA-730 kg/m^3	Binder-500 kg/m^3 W/B- 0.45, 1.7% SP, FA-940 kg/m^3 and CA-730 kg/m^3
	Coefficient of variation (%)	25.20	22.85	20.04	17.58
	Rank	4	3	2	1
Compressive strength	Mean	28.92	29.86	24.76	25.83
	Standard deviation	3.750	3.926	3.407	3.488
	Coefficient of variation (%)	12.97	13.15	13.76	13.51
	Rank	1	2	4	3

Table 7 Standard deviation of trial mixes with FA and CA constant at 960 kg/m^3 and 700 kg/m^3 respectively and variation in binder content, W/B and SP dosage

Test	Parameter	Trial MIXES			
		Binder-455 kg/m^3 W/B-0.42, 1.1% SP, FA-960 kg/m^3 and CA-700 kg/m^3	Binder-470 kg/m^3 W/B-0.42, 1.7% SP, FA-960 kg/m^3 and CA-700 kg/m^3	Binder-485 kg/m^3 W/B-0.45, 1.1% SP, FA-960 kg/m^3 and CA-700 kg/m^3	Binder-500 kg/m^3 W/B-0.45, 1.7% SP, FA-960 kg/m^3 and CA-700 kg/m^3
Slump flow	Mean	615.3	650	644.5	679.5
	Standard deviation	66.25	63.2	61.72	57.3
	Coefficient of variation (%)	10.77	9.723	9.577	8.432
	Rank	4	3	2	1
T_{50}	Mean	8.250	5.950	6.275	4.425
	Standard deviation	4.735	2.950	3.554	2.165
	Coefficient of variation	57.39	49.58	56.63	48.94
	Rank	4	2	3	1
V-funnel	Mean	14.75	11.28	11.60	9.000
	Standard deviation	8.525	5.395	5.832	4.200
	Coefficient of variation (%)	57.80	47.85	50.27	46.67
	Rank	4	2	3	1
L-box	Mean	0.7175	0.7900	0.7800	0.8525
	Standard deviation	0.1601	0.1257	0.1364	0.08808
	Coefficient of variation	22.31%	15.91%	17.48%	10.33%
	Rank	4	2	3	1

(continued)

Table 7 (continued)

Test	Parameter	Trial MIXES			
		Binder-455 kg/m^3 W/B-0.42, 1.1% SP, FA-960 kg/m^3 and CA-700 kg/m^3	Binder-470 kg/m^3 W/B-0.42, 1.7% SP, FA-960 kg/m^3 and CA-700 kg/m^3	Binder-485 kg/m^3 W/B-0.45, 1.1% SP, FA-960 kg/m^3 and CA-700 kg/m^3	Binder-500 kg/m^3 W/B-0.45, 1.7% SP, FA-960 kg/m^3 and CA-700 kg/m^3
Compressive strength	Mean	27.34	28.22	22.56	23.60
	Standard deviation	3.980	3.829	2.907	2.931
	Coefficient of variation	4	3	2	1
	Rank	14.56%	13.57%	12.88%	12.42%

Table 8 Standard deviation of trial mixes with FA and CA constant at 940 kg/m^3 and 730 kg/m^3 respectively and variation in binder content, W/B and SP dosage

Test	Parameter	Trial mixes			
		Binder-455 kg/m^3 W/B-0.42, 1.1% SP, FA-960 kg/m^3 and CA-730 kg/m^3	Binder- 470 kg/m^3 W/B-0.42, 1.7% SP, FA-960 kg/m^3 and CA-730 kg/m^3	Binder- 485 kg/m^3 W/B-0.45, 1.1% SP, FA-960 kg/m^3 and CA-730 kg/m^3	Binder- 500 kg/m^3 W/B-0.45, 1.7% SP, FA-960 kg/m^3 and CA-730 kg/m^3
Slump flow	Mean	598.3	632.3	627.5	651.3
	Standard deviation	60.7	59.73	57.7	52.39
	Coefficient of variation (%)	10.15	9.447	9.195	8.044
	Rank	4	3	2	1
T_{50}	Mean	9.000	6.800	6.850	4.925
	Standard deviation	4.706	3.452	3.771	2.472
	Coefficient of variation (%)	52.29	50.76	55.06	50.19
	Rank	3	2	4	1
V-funnel	Mean	16.63	12.45	12.50	9.250
	Standard deviation	8.370	5.951	6.124	4.662
	Coefficient of variation (%)	50.34	47.80	48.99	50.40
	Rank	3	1	2	4
L-box	Mean	0.6950	0.7650	0.7450	0.8375
	Standard deviation	0.1427	0.1310	0.1642	0.09106

(continued)

Table 8 (continued)

Test	Parameter	Trial mixes			
		Binder-455 kg/m^3 W/B-0.42, 1.1% SP, FA-960 kg/m^3 and CA-730 kg/m^3	Binder- 470 kg/m^3 W/B-0.42, 1.7% SP, FA-960 kg/m^3 and CA-730 kg/m^3	Binder- 485 kg/m^3 W/B-0.45, 1.1% SP, FA-960 kg/m^3 and CA-730 kg/m^3	Binder- 500 kg/m^3 W/B-0.45, 1.7% SP, FA-960 kg/m^3 and CA-730 kg/m^3
	Coefficient of variation (%)	20.53	17.13	22.04	10.87
	Rank	3	2	4	1
Compressive strength	Mean	28.84	29.83	24.79	25.93
	Standard deviation	3.733	3.878	3.191	3.153
	Coefficient of variation (%)	12.95	13.00	12.87	12.16
	Rank	3	4	2	1

constant in a fixed range as in rational method of mix design. The results demonstrated that trial mixes TM49 to TM64 were the optimum mixture design which satisfied all the fresh as well as hardened properties. This was accomplished due to the modification of binder content, w/b, coarse aggregate, fine aggregate and SP content.

Acknowledgements The authors highly acknowledge to the Vellore Institute of Technology, Vellore for their support through SEED GRANT fund in carrying out the research project. The authors are also grateful to BASF India Limited, Construction Chemicals Division, Kancheepuram (Tamil Nadu) for providing chemical admixtures to undertake the experimental study reported in this paper.

References

1. Nagamoto, N., Ozawa, K., Mixture properties of self-compacting, high-performance concrete. In: Proceedings, Third CANMET/ACI International Conferences on Design and Materials and Recent Advances in Concrete Technology, SP-172, V. M. Malhotra, American Concrete Institute, Farmington Hills, Mich., pp. 623–637 (1997)
2. Okamura, H., Ozawa, K.: Mix Design for Self-Compacting Concrete. Concrete Library of Japanese Society of Civil Engineers, June 25, 1995, pp. 107–120.
3. Okamura, H.: Self-compacting high-performance concrete. Concr. Int. Des. Constr. **19**, 50–54 (1997)
4. Okamura, H., Ouchi, M.: Self-compacting concrete: development, present use and future in self compacting concrete. In: Proceedings of the First International RILEM Symposium. RILEM Publications, Cachan Cedex, France, pp. 3–14 (1999)
5. Okamura, H., Ouchi, M.: Self-compacting concrete. J. Adv. Concr. Tech. **1**, 5–15 (2003)
6. Dinakar, P.: Design of self-compacting concrete with fly ash. Mag. Concr. Res. **2012**, 401–409 (2012)
7. Sonebi, M.: Medium strength self-compacting concrete containing fly ash: Modelling using factorial experimental plans. Cem. Concr. Res. **34**, 1199–1208 (2004)
8. Bouzoubaa, N., Lachemi, M.: Self-compacting concrete incorporating high volumes of class F fly ash. Cem. Concr. Res. **31**(3), 413–420 (2001)
9. Xie, Y., Liu, B., Yin, J., Zhou, S.: Optimum mix parameters of high-strength self-compacting concrete with ultra-pulverized fly ash. Cement Concr. Res. **32**(3), 477–480 (2002)
10. Ouchi, M., Nakamura, S., Osterson, T., Hallberg, S.E., Lwin, M.: Applications of self-compacting concrete in Japan, Europe and the United States. In: Proceedings of 5th International Symposium on High Performance Computing, Federal Highway Administration, Washington, D.C. (2003)
11. De Schutter, G., Bartos, P., Domone, P., Gibbs, J.: Self-compacting concrete. Whittles Publishing, Caithness, Scotland (2008)
12. Khaleel, O.R., Razak, R.A.: Mix design method of self-compacting metakaolin concrete with different properties of coarse aggregate. Mater Des. **53**, 691–700 (2014)
13. Ashish, D.K., Verma, S.K.: An overview on mixture design of self-compacting concrete. Struct. Concr. **20**(1), 371–395 (2019)
14. EFNARC: Specifications and Guidelines for Self-Compacting Concrete, EFNARC, UK, pp. 1–45 (2005)
15. IS 12269-1987 (reaffirmed in 2013): Ordinary Portland Cement 53 Grade-specification. Bureau of Indian Standards, New Delhi, India

16. IS 383-2016: Specifications for Coarse and Fine aggregates from Natural Sources for Concrete. Bureau of Indian standards, New Delhi, India.
17. IS 9013-1999 (reaffirmed 2004): Concrete Admixtures—Specifications. Bureau of Indian standards, New Delhi, India
18. IS 456-2000 (reaffirmed in 2007): Code of practice for plain and reinforced concrete. Bureau of Indian Standards, New Delhi, India
19. IS 516-1959 (reaffirmed 1999): Methods of Tests for Strength of Concrete. Bureau of Indian Standards, New Delhi, India
20. Ghoddousi, P., & Salehi, A.: Effect of mix proportion on robustness of self-compacting concrete. Gradevinar **67**, 1–9 (2015)
21. Naji, S., Hwang, S., Khayat, K., Robustness of self-consolidating concrete incorporating different viscosity-enhancing admixture. ACI Mate. J. (2011)

Corrosion Characteristics of Rebar Induced in Different Types of Fibre Reinforced Concrete

Ganesh Naidu Gopu and A. Sofi

Abstract Based on the early studies it is evident that fibre reinforced concrete exhibits good results in strength and workability compared to conventional concrete. This research work evaluates the practical corrosion resistance of rebar induced various forms of concrete reinforced with fibre. In this study, three multiple types of concrete reinforced with E-waste copper wire, steel fibre and E-Glass fibre at a constant percentage of 1% were used. To evaluate corrosion characteristics of rebar introduced in fibre reinforced concrete accelerated corrosion test and standard test method as per ASTM G109 were conducted. Test results were compared for all three types of fibre reinforced concrete. Results showed that E-Glass fibre reinforced concrete induced rebar is high corrosion resistivity compared to control concrete, steel fibre reinforced concrete and E-waste copper wire fibre reinforced concrete specimens.

Keywords Fibre reinforced concrete · Corrosion · Chloride environment · Rebar · Conventional concrete

1 Introduction

Concrete is among the most common building material for the formation of structural elements. Present technologies are using different types of concrete, among which fibre reinforced concrete has been accentuated due to its high strength exhibiting nature. Generally concrete is weak in restrict to crack initiation and crack propagation. Fibres plays important role to overcome these problem hence fibres introduced in concrete mix [1]. Different types of fibres are available in market such as steel fibres, glass fibres, basalt fibres etc. [2]. Increase the durability properties of concrete

G. N. Gopu · A. Sofi (✉)
Department of Structural and Geotechnical Engineering, Vellore Institute of Technology, Vellore, India
e-mail: asofi@vit.ac.in

G. N. Gopu
e-mail: gopuganeshnaidu@gmail.com

D. K. Ashish et al. (eds.), *3rd International Conference on Innovative Technologies for Clean and Sustainable Development*, RILEM Bookseries 29,
https://doi.org/10.1007/978-3-030-51485-3_26

by incorporated Fibres into the concrete and it increase chloride ion of the concrete [3]. In this research E-waste copper wire fibres introduced in concrete. It is the sources of electrical waste. It is not reacted with water (H_2O) and it is good conductor of electricity [4]. Initially, E-glass or electrical grid glass was developed for isolators for electrical wiring during the time it was provided to reinforce concrete as glass fibre. E-glass fibres are enhance the strength and durability properties of the concrete and also to produce concrete more acid resistance [5] and Steel fibres added in concrete to increase both the durability and the strength of the reinforced concrete. Steel fibre is good conductor of electricity [6]. Most basic tests to predict the corrosive resistance of the concrete are accelerated corrosion test and standard test for calculating corrosive resistance as per ASTM G109 [7]. Corrosion damage to concrete material is detected by using half-cell potential and LPR methods [8, 9]. Half-cell potential method gives only corrosion initiation and propagation but cannot explain corrosion rate [9]. AC impedance method (weight loss technique) is most viewed and suited method to estimate corrosion rate. Recently electrochemical techniques are coming into play [10]. Structures in severe carbonation concentrated areas are highly prone to premature durability problems [10]. Macro defect free cements were equipped with fibres to further increase the mechanical properties of concrete [11]. Corrosion in tension region causes more damage than corrosion in compression region in bride columns [12]. Frequency responses were used to study the corrosion characteristics of reinforcement embedded in concrete [13–15]. Corroded concrete specimen is taken under X-ray diffraction minute characteristics of concrete [16]. Zinc phosphate coatings were added along with the epoxy coating to control corrosion character [17–20]. Anion fabrication techniques are also tried to control corrosion of the reinforced concrete [21]. Most existing methods for controlling concrete corrosion in sewers are either very sluggish or harmful [22]. It was found from the tests that the specimens made with PPC displayed higher relative resistivity values and lower corrosion density values compared to those made with OPC in all structural formulations [23]. It has shown a tenfold decrease in the corrosion rate of it embedded in concrete under accelerated macrocell corrosion conditions in concrete using the engineered admixture method. Using the designed admixture model did not affect the compressive strength of mortars [24]. Anodic polarization, electron spectroscopy for chemical analysis (ESCA) and Auger electron spectroscopy (AES) were used to test the effect of inhibitors on corrosion of steel reinforcement in concrete [25]. Although the change in the time-to-cracking of concrete specimens integrating M2 was small, there was a significant improvement in the corrosion-resistant properties of concrete integrating R2 over the control specimens [26]. Steel corrosion inherent in concrete, two electrochemical measurements, namely the anodic Tafel slope and potential, were technically evaluated and semi-empirically adjusted by the intensity of chloride ions in concrete, and also the suggested model was used to analyze the concentration of chloride ions as well as the current density of corrosion around the cross-section and steel bar circumference in concrete members with one or more steel bars [26, 27]. This paper provides a detailed study on fiber reinforced concrete specimens' corrosion resistance.

2 Materials and Methods

2.1 Materials

53 grade of OPC was used as binding material. 12 mm coarse aggregate and natural river sand is used as fine aggregate. To ease the mix and to increase the flow-ability of concrete, water reducing agent is used. Table 1 was shows the properties of coarse and fine aggregate.

E-Glass fibres of 0.12 mm diameter and 6 mm length, E-waste copper wire fibres of 0.6 mm diameter and 30 mm length, steel fibres of 40 mm length and 0.8 mm diameter were mixed in concrete with a constant value of 1%. Figure 1a–c shows the image of fibres used in this study.

2.2 Mix Design and Proportion

Steel, E-waste copper wire and E-Glass fibres were mixed by 1% volume of concrete using hand mix and water reducing agent by volume of cement was added to achieve proper workability of concrete. Figure 2 demonstrates the preparation of concrete mix.

2.3 Accelerated Corrosion Test (ACT) Procedure and Experiment Setup

Concrete cylindrical specimens were prepared with a diameter 15 cm and a depth of 30 cm reinforced with 10 mm diameter steel rebar with a 25 mm of effective cover from the bottom of cylinder were prepared to conduct this test (ASTM G 109). Rebar embedded in such a way that 5 cm of bar length is protruded outside from the top surface of the cylinder as shown in Fig. 3.

Table 1 Properties of aggregates

Physical tests	Coarse aggregate	Fine aggregate
Specific gravity	2.58	2.66
Fineness modulus	4.32	2.32
Bulk density (kg/m^3)	1540	1780

Fig. 1 **a** E-glass fibres; **b** steel fibres; **c** E-waste copper wire fibres

Fig. 2 Preparation of concrete mix

Fig. 3 Preparation of test specimen

2.4 *Experimental Set up*

Concrete cylinders were kept in a transparent plastic E-Glass tank and test is carried out using anodic current method. A D.C power supply is used to transfer power to the reinforced bars in the cylinder. 3.5% of NaCl is poured in to the tank and cylinders were filled were immersed into the tank and care is taken to fill the NaCl solution is filled upto a depth of 50 mm from the top. The concrete embedded reinforced steel bar is used as anode and as a cathode, a steel plate is used. Figure 4 demonstrates the laboratory set-up of the accelerated corrosion test. Multimeter was used to check the voltage generated in anode.

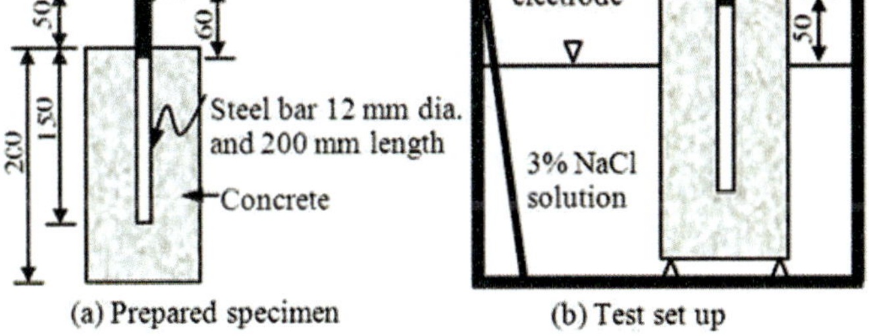

Fig. 4 Accelerated corrosion test setup

2.5 Examination of Chloride Ion Resistance as Per ASTM G109 Standard

To begin the test, steel rebar of 10 mm diameter are power wire brushed to maximum white and bars were cleaned by soaking them in sulfuric acid [28]. One end of the bar is tapped and drilled to form threads and size of threading is nearly equal to screw size. Each tapped end is bolted for two nuts. Length of bar should be 38 cm for which; both sides are wrapped with electroplaters tape for a length 100 mm on each side. Cover electroplaters tape with neoprene tubing on both sides for a length of 50 mm on each side. Bars were placed equidistant and 40 mm of bar is exposed outside, space between tape and tube are filled with two parts epoxy. Arrangement of bars in specimen was shown in Fig. 5. Care is taken while casting, that bars are clearly protected from moisture. Specimens are cured for 28 days. After removing from curing, top surface of the specimen is wire brushed and specimens are allowed to dry for 12 days. A flexi E-Glass dam is installed on the wire brushed side of the specimen. Sides of the dam are sealed using silicon sealant. Dam is filled with 3.5% of NaCl to a depth of 20 mm.

Fig. 5 Experimental setup of ASTM G 109

3 Experimental Results and Discussions

3.1 Accelerated Corrosion Test (ACT)

Table 2 shows the steel bar mass loss (in grams) for the concrete specimens incorporated by different types of fibres. Starting from an actual, undamaged 250 g per 225 mm long 10 mm diameter bar, after having been subjected to acceleration corrosion test bar weight was reduced. E-Glass fibre reinforced concrete samples performance batter than other concrete mix samples.

3.2 Corrosion Potential Rates

The top and bottom rebar corrosion potential was measured using a half-cell potential study as per ASTM C876 and copper–copper sulfate (CSE) half-cell reference electrode was used [29]. Potential graphs for the test were shown in Figs. 6 and 7. For top rebar, an average of corrosion values of three specimens of conventional concrete

Table 2 Loss of weight in rebar in accelerated corrosion test

Type of member	Clear cover (mm)	Voltage	Loss in weight (g)
Control concrete	25	20	30.7
Steel fibre reinforced concrete	25	20	20.8
E-waste glass fibre reinforced concrete	25	20	15
E-waste copper wire fibre reinforced concrete	25	20	22.4

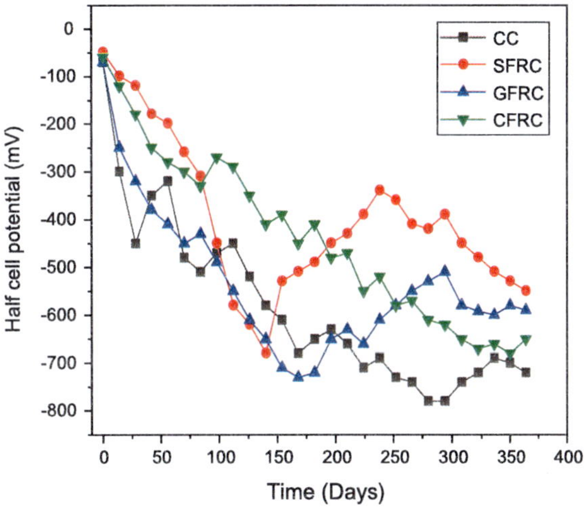

Fig. 6 Measurements of half cell potential for the top rebar

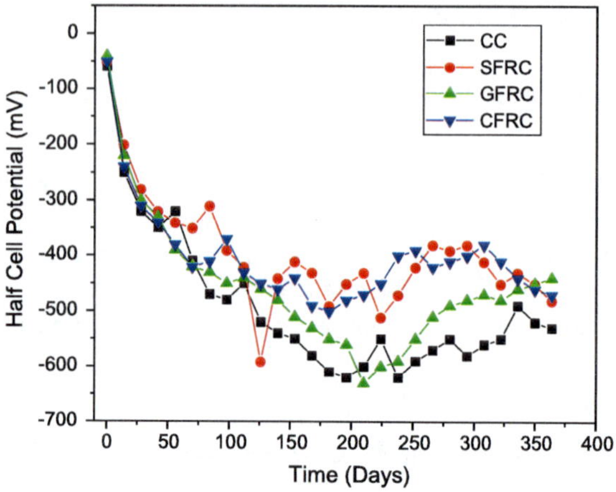

Fig. 7 Measurements of half cell potential for the bottom rebar

member and 3 types of fibre reinforced concrete was obtained. After 12 weeks of wet and dry ponding of the specimen fibre reinforced concrete specimens has a potential for corrosion of −399, −450, −437 mV, whereas conventional concrete member exhibited a potential value of −582 mV. For fiber reinforced concrete, an increase in the corrosion potential is demonstrated compared to conventional concrete. Individually corrosion potential values of steel fibre reinforced concrete were high, followed by E-waste glass fibre reinforced concrete and E-waste copper wire fibre reinforced concrete.

Bottom rebar corrosion potential values were plotted and shown in Fig. 8. Ponding and drying of about 12 weeks was carried out and the corrosion potential values of bottom rebar of steel fibre, E-waste glass fibre and E-waste copper wire fibre specimens has a potential for corrosion of −454, −395, −402 mV, whereas conventional concrete specimens has a potential for corrosion of −490 mV. Corrosion potential

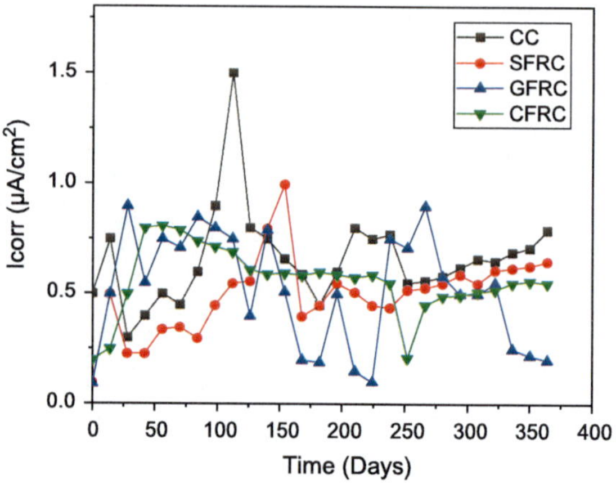

Fig. 8 Measurements of corrosion current density (I_{corr}) for the top rebar

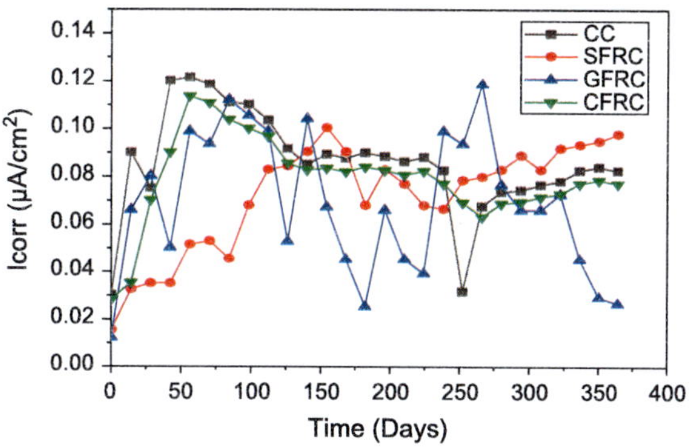

Fig. 9 Measurements of corrosion current density (I_{corr}) for the bottom rebar

rates of conventional concrete were more in bottom rebar compared to fibre reinforced concrete. Increase in corrosion rate of conventional concrete may be due cycling process of wetting and drying of sample in chloride environment, which lead to freezing and thawing action. This phenomenon may not occur is fibre reinforced concrete.

3.3 *Corrosion Current Density*

The top and bottom rebar corrosion current density was measured using a linear polarization resistance (LPR) as per ASTM G59 and a linear polarization resistance experiments were done using a Solartron potentiostat (model no. 1287, Roxboro Group) [30]. The specimens were allowed for 2 weeks of wet-dry cycling to increase the rate corrosion is induced [7]. For top rebar, Corrosion current density of conventional concrete is at 112 days is 0.66 mA/cm^2, where as for steel fibre, E-waste copper wire fibre and E-glass fibre the corrosion density were found to be 0.52, 0.56, and 0.49 mA/cm^2. For bottom rebar I_{corr} readings for conventional concrete is 0.114 mA/cm^2 and for steel fibre, E-glass fibre and E-waste copper wire fibre the values were 0.071, 0.068 and 0.078 mA/cm^2. NaCl solution of 3.5% is used to accelerate the corrosion rate and corrosion characteristics of top and bottom rebar are finalized through corrosion current density. Corrosion current densities of top and bottom rebar was plotted and shown in Figs. 8 and 9, respectively.

3.4 *Visual Examination for Crack Propagation*

Crack propagation for conventional concrete and fibre reinforced concrete clearly shown in Fig. 10a–c where as in fibre reinforced concrete members splitting crack formed was very minute compared to conventional.

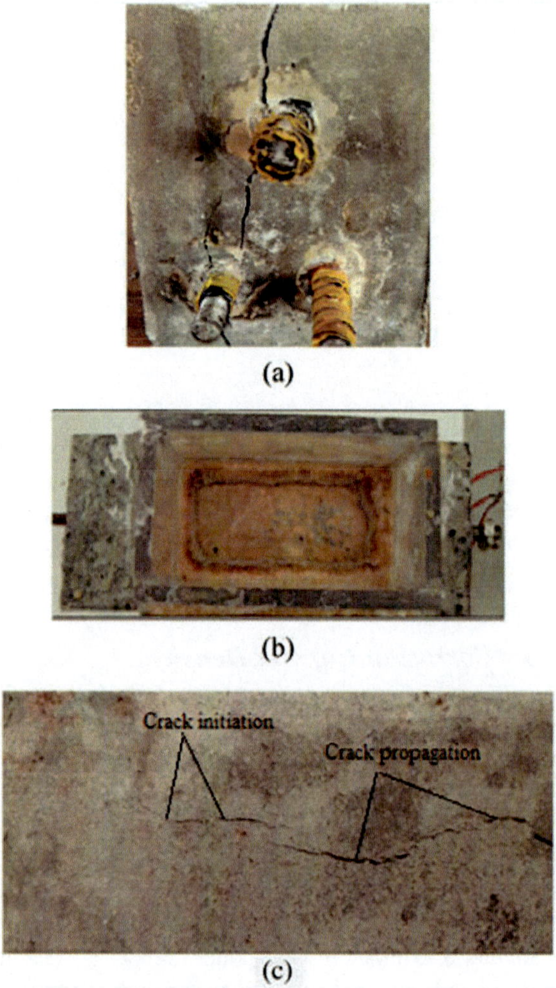

Fig. 10 **a** Splitting crack due corrosion in conventional concrete. **b** E-glass fibre reinforced concrete specimen after 365 days. **c** Splitting of crack due to corrosion in steel fibre reinforced concrete specimen

However specimens with steel and E-waste copper wire fibres were given crack but specimen with E-glass fibres does not cause any crack. This may be due to corrosion current density is good on E-glass fibre and crack values were automatically get reduced. Rebars from each type of concrete were taken and bars were cleaned with sulphuric acid. After cleaning surface of bar was hand brushed to the maximum white. Mass in loss of bars embedded in both conventional and fibre reinforced concrete members were measured as per ASTM G1 [31] and the final mass loss values given in Table 3.

Table 3 Mass loss of reinforcing bars from standard test as per ASTM G109

Specimen	Mass loss in top rebar (g)	Mass loss in bottom rebar (g)
Control concrete	12	2.5
Steel fibre reinforced concrete	5	1.0
E- glass fibre reinforced concrete	3	0.85
E-copper wire fibre reinforced concrete	3.22	1.1

4 Conclusion

Based on the conducted test results, the following conclusions can be drawn.

- From the ASTM standard test of cycling ponding and drying, current density values are less for fibre reinforced concrete compared to conventional concrete.
- Reinforcing bars induced in conventional concrete is has less corrosion potential compared to fibre reinforced concrete member.
- Crack initiation duration is increased due to the inclusion of fibre which also resists the corrosion induced damage (increase in durability).
- Visual examination of crack propagation or splitting crack values were very less for steel fibre reinforced concrete and E-waste copper wire fibre reinforced concrete specimen compare to conventional concrete. E-glass fibre reinforced concrete specimen surface not appeared any cracks.
- E-glass fiber reinforced concrete induced rebar has less loss in weight compared to control concrete, steel and E-waste copper wire fiber reinforced concrete during ACT test, from which it is clear that E-glass fiber reinforced concrete induced rebar has high corrosion resistivity.

References

1. Sri Durga Vara Prasad, M., Mani, E.: Mechanical behaviour of fibre reinforced concrete using shape memory alloys. Int. J. Innov. Technol. Explor. Eng. **9**(1), 230–232 (2019)
2. Alsaif, A., Bernal, S.A., Guadagnini, M., Pilakoutas, K.: Freeze-thaw resistance of steel fibre reinforced rubberised concrete. Constr. Build. Mater. **195**, 450–458 (2019)
3. Michel, A., Solgaard, A.O.S., Pease, B.J., Geiker, M.R., Stang, H., Olesen, J.F.: Experimental investigation of the relation between damage at the concrete-steel interface and initiation of reinforcement corrosion in plain and fibre reinforced concrete. Corros. Sci. **77**, 308–321 (2013)
4. Patel, S., Rana, R.S., Singh, S.K.: Study on mechanical properties of environment friendly Aluminium E-waste Composite with Fly ash and E-glass fibre. Mater. Today Proc. **4**(2), 3441–3450 (2017)
5. Kittl, P., Galleguillos, E., Diaz, G.: Properties of compacted copper fibre reinforced cement composite. Int. J. Cem. Compos. Light. Concr. **7**(3), 193–197 (1985)

6. Chen, C.H., Huang, R., Wu, J.K., Yang, C.C.: Waste E-glass particles used in cementitious mixtures. Cem. Concr. Res. **36**(3), 449–456 (2006)
7. G109-07: Standard Test Method for Determining the Effects of Chemical Admixtures on the Corrosion of Embedded Steel Reinforcement in Concrete Exposed to Chloride Environments. In: ASTM Book of Standards (2000)
8. Ji, Y., Zhan, G., Tan, Z., Hu, Y., Gao, F.: Process control of reinforcement corrosion in concrete. Part 1: Effect of corrosion products. Constr. Build. Mater. **79**, 214–222 (2015)
9. Gupta, T., Siddique, S., Sharma, R.K., Chaudhary, S.: Behaviour of waste rubber powder and hybrid rubber concrete in aggressive environment. Constr. Build. Mater. **217**, 283–291 (2019)
10. Haque, M.N., Al-Khaiat, H.: Carbonation of concrete structures in hot dry coastal regions. Cem. Concr. Compos. **19**(2), 123–129 (1997)
11. Di Maggio, R., Franchini, M., Guerrini, G., Poli, S., Migliaresi, C.: Fibre-matrix adhesion in fibre reinforced CAC-MDF composites. Cem. Concr. Compos. **19**(2), 139–147 (1997)
12. Tapan, M., Aboutaha, R.S.: Effect of steel corrosion and loss of concrete cover on strength of deteriorated RC columns. Constr. Build. Mater. **25**(5), 2596–2603 (2011)
13. Maalej, M., Chhoa, C.Y., Quek, S.T.: Effect of cracking, corrosion and repair on the frequency response of RC beams. Constr. Build. Mater. **24**(5), 719–731 (2010)
14. Fernandez, I., Bairán, J.M., Marí, A.R.: Mechanical model to evaluate steel reinforcement corrosion effects on σ–ε and fatigue curves. Experimental calibration and validation. Eng. Struct. **118**, 320–333 (2016)
15. Li, S., et al.: N-acylated isoindigo based conjugated polymers for n-channel and ambipolar organic thin-film transistors. Dye. Pigment. **109**, 200–205 (2014)
16. Askari, F., Ghasemi, E., Ramezanzadeh, B., Mahdavian, M.: Mechanistic approach for evaluation of the corrosion inhibition of potassium zinc phosphate pigment on the steel surface: Application of surface analysis and electrochemical techniques. Dye. Pigment. **109**, 189–199 (2014)
17. Mousavifard, S.M., Nouri, P.M., Attar, M.M., Ramezanzadeh, B.: The effects of zinc aluminum phosphate (ZPA) and zinc aluminum polyphosphate (ZAPP) mixtures on corrosion inhibition performance of epoxy/polyamide coating. J. Ind. Eng. Chem. **19**(3), 1031–1039 (2013)
18. Alibakhshi, E., Ghasemi, E., Mahdavian, M., Ramezanzadeh, B., Farashi, S.: Active corrosion protection of Mg-Al-PO43−LDH nanoparticle in silane primer coated with epoxy on mild steel. J. Taiwan Inst. Chem. Eng. **75**, 248–262 (2017)
19. Liu, X.R., Miao, M., Zhang, J.Y., Liu, J.K., Zhang, X.M., Wang, X.G.: Surface coordination and excellent anticorrosion performance of strontiumapatite nanocomposite. J. Ind. Eng. Chem. **80**, 656–666 (2019)
20. Li, W., et al.: Improved anti-corrosion performance of epoxy zinc rich coating on rusted steel surface with aluminum triphosphate as rust converter. Prog. Org. Coatings **135**, 483–489 (2019)
21. Hou, L., Li, Y., Sun, J., Zhang, S.H., Wei, H., Wei, Y.: Enhancement corrosion resistance of MgAl layered double hydroxides films by anion-exchange mechanism on magnesium alloys. Appl. Surf. Sci. **487**, 101–108 (2019)
22. Sun, X., Jiang, G., Bond, P.L., Wells, T., Keller, J.: A rapid, non-destructive methodology to monitor activity of sulfide-induced corrosion of concrete based on H2S uptake rate. Water Res. **59**, 229–238 (2014)
23. Pradhan, B.: Corrosion behavior of steel reinforcement in concrete exposed to composite chloride–sulfate environment. Constr. Build. Mater. **72**, 398–410 (2014)
24. Saraswathy, V., Muralidharan, S., Kalyanasundaram, R., Thangavel, K., Srinivasan, S.: Evaluation of a composite corrosion-inhibiting admixture and its performance in concrete under macrocell corrosion conditions. Cem. Concr. Res. **31**(5), 789–794 (2001)
25. Saricimen, H., Mohammad, M., Quddus, A., Shameem, M., Barry, M.: Effectiveness of concrete inhibitors in retarding rebar corrosion. Cem. Concr. Compos. **24**(1), 89–100 (2002)
26. Cramer, S., et al.: Corrosion prevention and remediation strategies for reinforced concrete coastal bridges. Cem. Concr. Compos. **24**(1), 101–117 (2002)
27. Xia, J., Li, T., Fang, J.X., Jin, W.: Numerical simulation of steel corrosion in chloride contaminated concrete. Constr. Build. Mater. **228**, 116745 (2019)

28. Yuenyongsuwan, J., Sinthupinyo, S., O'Rear, E.A., Pongprayoon, T.: Hydration accelerator and photocatalyst of nanotitanium dioxide synthesized via surfactant-assisted method in cement mortar. Cem. Concr. Compos. **96**, 182–193 (2019)
29. ASTM C876 Standard Test Method for Corrosion Potentials of Uncoated Reinforcing Steel in Concrete
30. ASTM G59 Standard Test Method for Conducting Linear Polarization Resistance Measurements
31. ASTM G1 Standard Practice for Preparing, Cleaning, and Evaluating Corrosion Test Specimens

Physical and Microstructural Properties of Construction and Demolition Waste Based Masonry Units

Vivian Lawrence Sequeira, Ashwin M. Joshi, Meghashree D. Kerekoppa, and Namratha Bharadwaj

Abstract The aim of this study was to employ crushed Demolished Brick Masonry, a variety of Construction and Demolition Waste as a replacement for conventional construction materials (natural soil and aggregates) primarily in the production of three types of masonry units, namely Solid Concrete Blocks, Stabilized Adobe Blocks and Stabilized Mud Concrete Blocks. In addition Controlled Low Strength Material was also studied. Long term strength of these specimens was studied at different ages to determine their performance. Scanning Electron Microscopy in addition to Energy Dispersive Spectroscopy and X-Ray Diffraction were used to study the microstructure and identify the phases present in it. SEM revealed the presence of Calcium-Silicate-Hydrate Crystals and voids in the microstructure. In Solid Concrete Blocks the distribution of C-S-H was sparse compared to commercial concrete blocks due to the lower cement content employed. Significant differences could not be observed in the morphology of C-S-H among different proportions of SAB studied. The microstructure of Controlled Low Strength Material was quite similar to that of Stabilized Adobe. Stabilized Mud Concrete Blocks had a particularly dense network of C-S-H compared to the other types of units. XRD indicated the presence of Anorthite, which was highest in Stabilized Mud Concrete Blocks and lowest in Stabilized Adobe Blocks.

V. L. Sequeira · M. D. Kerekoppa
Former Student, ICM, RASTA—Centre for Road Technology, Bangalore 560058, India
e-mail: vivian.ls14@gmail.com

M. D. Kerekoppa
e-mail: meghakerekoppa@gmail.com

A. M. Joshi (✉)
Research Scholar, Civil Engineering, BMS College of Engineering, Bangalore 560019, India
e-mail: ashwinmjoshi@gmail.com

N. Bharadwaj
Assistant Professor, BMS College of Architecture, Bangalore 560019, India
e-mail: namratha.b@bmsca.org

D. K. Ashish et al. (eds.), *3rd International Conference on Innovative Technologies for Clean and Sustainable Development*, RILEM Bookseries 29,
https://doi.org/10.1007/978-3-030-51485-3_27

Keywords Construction and Demolition waste · Demolished Brick Masonry waste · Concrete waste · Stabilized Adobe · Concrete Blocks · Stabilized Mud Concrete · Controlled Low Strength Material · Scanning Electron Microscopy

1 Introduction

The construction industry is one of the largest sectors in India, contributing about 7% to the national GDP and generating employment [1]. It is also one of the biggest producers of solid waste. Construction and Demolition Waste (CDW) constitutes up to one third of total Municipal Solid Waste (MSW), the other constituents being excavated and processed soil (36%), demolished masonry waste (31%) and concrete (23%) [2]. Another study estimates that the city of Bangalore alone generates 3600 T of CDW daily [1]. This waste is often improperly disposed by dumping into lakes and landfills, which causes both environmental and health problems. Severe environmental issues have been caused due to high consumption of natural resources such as river sand [3], which is the most common and extensively used material in construction. It has become important to explore alternative fine aggregates.

Construction and demolition waste, when processed properly, has the potential to be used as a replacement for natural sand as fine aggregates. Considerable number of researchers have worked on production of concrete based products or mixes by recycling concrete debris, demolished clay debris etc. [4]; some have used it in the preparation of masonry products [4] and paver blocks [5]. Joshi et al., have used Demolished Brick Masonry Waste (DBMW) in stabilized adobe blocks as a partial and total replacement for natural soil to identify the optimum percentage of replacement for the best performance [6]. It was noted about 60% of CDW comprises of excavated natural soil and demolished masonry [6]. Few researchers have worked on utilizing these materials in masonry units, which form a major portion of CDW generated. The data availability in this regard is also limited. Furthermore, the physical properties of such products have not been evaluated over the long term.

The objective of this study was to utilize the combinations of excavated natural soil (NS) and crushed DBMW in preparation of three types of masonry units, namely Solid Concrete Blocks (SCB), Stabilized Adobe Blocks (SAB) and Stabilized Mud Concrete Blocks (SMCB) and Controlled Low Strength Material (CLSM). The strength characteristics of these products were studied at different ageing to identify the appropriate proportions to achieve satisfactory results. Scanning Electron Microscopy (SEM), Energy Dispersive X-Ray Spectroscopy (EDS) and X-Ray Diffraction (XRD) techniques were employed to analyze the microstructure.

2 Methodology

Locally available excavated Natural Soil possessing about 20% clay content was procured and processed. DBMW was sourced from a local dump yard and deleterious contents were removed. On a flat surface the material was spread and crushed using rollers. NS and DBMW were sieved through 4.75 mm sieves and stored in barrels. OPC 53 Grade conforming to IS 12269: 2013 [7] was used in all aforementioned products. M-Sand conforming to IS 383: 2016 [8] was used in SCB and CLSM. Brick Waste (BW) and Mortar Waste (MW) were procured from a nearby CDW dump site. The crushed concrete waste (CW) was procured from a local processing plant. They were processed similarly to DBMW. Mortar waste consisted of only fractions of cement mortar without brick or concrete fractions.

2.1 Properties of the Products

Concrete Blocks—Concrete masonry units are commonly used in construction. They can be solid, hollow or of low density. In this study, blocks of three sizes were prepared using 4 mixes with different combinations of M-sand and DBMW. They were produced using egg-laying machines. Commercially procured concrte blocks, referred to as Commercial Blocks conforming to Indian Standards (IS) specifications were used for comparison under similar testing conditions. Blocks of length 400 mm and height 200 mm, and three thicknesses were used for the study:

- Thickness of the blocks
  - Type A – 4 inches or 100 mm
  - Type B – 6 inches or 150 mm
  - Type C – 8 inches or 200 mm

Mix Proportions—Concrete of grade M10 was designed using IS 10262: 2009 [9] with a constant free W/C ratio of 0.55. Four mix proportions were used to produce the blocks by varying the degree of replacement of M-Sand (MS) by DBMW. The different proportions and the designations used are presented in Table 1.

Stabilized Adobe Blocks "*Adobe—sun dried mudbrick*" is traditionally used in building construction. Adobe blocks are composed of natural soil dried under sunlight

Table 1 Solid concrete blocks—mix proportions

Designation	Fine Aggregate Proportion
R25	25% DBMW; 75% M-Sand
R50	50% DBMW; 50% M-Sand
R75	75% DBMW; 25% M-Sand
R100	100% DBMW; 0% M-Sand

and are low cost, eco-friendly materials which have lower embodied energy than conventional building materials. Cement was used as stabilizer. Dimensions of SAB was 230 × 145 × 100 mm. They were produced from a steel mould with five sides and ejectable bottom plate (Fig. 1). Jagadish et al. [10] had developed and used a similar mould previously.

Mix Proportions—Seven series of blocks were prepared with the constant cement content of 9% and varying the percentage replacement of NS by DBMW. The different combinations of NS and DBMW and the corresponding water content used in the preparation of various series of SAB is presented in Table 2.

Stabilized Mud Concrete Blocks—Stabilized Mud Concrete (SMC) is a composite mix of conventional concrete wherein fine aggregate is completely or partially replaced by natural soil and the coarse aggregate fraction may be natural or recycled. Blocks of dimension 230 × 150 × 100 mm were prepared using wooden moulds and were cured under wet burlap.

Fig. 1 SAB mould

Table 2 Stabilized adobe blocks—mix proportions

Series	Proportion	Water content (%)
A	NS—100%	28
B	NS—80%, DBMW—20%	27
C	NS—60%, DBMW—40%	24
D	NS—40%, DBMW—60%	23
E	NS—20%, DBMW—80%	22
F	DBMW—100%	21
G	NS—30%, DBMW—70%	23

Table 3 Stabilized mud concrete blocks—mix proportions

Series	Fine aggregate proportion
V1	50% BW + 50% MW
V2	25% BW + 75% NS
V3	40% BW + 60% NS
V4	100% Brick Waste

Mix Proportions—Four series of blocks were prepared using a fixed mix proportion—1 Cement: 3 Fine Aggregate (FA): 4.5 Coarse Aggregate (CA). FA comprised of a combination of crushed brick waste (BW) and excavated NS of particle size 4.75 mm passing. Series V1 comprised of crushed brick waste and crushed mortar waste in equal proportions as replacement to FA. CA fraction for all proportions consisted of crushed BW (passing 12 mm). The water content required to achieve a slump of 100 mm was determined through trial mixes. The different series designations and proportions of the constituents used in their preparation is presented in Table 3.

Controlled Low Strength Material—Defined as a "self-compacted cementitious material used primarily as backfill as an alternative to compacted fill" by ACI 229R [11]. 150 mm cubes specimens were cast and air dried before curing with water. The specimens were tested at specific predetermined time periods.

Mix Proportions—Three mixes were prepared by keeping the cement content at 5% and varying the proportions of other constituents to achieve flowability as per ASTM D 6103 [12]. Controlled mix was prepared with M-Sand as fine aggregate, while 4 other mixes were made by replacing M-Sand entirely by Concrete Waste (CW), Demolished Brick Masonry Waste (DBMW), combination of CW and DBMW (CMW) and excavated natural soil (SW) respectively.

- Controlled Mix (MS): Cement: GGBS: Fly ash: M-Sand: Water
- Mix 1 (CW): Cement: GGBS: Fly ash: Concrete waste: Water
- Mix 2 (DBMW): Cement: GGBS: Fly ash: Masonry waste: Water
- Mix 3 (CMW): Cement: GGBS: Fly ash: Concrete and Masonry Waste: Water
- Mix 4 (SW): Cement: GGBS: Fly ash: Soil Waste: Water

2.2 *Testing*

Physical properties (material characterization) of the constituent materials used in this study were evaluated according to the relevant Indian Standards (IS) guidelines. All products were cured with water for a period of 28 days after casting.

Compressive Strength—The specimens to be tested were cured by wet gunny bags covering the specimen entirely. Wet compressive strength (after soaking specimen for 24 hours) was determined at different ages. Concrete Blocks, Stabilized Mud

Concrete Blocks and Controlled Low Strength Material Cubes were tested in accordance with IS 516: 1959 [13], while Stabilized Adobe Blocks were tested in accordance with IS 3495: 1992 [14]. The average crushing strength of five specimens from each proportion were recorded. Water absorption characteristics were also checked.

Microstructural Analysis—The microstructure was studied by extracting samples from the failure surface of specimens tested for compressive strength. The samples were oven dried for 24 hours prior to testing. The morphology was examined through SEM. To minimize the effects of charging, test specimens were sputter coated with a 10 nm gold coating for 76 s prior to insertion in the SEM chamber. Images of the microstructure were captured at different magnifications to help observe various features. SEM imagery had been concluded to obtain the elemental composition of the microstructure. Then, Energy Dispersive X-Ray Spectroscopy (EDS) was conducted on the same samples. Another batch of oven dried samples were used for X-Ray Diffraction (XRD) analysis to obtain the mineralogical composition.

3 Results and Discussion

3.1 Material Properties

The particle size distribution of the materials used in the preparation of CB, SAB, SMCB and CLSM have been presented in Fig. 2. The properties of the fine aggregates used in CB are shown in Table 4. The physical characteristics of the various proportions of NS and DBMW used to prepare different series of SAB have been shown in Table 5.

3.2 Compressive Strength

Concrete Blocks—Test results of compressive strength (wet compressive strength) of SCB across different ages are reported in Table 6. SCB with higher proportions of DBMW exhibited increased strength values with time. Block crushing strength of all series of SCB satisfied the strength requirements specified by the BIS (Bureau of Indian Standards) [15]. Blocks of R75 proportion had the highest long term strength across different types overall. This suggests that the percentage of replacement of fine aggregates by DBMW can go beyond 50% (which is commonly considered as the limiting replacement value), in fact the complete replacement also yielded satisfactory results. Water absorption values at 28 day ageing was found to lie between 2.23–5.38%, which is acceptable. Graph of compressive strength versus age of SCB at testing is presented in Fig. 3.

Fig. 2 **a** Particle size distribution of CLSM and CB constituents, **b** particle size distribution of SAB proportions, **c** particle size distribution of SAB proportions

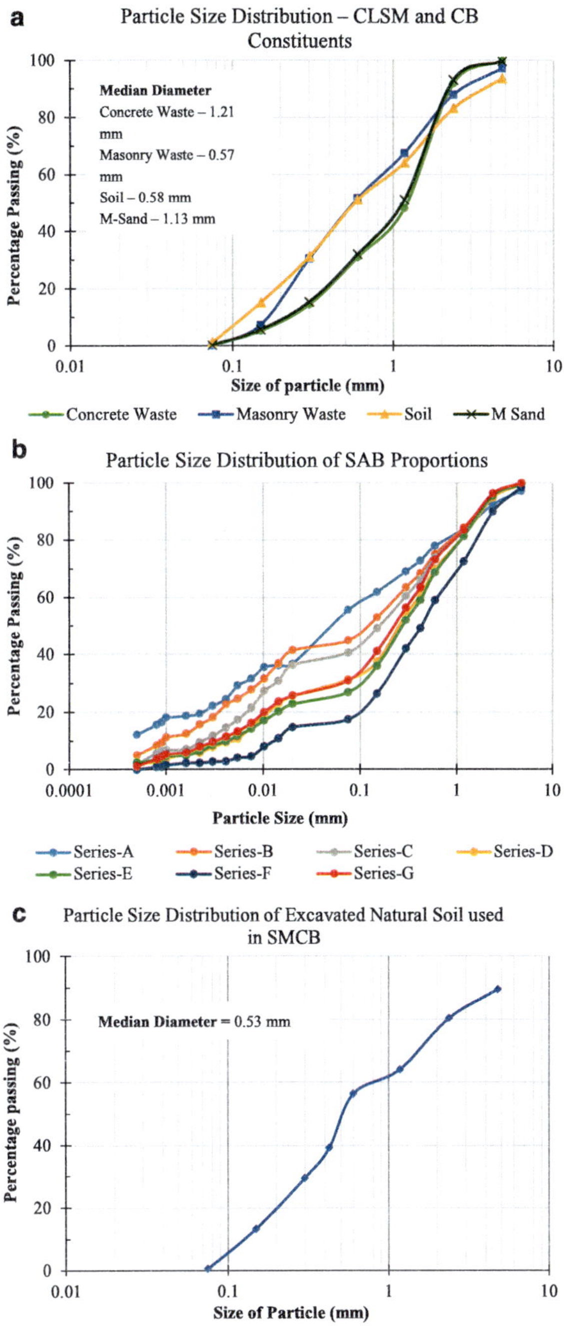

Table 4 Properties of fine aggregates used in concrete blocks [1]

Designation	Specific gravity	Fineness modulus
M-Sand	2.61	1.13
CB R25	2.62	3.78
CB R50	2.62	3.67
CB R75	2.59	3.73
CB R100	2.57	3.56

Table 5 Properties of natural and modified soil used in stabilized adobe blocks [6]

Series	Gs	Gravel (%)	Sand (%)	Silt (%)	Clay (%)	Median diameter (mm)	MDD (g/cc)	OMC (%)
A	2.64	2.92	41.51	36.07	19.50	0.05	1.91	16.40
B	2.40	0.66	54.56	30.58	14.20	0.12	1.94	15.20
C	2.43	0.56	61.78	29.66	8.00	0.16	1.99	12.60
D	2.46	0.62	68.18	25.20	6.00	0.25	2.03	12.20
E	2.48	0.52	72.56	21.92	5.00	0.28	1.98	12.80
F	2.53	1.58	80.96	15.46	2.00	0.44	1.90	14.20
G	2.47	0.12	69.08	24.80	6.00	0.23	2.05	11.60

Stabilized Adobe Blocks—The values of compressive strength of stabilized adobe blocks have been reported in Table 7. The compressive strength characteristics of SAB tested fulfilled the strength requirements specified by the BIS (3.50 MPa) in IS 3495 [14]. Blocks from Series G, having a mix proportion of 70% DBMW and 30% NS, exhibited the highest strength at an age of 30 days. Blocks from the same series were tested after 1 year, and the compressive strength was observed to increase from 5.36 up to 6.40 MPa. This suggests that up to a 70% replacement of natural soil by DBMW may yield the best performance characteristics. The 30 day water absorption values was found to lie in the range of 6.51–11.10%, well under the limit of 20% as prescribed in IS 3495 (Part 2) [14]. Graph of compressive strength versus age of SAB at testing is presented in Fig. 4.

Stabilized Mud Concrete Blocks—The values of compressive strength of stabilized mud concrete blocks have been reported in Table 8. The highest strength was given by the blocks of series. This suggests that the optimum extent of replacement of natural soil by DBMW is around 40%. A similar trend of increase in compressive strength in the long term was observed when higher proportions of DBMW were used in CB and SMCB. This strength increase is atypical as percentage strength gain over two year ageing is comparatively quite high when compared to behavior of a typical non-pozzolanic concrete. The water absorption values (at 28 day ageing) were observed in range of 1.77–4.88% which is least amongst the three categories of masonry blocks studied. Figure 5 presents the graph of compressive strength versus age of SMCB at testing.

Table 6 Compressive strength of concrete blocks across different ages

Series	Average compressive strength (MPa)								
	14 days			28 days			1 year		
	Type: A	Type: B	Type: C	Type: A	Type: B	Type: C	Type: A	Type: B	Type: C
R25	4.60	5.11	5.38	4.76	5.47	5.85	6.99	7.12	6.83
R50	5.92	6.14	5.72	6.49	6.33	6.55	8.02	6.04	6.58
R75	4.55	5.05	4.70	4.62	5.42	4.82	7.39	7.26	7.23
R100	3.35	4.82	4.46	3.55	4.88	4.62	6.94	5.43	6.27

Table 7 Compressive strength of stabilized adobe blocks across different ages

Block series	Average compressive strength (MPa)						
	A	B	C	D	E	F	G
7 days	2.16	2.28	3.04	3.60	4.35	3.63	4.48
30 days	3.04	3.36	4.34	5.23	5.15	3.80	5.36
1 year	–	–	–	6.07	6.23	4.63	6.40

Table 8 Compressive strength of stabilized mud concrete blocks across different ages

Block series	Average compressive strength (MPa)			
	3 days	7 days	28 days	2 years
V1	7.34	10.41	15.88	25.38
V2	8.08	12.20	17.36	27.36
V3	8.15	13.63	18.03	30.23
V4	6.89	10.08	13.05	27.03

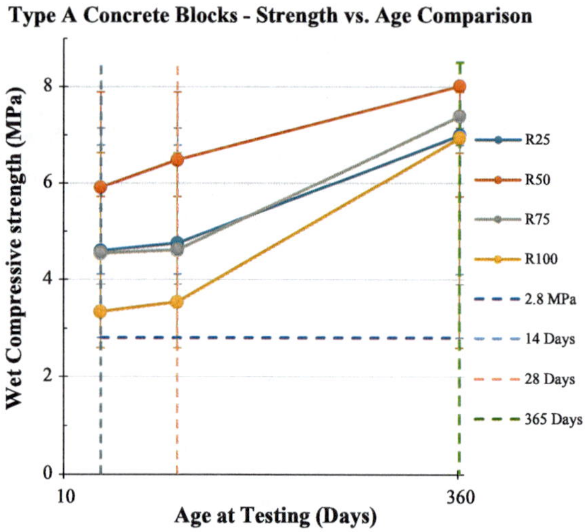

Fig. 3 Comparison of compressive strength of concrete blocks

Controlled Low Strength Material—Cube compressive strength values of CLSM specimens at different ageing has been reported in Table 9. There is a general increase in strength with ageing with MS proportion showing highest percentage increase. DBMW and CMW have also shown increased strength values with ageing. However there is a reduction in the strength of SW cubes after 28 days. The compressive strengths of CB, SAB, SMCB and CLSM have been compared in Fig. 6.

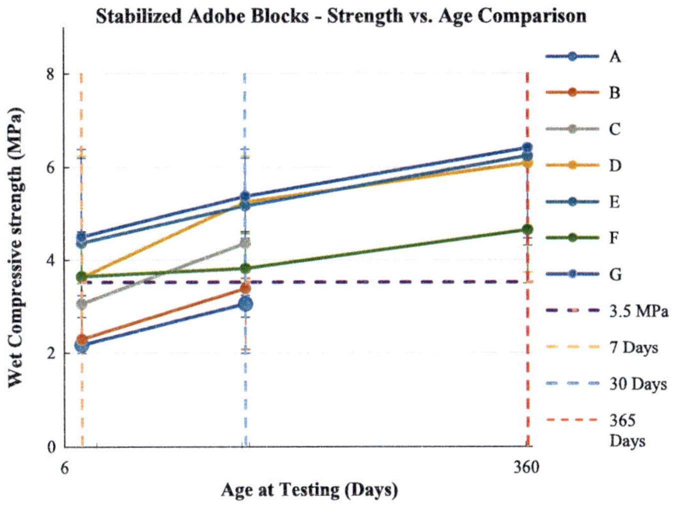

Fig. 4 Comparison of compressive strength of stabilized adobe blocks

Fig. 5 Comparison of compressive strength of stabilized mud concrete blocks

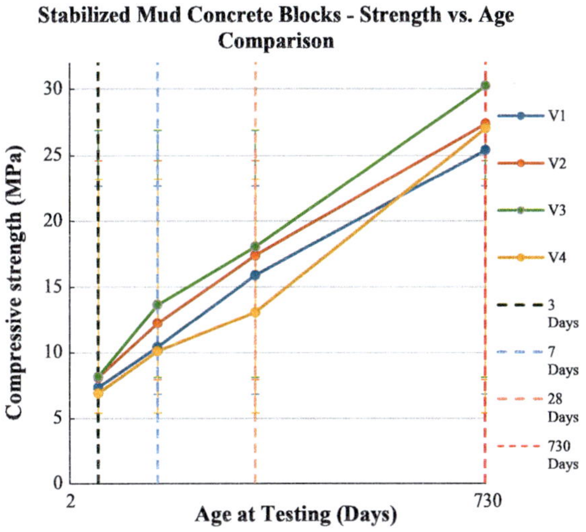

Table 9 Compressive strength of controlled low strength material cubes across different ages

Block series	Average compressive strength (MPa)				
	MS	CW	DBMW	CMW	SW
7 days	7.56	5.82	2.03	3.84	1.81
28 days	10.23	7.28	2.77	4.68	3.32
90 days	12.73	10.82	2.86	4.51	2.98
180 days	14.50	11.12	3.22	7.41	1.65

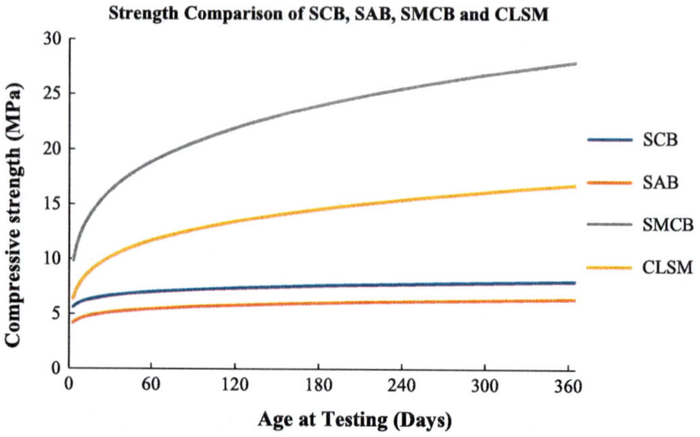

Fig. 6 Strength comparison of SCB, SAB, SMCB and CLSM

3.3 *Microstructure Analysis*

The results of SEM and EDS have been reported in Figs. 7 and 8 and Tables 10, 11, 12 and 13. The microstructure of concrete blocks of Series R75 (Fig. 7a.) showed a distribution of amorphous deposits, characteristic of Calcium Silicate Hydrate (C–S–H). EDS data reveals the Ca/Si ratio ranges between 0.86 and 2.04 indicating the presence of C–S–H, among other phases [16]. The presence of cracks was minimal, but interlayer spaces and voids (possibly air and capillary voids) were distributed randomly. The microstructure of blocks of Series R50 and R75 were quite similar in nature. Hexagonal crystals of Calcium Hydroxide were observed in Series R25 and R100.

The microstructure of SAB of Series G (Fig. 7b.) showed a dense distribution of C–S–H. The Ca/Si ratio ranges from 0.27 to 2.01. Some samples of Series G also had a high concentration of Fe (30.67%), although it is likely that this is due to the composition of the aggregates, rather than the cement used. Interlayer spaces and voids (capillary and air) were distributed throughout the sample. Cracks could be observed in the microstructure, but it is not clear if they formed due to sampling or other reasons. Blocks of Series A and B were similar, having a rather sparse distribution of C–S–H. However, the morphology of blocks of Series C, D, E, F and G blocks were alike, having a comparable distribution of C–S–H and voids. These blocks also exhibited better strength characteristics compared to Series A and B.

The microstructure of SMCB of Series V3 (Fig. 7c.) exhibited a very dense distribution of C–S–H. The Ca/Si ratio ranged from 0.85 to 1.98. Ettringite and Calcium Hydroxide crystals could also be seen. The microstructure of all stabilized mud concrete blocks were similar in nature, with the same morphological features and a similar distribution of voids. C–S–H formations were more prominently visible as compared to the other blocks.

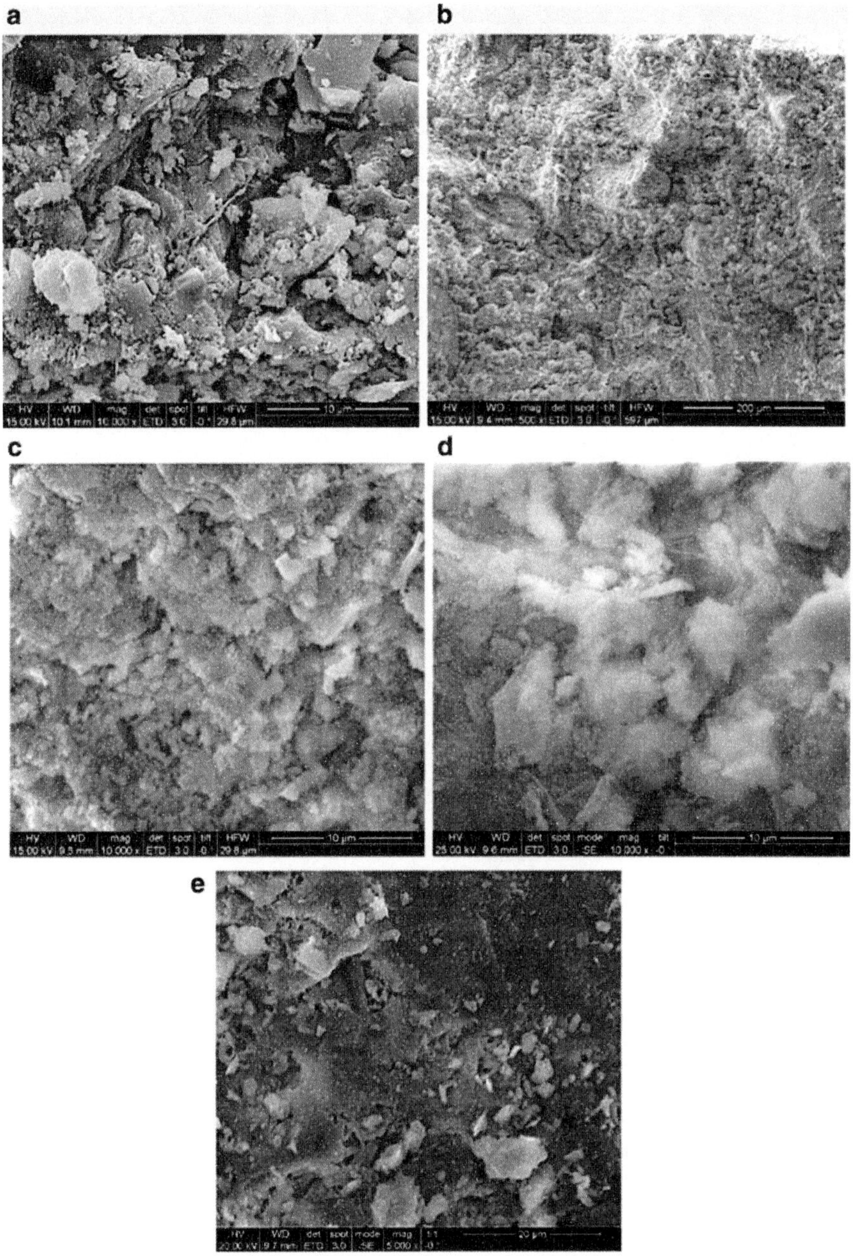

Fig. 7 **a** Concrete block—R75,**b** stabilized adobe block—series G, **c** stabilized mud concrete block—V3, **d** controlled low strength material—DBMW, **e** demolished brick masonry waste

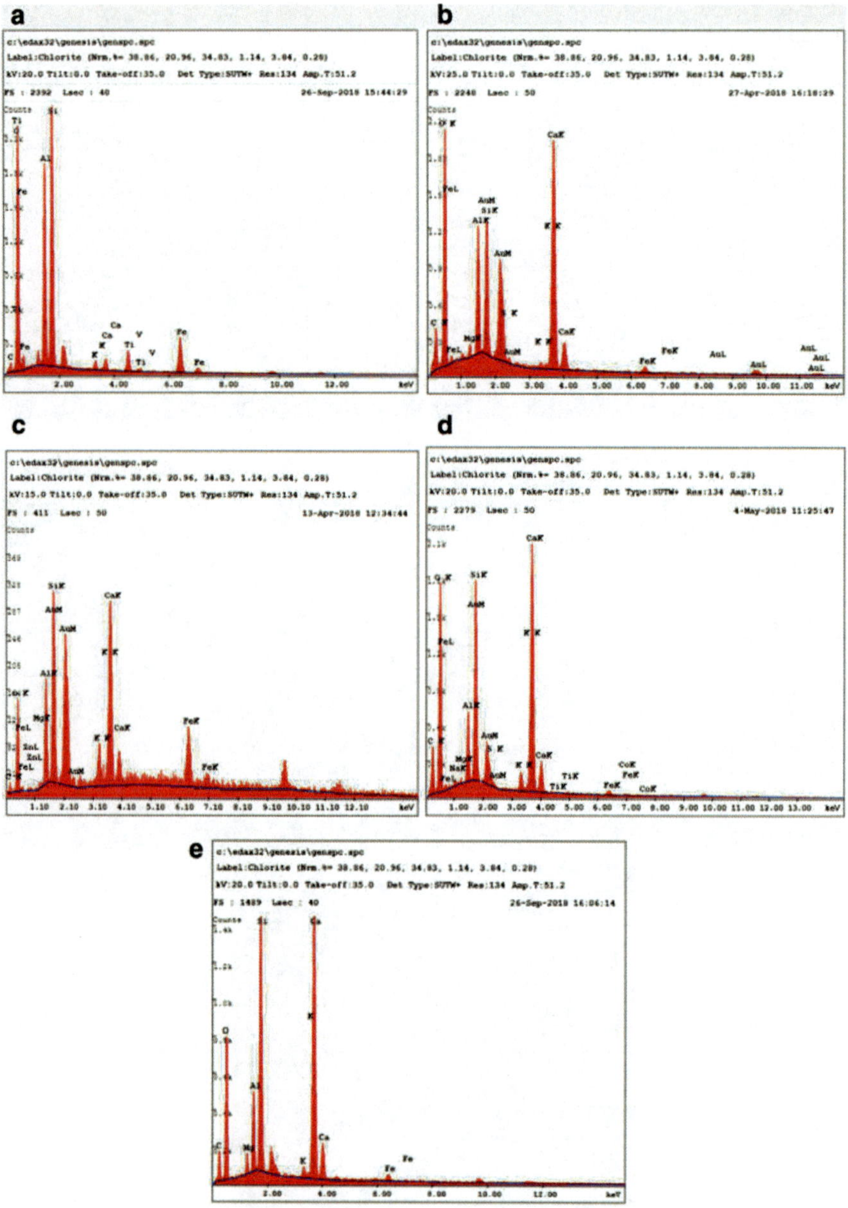

Fig. 8 **a** Demolished brick masonry waste—EDS data, **b** CB R75—EDS data, **c** SAB series G—EDS data, **d** SMCB series V3—EDS data, **e** CLSM series CW—EDS data

Table 10 Elemental composition of concrete blocks

Block series	Elemental concentration (percentage weight)				
	Calcium (Ca)	Silicon (Si)	Aluminium (Al)	Iron (Fe)	Oxygen (O)
R0	13.24	28.92	10.00	1.19	42.51
R25	23.96	9.09	2.30	0.43	58.91
R50	5.32	18.78	5.29	1.00	44.78
R75	13.13	6.44	6.36	0.88	49.07
R100	42.18	10.15	3.88	5.25	22.18

Table 11 Elemental composition of stabilized adobe blocks

Block series	Elemental concentration (percentage weight)				
	Calcium (Ca)	Silicon (Si)	Aluminium (Al)	Iron (Fe)	Oxygen (O)
A	3.18	14.29	12.67	3.03	48.51
B	13.52	5.48	4.54	2.39	22.83
C	5.76	9.73	5.84	3.88	22.83
D	10.12	15.65	11.49	2.31	57.46
E	13.61	12.40	7.00	4.46	31.40
F	23.84	5.51	3.10	5.70	37.02
G	18.81	9.36	5.16	19.84	12.42

Table 12 Elemental composition of stabilized mud concrete blocks

Block series	Elemental concentration (percentage weight)				
	Calcium (Ca)	Silicon (Si)	Aluminium (Al)	Iron (Fe)	Oxygen (O)
V1	46.15	14.75	4.27	4.84	14.26
V2	11.22	19.00	2.58	1.03	35.14
V3	18.64	9.40	3.51	0.99	41.42
V4	21.67	7.04	2.61	0.78	33.71

Table 13 Elemental composition of controlled low strength material

Block series	Elemental concentration (percentage weight)				
	Calcium (Ca)	Silicon (Si)	Aluminium (Al)	Iron (Fe)	Oxygen (O)
MS	17.14	23.53	4.77	0.00	38.58
CW	24.38	14.67	4.62	1.87	39.13
MW	14.43	19.76	9.77	6.64	29.11

The microstructure of CLSM (Fig. 7d) was similar to that of SAB, possibly because they are made of similar constituent materials. The Ca/Si ratio ranges from 0.73 to 1.66. SEM image of typical DBMW sample is shown in Fig. 7e.

Similar Ca/Si ratios were observed across the three types of masonry units proportions which developed the highest compressive strength.

The results of X-Ray diffraction analysis have been displayed in Fig. 9. The detailed analysis indicated the presence of Anorthite ($CaAl_2Si_2O_8$). Peaks were observed at 27.1°–28.3°, which corresponds to Quartz (Si).

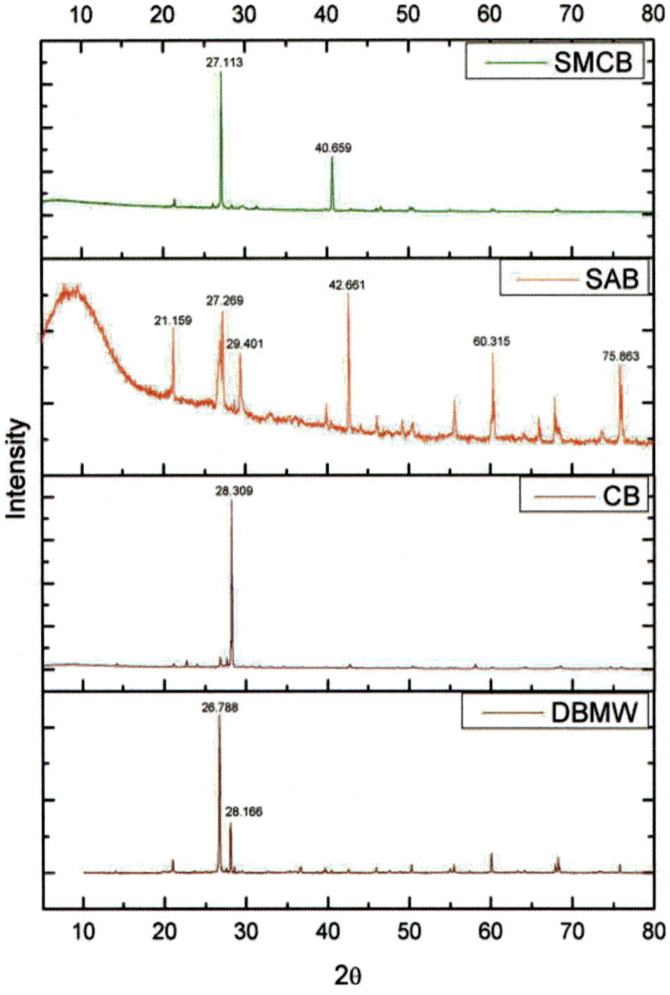

Fig. 9 XRD Data of DBMW, CB, SAB and SMCB

4 Conclusions

The conclusions drawn from the experimental research are listed below:

- Crushed demolished brick masonry waste with excavated natural soil constitute a major chunk of construction and demolition waste. It was effectively utilized to produce three types of masonry blocks and excavatable controlled low strength material.
- SCB and SAB with 70–75% DBMW replacement developed the highest compressive strength, which was corroborated by microstructure analysis through SEM and EDS.
- SMCB with 40% DBMW replacement developed the highest strength, and there was a 67.7% increase in compressive strength at 2 years ageing.
- Water absorption values at particular ageing was highest in SAB and least in SMCB; complementing the wet compressive strength characteristics which was highest in SMCB, amongst the three categories of masonry units.
- Amongst the different masonry units studied, SMCB developed the highest strength at 28 days and over the long term, followed by CB and SAB.

References

1. Joshi, A.M., Basutkar, S.M., Keshava, M., Raghunath, S., Jagadish, K. S.: Performance of masonry units prepared using construction and demolition waste as fine aggregates. In: 10th International Masonry Conference, Milan; Italy, pp. 1176–1185 (2018). https://doi.org/10.13140/RG.2.2.22838.55364
2. ICI—Guidelines: Recycling, use and management of C&D wastes. ICI-TC/05-01, 1st edn. Indian Concrete Institute, Chennai (2015)
3. Pillai, R.: Alternatives to river sand—a global perspective. In: Alternatives to river sand—a sustainable approach. Indian Concrete Institute, pp. 15–16, Bangalore (2013)
4. Joshi, A., Raghunath, S., Keshava, M., Jagadish, K.: Stabilized adobe using demolished brick masonry waste. In: National Seminar on emerging building materials and construction technologies, New Delhi (2018). https://doi.org/10.13140/RG.2.2.24096.84481
5. Report, T.: Experience of the first commercial scale pilot project for C&D waste management in India. IL&FS Environmental Infrastructure & Services Ltd., New Delhi (2014)
6. Joshi, A.M., Basutkar, S.M., Ahmed, M.I., Keshava, M., Rao, R.S., Jagadish, K.S.: Performance of stabilized adobe blocks prepared using construction and demolition waste. J. Build. Pathol. Rehabil. (2019). https://doi.org/10.1007/s41024-019-0052-x
7. Indian Standard (IS) 12269: Ordinary Portland Cement, 53 Grade—Specification (First Revision). Bureau of Indian Standards, New Delhi (2013)
8. Indian Standard (IS) 383: Coarse and Fine Aggregate for Concrete—Specification. Bureau of Indian Standards, New Delhi (2016)
9. Indian Standard (IS) 10262: Guidelines for Concrete Mix Design. Bureau of Indian Standards, New Delhi (2009)
10. Basu, S., Gumaste, K. S., Jagadish, K. S.: The stabilized adobe. In: Earth Construction Technologies, pp. 90–96, ASTRA and KSCST, Indian Institute of Science. Bangalore (1991)
11. ACI 229R-99: Controlled low strength materials, ACI Committee 229 (Reapproved 2005), American Concrete Institute (ACI), Farmington Hills, MI (1999).

12. ASTM D6103-97: Standard Test Method for Flow Consistency of Controlled Low Strength Material (CLSM), ASTM International, West Conshohocken, PA (1997)
13. Indian Standard (IS) 516: Method of test for strength of concrete. Bureau of Indian Standards, New Delhi (1959)
14. Indian Standard (IS) 3495 Part 1 & 2: Methods of tests of burnt clay building bricks. Bureau of Indian Standards, New Delhi (1992)
15. Indian Standard (IS) 2185 Part 1: Concrete Masonry Units—Specification. Bureau of Indian Standards, New Delhi (2005)
16. Mehta, P.K., Monteiro, P.J.M.: Concrete Microstructure, Properties, and Materials, 3rd edn. McGraw-Hill (2006)

Waste Recycled PET as a Binder in Polymer Mortar

Bhagyashri Sarde, Y. D. Patil, and Dhruv Jani

Abstract Recycled waste plastic (PET) use as binder, filler and aggregate in polymer concrete or mortar is one of the best solution for plastic waste disposal problem. In this work, waste plastic (PET) were converted into unsaturated polyester resin through glycolysis process. This resin was used as a binding material in polymer mortar. Three different resin content (10, 15, 18%) was used. The best physical and mechanical properties were showed by MIX C polymer mortar with 18 wt% resin, 32 wt% filler and 50 wt% sand. Results of this study reveal that feasibility of using recycled PET waste as resin or binding material in polymer mortar. Use of waste plastic as binder increases mechanical properties of concrete as well as reduces hazard to environment by utilizing plastic and reducing cement consumption. Polymer mortar prepared from recycled PET flakes can be used for many constructional applications such as for sewer pipes, repairing work etc.

Keywords Recycled plastic waste · PET · Polymer mortar · Red mud · Polymer concrete

1 Introduction

Central Pollution Control Board (CPCB), India in 2015 estimated that average plastic waste generation in the country was nearly 6.92% of municipal solid waste. The plastic processing industry predicted that polymer consumption from 2017 to 2022 is likely to grow at rate of 10.4%. Recycling this waste plastic and use as cement replacement in concrete is a new era of research. Due to versatile properties, easy to manufacture, low cost and lots of applications use of plastic in routine life is extensively increases [1, 2]. Massive production and consumption of plastic generates gigantic plastic waste. Plastic does not decompose naturally and therefore, alternative methods has to be implement to recycle it and use in routine life [3].

B. Sarde (✉) · Y. D. Patil · D. Jani
Sardar Vallabhbhai National Institute of Technolog, Surat, Gujarat, India
e-mail: sardebhagyashri2@gmail.com

D. K. Ashish et al. (eds.), *3rd International Conference on Innovative Technologies for Clean and Sustainable Development*, RILEM Bookseries 29,
https://doi.org/10.1007/978-3-030-51485-3_28

Recycling of plastic is renewable source of material and use of recycled plastic in routine life reduces the deterioration of environment, save energy and lessen the solid waste problems [4]. Polyethylene terephthalate (PET) is widely used in manufacturing of soft drink and mineral bottles, food containers. About 900 kilo tones of PET is made in India annually. Recycled PET can reduce energy up to 29–45% as compared to virgin PET and give important environmental benefits [5, 6]. High cost of traditional/virgin resin as binder in concrete is somewhat expensive as compared to cement based material. For that, resin prepared from recycled plastic waste use as binder in polymer concrete is acceptable solution [7, 8]. Addition of recycled plastic instead of virgin polymer are technically and economically feasible to improve performance and quality of concrete as well as to solve problem of waste disposal [8].

Polymer concrete is a cement less concrete made by mixture of mineral aggregate with polymerized monomer [9, 10]. Polymer concrete was recognized in 1970 and mostly used for repairing of building, thin covers for bridges etc. [11]. Polymer concrete does not contain hydrated cement phase, strength and hardening of concrete is occur due to polymerization reaction. Fast curing, good strength, stiffness, damping, good adhesion, long term durability, low permeability to water and aggressive solution, chemical resistant, light weight is the advantage of using PC as a building and construction material [10, 12]. Performance and application of polymer mortar or concrete is depending upon the type of resin and type of aggregate use. Epoxy resin, acrylic resin, polystyrene are the common types of resin used in manufacturing of polymer mortar/concrete.

Recycling of polymer is interesting aspect from environment and economic point of view but presence of moisture, longer exposure to high temperature, environment aging causes somewhat decrease in mechanical properties and thermal properties [13].

The paper represents the study of recycled polyethylene terephthalate waste prepared resin as a binder in polymer mortar. Polymer mortar was prepared with different resin percentage and effect of this on physical and mechanical properties of polymer mortar was observed.

2 Experimental Program

2.1 Material and Method

The material used for experiments are recycled polyethylene terephthalate (PET) flakes was obtained from JB Ecotex, Surat. Red mud (45μ) as fire retardant filler was and Silica sand (300μ) as fine aggregate was used in polymer mortar composition.

Thermoplastic PET flakes were converted into thermoset unsaturated polyester resin through glycolysis experiment in three necked bottom flask [14]. To carry out glycolysis experiment chemicals like Propylene Glycol, Maleic Anhydride, Phthalic

Anhydride, Zinc Acetate, Styrene are required. These chemicals are purchased from AASHKA Chemicals, Surat. Cement in concrete was fully replaced by unsaturated polyester resin, this unsaturated polyester resin act as binder to form polymer mortar/concrete.

Unsaturated polyester resin was mix with 40 wt% styrene to impart viscosity to it. Sand and filler was oven dried at 100 °C for 24 h. to minimize the moisture content up to 0.1% to form good bond between resin and sand [15, 16]. Resin is hydrophobic in nature, therefore, small amount of moisture can hinder the polymerization process Unsaturated polyester resin has low viscosity and good wettability, hence it can be use as binder in polymer mortar [17].

Apparent density of polymer mortar was measured according to ASTM C905-01 code [18]. Apparent density of mortar was measured by filling a mortar in 1-L vessel into two equal layers. Each layer is compacted by tilting vessel on alternate side approximately about 10 times, excess mortar was strike off and mass of mortar was measured on weighing machine. Apparent uncured density was calculated by formula.

$$D_u = W_m - W_a/V \tag{1}$$

where D_u is apparent uncured density, W_m is weight of mold with material, W_a is weight of unfilled mold and V is volume of mold.

Thermal conductivity was measured by using steady state guarded hot plate method according to ASTM C177 code. Disc of diameter 100 mm and height 50 mm was used for this purpose and thermal conductivity measured by using following formula:

$$K = QL/A\Delta T \tag{2}$$

Q stand for amount of heat conducted through samples, L stands for thickness of specimen, A is cross sectional area and ΔT is temperature difference.

Flow ability was carried out in slump cone to measure filling ability of polymer mortar. Polymer mortar was poured into slump cone and immediately uplifted to measure flow spread diameter. Compressive strength of polymer mortar were casted in cylinder with size 25 mm diameter and 25 mm height and performed according to ASTM C579-3018 Code [19]. The ASTM C307-18 code was used to carry out experiment on tensile strength of unsaturated polyester resin prepared polymer mortar [20]. For flexure test, samples with size 25 mm × 25 mm × 250 mm beam were casted and tested according to ASTM code C580-18 [21].

2.2 *Mixing*

Polymer Mortar was prepared by varying percentage of unsaturated polyester resin (10, 15, 18%), oven dried red mud filler and sand with 1.5 wt% MEKPO and 1.5 wt%

Table 1 Composition and mix proportion of mortar

Recipe	Resin content (%)	Filler content (%)	Sand content (%)
MIX A	10	40	50
MIX B	15	35	50
MIX C	18	32	50

Table 2 Chemical composition of red mud filler and sand

Compound (%)	Red mud filler	Sand
$Al(OH)_3$	98.8	–
Na_2O soluble	0.06	–
Humidity(105 °C)	0.30	–
SiO_2	–	97.3
Fe_2O_3	–	0.3
Loss on ignition	35	–

cobalt octate as a curing agent. Effect of variation of resin and filler content keeping sand content constant on the properties of polymer mortar was observed. The varying percentage of resin and filler is given in following Table 1. Three specimens for each mixture were prepared. For MIX A, six specimens for compressive strength test at 7 days and 28 days, six specimens for tensile strength test and six specimens for flexure strength were prepared and tested. Hundred specimens were prepared with following mix proportions.

The manufacture of mortars involves weighing, mixing of component into concrete mixer for 3 min., samples are casted into molds, vibrated and cured according to ACI 548 Code [22]. Mixing of polymer mortar was done by preparing dry mix of sand and filler firstly and then resin and curing agent was poured into dry mix in mixture. Chemical compositions of red mud filler are given Table 2.

3 Results and Discussion

Fresh/Rheological property and mechanical properties of mortar were performed after 7 and 28 days from casting according to standard code.

3.1 Fresh/Rheological Properties

Apparent density, thermal conductivity and Flow ability of mortar were determined for polymer mortars with different resin content. The density was increases was influenced by increase in resin content. Apparent density, thermal conductivity and

Table 3 Apparent density and thermal conductivity

Recipe	Apparent density (kg/m^3)	Thermal conductivity (W/mK)
MIX A	219	0.06523
MIX B	258	0.06686
MIX C	330	0.07500

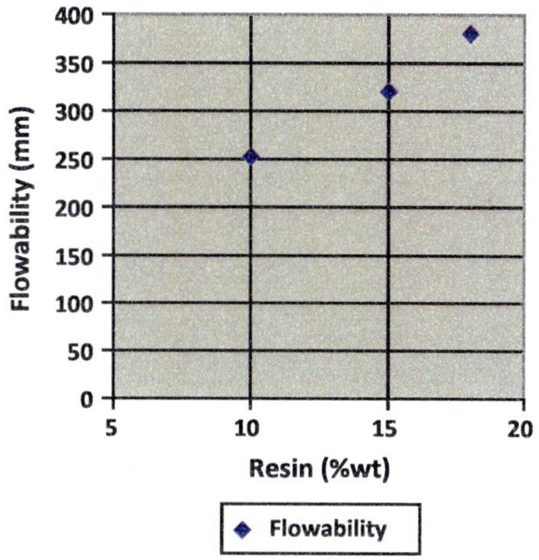

Fig. 1 Flow ability of polymer mortar with different wt% of resin

flow ability of MIX C was better than MIX A and MIX B. MIX C showed 68.8 and 78.18% increase in apparent density as compared to MIX A and MIX B. Thermal conductivity of MIX C was better, showed good resistance to temperature. Table 3 shows apparent density and thermal conductivity of different mix composition.

As shown in Fig. 1, flow spread diameter of mix A was 250 mm, for mix B 320 mm and for mix C 380 mm was observed. The flow spread diameter was increase due to increase in resin content. Acceptable flow ability of polymer mortar was 200 mm. MIX A exhibited a considerable increase in flow ability by 25% as compared to acceptable flow ability, MIX B showed 60% increase and MIX C showed 90% increase as compared to standard flow ability limit. This result indicates that, 18% resin content was sufficient to wet the sand and filler in the concrete mixture, resulting in good level compacting mortar with good thermal conductivity.

3.2 *Compressive Strength*

Compressive strength of mortar samples in function with resin percentage are shown in Fig. 2. It can be observed that mortar with 10 wt% resin showed 5 MPa compressive strength at 7 days and 6.9 MPa at 28 days. For 15 wt% resin, strength was 6.5 MPa at 7 days and 8.2 MPa at 28 days. Lowest values of compressive strength showed by

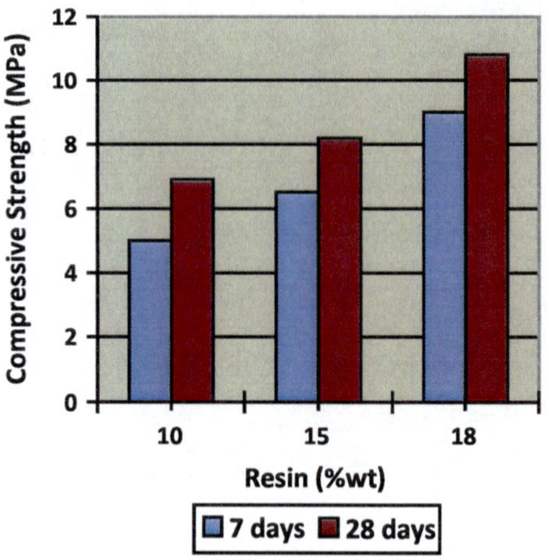

Fig. 2 Compressive strength of polymer mortar at 7 and 28 days

10 wt% resin indicate that, the lack of binding between sand, filler and resin, giving a weak polymer mortar. Increase in resin content (18 wt%) improves compressive strength in early days of curing. High resin content fill the gap between sand and filler result in good binding, good adhesion and high strength of polymer mortar. For cement mortar, the minimum compressive strength required was 7.5 MPa at the age of 28 days. MIX B showed 9% increase in strength at 28 days as compared to standard compressive strength. At 7 days of curing, the mortar with 18 wt% resin (MIX C) showed 9 MPa compressive strength and at 28 days of curing, 10.8 MPa strength was observed, which was 44% more than minimum compressive strength of cement mortar.

3.3 *Tensile Strength*

The tensile strength as function of resin content was shown in Fig. 3. It was observed that, tensile strength of mortar was improved slightly for MIX C. However, as compared to strength obtained for MIX C with MIX A and MIX B, increase in 35.71 and 73.21% strength at 7 days of curing was observed. Similarly, For MIX C, 51.47 and 88.23% increase in tensile strength at 28 days of curing was observed. By increasing the resin content, tensile strength of polymer mortar was improved for 18 wt% resin (MIX C).

3.4 *Flexure Strength*

Flexure strength of polymer mortar at seven and 28 days of curing with respect to resin content was shown in Fig. 4, The values of flexure strength of 18 wt% resin

Fig. 3 Tensile strength of mortar at 7 and 28 days

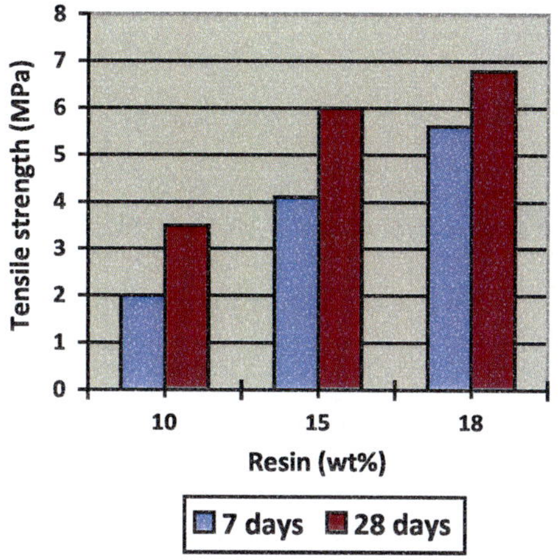

Fig. 4 Flexure strength of polymer mortar at 7 and 28 days

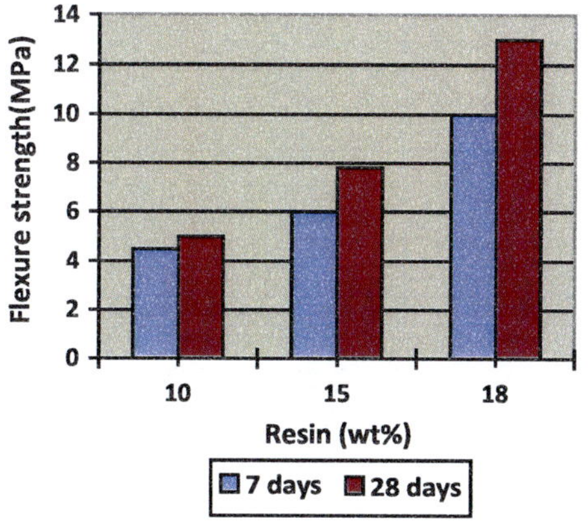

content mortar after 28 days of curing (13 MPa) rise slightly compared to values obtained at seven days (10 MPa). MIX C showed increase in 60% flexure strength as compared to MIX B at 7 and 28 days of curing. When MIX C was compared with MIX A, increase in flexure strength by 45% at 7 days of curing was observed. The mechanical behavior of MIX C showed good results, MIX C containing higher percentage of resin (18 wt.%), give better packing between phases and was sufficient to coated sand and filler particle in polymer mortar. Therefore, MIX C combination can be preferably use to prepare polymer mortar for precast application.

4 Conclusion

Use of virgin polymer and its effect on concrete is extensively studied by many researchers but the use of recycled polymer in concrete is less studied. Use of polymer prepared from recycled waste instead of virgin polymer enhances properties of polymer mortar or concrete. Many of waste polymers could be hazardous to environment if not properly use or recycle.

In this work, three different resin content and its effect on properties of mortar was studied. The unsaturated polyester resin was obtained from PET flakes through glycolysis process. This unsaturated polyester resin was used to manufacture polymer mortar. The physical and mechanical properties of polymer mortars are dependent on resin content.

Polymer mortar with 18 wt% resin, 32 wt% filler and 50 wt% sand (MIX C) showed good physical and mechanical properties as compared to 10 wt% (MIX A) and 15 wt% (MIX C). Increase in resin content gives a good wettability to sand and filler and subsequently improves compaction of mortar.

Addition of resin prepared from recycled PET flakes can be used for repairing purpose, bridge deck pavement, sewer pipes, road or airfield pavements, barriers, structural or decorative panels, drainage channels, swimming pools etc.

References

1. Yin, S., et al.: A life cycle assessment of recycled polypropylene fiber in concrete footpaths. J. Clean. Prod. **112**, 2231–2242 (2016)
2. Gu, F., et al.: From waste plastics to industrial raw materials: a life cycle assessment of mechanical plastic recycling practice based on real world case study. Sci. Total Environ. **601**, 1192–1207 (2017)
3. Tawfik, M.E., Eskander, S.B.: Polymer concrete from marble waste and recycled polyethylene terepthalete. J. Elastomers Plast. **38**, 65–78 (2006)
4. Mahdi, F., et al.: Physiochemical properties of polymer mortar composites using resins derived from post-consumer PET bottles. Cement Concr. Compos. **29**, 241–248 (2007)
5. Kalantar, Z.N., et al.: A review of using waste and virgin polymer in pavement. Constr. Build. Mater. **33**, 55–62 (2012)
6. Reis, J.M.L.: Fracture assessment of polymer concrete in chemical degradation solutions. Constr. Build. Mater. **24**, 1708–171 (2010)
7. Rebeiz, K.S.: Precast use of polymer concrete using unsaturated polyester resin based on recycled PET waste. Constr. Build. Mater. **10**, 215–220 (1996)
8. Wan Jo, B., et al.: Uniaxial creep behavior and prediction of recycled PET polymer concrete. Constr. Build. Mater. **21**, 1552–1559 (2007)
9. Ahn, S., et al.: Identification of stiffness distribution of fatigue loaded polymer concrete through vibration measurements. Compos. Struct. **136**, 11–15 (2016)
10. Son, S.W., et al.: Mechanical properties of acrylic polymer concrete containing methacrylic acid as an additive. Constr. Build. Mater. 669–679 (2012)
11. Jmal, H., et al.: Influence of the grade on the variability of the mechanical properties of polypropylene waste. Waste Manag. **75**, 160–173 (2018)
12. Tonet, KG: Polymer concrete with recycled PET: the influence of the addition of industrial waste on flammability. Constr. Build. Mater. **40**, 378–389 (2013)

13. Aciu, C., et al.: Recycling of polystyrene waste in the composition of ecological mortars. In: 8th International Conference Interdisciplinary in Engineering, INTER-ENG 2014, 9–10 Oct 2014, Trigu-Mures, Romania, vol. 19, pp. 498–505 (2014). Procedia Technology, Elsevier publication
14. Kin, J., et al.: Synthesis and applications of unsaturated polyester resin based on PET waste. Korean J. Chem. Eng. **24**(6), 1076–1083 (2007)
15. Konin, A.: Use of plastic waste as a binding material in the manufacture of tiles: case of eastes with a basis of polypropylene. Mater. Struct. **44**, 1381–1387 (2011)
16. Kanzhin, R.M., et al.: Conversion of polymer and resin waste into valuable product. Juniper Online J. Mater. Sci. **5**(2), 1–7 (2019)
17. Sekar, V.R., et al.: A technique to dispose waste plastics in ecofriendly way—application in construction of flexible pavements. Constr. Build. Mater. **28**(1), 311–320 (2012)
18. ASTM C905-01:2012 Standard test methods for apparent density of chemical resistant mortar, grout, monolithic surfacing and polymer mortar
19. ASTM C579-18 Standard test method for compressive strength of chemical resistant mortar, grout, monolithic surfacing and polymer concrete
20. ASTM C307-18: Standard test method for tensile strength of chemical-resistant mortars, grouts and monolithic surfacing
21. ASTM C580-18: Standard test method for flexure strength and modulus of elasticity of chemical-resistant mortars, grouts, monolithic surfacing and polymer concrete
22. ACI 548.1R-97: Guide for the use of polymers in concrete

A Study on Performance of Carbon Based Nano-enabled Cement Composites and Concrete

Mainak Ghosal and Arun Kumar Chakraborty

Abstract With the population expanding earth's limited resources have to be spread thin to meet our society's demand for sustainable development. Concrete consumes earth's finite resources infinitely but one of the certain and assured ways of promoting its sustainability is by improving its long-term durability. Contemporary research has shown that unconventional nanomaterials have the capability to address this issue. Our paper describes that with small addition of nanomaterials like Carbon Nanotubes (CNT) in optimized quantities the properties of concrete changes considerably. As cement hydration continues for years, it is observed that as the proportion of CNTs increased from the optimized level to its next level b.w.c, the mechanical strength of the cement composites gradually improved from 28 days to 1 year, with the addition of more Carbon Nanotubes. Also the durability properties of nano-enabled CNT added concrete showed vast improvements over the ages, when compared over ordinary control concrete or M-40 grade control concrete.

Keywords Cement · Concrete · Durability · Nanomaterials · Sustainability

1 Introduction

Concrete is the most consumed conventional building material consisting of cement and other natural raw materials at the macro level. With the cement production capacity of India likely to cross 600 million tonnes by 2030 and considering the burden on the natural resources, unfavorable environmental impacts would be on rise. Keeping in mind the Sustainability Development Goals (SDG) and also with the possibility of future carbon tax implementation, it's high time we become future equipped by creating more capabilities in the area of nanotechnology of cement and concrete. Major developments in cement- concrete performance were attained with application of super-fine particles viz. flyash, silica fume and now nanomaterials thus alleviating the problem of scarcity in natural materials and reducing green-house

M. Ghosal (✉) · A. K. Chakraborty
Indian Institute of Engineering Science and Technology, Shibpur, India
e-mail: mainakghosal2010@gmail.com

D. K. Ashish et al. (eds.), *3rd International Conference on Innovative Technologies for Clean and Sustainable Development*, RILEM Bookseries 29,
https://doi.org/10.1007/978-3-030-51485-3_29

gas (GHG) emissions. Due the emergence of new generation admixtures aided by advancement of visual technologies the studies of nanomaterials have exponentially increased in the last few years. This due to the fact that nanomaterials can also be used to minimize waste generation due to its recyclability in concrete [1].

The term 'nanomaterial' is a broad concept which means a material whose any external dimension is in nanoscale and nanomaterials are further subdivided into—nanoparticles, nanofibers and nanoplates. A nanoparticle has all the three dimensions in the nanoscale while nanofibers have two dimensions in the nanoscale, the other being in the microscale and nanoplates have only one dimension in the nanoscale, the other two dimensions being in the microscale. Nanomaterials especially oxide or carbon based nanomaterials like Nanosilica or Nanotubes of Carbon (CNT) or Carbon Nanofibers (CNF) which have been successfully synthesized from waste materials offer very smart solutions starting from structural health monitoring to improved properties to cement-concrete microstructural system, leading to more hydration products vis-à-vis strength and CNFs are considered a much cheaper nanomaterial than CNTs. Nanomaterials are also responsible for CO_2 absorption from the exposed hardened concrete surface by providing numerous nucleation sites in the cement-concrete matrix. Nucleation is the process whereby nanofillers' or nuclei (seeds) act as templates for crystal growth responsible for strength gains [2] and CO_2 absorption [3, 4]. So the influence of CNT/CNF enabled concrete on our environment through durability, carbon capture and waste utilization is huge thus leaving an everlasting impact on its sustainability.

Use of Nanotechnology and nanomaterials are already prevalent in the industry, world-wide, and new developments are taking place on a regular basis. Researchers have tried everything from 'micro' to 'nano' but, when we consider Nano materials in concrete, their perception takes time. Micro-materials have been widely used in construction during the last ten years but application of nanomaterials has surpassed that micro-effect in concrete. Though nano-application in cement-concrete is in a nascent stage and has been mainly confined to two areas, namely coatings and fabrications. With nano coatings market proposed to be saturated in a few years' time its time now to look up to the fabrication area which has the potential of providing a durable, innovative and green concrete, thus promoting sustainability in construction. All this leads to a new concrete called, 'Nano' concrete, whose behaviour may be interestingly novel w.r.to improved properties like increased bulk density/compressive strengths and durability. Identified breakthroughs in nano-fabrications include—globally in Edencrete's Colorado's Spillways/Georgia's Highways [5]; but there is very limited work on the use of nanotechnology or nanomaterials going in our country and this is an area where more emphasis is necessary—both in research field and in engineering use. Nanosilica concrete is also expensive but which may be nano-enabled is Nanodur UHPC (Dyckerhoff) [6].

Today the concrete end users demand long term performance and strength, durability and chemical resistance. Previously various researchers have used fibers but their actions were at the macro or micro (10^{-6} m) levels to which the approach of nano materials in concrete have replaced it at the nano or 10^{-9} m level. Now, the question arises is could nanomaterials be effective in present day complex concreting which is a heterogonous mixture of age old prescriptive codes, standards, specifications and new advanced materials like admixtures. However, to achieve results on optimized solutions, it is prudent to go for 'Performance Based' technique of these smart materials as against the common prescriptive approach for conventional materials in concrete. Once, Dr. A. P. J. Abdul Kalam [Ex-President of India (2002–07)] has aptly quoted that 'Four technologies are going to drive the future world and future India—biotechnology, nano-technology, information technology and environment technology'. With IT, India leads the world economy as the largest exporter, the other three (especially 'nano') also should not fall behind and it remains to see how India catches up fast with these magical three for manufacturing competitiveness and job generation.

1.1 Durability Vis-à-Vis Sustainability Aspects

India is a coastal country and the existing RC structures such as bridges, offshore platforms and multistoried buildings within 10 km from the coast comes under corrosion environment. Now, in concrete, environmental ingression deep into the microstructure affects the steel reinforcement beneath and the long term concrete strength is sacrificed thus affecting the sustainability of the construction. It is often said that what is durable is also sustainable. Durability clause has been enlarged in IS 456:2000 by thoroughly revising the previous 1978 code, though most of the foreign codes are on prescriptive lines in just specifying code-defined conditions for exposures. The Indian Standard IS 456:2000 also plays a prescriptive role in laying emphasis on concrete durability by mentioning requirement against exposure to aggressive chemical and sulphate attack and limiting the total acid soluble chloride and sulphate content in the mix expressed in kg/m^3 or % b.w.c. The present Paper utilizes a performance based approach in dealing with the mechanical and durability aspects of carbon nanomaterials in concrete. The durability parameters have been simulated in the IIESTS laboratory by testing cement concrete samples at longer term durations than the code prescribed 28 day strengths.

2 Experimental Investigations

The materials used were 43 Grade ordinary Portland cement, River sand, stone aggregates of 20 mm maximum size, plain potable drinking water, Superplasticizer and Nanomaterials like Carbon Nanotubes (industrial grade) (CNT).

Fig. 1 Mixing of CNT doped cement mortar composites

2.1 Dispersing the Nanomaterials

Nanomaterials exist in nature as entangled mass like form and when mixed with water they are non-dissolving. This is due to the presence of strong locally attractive *Van der Waals* forces. Thus, in concreting nano-addition faces problem in any medium be it aqueous or polymeric [7]. Ultrasonic energy has the power to tear off these local forces apart and render them functionally active. As far as our experimental programme was concerned, Carbon Nanotubes were dissolved in an aqueous solution of a 3rd generation suitable Superplasticizer (Poly Carboxylate Ether) after ultra-sonication for about 30 min in an external ultrasonicator bath (250 W Piezo-U-Sonic Ultrasonic Cleaner) in IIESTS laboratory. The time chosen for sonication is usually taken as the time required for 100% dissolution of the nanotubes otherwise without proper uniform dissolution the required strength will not at all be achieved. The weight of PCE added is about 1% by weight of cement. The dissolved CNTs in PCE were then added to the cement composites as shown in Fig. 1.

2.2 Tests on Cement Mortar Composites

As shown in Fig. 1, mortar cubes of 70 mm size were casted as per the Indian Standards. Industrial grade Carbon Nanotubes were added in proportions as per literature review i.e., 0.02, 0.05 and 0.1% b.w.c after proper dissolutions at w/c = 0.4. The cubes were vibrated both internally and externally by the respective internal and external vibrators during and after casting. The cubes were then ordinary cured under water and tested at 28 days, 90 days (3 months), 180 days (6 months) and 365 days (12 months or 1 year). Before testing in a compression testing machine, three specimens were converted to SSD (Saturated Surface Dry) condition by removing water from the surfaces. Then SSD weight of samples in air was measured along with their dimensions for Bulk Density calculations. The optimized proportions of the nanomaterials b.w.c. were which gave maximum strength at 28 days. These optimizations were repeated for performing strength & durability tests in concrete.

2.3 *Tests on M-40 Grade Standard Concrete*

For Compressive strength testing, 100 mm size concrete cubes were cast with cement, sand, stone aggregates and water in proportions as per the current IS: 10262 mix design code for M-40 Grade concrete for 100 mm slump, at water: cement = 0.4. The mix proportions were cement: sand: stone = 400 (kg/m^3): 687.54 (kg/m^3): 1293.04 (kg/m^3). Carbon Nanotubes were added in the optimized proportions as obtained in Test-result (2.2). The cubes were then ordinary cured under water in a curing tank, at room temperature of 27 ± 2 °C and tested for compressive strength at 28 days after natural surface drying and then immersed under aqueous solutions for strength testing at 3 months, 6 months, 12 months (1 year) respectively (Table 1).

3 Test Results and Discussion

The two most important properties of hardened concrete are its strength and durability. Table 2 shows the physical properties like bulk density, mechanical properties like compressive strength, potential durability and long term changes in the cement composites and concrete due to 'nano' addition.

Table 1 Sieve analysis results of river sand [13]

IS. sieve sizes	Cum % weight	% Passing		Remarks
(mm)	Retained			
4.75	–	100	90–100	Zone II, IS:383
2.36	6.77	93.23	75–100	
1.18	16.97	83.03	55–90	
600μ	44.95	55.05	35–59	
300μ	82.02	17.98	0–30	
150μ	98.28	1.72	0–10	
75μ	100	–		
Pan	–	–		

Table 2 Bulk density (kg/m^3) and compressive strength (N/mm^2) of both ordinary and nano-enabled Cement and concrete composites (% increase w.r.t. 0% additions)

% Nano additions in ordinary portland cement	28 day results		3 month results		6 month results		12 month results	
	Bulk density (% incr.)	Comp. str (% incr.)	Bulk density (% incr.)	Comp. str (% incr.)	Bulk density (% incr.)	Comp. Str (% incr.)	Bulk density (% incr.)	Comp str (% incr.)
0% CNT	2315.40	31.89	2290.12	31.20	2309.27	30.01	2229.10	30.01
0.02% CNT (optimized)	2356.24 (1.76%)	43.75 (37.19%)	2337.79 (2.08%)	35.59 (14.07%)	2372.71 (2.75%)	30.89 (2.93%)	2369.02 (6.28%)	28.53 (−4.93%)
0.05% CNT	2189.20 (−5.45%)	34.88 (9.38%)	2306.64 (0.72%)	31.85 (2.08%)	2348.61 (1.70%)	38.55 (28.46%)	2338.68 (4.92%)	41.69 (38.92%)
0.1% CNT	2267.23 (−2.08%)	24.83 (−22.13%)	2314.70 (1.07%)	31.50 (0.96%)	2353.40 (1.91%)	30.16 (0.50%)	2364.70 (6.08%)	50.78 (69.21%)
M-40 Grade Concrete	2559.34	43.03	2511.28	49.71	2589.38	48.34	2658.00	40.63
0.02% CNT enabled M-40 Concrete	2131.28 (−16.73%)	54.58 (26.84%)	2666.77 (6.19%)	72.37 (45.58%)	2667.45 (3.01%)	73.67 (52.40%)	2680.26 (3.83%)	82.00 (101.8%)

3.1 Detailed Discussions

The Results of the Experiments conducted tend to indicate that:

1. The physical properties like colour and density of the carbon nanotube enabled cement composites showed slight change when compared with the control samples of ordinary cement composites. The colour of the nanotube enabled mortar samples were little blackish in appearance and the samples specifically containing optimized 0.02% nanotube addition, were more denser that of control samples at all ages, as shown in Table 2. Also as per Fig. 3, the nano-enabled concrete was much more compact and denser than that of ordinary concrete or cement composites at all ages.

2. As per Table 2, the optimization for CNT is equal to 0.02% b.w.c for cement composites at 28 days. The mortar compressive strength shows that the compressive strength of the CNT cement composites shows a gain in strength of 38% at 28 days w.r.to control samples of ordinary cement composites but with the passage of time it fails to maintain that pace. Also, with the increased addition of carbon nanotubes (0.1% by cement wt.) the long term (1 year) strength gain increases appreciably to 69% with a denser composite. This attitude may be explained by the fact that as CNTs bond with C–S–H clusters [8] their strength increases but with the passage of time as more and more C-S–H clusters are generated due to later cement hydration, we need more percentage of CNTs to add up the strength.
3. As per Fig. 2, for 0.02% optimized CNT enabled M-40 concrete at air exposure, it is seen that as hydration proceeds it continued to be denser at latter ages as

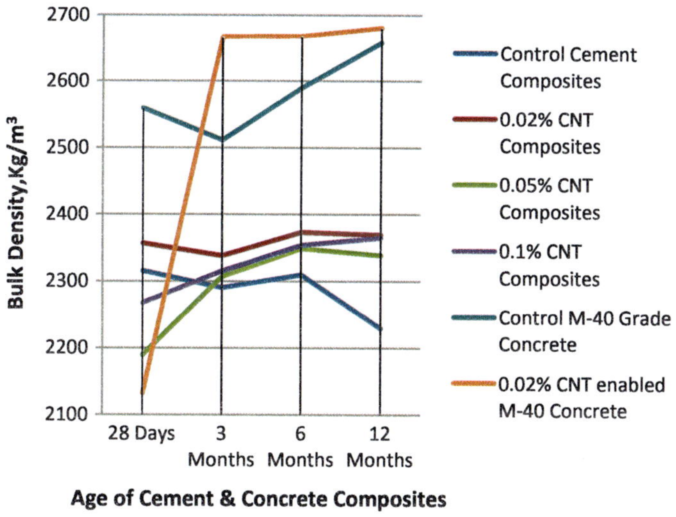

Fig. 2 Variations in density of nano-enabled cement composites and concrete over the ages

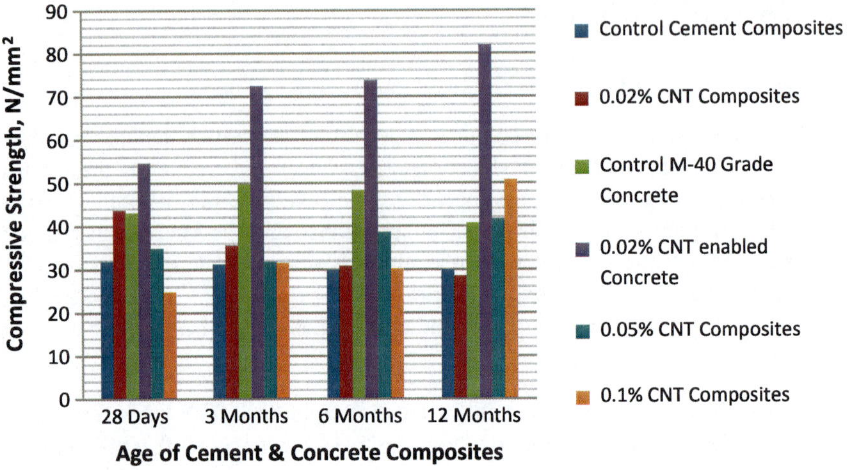

Fig. 3 Variations in strength of nano-enabled cement composites and concrete over the ages

reported in existing literatures for hardened concrete [9]. Also the density of nano-enabled cement composite for 0.02% CNT optimization attains maximum densification at all ages when compared to all other cement composites.

4. Carbon nano-concrete was fabricated with 0.02% CNT (optimum) and kept at air exposure after 28 days curing along with the controlled M-40 concrete samples. As per Table 2, the strength gain of the controlled samples got decreased after 3 months which may be due to the fact that the cement companies nowadays increase C_3S at the cost of C_2S. For nano-enabled concrete, it was observed that strength gain continued to abnormally increase at all ages, achieving more than 100% increase or doubling of strength from 40.63 N/mm^2 at 28 days(for controlled M-40 concrete) to 82.00 N/mm^2 for nano-enabled concrete at 1 years' time as shown in Table 2 and in Fig. 3.

4 Conclusion

The above results and discussion suggest that Carbon Nanotubes can improve the strength, density and durability of concrete (a widely used construction material), thereby reducing its total consumption and also deterioration after construction. This means that by using CNTs in concrete we can do away with large huge sections and instead can introduce thinner sections which need lesser formwork, reinforcement and detailing with lesser amount of energy, costing and finishes. Also more denser concrete will allow less moisture to pass through it resulting in more impervious concrete, making it a candidate for more durable structures. Thus, although carbon nanotubes-enabled concrete proves to be stronger, denser and durable than ordinary concrete, its applicability in construction sites in India, is negligible. Construction has lagged behind other industrial sectors, such as automobile, chemicals, electronics and biotech sectors, due to dearth in nanotechnology R&D and investments from large industrial corporations and venture capitalists [10]. Still, experimentation with CNT/CNF or other nanomaterials in cement concrete is a very active line of research in this decade with a need for more and more fruitful research. A review of literature on the durability of concrete reinforces the fact that though considerable research has been done on the effect of micro materials like fibers in concrete, only a few studies have been focused on the sustained use of nano-materials.

This fact is influenced by a multitude of factors and there are hardly any codes or standards which holistically address the performance evaluation of nanomaterials in cement concrete apart from the recent report from American Concrete Institute (ACI 241R-17). There exists a lack of coherence between the academic and the industry and the need for a strategy in the new industry creation in the nanotechnology field where not only public research institutes, but also existing industries (including SMEs) can participate. As seen in countries like Japan, China and Europe, to nullify the size disadvantage of nanomaterials and overcome diseconomies of scale, SMEs often utilize the process of technology transfer from public funded R&D institutions [11]. Though recently large corporations are investing in nanotechnology research (in-house), in India the research is mainly funded by Govt. and some of the challenges listed are its multidisciplinary nature, paucity of skilled labour and dependence on sophisticated equipments [12]. In this scenario and also looking at the climate change threat, there is a need for a paradigm shift to a performance based approach and probably influential and iconic institutions both public and private, may play a key role in fostering and standardizing the transformation from 'macro' to 'nano' within the construction industry. But for sustained development an active support from Indian SMEs is needed which then can further address the 'Make in India' initiatives. True to that line in creating a perfect ecosystem, the latest version of National Building Code of India, 2016 for example mentions new or alternate materials thus speaking of nanotechnology based advanced materials for the first time.

References

1. Khin, M.M.: A review on nanomaterials for environmental remediation. Energy Environ. Sci. **5**, 8075–8109 (2012)
2. Ouyang, X., Koleva, D.A., Ye, G., van Breugel, K.: Insights into the mechanisms of nucleation and growth of C–S–H on fillers. Mater. Struct. **50**(5) (2017). https://doi.org/10.1617/s11527-017-1082-y.
3. Lagerblad, B.: Carbon Dioxide Uptake During Concrete Life Cycle: State of the Art, 47 p. Swedish Cement and Concrete Research Institute (2005)
4. Hunt, C.M., Dantzler, V., Tomes, L.A., Blaine, R.L.: Reaction of Portland Cement with Carbon Dioxide. J. Res. Natl. Bureau Stand. **60**(5), 441–446 (1958)
5. https://www.edencrete.com/case-studies/colorado-spillways/
6. https://www.dyckerhoff.com/media/news/veranstaltung-emo-2019.
7. Li, H., Xiao, H.G., Yuan, J., Ou, J.: Microstructure of cement mortar with nano- particles. Compos. Part B **35**, 185–189 (2004)

8. Sobolkina, A., Mechtcherine, V., Khavrus, V., Maier, D., Mende, M., Ritschel, M., Leonhardt, A.: Dispersion of carbon nanotubes and its influence on the mechanical properties of the cement matrix. Cement Concr. Compos. **34**(2012), 1104–1113 (2012)
9. Iffat, S.: Relation between density and compressive strength of hardened concrete. Concr. Res. Lett. **6**(4), 182–189 (2015)
10. Zhu, W., et al.: Application of nanotechnology in construction. Mater. Struct. **37**, 649–658 (2004)
11. Purushotham, H.: Transfer of nanotechnologies from R&D institutions to SMEs in India. Asia-Pacific Tech Monitor **29**(4), 23–33 (2012)
12. Nanotechnology Development in India: Building Capability and Governing the Technology [TERI briefing paper. Supported by IDRC, Canada (2013). https://www.teriin.org/div/ST_BriefingPap.pdf.
13. Ghosal et al.: Application of nanomaterials on cement mortar and concrete: a study. IUP J. Struct. Eng. **X**(1), 7–15 (2017)

Cost and Feasibility Analysis of Chromium Removal from Water Using Agro and Horticultural Wastes as Adsorbents

Pushpendra Kumar Sharma, Sohail Ayub, and Bishnu Kant Shukla

Abstract Current study aims at revaluation and sustainable management of agro and horticultural solid wastes such as Almond (*Prunus dulcis*) Shells, Discarded Potato (*Solanum tuberosum*) and Petha (*Benincasa hispida*) reusing these as adsorbent aimed at low cost removal of chromium at source of wastewater generated from industries. The cost estimation and feasibility of adsorption of chromium from aqueous solutions was analyzed using Langmuir Adsorption Isotherms and varying thermodynamic parameters like enthalpy, Gibb's free energy and entropy changes at equilibrium state of adsorption respectively. At equilibrium state of Adsorption, the adsorbents, which were Almond (*Prunus dulcis*) Shells, Discarded Potato (*Solanum tuberosum*) and Petha (*Benincasa hispida*) showed positive change of enthalpy ΔH as 0.247, 0.288 and 0.2593 kJ/mole respectively proved endothermic nature of adsorption with strong association, the positive change of Gibb's free energy ΔG; 0.267, 0.267 and 0.3295 kJ/mole claimed the random feasibility. Binding of chromium adsorbate on to the surface of adsorbent resulted reduction in the degree of freedom with negative entropy ΔS as −0.0221, −0.0167 kJ/mole for Almond Shells and Petha where as positive change of entropy was observed for Discarded Potato 0.0013 kJ/mole, which showed a little lesser association.

Keywords Sustainable · Aqueous · Chromium · Pollution · Estimation

1 Introduction

There are various environmental concerns like ozone layer depletion, air soil and water pollution, emission limits due to which many projects have not been given

---

P. K. Sharma (✉) · B. K. Shukla
School of Civil Engineering, Lovely Professional University, Phagwara 144411, India
e-mail: p.sharmaji10@gmail.com

S. Ayub
Department of Civil Engineering, Aligarh Muslim University, Aligarh 202001, India

D. K. Ashish et al. (eds.), *3rd International Conference on Innovative Technologies for Clean and Sustainable Development*, RILEM Bookseries 29,
https://doi.org/10.1007/978-3-030-51485-3_30

environmental clearance or long lined despite of their importance in developing India.

Various industries are releasing heavy metals in the natural water bodies that cause serious environmental issues. Heavy metals are used in many industries, releasing traces into environment through air, water or soil. These heavy metals threaten all forms of life on globe. For instance chromium especially in macro amounts from electroplating and tanneries is severely contaminating our ground water and surface water sources making them toxic and carcinogenic.

So there is an urgent need of removing these contaminants at very sources before it joins our natural water bodies. The modern sustainable research has emphasized on cost, environment and society keeping in mind all the post and pre concerns. The multidimensional novel technology with a theme of '*aam ke aam, guthaliyon ke daam*' i.e. '7-Rs' stand for reuse, revalue, recovery, remediation, recirculation, reserve and right solid waste management of agrowastes and bye products which could otherwise be a serious matter to be seriously viewed upon.

This paper deals with the removal of very toxic, carcinogenic and dangerous hexavalent chromium from aqueous solutions though the trivalent chromium is life supporting in many ways [1]. Thousands of industries like tanneries, pigment and wood preservation, electroplating and leather manufacturing avoid the effluent treatment plants and removal of dangerous hexavalent chromium at their effluent points because of the costly removal through activated carbon and as a result these industries mostly in developing countries like India release chromium in natural streams thereby contaminating our drinking water resources increasing their pollution level high.

2 Literature Review

Many researchers have proved agro wastes; the potential adsorbents for chromium removal from wastewaters but none has estimated the cost of operations and the preparation cost of these agro waste adsorbent. The water so treated is of desired standards so as to be reused in multipurpose operations. These low cost agro and horticultural adsorbents are not only available in huge amount in agricultural country like India but also there is a big issue of their disposal. In the state of Punjab in India these agricultural wastes are burnt on spot at agricultural fields only called as 'Parali' burning creating lot of smoke, CO_2 emission and chemical fog in national capital range of Delhi. Many researchers have reported that instead of burning the agricultural wastes at spot, there may be an alternative to convert them in to value added wastes which can be the potential adsorbents for removal of heavy metals from aqueous solutions [2–16]. Various researchers have proved these horticultural and agro wastes to be potential adsorbents in last recent years [2–12, 17–21].

It is natural that the trivalent chromium which is very useful in many health problems is most found in macro amount and micro level amount of hexavalent chromium is observed in traces of effluent which can easily be treated with these cheaper adsorbents. In most of the times this dangerous hexavalent chromium is converted in to beneficial trivalent chromium. It has been observed that high molecule weight adsorbates can easily adsorbed on solid surfaces of adsorbents having good unevenness and cavities to catch the adsorbate like chromium. In physical adsorption an equilibrium of the process is achieved quickly through the slow diffusion process. It is well observed that the sorption depends on pairs of sorbate and sorbent along with the interference due to the other ions and adsorbates in complex nature of wastewater that could cover the attachment surface of main adsorbate to be adsorbed.

Nsami and Mbadcam [22] evaluated the sorption efficacy of chemically ready activated Carbon of Cola Nut Shells by Zinc Chloride on Methylene Blue plotting Langmuir Isotherms. They also studied various thermodynamic parameters for checking the feasibility of adsorption process at equilibrium.

Sharma and Ayub [23] first time estimated the cost of adsorbents for the removal of chromium from tannery wastewater using different agricultural and horticultural wastes like *Pisum sativum* peels, *Camellia sinensis* and *Zingiber officinale* mix and *Musa lacatan* peels to adsorb Cr (VI) and also observed the feasibility of chromium adsorption on to the various agro and horticultural wastes. The researchers obtained 65% removal in first 45 min and on increasing the contact time adsorption reached up to 90% but more contact time needs large volume of adsorption reactor thereby increasing the cost of construction of reactors making the process costly.

Sharma and Ayub [24] Studied economic feasibility of the three horticultural wastes like; *Mangifera indica* both mango seed and it's shell waste and *Carica papaya* adsorbing hexavalent chromium from wastewater of tanneries. The researchers also made first time a successful effort to find the cost of these adsorbents to adsorb 1 mg of chromium on to their surface in one layer only. The researchers proved these horticultural wastes potential and cost effective adsorbents.

Mahmood et al. [25] Performed thermodynamic study and equilibrium kinetics of adsorption process to remove Arsenic (III) from aqueous solutions using activated carbon prepared from Ferric Oxide mixed with used tea leaves. Conventionally they also plotted various Isotherms and Gibb's free energy curves for checking the feasibility and Arsenic adsorption efficiency but they didn't estimate the cost of adsorption.

Mohan et al. [26] removed arsenic by chemically treated low cost adsorbents using guava leaf waste, mango bark and sugarcane bagasse. Using Tempkin, Langmuir and Freundlich Isotherms they observed the adsorption efficiency and reported very cost effective and efficient adsorption of arsenic. But they also didn't determine the cost of adsorption.

Though there is sufficient existing literature on the low cost adsorption but none except [22, 24] has estimated the cost of adsorption of heavy metals by these agro and horticultural adsorbents, so that it could be proved that these are potential low cost adsorbents. So there has been a research gap on cost and feasibility analysis and keeping this in focus the authors first time analyzed the cost of these low cost agro and horticultural adsorbents to remove hexavalent chromium from aqueous solutions along with their feasibility at process equilibrium. This study is about the analysis of cost and feasibility of agro wastes like *Prunus dulcis* Shells, *Solanum tuberosum* discarded *potato* from Cold storages and *Benincasa hispida* from Petha Industries, used for the adsorption of hexavalent chromium from wastewaters. The research will lead knowledge of low cost and feasibility adsorption analysis and future researchers will be able to do this analysis for other adsorbents using thermodynamics and Langmuir Isotherms in equilibrium state of adsorption.

3 Materials and Methods

3.1 Sampling and Preparation of Sorbents

During batch experimentation, agro and horticultural wastes adsorbents such as *Prunus dulcis* Shells, *Solanum tuberosum* that were discarded from Cold storages and *Benincasa hispida* from Petha Industries were used to adsorb hexavalent chromium from aqueous solution. The optimized adsorbent amount was well mixed in a conical flask with 50 ml of standard synthetically diluted chromium solution and then left for attaining adsorption equilibrium. After adsorption equilibrium reached the holdup was sieved over Whatman No. 1 filter paper and then the filtrate was gauged for left over chromium content. This way the difference of initial and final residual chromium fraction of initial chromium content was the adsorbed chromium and when multiplied by 100 it was converted in percentage adsorption of chromium on to the surface of used adsorbent.

Adsorbents Preparation All the sorbents were first sun dried for 15 days, then oven dried at 95 °C for three days at a rate 5 h per day. Then was alternate mixed in 0.1 N NaOH and 0.1 N H_2SO_4, each washing followed by double distilled water (dd H_2O) rinse so as to get rid of any lignin and alkaline material. After oven drying at 95 °C for three days at rate of 5 h per day, prepared adsorbents were stored in desiccators so as not to absorb any humidity which reduces adsorption capacity occupying surfaces.

3.2 Batch Experimentation

Batch experimentation at equilibrium state of adsorption was run at absolute temperature of 298 K to optimize the affecting parameters like pH, dose of sorbent, initial chromium concentration, time of contact, sorbent size temperature and the initial mixing speed in rpm. Standard Solution of Chromium for different dilutions was used for all the above parameters optimization; to apply them for continuous flow experiment in a dynamic system. For all optimized values of process affecting parameters, the suspension through Whatman No. 1 filter paper, filtrates were assessed for residual chromium using Atomic Absorption Spectrometer (Make-Perkin Elmer, AAS, Pin AAcle 900 K Model) in the environmental engineering laboratory, civil engineering, ZHCET, AMU, Aligarh, India.

During batch experiments parameters, like adsorbent dose, pH, temperature, adsorbate initial concentration, grain size of adsorbent, contact time and mixing speeds in rpm were optimized so as to design a continuous flow system for the process of chromium removal. At the same time during optimization of affecting parameters both before and after use of adsorbents, Scanning Electron Microscopy (SEM) was used to understand the pattern and mechanism of adsorption along with thermodynamic characterization so as to check the feasibility of the low cost adsorption.

3.3 Feasibility Analysis

For thermodynamic feasibility analysis, Gibb's free energy equation (Eq. 1) was used;

$$\Delta G = -RT \ln K \tag{1}$$

where, ΔG; free energy change (joule/mole), R; universal gas constant 8.314 and T the Kelvin temperature 298 K. Plotting Gibb's free energy change (ΔG) on Y-axis and the absolute temperature (T) on X-axis for all adsorbents at equilibrium the linearization was mathematically simulated to the relation as shown by Eq. 2 [4, 5].

$$\Delta G = \Delta H - T \cdot \Delta S \tag{2}$$

where, ΔH; for change of enthalpy, ΔS; for change of entropy for activation [19].

3.4 Langmuir Isotherms

The Langmuir equation is given in Eq. 3

$$q_e = \frac{q_o K_L C_e}{1 + K_L C_e} \tag{3}$$

After simplification by linearization, we get Eq. 4

$$\frac{1}{q_e} = \frac{1}{q_o} + \frac{1}{q_o K_L} \times \frac{1}{C_e} \tag{4}$$

where, C_e; the sorbate concentration (mg/l), Q_e; adsorbate quantity stuck on unit gram of adsorbent (mg/gm), Q_o; single layer coverage of sorbent in (mg/g) which is intended and determined for the estimation of cost of adsorption. The Langmuir isotherm constants q_{max} and K_L calculated by intercept and slope of linearized

Langmuir curve (Eq. 5).

$$R_L = \frac{1}{[1 + \{1 + K_L C_o\}]} \tag{5}$$

The Equilibrium constant for an idea of feasibility of the adsorption such that its value if zero means irreversible, from zero to less than one favourable, at one linear and unfavourable if greater than unity. C_o is initial concentration of chromium which was taken fixed 30 mg/l at arte of optimized value and K_L is a pure number Langmuir constant of equilibrium which correlates energy associated with adsorption [27].

3.5 Scanning Electron Microscopy (SEM)

Scanning Electron Microscopy for *Prunus dulcis* Shells, *Solanum tuberosum* discarded from Cold storages and *Benincasa hispida* from Petha industries was also used to determine the surface textures and nature of adsorbents both prior and later used in process using JSM 6510 LV Make; Japan, at USIF Centre, AMU, Aligarh, India.

4 Results and Discussion

4.1 Optimization of Affecting Parameters

The maximum adsorption was observed in pH range 5.0–6.0 beyond which the efficiency of adsorption decreased [5, 28, 29].

The highest adsorption was observed at adsorbent dose ranging 10–20 g/L in all wastes and out of this range sorption was not satisfactory [9, 10, 16, 30]

Though there is no control on initial concentration in real conditions yet optimization of this parameter has a great role in designing of the dynamic systems. As per the conditions and concentrations in real effluents the dilution factor may be implemented as per requirement of designs of the systems. The best results were observed between 10 and 20 mg/l of initial concentration for process design purpose of dynamic reactors.

Chromium removed almost 60% in first 135 min. There after the process was very slow equilibrium acquiring but steady. Average grain size of adsorbent found optimum 1180 µm. The best suited temperature found 25 to 35 °C for adsorbents used in batch set ups. So all batch set ups were run at 25 °C i.e. 298 K. The optimum mixing speed at start was best observed in the range 80–100 rpm neither below nor above.

4.2 Thermodynamic Analysis for Feasibility

Determined enthalpy change (ΔH) all positive values 0.247, 0.288 and 0.259 3 kJ/mole respectively for each adsorbent Almond (*Prunus dulcis*)*Seed* Shells Waste (ASSW), Potato (*Solanum tuberosum*), discarded potato waste from Cold storages (DPWC) and Petha (*Benincasa hispida*) waste (PW) from Petha manufacturing Industries showed the endothermic nature of adsorption with strong binding; the entire positive changes in Gibb's free energy (ΔG) values 0.267, 0.267 and 0.3295 kJ/mole indicated the random feasibility and Binding of chromium on to the adsorbent surface reduced degree of freedom with negative entropy ΔS as −0.0221, −0.0167 kJ/mole for *Prunus dulcis*, Shells and *Benincasa hispida* from Petha Industries where as positive change of entropy was observed for *Solanum tuberosum* discarded from Cold storages 0.0013 kJ/mole which showed a little lesser association [19] as shown in Table 1 [11, 31–35].

Using Gibb's free energy plots as shown in Fig. 1, Thermo-dynamic Parameters for Almond seed shell waste (ASSWTP), Fig. 2 Thermo-dynamic Parameters for discarded potato waste from cold storage (DPWCTP) and Fig. 3 Thermo-dynamic Parameters for petha waste (PWTP) as adsorbent with linearized, simulated and modeled for the feasibility study of the process.

Values of ΔG, ΔH and ΔS used in the calculations were calculated at 298 K as shown in Table 1 in later section.

4.3 Langmuir Curves for Maximum Monolayer Coverage

The Langmuir curves for *Prunus dulcis* Shells, *Solanum tuberosum* discarded from Cold storages and *Benincasa hispida* from Petha Industries as in Fig. 4, Almond Seed Shell Waste Langmuir (ASSWL) Plot, Fig. 5, Discarded Potato Waste from Cold storage Langmuir (DPWCL) Plot and Fig. 6, Petha Waste Langmuir (PWL) Plot respectively were plotted from the observations. This helped the researchers to determine maximum monolayer coverage to estimate the cost of adsorption in following steps.

Various Langmuir constants values and Thermodynamic parameters of used adsorbents observed are shown in Table 1 as follows

4.4 SEM Analysis

The samples being non-conducting in electric nature; were gold coated before Scanning Electron Microscopy (SEM). SEM analysis was observed very self-explained; adsorbents found with sufficient surface with undulations and cavities to capture chromium on to these surfaces as shown in Fig. 7. Before adsorption there were

Table 1 Langmuir and thermo-dynamic parameters

At 298 K	Langmuir constants					Thermodynamic parameters		
Sorbent	Max monolayer coverage capacity, q_{max} (mg/g)	$1/q_0$	Langmuir constant (KL)	RL (favourable if 0 < RL < 1)	R^2 (Best error distribution or perfect correlation) (must be <1)	ΔG (KJ/mole)	ΔH (KJ/mole)	ΔS (KJ/mole)
ASSW	11.11	0.09	0.1627	0.1648	0.8702	0.267	0.247	-0.0221
DPWC	2.0	0.50	1.282	0.029	0.866	0.267	0.288	0.0013
PW	2.754	0.363	0.6269	0.05658	0.955	0.3295	0.259	-0.0167

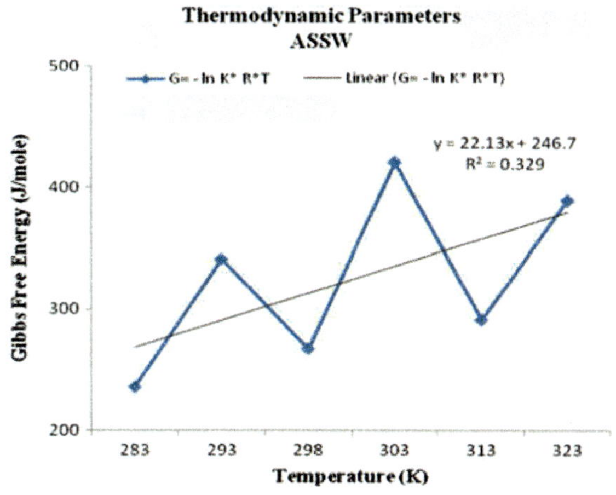

Fig. 1 Gibb's free energy versus temperature for almond seed shell waste (ASSW)

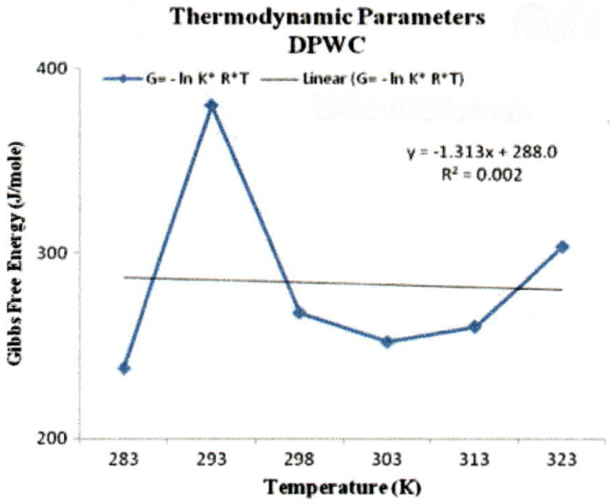

Fig. 2 Gibb's free energy versus temperature for discarded potato waste from cold storage (DPWC)

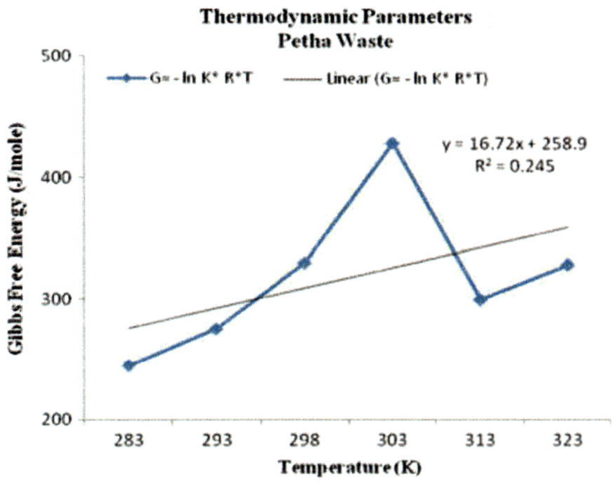

Fig. 3 Gibb's free energy versus temperature for Petha waste (PW)

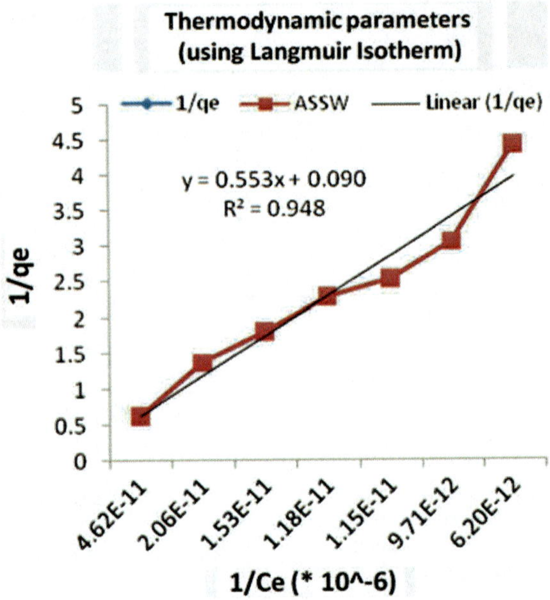

Fig. 4 Langmuir curve for almond seed shell waste (ASSWL)

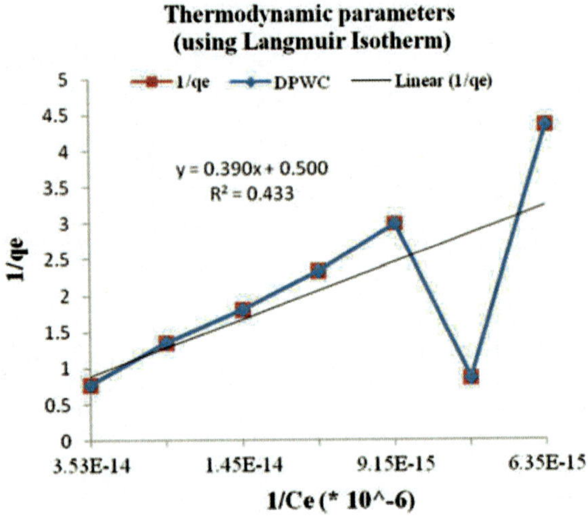

Fig. 5 Langmuir curve for discarded potato waste from cold storage (DPWCL)

many holes and layered wrinkles in almond seed shell case to adsorb the chromium on vast surfaces Fig. 7a and after the adsorption these holes and wrinkles are occupied by adsorbate making quite even surface without holes Fig. 7b. In case of discarded potato waste before adsorption there appears a cave like texture of the surface Fig. 7c and after the adsorption this cave like surface is seen even surface as shown in Fig. 7d. Similarly in Petha waste before the adsprption there are ripples like hollows

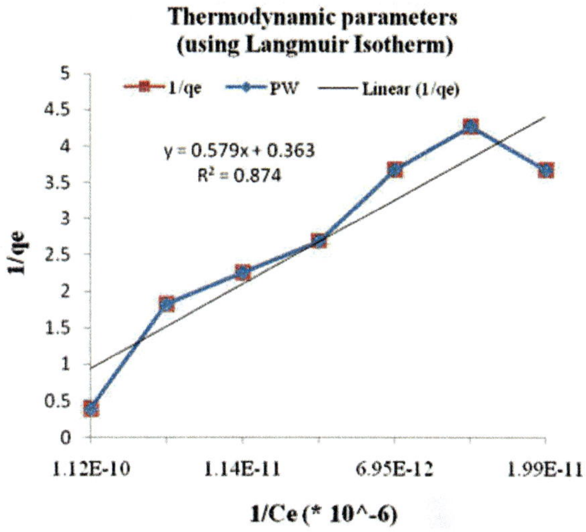

Fig. 6 Langmuir curve for Petha waste (PWL)

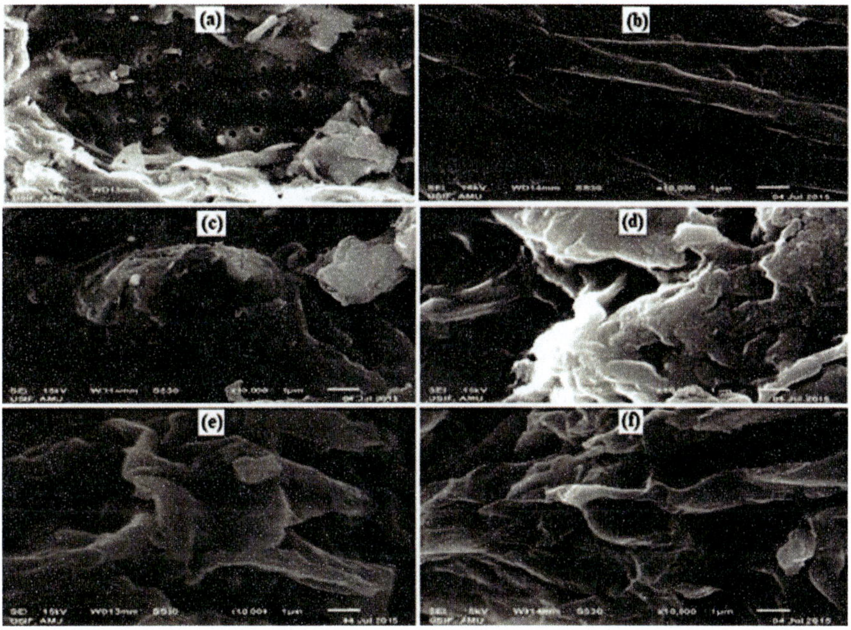

Fig. 7 SEM analysis for all adsorbents before and after use for **a** almond seed shell waste raw; **b** almond seed shell waste used; **c** discarded potato waste raw; **d** discarded potato waste used; **e** Petha waste raw; **f** Petha waste used

Fig. 7e but after the adsorption of chromium these gaps are quite shallow indicating adsorption as shown in Fig. 7f [32–35].

5 Cost Estimation

Along with feasibility study of adsorption estimation of cost is also very much essential to design the dynamic adsorption process reactors. No researcher has given a thought to this important idea. The process cost is directly proportional to the cost of adsorbent. From ancient times activated carbon has been most used for the chromium removal from solutions and it proved to be very costly and unaffordable by industrialists of developing countries. The problem could not be solved here it is hardly available so to meet out its day by day increasing demand as a result low cost materials become today's urgent need which are comparable to activated carbon with respect to local environmental aspects, availability, feasibility, efficiency and economy etc. The steps for the calculation of cost for the adsorbents preparation have been taken from the rarest literature available [3, 33–35]. The cost evaluation and economic feasibility of above agro and horticultural wastes adsorbents in this research work to adsorb chromium from leather manufacturing industrial effluents has been done and at a glance shown in Tables 2 and 3 for all the adsorbents used here.

Performing all steps of methodology like chemical and physical activation all the expenditures occurred were included in cost then the total cost of preparation of these Adsorbents is shown in Table 2. Each 1 kg of adsorbent prepared cost amounted as

Table 2 Estimating and costing of adsorbent per kg of agro wastes (ASSW, DPWC, PW)

Item	Unit cost (Rs.)	ASSW	Waste	DPWC	Waste	Petha	Waste
Cost		Amount	Price (Rs.)	Amount	Price (Rs.)	Amount	Price (Rs.)
H_2So_4	6.00/l	1.2 lit	7.20	3.0 lit	18.00	1.5 lit	9.00
NaOH	1.60/l	2.0 lit	3.20	5.0 lit	8.00	5.0 lit	8.00
Drying	5.80/kwh	0.6 kwh 9° 5 h	3.48	0.6 kwh 90° 5 h	3.48	0.6 kwh 90° 5 h	3.48
Heating	5.80/kwh	0.6 kwh 9° 5 h	3.48	0.6 kwh 90° 5 h	3.48	0.6 kwh 90° 5 h	3.48
Grinding	5.80/kwh	0.8 kwh1h	4.64	0.4 kwh 30 min	2.32	0.4 kwh 30 min	2.32
Net			22.00		35.28		26.28
Overhead 15% of net			3.30 + 50 (PC)		5.30		3.94
Total			75.30		40.58		30.22

Table 3 Unit gm chromium removal cost

Sorbent	Adsorptivity (mg/g)	Cost of adsorbent (Rs./kg)	Cost of adsorbent to remove 1 g chromium (Rs.)
ASSW	11.11	75.30	6.78
DPWC	2.00	40.58	20.29
PW	2.75	30.22	10.99
AC	4.221	600	142.15

Rs. 75.30, Rs. 40.58 and Rs. 30.22 respectively for ASSW, DPWC and PW whereas the market cost of popularly used adsorbent activated carbon is almost Rs. 600 per kg.

The estimated cost per gram of chromium adsorbed by *Prunus dulcis* Shells, *Solanum tuberosum* discarded from Cold storages and *Benincasa hispida* from Petha Industries was estimated as Rs. 6.78, Rs. 20.29 and Rs. 10.99 respectively whereas the same cost for conventional and common activated carbon is observed to be Rs. 142.145 which shows a great difference in cost. So, these agro and horticultural adsorbents have potential economic adsorption of chromium. The adsorbents were developed from agro and horticultural wastes otherwise these wastes require to be managed because their disposal is a big concern with environment. So the utilization of these wastes to adsorb hexavalent chromium from wastewater helps in agro waste management. So these agro and horticultural wastes not only removed toxic chromium but also reused and revalued indirectly benefitting the society and environment otherwise farmers specially in Punjab, India, usually burn in fields due expensive transport. The burning of these wastes at fields not only increases the greenhouse gases emission being an environmental and climatic issue but also degrades the ambient air quality. This low cost removal of chromium is easily affordable by the industrialists so their final product cost is lowered ecofriendly. Hence tanneries can play an important role for employment, environment and natural streams purification economically. The low cost potential adsorbents will undoubtedly be used by the industrialists to adsorb the heavy metals from the industrial effluents frequently. This practice will not only produce economic products from industry but also reduce the heavy metal contamination of natural streams which are the main sources of our water reducing water treatment cost. In addition to the environmental pollution control there will be the best agricultural waste management also for our society. So, this research offers Sustainable development of industry, society and nation simultaneously which is its beauty.

6 Conclusion

The studied agro and horticultural wastes; *Prunus dulcis* Shells, *Solanum tuberosum* discarded from Cold storages and *Benincasa hispida* from Petha Industries are observed potential adsorbents to adsorb toxic chromium. The process is not only economic and ecofriendly but also a kind of solid waste management of the agro and horticultural wastes; saving fields and environment which would have been damaged due to conventional burning. These agro and horticultural wastes being available in plenty amount in India, so their regeneration is not very necessary; otherwise in the countries where these low cost agro and horticultural adsorbent are not available in abundance can be regenerated also.

References

1. Katz, S.A.: The analytical biochemistry of chromium. Environ. Health Perspect. **92**, 13–16 (1991)
2. Sharma, P.K., Ayub, S., Tripathi, C.N.: Agro and horticultural wastes as low cost adsorbents for removal of heavy metals from wastewater: a review. Int. Ref. J. Eng. Sci. **2**(8), 18–22 (2013)
3. Sharma, P.K., Ayub, S., Tripathi, C.N.: Sorption Studies of synthesized horticultural waste seeds for removal of hexavalent chromium from aqueous solutions. Int. J. Environ. Sci. Toxic. Res. **3**(7), 101–114 (2015)
4. Sharma, P.K., Ayub, S., Tripathi, C.N.: Isotherms describing physical adsorption of Cr(VI) from aqueous solution using various agricultural wastes as adsorbents. Civil Environ. Eng. **3**, 1–20 (2016)
5. Sharma, P.K., Ayub, S., Tripathi C.N. and Ajnavi, S.: Removal of chromium (VI) From aqueous solutions using discarded solanum tuberosum as adsorbent. Res. Inventy Int. J. Eng. Sci. **6** (11), 5–15 (2016).
6. Sharma, P.K., Ayub, S., Tripathi, C.N.: Physisorption of chromium from aqueous solutions using agro and horticultural wastes as adsorbent. Int. J. Appl. Environ. Sci. **4**, 607–631 (2017)
7. Ayub, S., Ali, S.I., Khan, N.A.: Treatment of wastewater by agricultural wastes. Environ. Pollut. Cont. J. **2** (1) (1998)
8. Ayub, S., Ali, S.I., Khan, N.A.: Extraction of chromium from the wastewater by adsorption. Environ. Pollut. Cont. J. 2(5) (1999)
9. Ayub, S., Ali, S.I., Khan, N.A.: Efficiency evaluation of Neem bark (Azadirachtaindica) bark in the treatment of industrial wastewater. Environ. Pollut. Cont. J. **4**(4), 34–38 (2001)
10. Ayub, S., Ali, S.I., Khan, N.A. Adsorption studies on the low cost adsorbent for the removal of Cr (VI) from the electroplating industries. Environ. Pollut. Cont. J. **5**(6) (2002)
11. Ayub, S., Ali, S.I., Khan, N.A.: Chromium removal by adsorption on coconut shell. J.Indian Assoc. Environ. Manag. **30**, 30–36 (2003)
12. Ayub, S., Khorasgani, F.C.: Adsorption process for wastewater treatment by using coconut shell. Int. J. Civil Struct. Env. Infrastruct. Eng. Res. Dev. **4**(3), 21–34 (2014)
13. Ajmal, M., Rifaqat, A.K.R., Bilquees, A.S.: Adsorption and removal of dissolved metals using pyrolusite as adsorbent. Environ. Monit. Assess. **38**, 25–35 (1995)
14. Ajmal, M., Rao, R.A.K., Ahmad, R., Ahmad, J.: Adsorption studies on *Citrus reticulata* (fruit peel of orange): removal and recovery of Ni(II) from electroplating wastewater. J. Hazd. Mater. **49**(1–2), 117–131 (2000)
15. Raji, C., Anirudhan, T.S.: Sorptive behavior of Chromium (VI) on saw dust carbon in aqueous media. Ecol. Env. Cons. **4**(1), 2–33 (1998)
16. Ayub, S., Sharma, P.K., Tripathi, C.N.: Removal of hexavalent chromium using agro and horticultural wastes as low cost sorbents from tannery wastewater: a review. Int. J. Res. Civil Eng. Archit. Des. **2**(3), 21–35 (2014)
17. Sharma, P.K., Dubey, S.K., Rehman, Z.: A study on arsenic removal from ground water by adsorption process. In International Conference on Emerging Technologies for Sustainable

Environment, (ETSE-2010), pp. 164–167, Department of Civil Engineering, A.M.U. Aligarh, India (2010)
18. Camino, G., Paserini, A., Toscano, G.: Removal of toxic and Cr(VI) from aqueous solution by hazelnut shell. Water Res. **34**(11), 2955–2962 (2000)
19. Ayub, S., Ali, S.I., Khan, N.A.: Comparative study of different agro based adsorbents for the treatment of wastewater. Curr. World Environ. **1**(2), 109–116 (2006)
20. Baskaran, P.K., Venkatraman, B.R., Hema, M., Arivoli, S.: Adsorption studies of copper ion by low cost activated carbon. J. Chem. Pharm. Res. **2**(5), 642–655 (2010)
21. Veena Devi, B., Jahagirdar, A.A., Zulfiqar Ahmed, M.N.: Adsorption of chromium on activated carbon prepared from coconut shell. Int. J. Eng. Res. Appl. **2**(5), 364–370 (2012)
22. Nsami, J.N., Mbadcam, J.K.: The adsorption efficiency of chemically prepared activated carbon from cola nut shells by $ZnCl_2$ on methylene blue. J. Chem. **2013**, 1–7 (2013)
23. Sharma, P.K., Ayub, S.: The Cost Analysis and Economic feasibility of AGRO wastes to adsorb chromium(VI) from wastewater. Int. J. Civil Eng. Technol. (IJCIET). **10**(2), 2387–2402 (2019a)
24. Sharma, P.K., Ayub, S.: Economic feasibility study of horticultural wastes for chromium adsorption from tannery wastewater. Int. Water Resour. Assoc. (India) J. **8**(2), 51–58 (2019)
25. Mahmood, T., Aslam, M., Naeem, A., Siddique, T., Din, S.U.: Adsorption of As(III) from aqueous solution on to iron impregnated used tea activated carbon: equilibrium, kinetic and thermodynamic study. J. Chil. Chem. Soc. **63**(1), 3855–3866 (2018)
26. Mohan, D., Markandeya, Dey, S., Dwivedi, S.B., Shukla, S.P.: Adsorption of arsenic using low cost adsorbents: guava leaf biomass, mango bark and bagasse. Curr. Sci. **117**(4), 649–660 (2019)
27. Dada, A.O., Olalekan, A.P., Olatunya, A.M., Dada, O.: Langmuir, Freundlich, Temkin and Dubinin–Radushkevich isotherms studies of equilibrium sorption of Zn^{2+} unto phosphoric acid modified rice husk. IOSR J. Appl. Chemi. (IOSR-JAC) **3**(1), 38–45 (2012)
28. Mahesh, S., Rama, B.M., Praveen, K.N.H., Usha, L.K.: Adsorption kinetics of dihydric Phenol Hydroquinone on Activated Carbon. Indian J. Environ. Health. **41**(4), 317–325 (1999)
29. Sharma, D.C., Forster, C.F.: Removal of Hexavalent chromium using Sphagnummors peat. Wat. Res **27**(7), 1201–1208 (1993)
30. Rao, M., Parwate, A.V., Bhole, A.G.: Utilization of low cost adsorbents for the removal of heavy metals from wastewater. Env. Poll. Cont. J. **5**(3), 12–23 (2002)
31. Baisakh, P.C., Patnaik, S.N.: Removal of Hexavalent chromium from aqueous solution by adsorption on coal char. Ind. J. of Env. Hlth **4**(3), 189–196 (2002)
32. Gupta, S., Babu, B.V.: Adsorption of Cr(VI) using activated neem leaves: kinetic studies. Adsorption **14**(1), 85–92 (2008)
33. Sharma, P.K., PhD Thesis. Performance evaluation and feasibility study of the agro and horticultural wastes as adsorbent for the removal of chromium from tannery wastewater. July 2016 submitted to Dr. APJ AKTU, Lucknow, UP, India. (2016)
34. Sharma, P.K., Ayub, S.: The cost analysis and economic feaibility of AGRO wastes to adsorb chromium (VI) from wastewater. Int. J. Civil Eng. Technol. **10**(2), 2387–2402 (2019b)
35. Sharma, P.K., Ayub, S.: Economic feasibility study of horticultural wastes for chromium adsorption from tannery wastewater. Int. Water Resour. Assoc. (India) J. **8**(2), pp. 1301–2277 (2019)

Influence of the Packing Density of Fine Particles in Ternary, Quaternary and Quinary Blends on High Performance Concrete

Bhawani Singh, Akanksha Pathania, Mitali Gupta, Ankush Saini, and Abhilash Shukla

Abstract Concrete is a widely used construction material all over the world. Increasing the cement content in concrete leads to increase in the CO_2 emission which ultimately contributes to global warming. In order to increase the strength and durability of concrete without increasing the cement content becomes important. The solution to this problem lies in the packing density. According to the previous researches, for reduction in voids, ultra-fine pozzolanic materials should be added to form a dense cement paste to achieve a higher packing density. Addition of mineral admixtures can minimize the porosity of concrete and improve the microstructure, which leads to improve the mechanical properties of concrete. This paper presents an experimental approach for packing density measurement of ternary, quaternary and quinary mix. Mixtures are made with the inclusion of rice husk ash, fly ash, metakaolin, and ultra-fine slag as a replacement of Ordinary Portland Cement. Packing density of the ternary, quaternary and quinary cementitious blends were studied using the Punkte test method, wet packing test method and Relative Density Index (β) method respectively. Results indicated maximum packing density obtained by the replacement of cement by mineral admixtures varied between 25 and 30% in ternary, quaternary and quinary mixtures.

Keywords Particle packing · Relative density · Solid concentration · Wet packing density

B. Singh · A. Shukla (✉)
Department of Civil Engineering, JUIT Waknaghat, Solan, Himachal Pradesh, India
e-mail: abhilash.shukla29@gmail.com

B. Singh
e-mail: bhawani00015@gmail.com

A. Pathania
CHC Consultant Asia Pacific LLP, Mohali, India

M. Gupta
Ramboll India PVT. Ltd, Gurugram, India

A. Saini
Architectural and Structural Consultant, Panchkula, India

D. K. Ashish et al. (eds.), *3rd International Conference on Innovative Technologies for Clean and Sustainable Development*, RILEM Bookseries 29,
https://doi.org/10.1007/978-3-030-51485-3_31

1 Introduction

Concrete is the most utilized resource in the construction of buildings and infrastructure worldwide. "For achieving its high mechanical properties" researchers have used an increased amount of cement up to 1700 kg/m^3 [1]. Such increase in the amount of cement the production of cement increases rapidly, which causes environmental issues. The material is used so widely that the world cement production now contributes 20% of anthropogenic global CO_2 emission which causes global warming. For this reason cement industries ranked third for the emission of CO_2 [2, 3]. On the other hand, high cement content causes higher shrinkage which leads to increase the dimensional instability [4]. So to overcome its adverse effects, it turns necessary to reduce the amount of cement in concrete with an alternate that has cementitious property and also fills the voids between the particle without compromising the strength of concrete. Here comes the role of mineral admixture and fillers which possesses the cementitious property and also fills the voids. Another advantage of using mineral admixtures is that they undergo a pozzolanic reaction which increases the pore and interface refinement [5–7]. Packing density gives an alternate way to utilizing the potential of mineral admixtures by selecting the appropriate percentage replacement of cement rather than applying the hit and trial methods for the selection of material. Particle packing involves the selection of appropriate sizes and proportions of particulate materials to get a suitable combination for optimal packing. Packing density gives one possible way to reduce the adverse effect of cement by producing the Ultra-High-Performance concrete. Ultra-High-Performance concrete becomes more popular in the construction world, due to its extraordinary benefits, like high compressive strength, high tensile strength, high tensile ductility, high durability, and high dimensional stability [8–10]. It has been observed in the literature Lange et al. [11], Ming et al. [12] extensively investigated that the packing density plays an important role in the properties of the hardened paste. According to de Larrard et al. [13] it also affects the hydration process which leads to improving the microstructure and ultimately increases the performance of the concrete. This issue has been considered by most of the early researchers. They have worked on the particle size distribution and its effect on the packing density of cementitious material. For instance, de Larrard et al. [13] used limestone fillers, silica fume, cement and analyze the density of the mixture by using a linear model. Zhang et al. [14], Wang et al. [15], Aiqin et al. [16] investigated the relationship between cement and particle size distribution of pozzolanic material. They performed the test on the binary system and studied the water requirement and hydration rate of cement. Similarly, Fuller et al. [17] have investigated the importance of the particle size distribution of the aggregates and the properties of the concrete based on the packing of constituent materials. Youjun et al. [18] and Long et al. [19] analyzed the effectiveness of supplementary cementitious material in filling up the voids and volume of hardened paste with the help of Aim and Goff model. Peng et al. [20] and Niu et al. [21] explored the packing effect of ultra-fine powder with the help of the Andreasen equation. Several studies have examined the packing effect like Lu [22] and Long et al. [19], they investigated the packing effect of ultrafine fly ash on cementitious material paste with very low water to binder ratio. It has been concluded that the incorporation of ultra-fine powders with a variety of sizes helps to improve the packing property and also help to maximize the benefit of cementitious material [23, 24]. A comprehensive overview of various particle packing theories and methods can be found in [25–27]. Wong et al. [26] classified the packing method into the direct and indirect method. In indirect method void content and packing density was measured by calculating the bulk density of aggregate. This method conforms to British Standard BS 812: Part 2: 1995 and in case of indirect method void content is calculated concerning the water demand. Fennis [27] demonstrate the various methods which calculate the void content in term of water demand like Water demand (France), Puntke test (Germany), Mixing energy method, Proctor test, Centrifugal consolidation, Water demand method (Japan), Rheology—Krieger, and Dougherty. Li et al. [28] performed experiments on the dry and wet packing densities of concrete mixes for different combinations under different levels of compactions using bulk density method and evaluated the solid concentration of particles. Kumar et al. [29], Wille et al. [30], Dhinakaran et al. [31] optimized the mineral admixture content to get the optimal packing density and investigated the durability parameter and calculated the compressive strength without using the heat treatment for curing. Rashad [32] reviewed the previous studies carried out on the effect of quartz powder on properties of traditional cementitious materials and geopolymers. It can be concluded that in the development of the Ultra-High-Performance concrete, packing density is playing a more and more important role to produce sustainable concrete. This study aims to determine the replacement percentage of cementitious material by packing density methods to develop Ultra-High-Performance concrete with less amount of cement and to achieve high compressive strength. High-performance cementitious material containing ultra-fine slag, metakaolin, rice husk ash, fly ash has been used in this study. The effect of ultrafine slag, metakaolin, rice husk ash, fly ash on packing densities of ternary, quaternary and quinary cementitious material was investigated by Puntke test, Wet packing method, and Relative density index (β) method respectively. The aim of obtaining packing density is to fill the gap in particle size gradation for minimizing the voids content. Secondly, to determine optimized proportion of mineral mixture required to get high packing density. Influence of these mineral admixtures on the relative density of corresponding pastes with low water/binder ratios was also studied. Mixtures showing maximum packing density in the ternary, quaternary and quinary mixes were used to investigate the compressive strength.

Table 1 Specific gravity and chemical composition of materials

Material	SiO_2	Al_2O_3	Fe_2O_3	CaO	MgO	Na_2O_3	K_2O	SO_3	Specific gravity
OPC53	22.60	4.30	2.40	64.40	2.10	0.12	0.40	2.30	3.15
UFS	22	5.4	4.2	63	1.1	–	–	2.3	2.83
Fly ash	57.96	25.86	5.31	3.98	1.58	0.9	1.15	0.34	2.17
Metakaolin	57.10	34.46	3.94	1.24	1.28	0.3	0.08	–	2.5
Rice husk ash	93.70	0.30	0.20	0.60	0.40	0.20	1.40	0.10	2.53

2 Experimental Details

2.1 Material Used

Materials used: OPC 53 grade cement conforming to IS 12269 (2016). Fly Ash of class F confirming to IS 3812 (2013) obtained from Guru Gobind Singh super thermal power plant, Rupnagar, Punjab. Metakaolin confirming to IS 1727 (2018) procured from the Kaomin Industries LLP, Vadodra, Gujarat. Ultra-fine slag by the commercial name of ALCCOFINE 1203 was obtained from Counto Microfine Products Limited Goa conforming IRC:SP:70 (2016), IS 456 (2000), IS 12089:1987. Rice husk ash confirming to IS 3812 (2013) procured from KGR fusions Private Limited, Ludhiana Punjab. Quartz powder and Quartz sand was acquired from Surya Min Chem, Barwala, Delhi. Crushed stone sand has been used which was locally from Nangal, Punjab. Particle size distribution has been determined using laser particle size analyzer. Oxide composition of materials were determined by X ray fluorescence at IIT Bombay. Specific gravity was calculated by the Le-chatelier apparatus confirming to IS 5514 (2016). "Table 1 shows the specific gravity and chemical composition of materials" and in "Fig. 1 shows the particle size distribution of materials". "Table 2 shows the Blain surface area of cementitious material" and "Table 3 shows the properties of fine aggregate".

2.2 Experimental Program and Test Methods

The experimental program consists of two phases. The first phase involves determining the packing density for different combinations. In the second phase, compressive strength was determined for mixtures which showed maximum packing density. In the first phase ternary, quaternary and quinary mixes were finalized. Puntke method, Wet packing method, and Relative density index (β) method were adopted for the mixtures. In the case of ternary, quaternary and quinary blend maximum replacement of mineral admixture varied from 5 to 25%, 10% to 40% and 10% to 40% respectively. However, the mixtures which have shown the highest packing density are discussed. For ternary mixtures, Puntke method was adopted [33]. The

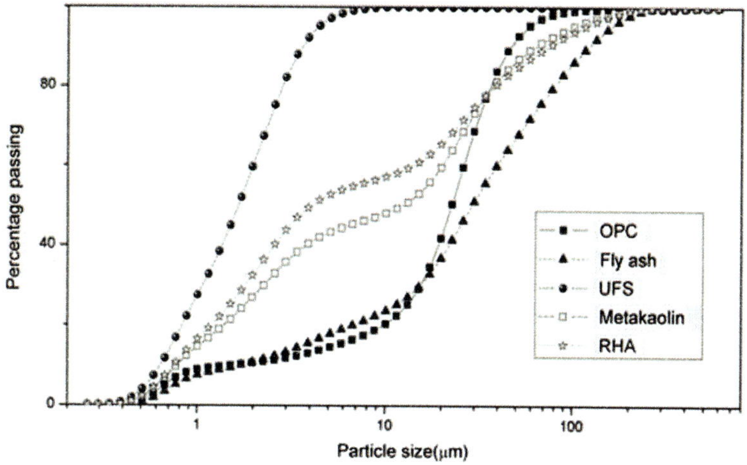

Fig. 1 Particle size distribution

Table 2 Blain surface area of cementitious material

Material	Blain surface area (m^2/kg)
OPC53	360
UFS	>1200
Fly Ash	410
Metakaolin	350
Rice husk ash	1315

Table 3 Properties of fine aggregate

Material	Specific gravity	Water absorption %
Quartz powder	2.65	Nil
Quartz sand	2.34	Nil
Crushed stone sand	2.6	1

basic principle of Puntke test is that the water added to the dry mixture fills the voids between the particles of the mixture and acts as a lubricant which leads to improve the compactness of the mixture. After filling all the voids excess water starts to appear on the surface which indicates the saturation limit. But the limitation of this method is that it relies on the water demand, but not on the apparent density and void content. For quaternary mixtures, method demonstrated by Wong et al. [26] for calculating the particle density, void ratio and void content of cementitious materials was adopted. This method does not depend upon the consistency or water demand. In this method, the calculations have been performed for various water to cementitious ratio. Solid concentration and void content were measured by determining the bulk density. The main advantage of this method is that air content is also taken into consideration and it also takes less time for mixing as compared to the indirect methods (like consistency method). In this case, the experimental investigation was carried out by increasing the percentage replacement of mineral admixtures from 5 to 40% and an increment of 5% in each step. In this study fly ash has been kept at lower percentages due to its higher particle size than other mineral admixtures. For quinary mixtures, the relative density index method was adopted. This method is based on the assumption that the air content of fresh paste is neglected and the minimum water required is determined which just fills the interspaces. In other words, it can be said that the volume of minimum water is equal to the volume of the interspaces. It is also based upon the relative density of the mixtures and the method of mechanical compaction [4, 20].

3 Results and Discussions

3.1 *Ternary Mixtures*

Partial replacement of cement was done on the mass basis with fly ash, rice husk ash, metakaolin and ultrafine slag respectively. The replacement was varied from 5 to 40%. About 230 combinations were assessed for achieving maximum packing density. Minimum and maximum packing density obtained was 0.53 and 0.58 respectively. "The packing density ranges with the maximum replacement of mineral admixture are shown in (Table 4)".

Results

Cement + Rice Husk Ash + Fly Ash (CRF)

Thirty-seven mixes combinations were assessed. The replacement of cementitious materials was varied from 5 to 40%. It has been observed that the packing density increased up to 25% replacement levels and then started to decrease as "Fig. 2a shows the variation in packing density with different replacement levels of cement". It is noted that the increase in fine content in the mix increased the specific surface area and as a result water demand increased. In other words, the interstitial water arrives at its minimum, the overall adsorbed water rises due to the large specific surface

Table 4 Packing density ranges with the maximum replacement of mineral admixture

Mix combination	% replacement of mineral admixtures at maximum packing density achieved	Range of packing density
CRF	25	0.54–0.57
CMU	10	0.55–0.58
CFU	25	0.55–0.58
CFM	5	0.54–0.56
CUR	15	0.53–0.57
CMR	5	0.53–0.56

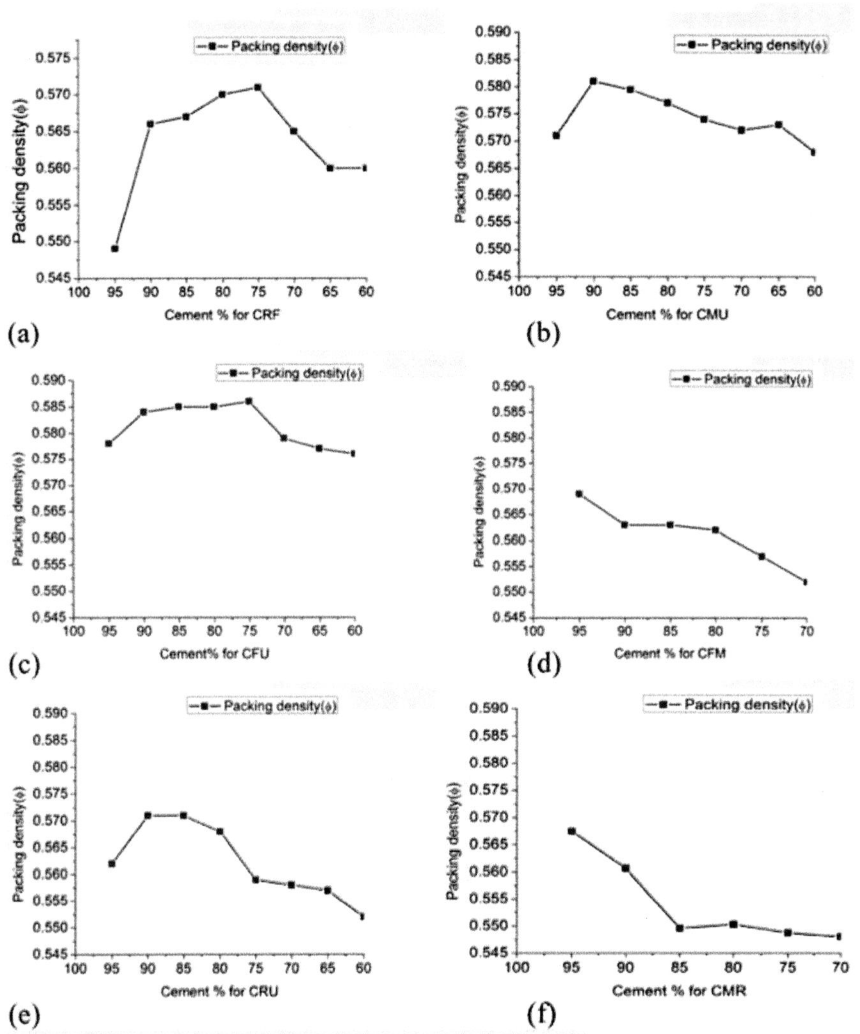

Fig. 2 Variation in packing density with different replacement levels of cement **a** CRF, **b** CMU, **c** CFU, **d** CFM, **e** CRU, **f** CMR

area of mineral admixtures. In this case, the w/b ratio ranged between 0.25 and 0.30. Increase in the water demand leads to the decrement in packing density. In this case, the packing density was observed in the range of 0.54–0.57.

Cement + Metakaolin + Ultrafine Slag (CMU):

Seventy combinations were studied with different percentages of metakaolin and ultrafine slag. Replacement levels up to 40% were investigated. Since the particle size of both metakaolin and ultrafine slag was similar, packing density was gradually decreased after 10% cement replacement as "Fig. 2b shows the variation in packing density with different replacement levels of cement". Here the w/b ratio ranged between 0.23 and 0.28 for the replacement levels up to 40%. The increased amount of water causes a decrease in packing density due to dispersion between the particles of the mix, causing the solid concentration to decrease. In this case, the packing density varied from 0.55 to 0.58.

Cement + Fly Ash + Ultrafine Slag (CFU):

Thirty-five combinations were investigated and the results indicated that the minimum water requirement gradually declines as the fine content increased to 25%. Since the void content decreased, resulting in increasing levels of compactness. With the further increase in fine particles resulted in increase of water demand and the packing density decreased as as "Fig. 2c shows the variation in packing density with different replacement levels of cement". Owing to the packing effect of mineral admixture, the packing density of (CFU) system increased and the interstitial water declines with increasing fly ash and ultrafine slag content. Hence, decreasing the minimum water requirement. w/b ratio range varied from 0.23 to 0.24 as the replacement of cement was varied from 5 to 40%. The range of packing density observed was in between 0.55 and 0.58.

Cement + Fly Ash + Metakaolin (CFM):

Thirty combinations were assessed and it was observed that in this case, particle packing density was decreased as the replacement level increased to 30%. Metakaolin has a very small particle size as compared to fly ash and cement. So, they did not pack well (increased water demand) and particle packing density was gradually decreased after 5% replacement of cement as "Fig. 2d shows the variation in packing density with different replacement levels of cement". The w/b ratio range was in the range of 0.24–0.29 as the replacement of cement varied from 5 to 30%. Packing density range observed was 0.54–0.56.

Cement + Rice Husk Ash + Ultrafine Slag (CRU):

It was noted that the particle packing density was increased up to 15% replacement of cement. It was due to the filling effect which decreased the void content and consequently, the water demand (i.e. amount of water needed to fill up the voids). After 15% replacement of cementitious material with cement, the packing density started to decrease as "Fig. 2e shows the variation in packing density with different replacement levels of cement". More water requirement causes a suspension in the system. In this case, the range of w/b ratio and packing density was in the range of 0.24–0.28 and 0.53–0.57 respectively.

Cement + Metakaolin + Rice Husk Ash (CMR):

Similarly, it was observed that packing density was gradually decreased as replacement of cement varied from 5 to 30%. After 5% replacement of mineral admixture with cement, packing density decreased as "Fig. 2f shows the variation in packing density with different replacement levels of cement". Metakaolin did not pack well to

Table 5 Packing density of different combinations

Combination	OPC 53 (%)	RHA (%)	Fly ash (%)	MK (%)	UFS (%)	w/b	Packing density (Φ)
CRF	75	16.25	8.75	0	0	0.25	0.57
CMU	90	0	0	2.5	7.5	0.22	0.58
CFU	75	0	12.5	0	12.5	0.24	0.58
CFM	95	0	2.5	2.5	0	0.24	0.56
CRU	85	7.5	0	0	7.5	0.24	0.57
CMR	95	2.5	0	2.5	0	0.24	0.56

give maximum packing density due to its small particle size as compared to rice husk ash. Effective packing can be attained by selecting proper proportions and particle size gradation to fill the voids between the larger particles. In this case, the w/b ratio range varied within 0.24–0.31 and the packing density varied within 0.53–0.56. "The combinations which yielded maximum packing density are shown in (Table 5)".

Discussion

"Packing density ranges with the maximum replacement of mineral admixtures as shown in (Table 4)". The maximum range of replacement was up to 25% in the case of the CRF and CFU and minimum 5% in the case of the CFM and CMR as "the variation of packing density with different replacement levels of cement (1) CFU, (2) CRF, (3) CMU, (4) CFM, (5) CRU, (6) CMR is shown in (Fig. 2)". This effect is seen due to the fact that water was not able to fill all the voids of the mix due to the agglomeration of fine particles and they also showed very less packing density as compared to CRF and CFU and on the other hand, combination CFU showed the maximum packing density of 0.58 with w/b ratio of 0.22, "Packing density of different combinations are shown in (Table 5)". CMU also showed similar values of packing density at lower replacement levels. Possible explanation is the fineness and ball-bearing effect of fly ash and lower particle size distribution of ultra-fine slag which leads to filling all the voids between the particle and constituted a dense paste.

3.2 *Quaternary Mixtures*

Partial replacement of ordinary Portland cement with ultrafine slag (UFS), metakaolin and fly ash was done with respect to mass. Packing densities were calculated in different percentages. Amongst all the combinations experimented, maximum packing densities were achieved at 30% replacement of cementitious material with cement. "Table 6 shows the mix proportion with the highest packing density". The water requirement (optimum water) for achieving highest packing density was different. The amount of fine material becomes important. The water requirement increased with an increase in percentage replacement of OPC and it

Table 6 Mix combinations (with maximum packing density)

Combination	OPC 53 (%)	UFS (%)	MK (%)	FA (%)	Packing density
CUMF34	70	12	10	8	0.694
CUMF36	70	10	11	9	0.689
CUMF37	70	11	10	9	0.687
CUMF35	70	13	9	8	0.687
CUMF32	70	10	12	8	0.686

Note C Cement, *U* Ultra-fine slag, *M* Metakaolin, *F* Fly ash and 34, 36, 37, 35, 32 show the sample number

was constant for 20–35% replacement of OPC as shown in "Fig. 3 packing density and water to solid ratio variation with percentage replacement of the OPC". In 85% of cement content, paste formation was observed at 0.26 water to solid ratio (15% replacement of cement was achieved at 0.26 water to solid ratio). Further decrement in cement content, the paste formation was noted 0.28 water to solid ratio. Packing density gradually increases till 30% replacement and then decreases as replacement percentage increases as shown in "Fig. 3 packing density and water to solid ratio variation with percentage replacement of OPC". Sixty combinations were tried for different replacement levels with mineral admixtures.

Discussion

"Highest packing densities of mix proportions are shown in (Table 6)". Results indicates that the maximum packing density of 0.69 was obtained for the combination which contains 70% cement, 12% ultrafine slag, 10% metakaolin, and 8% fly ash.

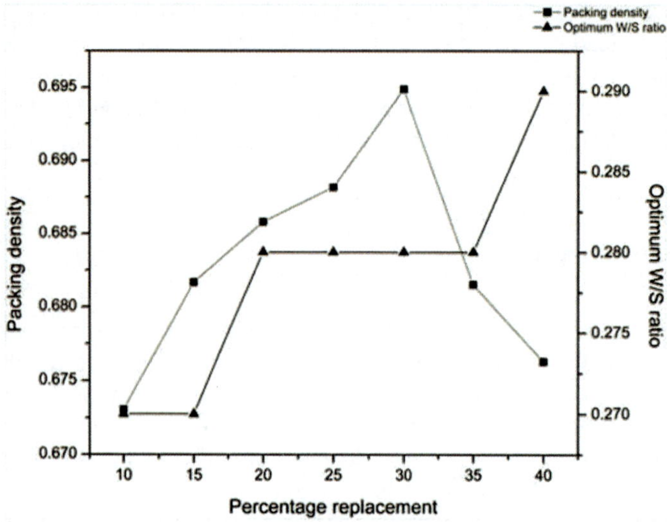

Fig. 3 Packing density and w/s ratio variation with percentage replacement of OPC

Packing density observed to increase as the percentage replacement of cement with ultrafine slag, metakaolin and fly ash increased. At low water to solid ratio (w/s), paste formation was difficult due to less water which did not provide enough lubrication layer between the particles. It is important to note that water added must be sufficient to fill up the voids for avoiding air entrapment in voids. On the other hand, at high water to solid ratio (w/s) the paste formed has a higher water content which formed a suspension and the particles are unnecessarily separated in the water, causing the solid concentration to decrease as the w/s ratio increases. In the case of optimum water to solid ratio (w/s) the packing density is maximum because it reaches its saturation point. At saturation point, particles are more closely packed, the voids are at the minimum level and the solid concentration reaches its maximum value. To obtain the maximum solid concentration, it is recommended to start with a low w/s ratio to high w/s ratio successively until the solid concentration has reached a maximum value. In this case, trials were started from the lowest water to solid ratio 0.25 for mix proportion with 90% cement content. To obtain a paste, the lowest water to solid ratio increased as the percentage of cement in the mix decreased. This is attributed to the increase in the fine content of the mix as they increase the water requirement, because of the increase in the specific surface area of the mixture.

3.3 *Quinary Mixtures*

Various trial mixes were prepared for the quinary mix by changing the proportions of the materials in the mixes. The replacement of the cement was done starting from 10% and up to 40% with the gradual decrease in the cement content of 5% in each step. In this case except cement, the percentages of the metakaolin and UFS were kept higher than the RHA because they have a smaller particle size which helps in achieving continuous particle-size gradation. Supplementary cementitious material with higher fineness is more effective because it would produce a broader range of particle size distribution [34, 35]. A total of 70 trials mixes was prepared and, five mixes which provided maximum packing densities are shown in the (Table 7)".

Table 7 Maximum packing density of trial combinations

Mix	OPC (%)	MK (%)	UFS (%)	RHA (%)	FA (%)	Relative density (β)
CMURF 44	70	8	9	7	6	0.720
CMURF 45	70	8	10	7	5	0.719
CMURF 46	70	8	11	6	5	0.718
CMURF 58	70	11	9	6	4	0.717
CMURF 63	65	11	11	7	6	0.717

Discussion

Particle packing test for one combination shows that the packing density initially increases with the increase in the water to binder ratio and reaches to a maximum and with a further increment increase in water to binder ratio a dip is observed. It is due to the increase in solid concentration. Volume of voids decreases filling all the voids and giving maximum value of the packing density. “Fig. 4 shows the variation in the relative density with the w/b ratio for the combination CMURF 44” and “Fig. 5 shows the variation of the Percentage cement replacement versus Relative density”. The packing density of the combinations with the 10% replacement of the cement gives the minimum value of 0.692. 30% replacement levels provided the maximum packing density of 0.72 which is highest among all combinations and further replacement showed the slump in the density values. “Table 7 shows the values of the maximum packing density obtained from all the combinations tested”.

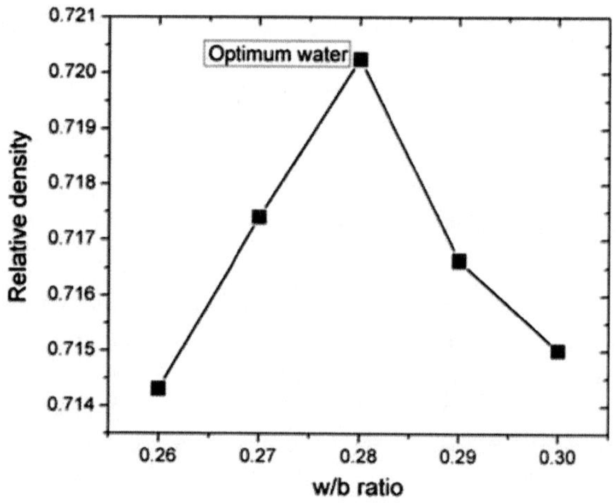

Fig. 4 Variation in the relative density to w/b ratio for CMURF44

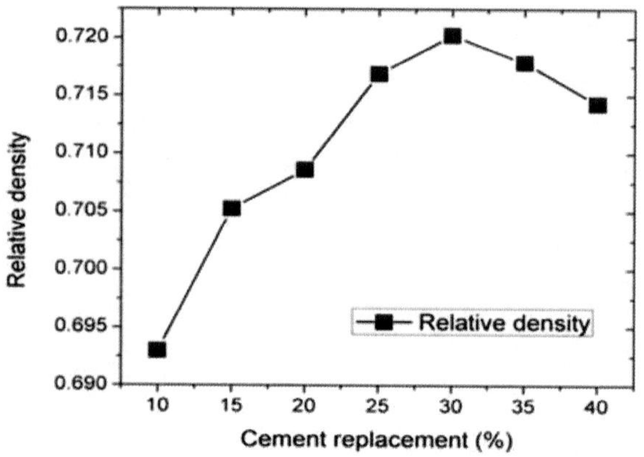

Fig. 5 Percentage cement replacement versus relative density for CMURF44

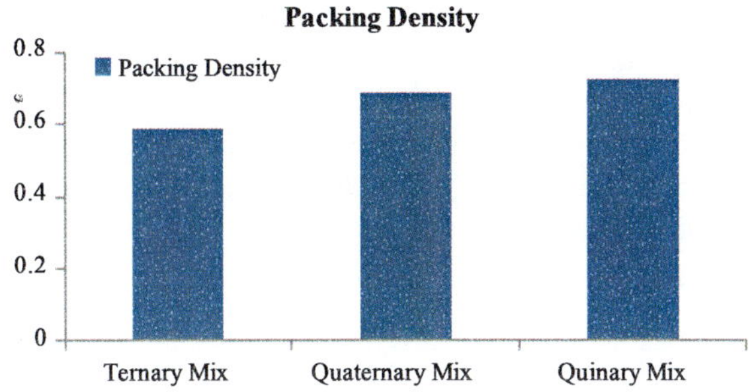

Fig. 6 Comparison of ternary, quaternary and quinary mix to packing density result

The water requirement increased slightly with the increase in the quantity of the fines in the combination corresponding to the maximum relative density for a combination. Relative density showed an increment trend up to 30% replacement levels. The maximum value of the relative density obtained was 0.72 of the mix which contains 70% cement, 8% metakaolin, 9% ultra-fine slag, 7% RHA, and 6% fly ash content. With the increase in the fine content, the requirement of water has also increased by a small amount. Result emphasized that the "packing density of the quinary mixtures showed higher than the ternary and quaternary mixtures as shown in (Fig. 6)". Ternary mixtures have shown the lowest packing density among all mixtures. Packing density is increased as the number of fine particles increases as they cover the wide range of the particle sizes which leads to decrease in the number of voids in the mixture. It is due to the fact that with a broader range of particle size distribution, medium sized particles would fill up the voids between the larger particles, the fine particles would fill up the voids between the medium sized particles and very fine particles would fill up the voids between the fine particles and so on. This phenomenon results in the removal of voids by the successive filling effect. It can be said that by blending fine materials continuous particle size gradation can be obtained which results in increment of packing density. The same trend has been observed by Kwan et al. [36]. So maximizing the packing density of aggregate particles will reduce the amount of cement in concrete production and it supports sustainable development.

3.4 Mix Proportions and Compressive Strength

In the second phase, compressive strength was calculated for each accepted trial mix which showed the maximum packing density in ternary, quaternary and quinary blends. The procedure followed for mixing of the materials is based on the traditional techniques and recommendation, steps are as follows (1) Materials were weighed according to the mix design prepared, (2) All the constituents were mixed in the dry state with the help by using electrically driven high energy mechanical mixer IS 10890 (2018) of the epicyclic type (imparting both planetary and a revolving motion to the mixer paddle) for approximately 2 min, (3) 50% water (calculated water according to water to cementitious ratio) and 50% superplasticizer were added to the mixer and mixed for another 2 min, (4) Remaining portion of superplasticizer and water were added and mixed for another 2 min at high speed (5) Materials were mixed at for 2.5 min, (6) Further, mixing was carried out for about 1.5 min at the higher speed. Till this point of time, thick paste was formed for casting. Concrete compressive strength was measured using $70.7 \times 70.7 \times 70.7$ mm^3 cube specimens, conforming to Indian Standard Specifications IS 4031-part 6 (2019). Compressive strength was taken average of three specimens. Moulds were placed on the vibrating table. Addition of concrete in mould was done in three phases by filling one third of the mould and vibrating. This process continued till the filling of moulds with concrete. Vibration for the expulsion of air continued for minimum two minutes. After casting of specimens, the specimens were covered with plastic sheet to avoid the loss of moisture. Cubes were taken out from the moulds after 24 h. Specimens were cured in a water tank at temperature of 20 ± 5 °C. Compressive strength was determined at the end of 28 days of curing. The main components consist of cementitious paste and sand. The sand used in this study is of two types, quartz sand and crushed stone sand. Quartz powder is used to further increase the density of the mix. The mix design has been prepared for the mixes having the highest packing densities by keeping cement content at 1100 kg/m^3. Coarse aggregates were not used in the mixes because the basic concept for obtaining HPC is to make the matrix as dense as possible and avoid weak zones (ITZ). The densification is possible by the complete elimination of coarse aggregates. Although the maximum size of the particles is ~600 µm. "Mix ratio by mass shown in Table 8 for the ternary, quaternary and quinary mix". "The compressive strength results for the ternary, quaternary and quinary mix are presented in Table 9".

Component of concrete mix design and mix ratio by mass (Cement content 1100 kg/m^3)

Mix. No	C	UFS	MK	FA	RHA	QS	CSS	QP	SP	W
CFU	1	0.166	0	0.166	0	0.187	0.212	0.022	0.013	0.275
CUMF 34	1	0.171	0.142	0.114	0	0.167	0.185	0.019	0.011	0.266
CMURF 44	1	0.128	0.114	0.085	0.099	0.167	0.185	0.019	0.013	0.265

Note *C* Cement, *UFS* Ultrafine slag, *MK* Metakaolin, *FA* Fly ash, *RHA* Rice husk ash, *QS* Quartz sand, *CSS* Crushed stone sand, *QP* Quartz powder, *SP* Superplasticizer, *W* water

Table 9 Compressive strength and curing age

Mix. No	Packing density	Mix. type	Compressive strength (N/mm^2)	Curing age (Days)
CFU	0.58	Ternary	85	28
CUMF 34	0.69	Quaternary	94.13	28
CMURF 44	0.72	Quinary	100.45	28

Compressive strength of the cubes was lower than as expected because the temperature of the curing tank was quite low, resulting in incomplete hydration and underutilized potential of the mineral admixture as lower the hydration of cement, the lower is the production of $Ca(OH)_2$ which is necessary for the pozzolanic reaction of mineral admixtures. Therefore, during the hydration process, the hydration products of the cement particles need to bridge a larger distance, eventually leading to lower strengths and strength enhancement criteria is also depends upon the contribution of the pozzolanic reaction effect of mineral admixtures. So if the proper curing is employed, these materials can react with the products of cement hydration to produce additional gels which would contribute to strength enhancement and durability. Although the maximum compressive strength of 100.45 MPa was achieved by the quinary sample which posses the packing density of 0.72. In the case of the quaternary sample, it showed the maximum strength of 94.13 MPa and in the case of the ternary sample, it showed the least strength of 85 MPa. It may be because of the number of voids may remain in the mix. Failure of the cubes observed was more of brittle failure than the D shaped failure as observed in the conventional concrete.

Form the first phase it can be observed that, it is possible to reduce the cement content by up to 30% by using particle packing methods. Alternative pozzolanic materials are required to be used in order to improve the properties of concrete and reduce the amount of cement used in concrete production. In the second phase, it can be observed that by using the same amount of reduced cement in the concrete mix showed higher compressive strength. Higher concrete strength can minimize the required member size, which contributes to a decrease in CO_2 emissions by using a smaller amount of concrete. As mentioned by KORE SD [37] adopting a packing density approach for the development of concrete mixes would reduce the annual global cement production from 4.2 billion tons by 0.51 billion tons and CO_2 release from 3.95 billion tons to 3.47 billion tons. From the above study, it can be concluded that adopting a packing density approach for the development of concrete mixes would reduce the annual global cement production and help to produce sustainable concrete, which ultimately mitigating environmental pollution to a large extent.

4 Conclusions

The experimental results presented in this paper verified that the dense packing effect of fly ash, ultrafine slag, metakaolin, rice husk ash is highly important for high strength of concrete and also helps to achieve the sustainable approach to the development of concrete. Specific findings of this research include the following:

1. The indirect method for quaternary and quinary paste is not reliable. Since it focuses on water demand and not on the void content, bulk density, and compaction. Instead of using the indirect method, the wet packing method and relative density index methods are better options for quaternary and quinary mixes respectively.

2. Ternary samples showed the least packing density values as compared to quaternary and quinary samples. In the ternary sample, maximum packing density was achieved 0.58 for a combination in which Cement (75%) + Fly ash (12.5%) + Ultrafine Slag (12.5%) was used.
3. In the quaternary case, replacement results emphasized that when the cement and ultra-fine slag are constant packing density increases as fly ash content increases. The maximum packing density was achieved 0.69. It obtained from a combination in which Cement (70%) + Ultrafine slag (12%) + Metakaolin (10%) + Fly ash (8%) was used.
4. In the case of quinary samples, 30% replacement of the cement with additional cementitious material showed the maximum relative density and the maximum value of the relative density obtained was 0.72 of the mix which contains 70% cement, 8% metakaolin, 9% ultrafine slag, 7% RHA and 6% fly ash content.
5. Mixture combinations that cover the wide range of particle sizes achieved the maximum packing density.
6. The particle packing of the material followed a trend in which initially the packing increased with the increases in the replacement of cement with and after that packing density started to decrease.
7. Results emphasized that the maximum replacement of mineral admixture with cement up to 30% in the case of quinary and quaternary and 25% in the case of ternary is sufficient to fill the maximum voids and further increase or decrease results in the lower packing.
8. An increase in the percentage of fine content leads to an increase in the water demand due to an increase in the specific surface area of the mixture.
9. Maximum compressive strength of 100.45 MPa was achieved by the quinary sample and in the case of the quaternary sample, it showed the maximum strength of 94.13 MPa. The ternary sample showed the least compressive strength of 85 MPa as compare to quaternary and quinary samples. 1100 kg/m^3 of cement content was used for ternary, quaternary and quinary samples for the mix design.
10. Maximum packing of particles formed a dense structure which leads to improving the packing density of structure and ultimately gives the high compressive strength.
11. If the curing of the specimen is carried out at an elevated temperature, these concrete mixes are expected to provide higher strength.

References

1. Neville, A., Aitcin, P.C.: High performance concrete—an overview. Mater. Struct. **31**(2), 111–117 (1998)
2. Anand, S., Vrat, P., Dahiya, R.P.: Application of a system dynamics approach for assessment and mitigation of CO_2 emissions from the cement industry. J. Environ. Manag. **79**(4), 383–398 (2006)

3. Andrew, R.M.: Global CO_2 emissions from cement production. Earth Syst. Sci. Data **10**(1), 195 (2018)
4. De Larrard, F.: Concrete Mixture Proportioning–A Scientific Approach. E & FN Spon. New York (1999)
5. Feng, N., Shi, Y., Hao, T.: Influence of superfine mineral powder on the fluidity and strength of cement paste. J. Shandong Inst. Build. Mater. **12**(S1), 103–109 (1998)
6. Goldman, A., Bentur, A.: The influence of microfillers on enhancement of concrete strength. Cem. Concr. Res. **23**(4), 962–972 (1993)
7. Tan, K.-f., Pu, X.-c.: Strengthening mechanism of mineral admixtures to concrete. J. Southwest Univ. Sci. Technol. **3** (2007)
8. Neville, A.M. Properties of Concrete, 4th edn. Longman (1995)
9. Wong, Henry, H.C., Kwan, A.KH: Packing density: a key concept for mix design of high performance concrete. In: Proceedings of the Materials Science and Technology in Engineering Conference, HKIE Materials Division, Hong Kong, pp. 1–15 (2005)
10. Vejmelková, E., Pavlíková, M., Keršner, Z., Rovnaníková, P., Ondráček, M., Sedlmajer, M., Černý, R.: High performance concrete containing lower slag amount: a complex view of mechanical and durability properties. Constr. Build. Mater. **23**(6), 2237–2245 (2009)
11. Lange, F., Mörtel, H., Rudert, V.: Dense packing of cement pastes and resulting consequences on mortar properties. Cem. Concr. Res. **27**(10), 1481–1488 (1997)
12. Ming, T., Jiachun, W., Lianjun, Li.: Research on fractal characteristics of concrete materials pore with MIP. J. Shenyang Archit. Civil Eng. Univ. (Nat. Sci.) **17**(4), 272–275 (2001)
13. de Larrard, F., Thierry, S.: Optimization of ultra-high-performance concrete by the use of a packing model. Cement Concr. Res. **24**(6), 997–1009 (1994)
14. Zhang, C., Wang, A., Tang, M., Liu, X.: The filling role of pozzolanic material. Cem. Concr. Res. **26**(6), 943–947 (1996)
15. Wang, A., Zhang, C., Zhang, N.: Study of the influence of the particle size distribution on the properties of cement. Cem. Concr. Res. **27**(5), 685–695 (1997)
16. Aiqin, W., Chengzhi, Z., Ningsheng, Z.: The theoretic analysis of the influence of the particle size distribution of cement system on the property of cement. Cem. Concr. Res. **29**(11), 1721–1726 (1999)
17. Fuller, W.B., Thompson, S.E.: The laws of proportioning concrete. Trans. Amer. Soc. Civ. Engrs. Bd **59** (1907), 67
18. Youjun, X., Baoju, L., Guangcheng, L.: Study on dense packing properties of cementitious materials. J. Chin. Ceram. Soc. **29**(6), 512–517 (2001)
19. Long, G.-c., Wang, X.-y., Xiao, R.-m.: Research of filling role of mineral blends in C3S cementitious system. J. Build. Mater. **3** (2002)
20. Peng, Y., Shuguang, Hu., Ding, Q.: Dense packing properties of mineral admixtures in cementitious material. Particuology **7**(5), 399–402 (2009)
21. Niu, Q., Naiqian, F., Jing, Y.: Packing of superfine mineral powder in cement. Guisuanyan Xuebao (J. Chin. Ceram. Soc.) (China) **32**, 102–106 (2004)
22. Lu, H.G.: Introduction to Powder Technology. Tongji University Publisher, Shanghai (1998)
23. Richard, P., Cheyrezy, M.: Reactive powder concretes with high ductility and 200–800 MPa tensile strength. In: San Francisco: ACI Spring Convention, SP, pp. 144–24 (1994)
24. Richard, P., Cheyrezy, M.H.: Reactive powder concrete. Cem. Concr. Res. **25**(7), 1501–1511 (1995)
25. Kumar, S.E.N.T.H.I.L., Manu S.: Particle packing theories and their application in concrete mixture proportioning: a review. Indian Concr. J. **77**(9), 1324–1331 (2003)
26. Wong, Henry, H.C., Kwan, A.K.H.:. Packing density of cementitious materials: part 1—measurement using a wet packing method. Mater. Struct. **41**(4), 689–701 (2008)
27. Fennis, S.A.A.M.: Measuring water demand or packing density of micro powders: comparison of methods (2008)
28. Li, L.G., Kwan, A.K.H.: Packing density of concrete mix under dry and wet conditions. Powder Technol. **253**, 514–521 (2014)

29. Kumar, S., Acharya, G., Mhamai, S.: Reactive powder concrete with mineral admixtures. J. Emerg. Technol. Innov. Res. **2**(6), 1749–1757 (2015)
30. Wille, K., Naaman, A.E., Parra-Montesinos, G.J.: Ultra-high performance concrete with compressive strength exceeding 150 MPa (22 ksi): a simpler way. ACI Mater. J. **108** (1) (2011)
31. Dhinakaran, G., Thilgavathi, S., Venkataramana, J.: Compressive strength and chloride resistance of metakaolin concrete. KSCE J. Civil Eng. **16**(7), 1209–1217 (2012)
32. Rashad, A.M.: Effect of quartz-powder on the properties of conventional cementitious materials and geopolymers. Mater. Sci. Technol. **34**(17), 2043–2056 (2018)
33. Puntke, W.: Wasseranspruch von feinen Kornhaufwerken. Beton-Dusseldorf- **52**(5), 242–249 (2002)
34. Kwan, A.K.H., Wong, H.H.C.: Packing density of cementitious materials: part 2—packing and flow of OPC+ PFA+ CSF. Mater. Struct. **41**(4), 773 (2008)
35. Chen, J.J., Kwan, A.K.H.: Triple blending with fly ash microsphere and condensed silica fume to improve performance of cement paste. J. Mater. Civ. Eng. **25**(5), 618–626 (2013)
36. Kwan, A.K.H., Fung, W.W.S.: Packing density measurement and modelling of fine aggregate and mortar. Cement Concr. Compos. **31**(6), 349–357 (2009)
37. Kore, S.D., Vyas, A.K.: Packing density approach for sustainable development of concrete. J. Mater. Eng. Struct. «JMES» **4**(4), 171–179 (2017)

Experimental Investigation of Rheological Properties of Recycled Aggregate Concrete

B. Suguna Rao, Sayyed Ibrahim uz Zaman, and Srikanth M. Naik

Abstract Increasing construction activity has increased the demand for construction materials, and concrete is one of the most widely used construction materials. Concrete is popular for its property of mold-ability and fluidity at early stages. The fluidity of concrete can be characterized by the study of its flow behavior in fresh state. The rheological behavior of recycled aggregate concrete is affected by superplasticizer and mineral admixtures. Hence, this study is carried out with the inclusion of mineral and chemical admixtures. Plastic viscosity of concrete is predicted using models which are essential for concrete production. An attempt is done to study the compressive strength of concrete by replacing the natural coarse aggregates by recycled coarse aggregates. After understanding the compressive strength of various replacement ratios, an optimum replacement ratio is identified using Taghuchi method. Using the Taghuchi method—which is based on principles of design of experiments—the optimum replacement ratio was found to be 45% [2]. The rheological properties were studied using direct shear box test to obtain the corresponding yield stress and plastic viscosity of concrete made from natural coarse aggregates and concrete made from recycled coarse aggregates.

Keywords Recycled aggregate concrete (RAC) · Yield stress (τ0) · Plastic viscosity (μ) · Rheology

B. S. Rao (✉) · S. I. Zaman · S. M. Naik
Department of Civil Engineering, MSRIT (Autonomous Institute, Affiliated To VTU), Bangalore, India
e-mail: suguna_rao@msrit.edu

S. I. Zaman
e-mail: sayyed.zaman@gmail.com

S. M. Naik
e-mail: srikanth_naik@yahoo.com

D. K. Ashish et al. (eds.), *3rd International Conference on Innovative Technologies for Clean and Sustainable Development*, RILEM Bookseries 29,
https://doi.org/10.1007/978-3-030-51485-3_32

1 Introduction

1.1 General

With tremendous advancements in the field of construction, it has become a necessity to maintain sustainability. Most suitable way to achieve this would be to use alternative building materials obtained from construction and demolition waste. The population explosion has increased the construction activities, resulting in an increased requirement of construction materials, concrete being one of the important construction material which is in high demand. Meanwhile, this increased activity has also increased the quantity of construction and demolition waste, creating a threat to the environment. The recycle and reuse of such construction and demolition waste materials will yield a tangible impact in protecting the environment [2]. To meet the challenge of rapid growth in urbanization and scarcity of natural aggregates, demolition waste can be utilized without compromising the structural strength of the concrete.

Most of the governments across the world have made it compulsory to assess the impact of construction activities on the environment. Hence it becomes our responsibility as civil engineers to ensure safety and satisfy all the required design criteria without causing a negative impact on the environment [4]. One of the best ways to achieve sustainable growth in construction would be to use alternative building materials in the place of conventional materials. Continuous construction activities and demolition of structures leads to accumulation of construction and demolition waste, which in addition to causing the depletion of land and destroying its landscape also generates debris which must be disposed of somewhere safely. Also, as the natural aggregates required for concrete production is depleting, we can use demolished concrete wastes as aggregates in new concrete.

At present, a wide range and variety of materials are being used to produce conventional concrete due to increase in the demand on natural aggregates [5]. Earlier researchers have worked on understanding the behavior of recycled aggregate concrete in its hardened state, but not much research has focused on understanding the rheological properties of recycled aggregate concrete. Hence the behavior of recycled aggregate concrete needs to be studied to understand specific parameters that influence its rheological behavior [14]. Rheological study includes research on the flow property of fresh concrete and deformation of hardened concrete. Advancement in the field of concrete technology has altered the requirement of basic slump test on concrete for measuring its workability while using concrete for special purposes [1]. Water absorption of recycled coarse aggregates are higher compared to natural coarse aggregates. The characterization of rheological parameters are based on the correlation between shear stress and shear strain. The key parameters that play a vital role in obtaining workable concrete are plastic velocity and yield stress. The previous literature proposes models and equations to present the relationship between shear stress and shear strain of flow materials, among them the Bingham model is the most popular one [7]. The yield stress obtained describes the minimum stress needed to initiate the flow. Plastic viscosity signifies the resistance to flow when there is increase in yield stress. As the plastic viscosity increases there is a greater resistance to flow.

Rheometers are the most popular devices used to evaluate the rheological parameters which work on the response to applied forces which are dynamic in nature. The rheometers are classified into two types: one which controls the applied shear stress or shear strain known as rotational or shear rheometers and the other which applies extensional stress or extensional strain known as extensional rheometers. But these are expensive devices and difficult to obtain [8]. Hence this study is carried out using recycled coarse aggregates to explore the rheology of concrete to make a comparison with that of natural coarse aggregates in a more sustainable and economic manner. The aim of this study is to arrive at a standard mix proportion for specialized use of recycled coarse aggregates in concrete to produce workable concrete with desired strength and durability. Due to the various parameters involved and their variability, different procedures are developed to arrive at the right mix. The conventional mix design cannot be directly applied for mix proportioning, hence ACI method of mix design is adopted with a number of trial mixes to arrive at a suitable mix proportion through experimentation. Based on the rheological parameters it will be easier to characterize the recycled aggregate concrete to be suitable for desired strength and flow properties.

1.2 Concrete Shear Box

Rheological parameters are usually studied using rheometers. Rheometers are expensive and are not easily available [1]. Hence concrete shear box was found to be the most suitable device to study rheological parameters. The characterization of rheology is based on the correlation between shear stress and shear rate. Concrete shear box is a device developed to understand the rheological properties of concrete and gives similar results as rheometers [1]. Shear box tests are static tests which are strain controlled. Basically, it is a strain-controlled device in which normal stress is applied initially. The three set of normal stresses are subjected to fixed value of displacement rate (shear strain) for which shear stresses are obtained. Peak value of shear stress for the respective shear stress and shear strain are noted. The tests were repeated for three different displacement rates, keeping the normal stress constant. The value of shear stress is evaluated for zero normal stress. From the graph of Bingham model, we arrive at relative yield stress and relative plastic viscosity. Concrete shear box apparatus used to study rheological parameters is shown in Fig. 1

Fig. 1 Concrete shear box apparatus

2 Experimental Program

Concrete mixes were prepared using natural coarse aggregate and recycled coarse aggregate to arrive at high strength recycled aggregate concrete. From the previous literature it was noted that extensive research has been carried out on the utilization of recycled aggregates for normal strength concrete, hence in this research an effort is made to focus on achieving high strength concrete accommodating recycled coarse aggregate as partial replacement of natural coarse aggregate to arrive at structural concrete [2, 16, 18]. The specimens cast were tested and cured for 7 days and 28 days. A total of 72 cube specimens were cast to the size of 150 mm × 150 mm × 150 mm. Taguchi method based on principles of design of experiments was applied to arrive at optimum values, and it was found that optimum replacement ratio of recycled coarse aggregate was 45% [2]. Slump test was performed on both conventional concrete as well as optimum replacement recycled aggregate concrete. Since slump test gave limited values which was almost the same for both mixes, concrete shear box test was performed to evaluate rheological properties of recycled aggregate concrete.

2.1 Materials Specifications

Cement: OPC Ordinary Portland Cement of 43 grade.

Fine Aggregate: Manufactured sand of 4.75 mm downsize is used due to scarcity of natural river sand.

Coarse aggregate: Natural coarse aggregate and recycled coarse aggregate of 12.5 mm downsize.

Chemical Admixture: Conplast SP 430 DIS superplasticizer.

Mineral Admixture: Silica fume with specific gravity of 2.2.

Table 1 Properties of cement

Property	Result
% fineness	5%
Specific gravity	3.05
Standard consistency	33%
Setting time	**Initial set—40 min** **Final set 5 h 45 min**

Table 2 Test results of normal coarse aggregates

Properties		Result
Fineness modulus		6.25
Impact		18.57%
Crushing		22.21%
Water Absorption		2.47%
Specific gravity		2.65
Abrasion value		23.50%
Soundness		8.90%
Shape		Angular
Size		12.5 mm
Bulk density	Compact state	1.58 kg/l
	Loose state	1.47 kg/l

Water: Portable water.

Cement was tested as per the provisions of per IS 12269–1987. The tests conducted on Cement, coarse aggregate are shown in Tables 1, 2, and 3. The tests were conducted to understand the engineering properties of materials and are utilized in this study.

Table 3 Test results of recycled coarse aggregates

Properties		Result
Fineness modulus		5.45
Impact		24.90%
Crushing		24.60%
Water Absorption		4.42%
Specific gravity		2.46
Abrasion value		32%
Soundness		11.60%
Shape		Angular
Size		12.5 mm
Bulk density	Compact state	1.512 kg/l
	Loose state	1.428 kg/l

The recycled coarse aggregates used in this project were obtained from Rockwell Crystal Crushing unit, Bangalore and were tested for engineering properties. The properties of recycled aggregate concrete tested are shown in Table 3.

2.2 *Mix Design Methodology*

The mix design methodology adopted was based on Perumal's method of mix design for M60 grade of concrete [16]. The mix proportions obtained according to the mix design are tabulated in Table 4. Perumal's mix design is based on ACI method.

2.3 *Slump Test*

Optimum dosage of superplasticizer was arrived at by conducting slump test for trial mixes. Initially, slump tests were carried out by increasing the dosage up to 5% and observing their slump value. For a dosage of 5% it is observed that slump value of 40 mm is arrived at. From the further study on recycled aggregate concrete it was noted by the authors that optimum replacement of recycled coarse aggregate was 45% [2]. Further, this study aims to understand the rheological behavior of recycled aggregate concrete and arrive at a good correlation with the properties of recycled coarse aggregates. From this study the authors have attempted to study the rheological behavior of recycled aggregate concrete for the optimum replacement ratio of recycled coarse aggregate. Optimum dosage of super plasticizer is arrived for 0 and 45% replacement ratio of recycled aggregate concrete by conducting trial mixes.

Slump test was performed on both soaked and unsoaked condition of recycled aggregate concrete on both 0 and 45% replacement of coarse aggregates to determine the optimum dosage of super plasticizer and the condition of recycled aggregates to be used in the project. Both 0 and 45% replacement of coarse aggregates gave the same results and are shown in Tables 5 and 6.

This work was carried out by considering the pre-soaking condition of recycled aggregates as the water absorption of recycled aggregates are more and hence consumption of super plasticizer is also high. The aggregates were soaked by keeping it in water for a period of not more than 24 h. Before adding it to the mix, the aggregates were surface dried (with the help of a cloth) to maintain the water content required for the mix.

Table 4 Mix proportions for various replacement percentages of coarse aggregates

Material (kg/m^3)	Percentage replacement of NCA by RCA (%)											
	0	10	20	30	40	45	50	60	70	80	90	100
Cement	500	500	500	500	500	500	500	500	500	500	500	500
Silica fume	55.55	55.55	55.55	55.55	55.55	55.55	55.55	55.55	55.55	55.55	55.55	55.55
Sand-FA	645.6	635.6	624.7	613.7	602.7	597.3	591.8	580.8	569.9	559.0	548.0	537.1
NCA	1000	900	800	700	600	550	500	400	300	200	100	0
RCA	0	100	200	300	400	450	500	600	700	800	900	1000
Water	150	150	150	150	150	150	150	150	150	150	150	150
SP (2.5%)	12.5	12.5	12.5	12.5	12.5	12.5	12.5	12.5	12.5	12.5	12.5	12.5

Table 5 Slump values of recycled aggregates for unsoaked condition

Dosage of SP in %	1	2	3	4	5
Slump (mm)	0	0	0	20	40

Table 6 Slump values for soaked condition of recycled aggregates

Dosage of SP in %	1	1.5	2	2.5
Slump value (mm)	0	0	30	40

2.4 Compression Test

The concrete mixes were cast into cubes of size 150 mm × 150 mm × 150 mm and were kept for curing for 7 days and 28 days. All the ingredients were mixed with the help of a pan mixer. Compression test was performed on all the cubes after curing for 7 days and 28 days and are as shown in Fig. 2.

Based on the strength test results the replacement percentage of 45% was found to be optimum [2]. Further to understand the rheological behavior of recycled aggregate concrete and predict its plastic viscosity and yield stress, concrete shear box test was proposed for concrete with 0% replacement and concrete with 45% replacement. This test will help in understanding the performance of recycled aggregate concrete for the optimum replacement ratio.

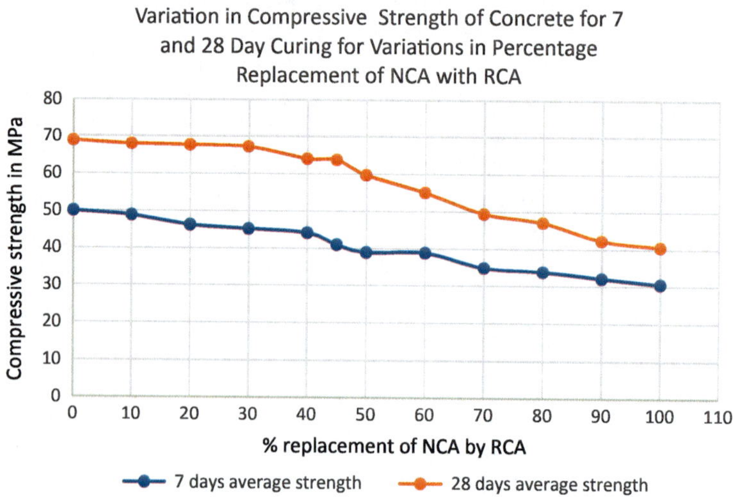

Fig. 2 Compressive strength for 7 days and 28 days with Variations in percentage replacement of coarse aggregate

2.5 Concrete Shear Box Test

Slump tests conducted for 0% replacement and 45% replacement achieved almost the same value with the optimum dosage of super plasticizer. Recycled aggregates were used by pre-soaking them, to avoid the stiffness of the mix. Further slump test is carried out for proposed M60 grade of concrete with 0 and 45% replacement. Shear box tests were conducted by maintaining constant shear strain and for various normal stresses. The tests were carried out for different normal stresses keeping the displacement strain constant. Ultimate shear stress value with respect to normal stress is studied. These shear stress value will help to arrive at yield stress and plastic viscosity which are Bingham parameters. In this study three different displacement rates 0.25, 0.62 and 1.25 mm/min in combination with three different sets of normal stresses 0.025, 0.05 and 0.10 MPa were used. Concrete mix with 0% replacement is considered as Mix 1 and 45% replacement is considered as Mix 2. Concrete mixes were prepared and are poured in shear box of size 150 mm × 150 mm × 150 mm in three layers, each layer being tamped 25 times using tamping rod. A typical plot of shear stress versus shear strain was plotted that indicates increase in shear stress increases shear strain but after reaching a peak value there is decline. From this peak stress is arrived for a specific strain rate and normal stress. Refer to Figs. 3, 4, 5 and 6 and for Mix 1 and Figs. 7, 8, 9 and 10 for Mix 2. Further tests were conducted for different displacement rates keeping normal stress constant for which peak shear stresses were obtained. These peak shear stresses and normal stresses are considered to plot graph for each displacement rate. The y-axis intercept of the graph indicates intrinsic shear strength at zero normal stress [1]. Further peak shear stress was obtained for zero normal stress for different displacement rates. The Y-axis intercept gives the yield stress at zero normal stress and the slope of the graph gives

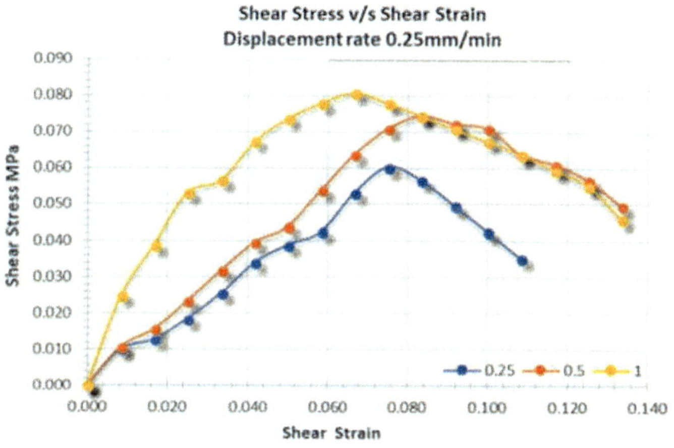

Fig. 3 Shear stress distribution for Mix 1 displacement rate of 0.25 mm/min

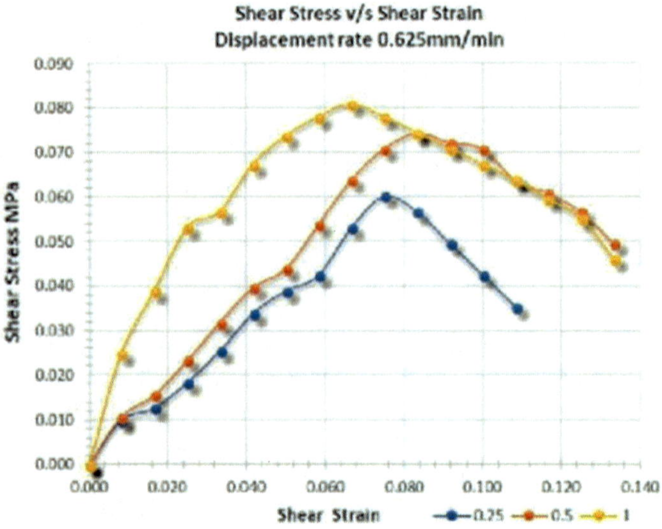

Fig. 4 Shear stress distribution for Mix 1 for displacement rate of 0.625 mm/min

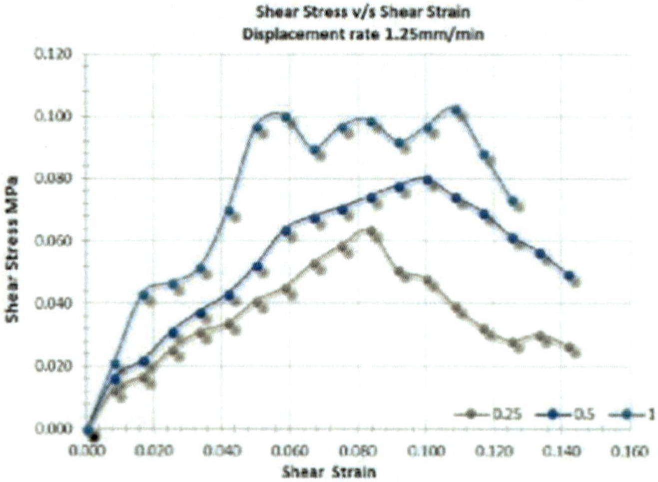

Fig. 5 Shear stress distribution for Mix 1 for displacement rate of 1.25 mm/min

the plastic viscosity of the mix. The graph plotted in Fig. 6 represents the Bingham parameter for Mix 1 and Fig. 10 represents Bingham parameter for Mix2.

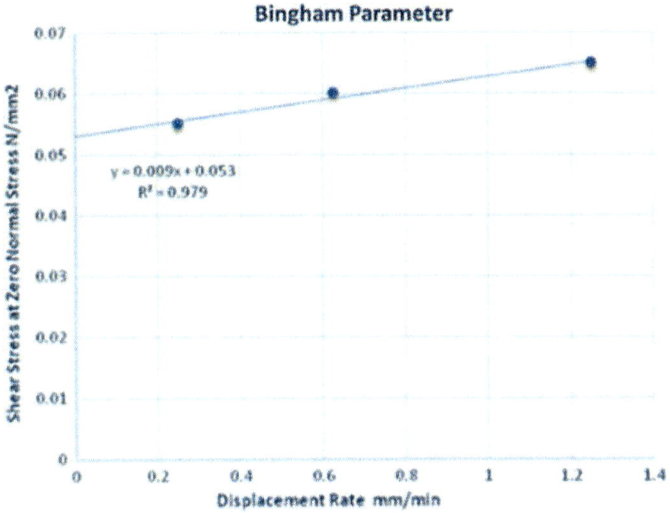

Fig. 6 Peak shear stress at zero normal stress versus displacement rate for Mix 1 Yield stress 0.0531 MPa; Plastic viscosity = 35 MPa-s

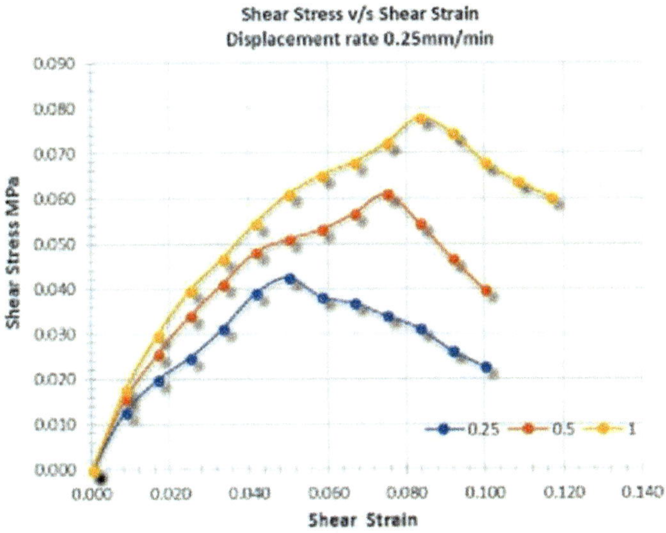

Fig. 7 Shear stress distribution for Mix 2 displacement rate of 0.25 mm/min

3 Results

Results obtained by conducting concrete shear box test for Mix.1 and Mix. 2 were plotted as initial shear stress versus shear strain and are represented in Figs. 3, 4

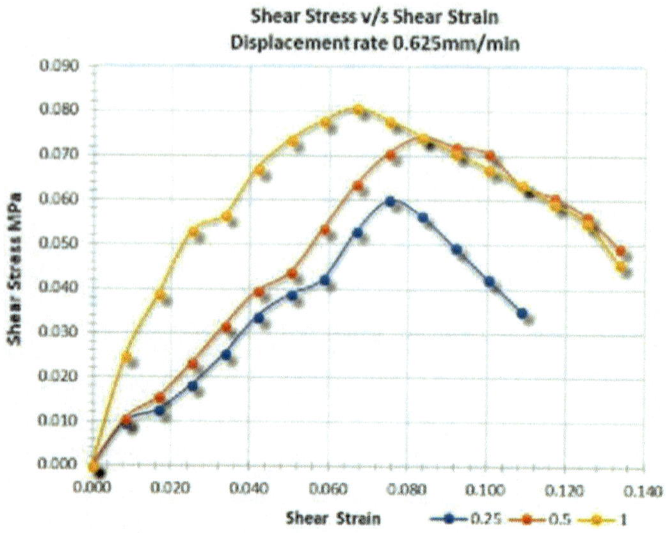

Fig. 8 Shear stress distribution for Mix 2 displacement rate of 0.625 mm/min

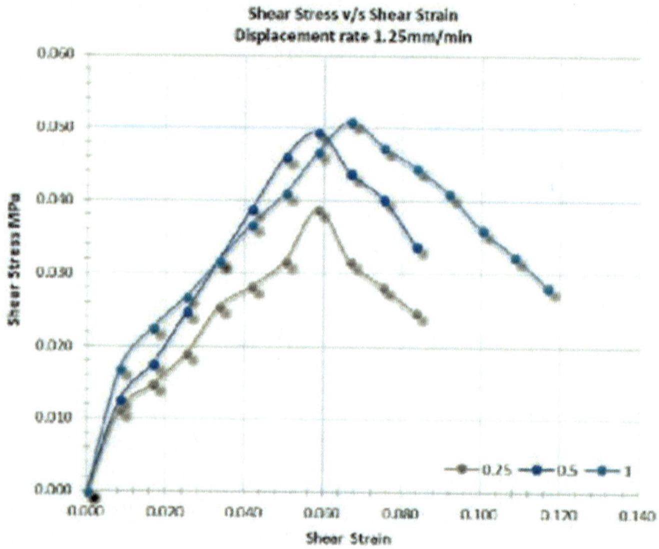

Fig. 9 Shear stress distribution for Mix 2 displacement rate of 1 0.25 mm/min

and 5 for displacement rates of 0.25, 0.62 and 1.25 mm per minute. The peak stress obtained for various displacement rates are as shown in Table 7. The variation of shear stress and shear strain subjected to normal stress of 0.02, 0.04 and 0.09 MPa were observed. It is observed that as the normal stress increased for a particular

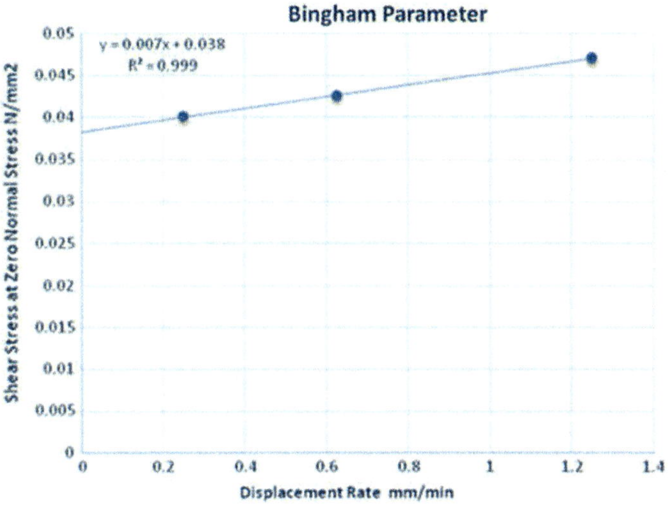

Fig. 10 Peak shear stress at zero normal stress vs displacement rate for Mix 2 Yield stress = 0.0382 MPa; Plastic viscosity = 29 MPa-s

Table 7 Peak shear stress at zero normal stress for all three displacement rates of Mix 1 & Mix 2

Displacement rate (mm/min)	Normal stress (MPa)	Peak shear stress values (MPa) MIX 1	Peak shear stress values (MPa) MIX 2
0.25	0.02	0.07	0.05
	0.05		
	0.10		
0.62	0.02	0.07	0.06
	0.05		
	0.10		
1.25	0.02	0.06	0.06
	0.05		
	0.10		

displacement rate, the peak stress also increased for Mix. 1 compared to Mix. 2 as shown in Figs. 6, 7 and 8. The increase in peak stress and normal stress in Mix. 1 could be due to increased particle packing of aggregates leading to better interlock compared to Mix. 2. It is also observed that at lower displacement rates peak stresses are low compared to higher displacement rate, this may be due to availability of time for the particle re-orient. It is likely that as displacement rate influences the Bingham parameters, it can be considered as factor for influencing the results. It was observed that for a given water content and the yield stress, value decreased for Mix. 2. This can be attributed to that fact that the presence of adhered mortar on the surface of recycled aggregate may reduce the inter-particle friction. Whereas in Mix.

1 there is a better inter-particle bonding between aggregate phase and cement paste. Values obtained using concrete shear box are of a comparatively higher value than those obtained using rheometers. This may be due to particle interference or friction during shearing at lower rates. But the trends reported are like the results obtained using rheometers [3]. In this study it can be observed that yield stress and plastic viscosity are affected by the grading of aggregates and its properties.

4 Conclusions

1. Based on the results of the slump test, it can be predicted that the effect of Super plasticizer increases in case of recycled aggregate concrete for soaked conditions.
2. There is a decrease in cube strength of recycled aggregate concrete with increase in percentage replacement of natural coarse aggregates by recycled coarse aggregates.
3. The strength decreases for certain replacement levels and optimum strength is achieved for 40–45% replacement levels.
4. To correlate and have a better understanding of the rheological behavior of recycled aggregate concrete, shear box test is conducted to know the rheological parameters in static condition.
5. Based on further study on recycled aggregate concrete, a model can be predicted to know the performance of recycled aggregates in concrete. From the arrived results it can studied that Mix. 1 has higher yield stress values compared to Mix. 2.

5 Scope for Further Investigation

The research work can be continued further for different water cement ratios while keeping other design parameters constant to evaluate the rheological parameters. Studies can be conducted by keeping the water cement ratio constant and changes can be made to other parameter in the mix design and then the rheological parameters can be evaluated. It was also realized by the authors that percentage of superplasticizer added was at the higher end and for future work better super plasticizer with lesser dosage can be tried to arrive at high strength recycled aggregate concrete. Based on this, the best mix design procedure can be prepared.

References

1. Girish, S., Santhosh, B.S.: Concrete shear test: a new tool for determining rheological properties of fresh Portland cement concrete. In: Proceedings of International Conference in Civil Engineering "Civil Engineering and Building Materials"(CEBM-2012), pp. 29–294. Taylor & Francis Group, London (2012)
2. Suguna Rao, B. Govindagowda, G., Naik, S.M.: Optimization of recycled aggregate concrete May 2017. IRE J. **1**(1). ISSN: IRE1700015
3. Lakshmi Sravanthi, B., Suguna Rao, B.: Strength Properties of High Strength Concrete Using Recycled Concrete Aggregate. PG RIT (2016)
4. Faleschini, F., Jiménez, C., Barra, M., Aponte, D.: Enric Vázquez. Carlo Pellegrin, Rheology of fresh concrete with recycle aggregates (2014)
5. Bizinotto, M.B., Faleschini, F., Gonzalo Jiménez Fernández, C., Hernández, D.F.A.: Effects of chemical admixtures on the rheology of fresh recycled aggregate concretes (2017)
6. Neophytu, M.K.A., Porgouri, S., Kanellopoulus, A.F., Petrou, M.F., Ioannou, I., Georgiou, G., Alexandrou, A.: Determination of the rheological parameters of self-compacting matrix using slump flow test (2010)
7. Iris, G.T., Belén, G.F., Fernando, M.A., Diego, C.L.: Self-compacting recycled concrete: Relationships between empirical and rheological parameters and proposal of a workability box (2017)
8. Knaack, A.M., ASCE, S.M., Kurama, Y.C.: ASCE, Rheological and Mechanical Behavior of Concrete Mixtures with Recycled Concrete Aggregates (2012)
9. M. Bravo, J. de Brito, L. Evangelista, J. Pacheco, Superplasticizer's efficiency on the mechanical properties of recycled aggregates concrete: Influence of recycled aggregates composition and incorporation ratio.
10. Malešev, M., Radonjanin, V., Broćeta, G.: Properties of Recycled Aggregate Concrete, Contemporary Materials, V−2 (2014)
11. González-Taboada, I., González-Fonteboa, B.: Fernando Martínez-Abella. Sindy Seara-Paz, Analysis of rheological behavior of self-compacting concrete made with recycled aggregates (2017)
12. Tam, C.T., Ong, K.C.G., Akbarnezhad, A. and Zhang, M.H., Research on Recycled Concrete Aggregates at NUS (2013)
13. Girish, S., Indumathi, C., Vengala, J., Ranganath, R.V.: Rheological properties of self compacting concrete using direct shear box. Indian Concr. J. **83**(8), pp. 47–53 (2009)
14. Amario, M., Pepe, M., Martinelli, E., Dias Toledo Filho, R.: Rheological behavior at fresh state of structural recycled aggregate concrete. In: Hordijk, D.A., Luković, M. (eds.), High Tech Concrete: Where Technology and Engineering Meet. © Springer International Publishing AG. 10.1007/978-3-319-59471-2_27
15. Koehler, E.P., Fowler, D.W.: Development of Portland rheometer for fresh Portland cement concrete. ICAR Report 105–3F (2004)
16. Koehler, E.P., Fowler, D.W.: "Summary of Concrete Workability Test Methods", RESEARCH REPORT ICAR-105-1. The University of Texas at Austin August, International Center for Aggregates Research (2003)
17. Bairagi, N.K., Vidhyadhara, H.S., Ravande, K.: Mix Design Procedure for Recycled Aggregate Concrete (1990)
18. Jony, B et al.: Study of Properties of Sustainable Concrete using Slag and Recycled Concrete Ag aggregate-International Journal of Engineering Research & Technology (2014)
19. Suguna Rao, B., Suresh, A., Naik, S.: Shrinkage behaviour of high strength concrete using recycled concrete aggregate. In: As Proceeding of ICSCBM Springer book chapter doi.org/10.1007/978–981–13–3317–0_74, pp. 829–837.

Management of Sustainable Infrastructure Projects: A Scientometric Analysis

Abid Hasan and Arka Ghosh

Abstract Infrastructure projects play a vital role in the social and economic developments in both developed and developing countries. In the last few years, many studies have examined the application of project management theories and practices in the context of sustainable infrastructure projects. Consequently, considerable research has been performed to devise new strategies and refine existing project management practices to enhance sustainability in different aspects of project delivery. Previous studies reveal various social, economic, and environmental issues encountered over the life cycle of infrastructure projects. This paper undertakes a 10-year (2010–2019) scientometric analysis of 51 relevant journal articles found in the Scopus database to identify the research trends in the management of sustainable infrastructure projects. It provides insights into the key research themes gaining the interest of the researchers globally. For instance, the review identified four major research clusters viz. sustainability metrics, processes and factors, community impact, and sustainability triangle. An enquiry into the countries and authors with the maximum number of publications and the collaborative network of the authors reveals a lack of research on project management in sustainable infrastructure projects in the context of developing countries. The findings are expected to inform future studies in this research area.

Keywords Sustainable infrastructure · Project management · Infrastructure projects · Scientometric analysis

1 Introduction

Infrastructure projects drive social and economic growth during both construction and post-construction phases of their life cycle. However, they also contribute heavily to greenhouse gas emissions. While sustainability has captured extensive attention

A. Hasan (✉) · A. Ghosh
School of Architecture and Built Environment, Faculty of Science, Engineering and Built Environment, Deakin University, Geelong, VIC 3220, Australia
e-mail: abid.hasan@deakin.edu.au

D. K. Ashish et al. (eds.), *3rd International Conference on Innovative Technologies for Clean and Sustainable Development*, RILEM Bookseries 29,
https://doi.org/10.1007/978-3-030-51485-3_33

of practitioners and policymakers in the construction industry globally [1], there has been a lack of focus on the sustainability of infrastructure projects. As per the Rio Declaration (1992), sustainable development should be the theme for all development projects. Infrastructure projects have significant effects on implementing the principles of sustainable development [2, 3]. Cost-effective and environmentally sound infrastructure can underpin a country's economic development [4]. Additionally, resilient, robust, and adaptive infrastructure projects can meet the long-term social, economic and environmental goals of sustainability and overcome the threats associated with unsustainable development [5].

Previous studies show that infrastructure projects are increasingly embracing the principles of sustainable development or at least incorporating the perspective of sustainable development in the traditional approach of project design and execution. In the last few years, the sustainability criterion has become increasingly important in the delivery of infrastructure projects as stakeholders are focussing more on eco-friendly and economically efficient construction methods and practices during the life cycle of the project [6]. Similarly, a noticeable increase can be observed in research on sustainable infrastructure projects. However, most of the previous studies focus on the technical aspects of project delivery to improve sustainability and consequently, research efforts directed at project management practices in sustainable infrastructure projects are scarce [6].

Since the number of infrastructure projects is likely to grow in the future, especially in developing economies [2], the proper management of sustainable development shall play a crucial role in improving the sustainability performance in these projects. For instance, Ibrahim et al. [7] proposed a set of risk-related factors to be included in the earned value management calculation to assess and evaluate the progress of modern sustainable infrastructure construction projects. Meng et al. [8] found that top managers' leadership competence, with managerial competence being the primary determinant, followed by intellectual competence, directly drives the entire life cycle of an infrastructure project to accomplish infrastructure sustainability.

This review paper undertakes a scientometric review of studies on project management in the context of sustainable infrastructure projects published during the last ten years (2010–2019). It is expected that the outcomes of this study will direct more research efforts in this area by providing the findings of previous studies in a concise manner. The review will also identify a few research gaps that offer opportunities for further research.

2 Review Approach

The present study utilises the scientometric mapping technique to analyse the extant literature on the management of sustainable infrastructure projects. The scientometric analysis is a generic process of domain analysis, which serves the purpose of identifying the intellectual structure of a scientific domain [9]. It provides a reliable and straightforward quantitative methodology comprising bibliometric tools and methods to analyse the literature and its outputs and to recognise the potentially insightful patterns and trends in research [10]. This technique is suitable for visualising significant patterns and trends in a large body of literature and bibliographic data [11]. The quantitative overview of the research landscape in the field of enquiry provides useful insights into the existing literature that would not be possible through other methods [12]. In the last few years, scientometric analysis approach has found growing acceptance among researchers in the built environment [13, 14].

A comprehensive desktop search was conducted under the **title/abstract/keyword** field of the Scopus database. Keywords for searching were 'sustainable infrastructure project', 'infrastructure project', and 'project management'. Project management is relevant across various sectors and disciplines, and it is not just limited to infrastructure projects. Therefore, the researchers limited the search results to construction literature to make meaningful and well-focused inferences from past studies as the fields of sustainability and project management are large and diverse. After the first round of search, 78 articles were identified. In the second round, both researchers reviewed the abstracts of the identified list of articles independently. It was found that project management in sustainable infrastructure projects was not the primary focus of 27 papers. As a result, these articles were removed from further analysis. Finally, the remaining 51 studies were included in this review. Previous studies show that a sample size of more than 50 papers provides a suitably large sample for using scientometric analysis [15].

Sustainability (Switzerland) has published the highest number of relevant publications, which is eight, followed by *International Journal of Project Management* (four publications), and *WIT Transactions on Ecology and the Environment* (three publications). Four other journals, namely, *Journal of Cleaner Production*, *Journal of Construction Engineering and Management*, *Journal of Management in Engineering* and *Construction Management and Economics* have two publications each. The search covered the period 2010–2019 and was not limited to a particular region. Table 1 presents an overview of the sources.

Scientometric analysis was performed on 51 journal articles using *VOSviewer*. In scientometric analysis, author keywords reflect the core content of scholarly publications and identify the main areas of research activity within any domain [16]. Probing a network of connected keywords can provide investigators with an accurate picture of scientific knowledge production, revealing patterns, relationships, and intellectual organisation of areas of research activity [17]. The analysis of the data comprised of importing the extracted and filtered Scopus dataset comprising title, abstract, and keyword fields into *VOSviewer* and selecting the map based on text data option. It created a term-co-occurrence map based on text data to identify the prominent research clusters.

Table 1 Sources of reviewed articles

Journal	Number of articles
Sustainability (Switzerland)	8
International Journal of Project Management	4
WIT Transactions on Ecology and the Environment	3
Journal of Cleaner Production	2
Journal of Construction Engineering and Management	2
Journal of Management in Engineering	2
Construction Management and Economics	2
Others	28
Total	51

3 Review Findings

3.1 *Wave of Research in the Management of Sustainable Infrastructure Projects*

Figure 1 shows that the wave of research has risen in the field of project management in sustainable infrastructure projects. Based on the Scopus database, 14 journal publications were published on this topic in 2019 alone, which is a sharp increase from four articles published in 2018. It can be argued that the research interest on

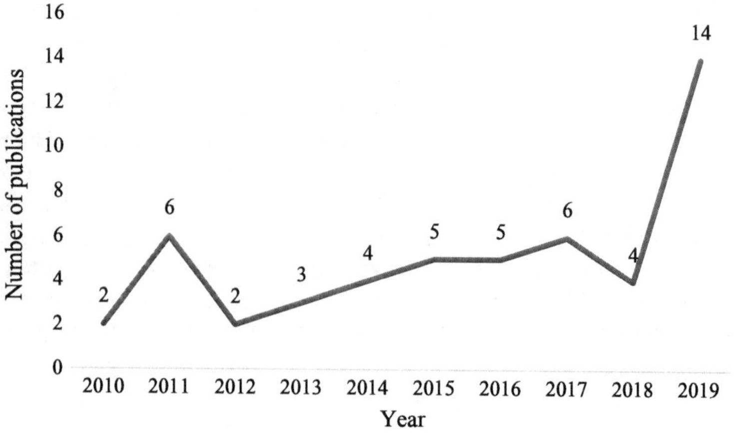

Fig. 1 Year-wise publications on project management in sustainable infrastructure projects

project management in infrastructure projects is catching the attention of researchers due to more emphasis on sustainability criterion in the construction industry.

3.2 Leading Researchers, Institutes and Countries

The awareness of the existing scientific collaboration networks could facilitate access to research funds, specialities, and research expertise. It also enhances research productivity and assists investigators in developing collaborative networks with each other to work together on future research and funded projects [18]. Consequently, these insights are useful in promoting scientific collaboration and scholarly communications among researchers working across different parts of the world [18]. The data on co-authorship is a useful metric for scientific collaboration and research productivity. Previous studies show that collaborative research is generally published in outlets with higher impact and receives more citations [19]. Therefore, knowledge of the current scientific collaboration networks in the research domain of project management in sustainable infrastructure projects can encourage further collaboration and access to research funds and expertise.

The analysis of the most influential countries based on research output and collaboration between researchers from these countries reveal the United States (US), United Kingdom (UK), Australia, and China lead research in the area of project management in sustainable infrastructure projects. An inquiry into the institutes leading the research output in this field shows that leading researchers in this area are affiliated to the University of Hong Kong, Hong Kong Polytechnic University, Arizona State University, University of Queensland, and University of Melbourne. An analysis of the various researcher networks in the area of project management in sustainable infrastructure projects was also performed to identify various leading researchers currently associated with this field and their collaborative networks. It was found that Shen, L, Kumaraswamy, M and Liu, B have produced the maximum number of journal articles in this research area.

For visualising the research collaboration network map of authors, the minimum number of documents for an author was set to one, which resulted in a network map of 144 authors. Figure 2 shows the prominent author research collaboration network of seven authors. The collaboration map of countries of affiliation of the authors is shown in Fig. 3.

3.3 Clusters of Research on Project Management in Sustainable Infrastructure Projects

A network of related keywords provides an accurate picture of scientific knowledge production in terms of patterns, relationships, and intellectual organisation of the

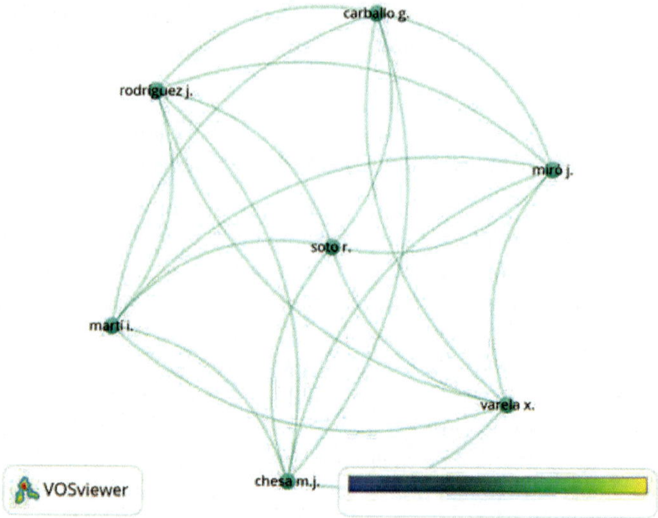

Fig. 2 Author research collaboration network (based on minimum one document)

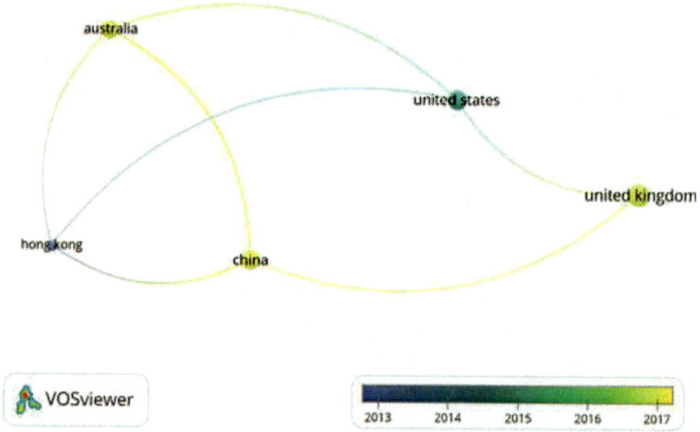

Fig. 3 Country collaboration map (based on the affiliation of authors)

topics covered [20]. Therefore, a co-occurrence network of keywords based on the closeness and strength of existing links was created using *VOSviewer*. The initial mapping of the extracted and filtered Scopus dataset comprising title, abstract, and keyword fields led to a discovery of 1847 terms. To refine it to observe the most prominent terms, we set the minimum number of occurrences of each term as seven. Out of 51 terms meeting this threshold, *VOSviewer* selected the top 60% of the terms with a high relevance score, i.e. 31 terms. In order to obtain a meaningful network map, redundant terms such as article, methodology, integration, study, industry, and way were filtered out from the network map. Finally, the network map shown in

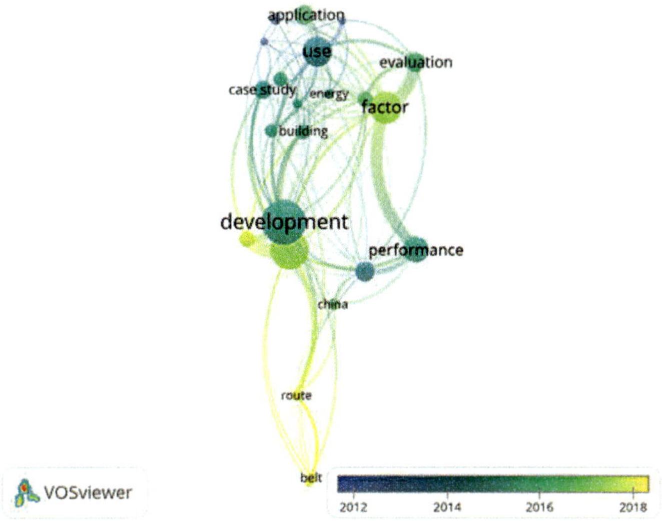

Fig. 4 Text data overlay map of co-occurrence of major research areas

Fig. 4 was obtained with the following four major research clusters: sustainability metrics, processes and factors, community impact, and sustainability triangle. The darker colours are indicative of matured research areas while the lighter shades depict emerging areas of research in this field. These findings on important research topics and research clusters help in the exploration of emergent trends within the specific field [21].

3.3.1 Cluster 1- Sustainability Metrics

The review shows that many researchers have focussed on developing frameworks and models for reliable sustainability assessment in infrastructure projects [22–25]. These frameworks and metrics help in evaluating sustainable practices quantitatively and thereby, assist in incorporating sustainable elements into projects. For example, Hasheminasab et al. [25] developed the sustainability indicators with regard to three pillars of sustainable development viz. environmental indicators (e.g. atmosphere, water, land and soil pollution, natural resource and biodiversity), social indicators (e.g. poverty and equality, health, safety and security, education and welfare) and economic indicators (e.g. energy consumption, financial, economic performance, occupation and earning). Fernández-Sánchez and Rodríguez-López [26] developed a methodology to identify, classify and prioritise sustainability indicators based on risk management standards. Whereas Shen et al. [27] introduced Key Assessment Indicators (KAIs) for assessing the sustainability performance of infrastructure projects. They identified five KAIs for the social dimensions (e.g. public safety and effects on local development), seven KAIs for the environmental dimensions (e.g. effect on water quality and ecological effect), and eight KAIs for the economic dimension (e.g. financial risk and life cycle cost) of infrastructure project sustainability. Aboushady and El-Sawy [28] proposed a qualitative assessment framework for identifying and prioritising sustainability indicators influencing projects. Draper et al. [29] suggested the use of the top-down-bottom-up methodology for defining and assessing sustainability criteria for maximising sustainability.

Yang et al. [30] presented a conceptual framework for managing sustainability knowledge and for raising the awareness and research efforts in the area of sustainable transport infrastructure. Similarly, Alnoaimi and Rahman [3] devised a sustainability assessment framework focusing on all aspects of sustainability throughout the infrastructure project life cycle to provide greater transparency to the stakeholders. Yu [31] proposed a project sustainability assessment system considering environmental, social, and economic criteria to achieve sustainability during the execution phase of the project life cycle.

3.3.2 Cluster 2- Processes and Factors

Previous researchers have identified factors and project management processes that affect sustainability in infrastructure projects. While some researchers suggest that large infrastructure projects and megaprojects can negatively affect biodiversity conservation and ecosystem [32], others argue that such projects could resolve increasing environmental challenges through individual and community empowerment and improvements in the living standards [5]. Zhang et al. [33] recommended that different types of political risks should be considered to strengthen the sustainable development of politically sensitive infrastructure projects. A typology of political risks and mapped different political risks in three-dimensional space could facilitate the formulation of different solutions to deal with various types of political risks [33]. Hughes [34] found that complete assessment of trade-offs between conservation and other values and investigation into environmental impacts is necessary to make good decisions for infrastructure projects based in eco-fragile regions and key biodiversity areas. The proposed road and rail routes of the Belt and Road Initiative were overlaid on key biodiversity areas and protected areas to predict biodiversity hotspots for over 4138 animal and 7371 plant species.

Martí et al. [35] in their study on Barcelona City Council's sustainable drainage systems found that non-standardised construction practices and lack of maintenance plan could ultimately lead to management problems in the long term. Therefore, it is important to consider all aspects of the project life cycle while making decisions in sustainable infrastructure projects. Similarly, Liu et al. [36] suggested an integrated management and design process for a holistic solution that combines stormwater management and urban space improvements. Previous studies have also stressed the use of infrastructure technologies and building information modelling during different phases of infrastructure project delivery [5, 37–39]. Eriksson et al. [40] found that industrialised construction could improve both short-term efficiency and long-term innovation and sustainability in projects.

3.3.3 Cluster 3- Community Impact

Previous studies show that decision making in sustainable construction is complicated by the involvement of multiple state and public stakeholders in projects [41]. Based on interviews with thirty community members, Colvin et al. [42] examined the social impacts of the TasWind proposal for the development of a wind energy facility. They recommended practitioners to use the consultation process for developing good engagement practice in infrastructure projects. In infrastructure projects such as energy infrastructure projects, where community involvement and support play an important role in project success, the consultation process around a proposed project before the social impact assessment process could help in identifying and analysing the anticipatory impacts of the proposed project [42].

Park et al. [43] found factors such as economic, construction-related, and safety-related characteristics are the most important causes of conflicts in energy infrastructure projects. The researchers proposed a conflict management strategy for more sustainable project execution. Kaminsky [44] studied the influence of national-level culture on construction permitting practices to understand how the culture shapes ways in which various approval processes govern projects. Sykes et al. [45] found that local politics and leadership as the most influential factors in community-sensitive projects. Wong et al. [46] suggested that the construction industry's interaction with the public should be a continuous process to promote safety, social and environmental responsibility. Therefore, stakeholder management and community support play a crucial role in the success of sustainable infrastructure projects.

3.3.4 Cluster 4- Sustainability Triangle

Previous studies show that economic performance receives the highest priority in the current practice of project feasibility study. Whereas social and environmental performance receives relatively less attention. Shen et al. [2] suggested the need for shifting the traditional approach of the project feasibility study to a novel approach based on the principles of sustainable development. Loosemore [47] concluded that traditional procurement practices need to be changed to encourage social procurement opportunities. Organisations should incorporate sustainability principles when making decisions on infrastructure investments to ensure long-term resilience [48].

Lenferink et al. [49] found that integrated Dutch Design-Build-Finance-Maintain projects could lead to more sustainable infrastructure development because of the life cycle optimisation incentives provided by the linked contract stages of design, construction, and maintenance. Zhang et al. [50] integrated relevant "people" into public–private partnerships to establish a public-private-people partnership (4P) approach for more sustainable post-disaster infrastructure projects. Hueskes et al. [51] found that sustainability considerations play a limited role in public–private partnerships resulting in the social dimensions of sustainability being neglected. Hueskes et al. [51] also noted that a "strong" sustainability perspective seems inherently incompatible with the contractual public–private partnerships project structure.

Kivilä et al. [6] found that the alliance contract promotes economic, environmental, and social sustainability. Therefore, the review shows that the sustainability triangle or principles and contract types are interrelated in sustainable infrastructure projects.

4 Conclusions and Future Research Directions

The need for delivering sustainable projects in recent years has led to the management in sustainable infrastructure projects translating into one of the emerging research areas of research in the built environment. Previous studies show that there is a strong need to develop new project management strategies to enhance sustainability in infrastructure projects. While the US, UK and Australia are leading research in this field, the review found growing research interest on this topic in Asian countries such as China. The main research areas in this field were identified as sustainability metrics, processes and factors, community impact, and sustainability triangle. The review also highlights a few research gaps for directing future research on this topic. While research suggests a broader approach to project management in sustainable infrastructure projects by considering the economic, environmental, and social dimensions of sustainability [52], the challenges in formulating measurable social sustainability criteria for measuring the social dimensions of sustainability have been highlighted by the previous researchers [51]. Therefore, developing robust project-level sustainability evaluation systems to evaluate the sustainability status of infrastructure projects during various phases of the project life cycle remains an important area of research. This review also found that socio-economic and political risks are the most common barriers to sustainable procurement in infrastructure projects. Future researchers could propose innovative approaches to project procurement to address these barriers. Similarly, research on the implementation of digital technologies in the delivery of sustainable infrastructure projects demands more attention. Additionally, future researchers should investigate the sustainability performance in megaprojects, which remains an unexplored area of research. It must be noted that the review only reflects the research published in journals indexed in the Scopus database. The articles published in conference proceedings and other publications such as books were excluded. Therefore, the use of different databases or inclusion and exclusion criteria in future studies can be expected to generate different results and new insights.

References

1. Xue, B., Liu, B., Sun, T.: What matters in achieving infrastructure sustainability through project management practices: a preliminary study of critical factors. Sustainability **10**(12), 4421 (2018)

2. Shen, L., Tam, V.W.Y., Tam, L., Ji, Y.: Project feasibility study: the key to successful implementation of sustainable and socially responsible construction management practice. J. Cleaner Prod. **18**(3), 254–259 (2010)
3. Alnoaimi, A., Rahman, A.: Sustainability assessment of sewerage infrastructure projects: a conceptual framework. Int. J. Environ. Sci. Dev. **10**(1) (2019)
4. Busari, O., Ndlovu, J.: Leveraging water infrastructure for transformative socio-economic development in South Africa. WIT Trans. Ecol. Environ. **162**, 435–446 (2012)
5. Kattel, G.R., Shang, W., Wang, Z., Langford, J.: China's South-to-North water diversion project empowers sustainable water resources system in the North. Sustainability **11**(13), 3735 (2019)
6. Kivilä, J., Martinsuo, M., Vuorinen, L.: Sustainable project management through project control in infrastructure projects. Int. J. Project Manage. **35**(6), 1167–1183 (2017)
7. Ibrahim, M.N., Thorpe, D., Mahmood, M.N.: Risk factors affecting the ability for earned value management to accurately assess the performance of infrastructure projects in Australia. Constr. Innov. (2019)
8. Meng, J., Xue, B., Liu, B., Fang, N.: Relationships between top managers' leadership and infrastructure sustainability: A Chinese urbanization perspective. Eng. Constr. Archit. Manage. **22**(6), 692–714 (2015)
9. Chen, C.: Science mapping: a systematic review of the literature. J. Data Inf. Sci.S **2**(2), 1–40 (2017)
10. Zheng, C., Yi, C., Lu, M.: Integrated optimization of rebar detailing design and installation planning for waste reduction and productivity improvement. Automat. Constr. **101**, 32–47 (2019)
11. Cobo, M.J., López-Herrera, A.G., Herrera-Viedma, E., Herrera, F.: Science mapping software tools: review, analysis, and cooperative study among tools. J. Am. Soc. Inform. Sci. Technol. **62**(7), 1382–1402 (2011)
12. Cheng, F.F., Huang, Y.W., Yu, H.C., Wu, C.S.: Mapping knowledge structure by keyword co-occurrence and social network analysis: evidence from Library Hi Tech between 2006 and 2017. Library Hi Tech **36**(4), 636–650 (2018)
13. Hosseini, M.R., Maghrebi, M., Akbarnezhad, A., Martek, I., Arashpour, M.: Analysis of citation networks in building information modeling research. J. Constr. Eng. Manage. **144**(8), 1–13 (2018)
14. Jin, R., et al.: A science mapping approach based review of construction safety research. Saf. Sci. **113**, 285–297 (2019)
15. Mahon, N.A., Joyce, C.W.: A bibliometric analysis of the 50 most cited papers in cleft lip and palate. J. Plastic Surg. Hand Surg. **49**(1), 52–58 (2015)
16. Su, H.N., Lee, P.C.: Mapping knowledge structure by keyword co-occurrence: a first look at journal papers in technology foresight. Scientometrics **85**(1), 65–79 (2010)
17. Van Eck, N.J., Waltman, L.: Visualizing bibliometric networks. In: Measuring Scholarly Impact, pp. 285–320. Springer (2014)
18. Ding, Y.: Scientific collaboration and endorsement: network analysis of coauthorship and citation networks. J. Informetr. **5**(1), 187–203 (2011)
19. Glänzel, W., Schubert, A.: Analysing scientific networks through co-authorship. In: Handbook of Quantitative Science and Technology Research, pp. 257–276. Springer, Dordrecht (2004)
20. Van Eck, N.J., Waltman, L.: Visualizing bibliometric networks. In: Measuring Scholarly Impact, pp. 285–320. Springer, Cham (2014)
21. Pollack, J., Adler, D.: Emergent trends and passing fads in project management research: a scientometric analysis of changes in the field. Int. J. Project Manage. **33**(1), 236–248 (2015)
22. Cass, D., Mukherjee, A.: Calculation of greenhouse gas emissions for highway construction operations by using a hybrid life-cycle assessment approach: case study for pavement operations. J. Constr. Eng. Manage. **137**(11), pp. 1015–1025 (2011)
23. Lee, J.C., Edil, T.B., Benson, C.H., Tinjum, J.M.: Evaluation of variables affecting sustainable highway design with BE2ST-in-Highways system. Transp. Res. Rec. **2233**(1), 178–186 (2011)
24. Zhou, J., Liu, Y.: The method and index of sustainability assessment of infrastructure projects based on system dynamics in China. J. Ind. Eng. Manage. **8**(3), 1002–1019 (2015)

25. Hasheminasab, H., Gholipour, Y., Kharrazi, M., Streimikiene, D.: A novel metric of sustainability for petroleum refinery projects. J. Cleaner Prod. **171**, 1215–1224 (2018)
26. Fernández-Sánchez, G., Rodríguez-López, F.: A methodology to identify sustainability indicators in construction project management—application to infrastructure projects in Spain. Ecol. Ind. **10**(6), 1193–1201 (2010)
27. Shen, L., Wu, Y., Zhang, X.: Key assessment indicators for the sustainability of infrastructure projects. J. Constr. Eng. Manage. **137**(6), 441–451 (2010)
28. Aboushady, A., El-Sawy, S.: Qualitative assessment framework to evaluate sustainability indicators affecting infrastructure construction projects in developing countries using the analytical hierarchy process (AHP). WIT Trans. Ecol. Environ. **179**, 1309–1320 (2013)
29. Draper, F., Oltean-Dumbrava, C., Kara-Zaitri, C., Newbury, B.: Individual learning on environmental vocational education and training courses does not always lead to the workplace application of knowledge and skills. J. Educ. Work **27**(6), 651–677 (2014)
30. Yang, J., Yuan, M., Yigitcanlar, T., Newman, P., Schultmann, F.: Managing knowledge to promote sustainability in Australian transport infrastructure projects. Sustainability **7**(7), 8132–8150 (2015)
31. Yu, W.D., Cheng, S.T., Ho, W.C., Chang, Y.H.: Measuring the Sustainability of construction projects throughout their lifecycle: a Taiwan lesson. Sustainability **10**(5), 1523 (2018)
32. Lehtonen, M.: Ecological economics and opening up of megaproject appraisal: lessons from megaproject scholarship and topics for a research programme. Ecol. Econ. **159**, 148–156 (2019)
33. Zhang, C., Xiao, C., Liu, H.: Spatial big data analysis of political risks along the belt and road. Sustainability **11**(8), 2216 (2019)
34. Hughes, A.C.: Understanding and minimizing environmental impacts of the belt and road initiative. Conserv. Biol. (2019)
35. Martí, I., Varela, X., Chesa, M.J., Rodríguez, J., Miró, J., Soto, R., Carballo, G.: The SUDS commission of the Barcelona City council as an entity that integrates the different perspectives. Rev. de Obras Publicas **166**(3607), 99–106 (2019)
36. Liu, L., Fryd, O., Zhang, S.: Blue-green infrastructure for sustainable urban stormwater management—lessons from six municipality-led pilot projects in Beijing and Copenhagen. Water **11**(10), 2024 (2019)
37. Ibrahim, N.H.: Reviewing the evidence: use of digital collaboration technologies in major building and infrastructure projects. J. Inf. Technol. Constr. (ITcon) **18**(3), 40–63 (2013)
38. Saldanha, A.G.: Applications of building information modelling for planning and delivery of rapid transit. In: Proceedings of the Institution of Civil Engineers-Municipal Engineer. Thomas Telford Ltd. (2017)
39. Subsomboon, K., Tantanee, B., Saratai, S., Buranajarukorn, P.: The 4DCAD In Project Planning and budgeting of the new urban infrastructure for the Phitsanulok Central Park, Thailand. Geogr. Tech. **14** (2019)
40. Eriksson, P.E., Olander, S., Szentes, H., Widén, K.: Managing short-term efficiency and long-term development through industrialized construction. Constr. Manage. Econ. **32**(1–2), 97–108 (2014)
41. Brooks, A., Rich, H.: Sustainable construction and socio-technical transitions in London's mega-projects. Geograph. J. **182**(4), 395–405 (2016)
42. Colvin, R., Witt, G.B., Lacey, J., Witt, K.: The community cost of consultation: Characterising the qualitative social impacts of a wind energy development that failed to proceed in Tasmania, Australia. Environ. Impact Assess. Rev. **77,** pp. 40–48 (2019)
43. Park, C.Y., Han, S.H., Lee, K.W., Lee, Y.M.: Analyzing drivers of conflict in energy infrastructure projects: empirical case study of natural gas pipeline sectors. Sustainability **9**(11), 2031 (2017)
44. Kaminsky, J.: The global influence of national cultural values on construction permitting. Constr. Manage. Econ. **37**(2), 89–100 (2019)
45. Sykes, P., Bell, M., Dissanayake, D.: Combined use of a backcast scenario and cross-impact matrix analysis to identify causes of uncertainty in a nascent transport infrastructure project. Transp. Res. Part B Methodol. **116**, 124–140 (2018)

46. Wong, K.K.W., Kumaraswamy, M., Mahesh, G., Ng, S.T.: Utilizing societal engagement as a vehicle for enhancing the image and sustainability of the construction industry. Sustain. Dev. **20**(3), 222–229 (2012)
47. Loosemore, M.: Social procurement in UK construction projects. Int. J. Project Manage. **34**(2), 133–144 (2016)
48. Woodward, J., Stoughton, K.M., Begley, L., Boyd, B.: Evaluating water assets using water efficiency framework for infrastructure intensive agency. J. Water Resourc. Plann. Manage. **145**(2), 04018090 (2018)
49. Lenferink, S., Tillema, T., Arts, J.: Towards sustainable infrastructure development through integrated contracts: experiences with inclusiveness in Dutch infrastructure projects. Int. J. Project Manage. **31**(4), 615–627 (2013)
50. Zhang, J., Zou, W., Kumaraswamy, M.: Developing public private people partnership (4P) for post disaster infrastructure procurement. Int. J. Disas. Resilience Built Environ. **6**(4), 468–484 (2015)
51. Hueskes, M., Verhoest, K., Block, T.: Governing public-private partnerships for sustainability: an analysis of procurement and governance practices of PPP infrastructure projects. Int. J. Project Manage. **35**(6), 1184–1195 (2017)
52. Parrish, K., Chester, M.: Life-Cycle assessment for construction of sustainable infrastructure. Pract. Period. Struct. Des. Constr. **19**(1), 89–94 (2013)

Utilization of Stone Dust as an Effective Alternative for Sand Replacement in Concrete

Pooja Jha, A. K. Sachan, and R. P. Singh

Abstract Construction activities are increasing day by day in different regions, and it requires many natural resources. Alternative materials have been searched for, which can fully or partially replace naturally available material in construction. Stone dust (SD) is one such alternative material which can be successfully used in construction and used as partial replacement of natural sand. The influence of the SD on fresh and hardened properties of concrete are examined and presented in this paper. In this study, sand is replaced by stone dust in different percentages ranging from 0 to 60%. The influence of replacement of sand by SD was analyzed through evaluation of fresh property such as workability (WKA), and mechanical properties, including compressive strength (CS) and splitting tensile strength (TS). Durability properties mainly water absorption (WA), acid resistance (AR) and sulphate resistance (SR) are also determined. The specimen cubes were subjected to 7 Days, 28 Days, and 90 Days and 180 Days moist curing as well as sulphate solution (10,000 ppm) curing. The results show that the optimum replacement level of SD is 40% based on the compressive strength and per cent loss of compressive strength. Sulphate attack is observed minimum at this replacement level. Therefore SD can be effectively utilized in concrete as a valuable alternative for sand and may prove beneficial in construction.

Keywords Concrete · Compressive strength · Optimum · Replacement · Stone dust · Workability

1 Introduction

Nowadays, concrete is the primary composite building material which is used in the construction industry. Usage of concrete as a construction material is more popular nowadays. Aggregates are one of the essential components of concrete which are available in the natural form. Due to the massive scale depletion of sand particles,

P. Jha (✉) · A. K. Sachan · R. P. Singh
Department of Civil Engineering, Motilal Nehru National Institute of Technology, Allahabad 211004, India
e-mail: rce1607@mnnit.ac.in

D. K. Ashish et al. (eds.), *3rd International Conference on Innovative Technologies for Clean and Sustainable Development*, RILEM Bookseries 29,
https://doi.org/10.1007/978-3-030-51485-3_34

there is a possibility of the shortage of these materials and also create environmental problems. For the conservation of these materials, alternative and supplementary cementitious materials which can be used as partial or full replacement of conventional material need to be investigated. SD may prove to be a promising alternative in preparation of concrete mix as fine aggregate with partial replacement. SD may be beneficial in concrete in terms of strength and economy when stone dust replaces sand partially [1]. The optimum replacement level of SD as fine aggregates was 40% [2].One of the waste materials such as quarry dust used as a good alternative for sand in mixtures which gives more strength at 50% replacement [3]. The CS, TS and flexural strength (FS) of concrete found an increase in strengths while using crushed SD as sand replacement in concrete [4]. The detailed study on the durability properties included resistance against acid attack of concrete when cubes cured in hydrochloric acid (HCI) and sulphuric acid (H_2SO_4) solutions [5].

In the present study, it is aimed to find out the optimum replacement level of SD used as fine aggregates by evaluating the compressive strength performance at 7 Days, 28 Days, 90 Days and 180 Days.

2 Experimental Works

2.1 Ingredients Used

Cement Ordinary portland cement (OPC) of 43 grade was used conforming to IS 269 specification. Table 1 shows the various physical properties of 43 grade OPC cement and these properties were determined as per IS: 269.

Coarse Aggregates Crushed gravel is used as coarse aggregate (20 and 10 mm) in the ratio 60:40, respectively. Surface dry and dust-free aggregates are used in the present work. The different physical properties were examined as per the IS code

Table 1 Different physical properties of OPC

Physical properties	Experimental results	Limits specified by BIS: 269
Compressive strength (N/mm^2)		
3 days	23.30	> 23
7 days	33.48	> 33
28 days	43.50	> 44
Consistency (%)	27	
Initial setting time (min)	37	30 (min)
Final setting time (min)	443	600 (max)
Specific gravity	3.14	3.15

Table 2 Various physical properties of coarse aggregates

Physical properties	Characteristics
Colour	Dark grey
Fineness modulus (FM)	6.5
Size (Maximum)	20 mm
WA	0.7%
Shape	Angular
Specific gravity	2.75 (20 mm) 2.60(10 mm)

383: 2016 in the laboratory at MNNIT Allahabad as given in Table 2. The gradation chart of coarse aggregates of the sieve analysis test results is shown in Fig. 1 as per the IS code recommendation.

Fine Aggregates River Sand is used as fine aggregates and obtained from the Allahabad city, UP conforming to zone II as per IS: 383-2016. Table 3 shows the various physical properties of river sand were determined in the laboratory of MNNIT Allahabad.

Stone Dust It was collected from the crusher plant of Lucknow, UP, India. In the present study was investigated due to its local availability in the vicinity of city Lucknow, which can be effectively utilized as an alternative for replacement of sand and hence it will be cost effective. Also, no efforts in the literature are made to assess the properties of locally available stone dust at Lucknow and its application in the

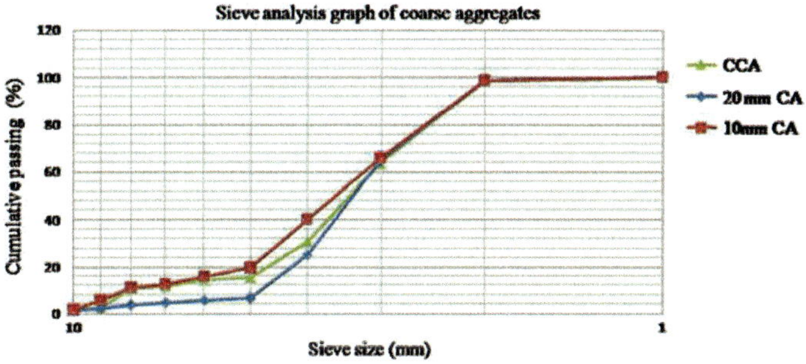

Fig. 1 Grading of coarse aggregates

Table 3 Various physical properties of river sand and stone dust

Physical properties	Characteristics of river sand	Characteristics of stone dust
Specific gravity	2.51	2.40
WA (%)	1	1.1
FM	2.7	2.90

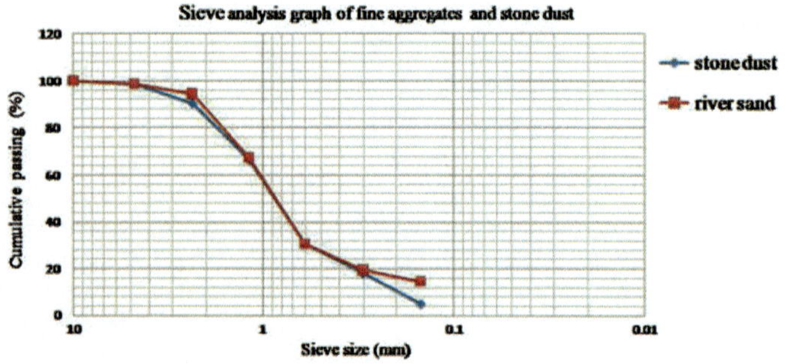

Fig. 2 Grading of river sand and stone dust

preparation of concrete. It was dry and dark grey colour and the test procedures described in IS: 383-2016. A comparison of physical properties of river sand and stone dust are shown in the above Table 3. The gradation chart of river sand and stone dust is shown in Fig. 2 as per the IS code recommendation.

Water For the preparation of the concrete mix and curing purposes, clean and potable water free from the harmful substances like organic matter, oil, acids as per code IS 456-2000, was used.

Superplastizer Master Rheobuild 817 RL was used as superplastizer in concrete mixes preparation.

2.2 Design Mix Proportions (DMP)

The concrete M25 with 0.45 water-cement ratio was prepared, and the standard deviation value for M25 grade concrete was used 4 N/mm^2 as per the procedure described in IS: 10262-2009. Locally available materials were used in the design mix proportion, as shown in Table 4.

Table 4 Details of mix design proportions

Ingredients used	Amount (kg/m^3)
Cement	326
Water	160.65
Coarse aggregates	
20 mm	753.5
10 mm	502.35
Fine aggregates - Sand	698
Superplasticizer	3.26

2.3 Mixing and Casting of Samples

For the preparation of mix proportions, locally available materials were firstly weighed and then mixed in a concrete mixer. The tap water was added after all materials are mixed properly in the mixer. By using the table vibrating machine, the cubes were filled and compacted using the tamping rod (25 times tamping). The samples were de-moulded for 24 h and immersed in water tank for curing purpose till the testing days. All samples were weighed and tested at 7, 28, 90 and 180 days. After 28 days of water curing, cubes were cured in acid (5% HCI, 5% H_2SO_4 solution) and salt solutions (10,000 ppm sodium sulphate solution (Na_2SO_4) for durability evaluation. Sand was replaced by stone dust in different percentage such as 0% (control mix), 10% (10SD), 20% (20SD), 30% (30SD), 40% (40SD), % (50SD), 60% (60SD) respectively.

2.4 Methods

WKA It was determined by using the slump test and compacting factor (CF) test as per the test procedure described in IS 1199: 1959.

CS Test Cubes (each of dimension 150 mm) were cast and used for testing the CS of concrete as per IS 516: 1959. Three specimens were tested, and their average value has been considered to determine the CS of concrete. The cubes were tested at 7, 28, 90 and 180 days of water curing.

TS Test Using cylinder having the dimension 300 mm length and 150 mm diameter to determine the TS of concrete at 28-day water curing as per the procedure given in IS 5816:1999.

WA Test The concrete samples were oven-dried in an oven at 110 ± 5 °C for 24 h after 28 days of water curing, and its weight was measured, say W_1. The concrete samples were immersed in water for 48 h, and the weight was measured, say W_2. The difference between the over dried mass and the saturated mass as a percentage of the oven-dried mass gives the percentage WA of concrete samples.

AR Test Concrete samples were weighed and immersed in 5% HCI solution and 5% H_2SO_4 solution for 90 and 180 days after 28 days of water curing. Samples were cleaned with the help of water to remove the impurities after 90 and 180 days. The CS and weight of concrete specimens were determined. The percentage loss in weight and percentage loss in CS were determined to assess the durability of concrete mixes under the acidic environment.

SR Test The concrete samples were immersed in 10,000 ppm sodium sulphate (Na_2SO_4) solutions for 90 and 180 days, and sulphate resistance test was performed similarly to that described for the acid resistance test.

3 Results and Discussion

3.1 *Effect of Stone Dust on Fresh and Hardened Properties of Concrete*

WKA of Concrete WKA of different mixes was determined by using the slump test and CF test as per the test procedure laid down in IS 1199-1959. To examine the low or medium WKA, both Slump and CF test were performed for workability. The effect of the addition of SD as partial replacement of natural sand on WKA of concrete is shown in Figs. 3 and 4.

WKA of concrete mixes decrease with increase in different percentages (0, 10, 20, 30, 40, 50 and 60%) of SD. Smooth texture and round shape make the concrete more workable, and it is observed higher in the control mix. The angular shape and rough texture of crushed SD enhanced the internal friction in concrete mixes. Therefore WKA decreases with an increase in the percentage of SD replacement. Pofale et al. [6] had shown similar observations while testing the WKA when stone dust replaced the natural sand partially.

CS of Concrete The CS of concrete at 7 days, 28 days, 90 days and 180 days were tested as per IS 516-1959 and Fig. 5 shows the variation of the CS with varying percentage of stone dust replacement. It is observed that the CS of concrete first

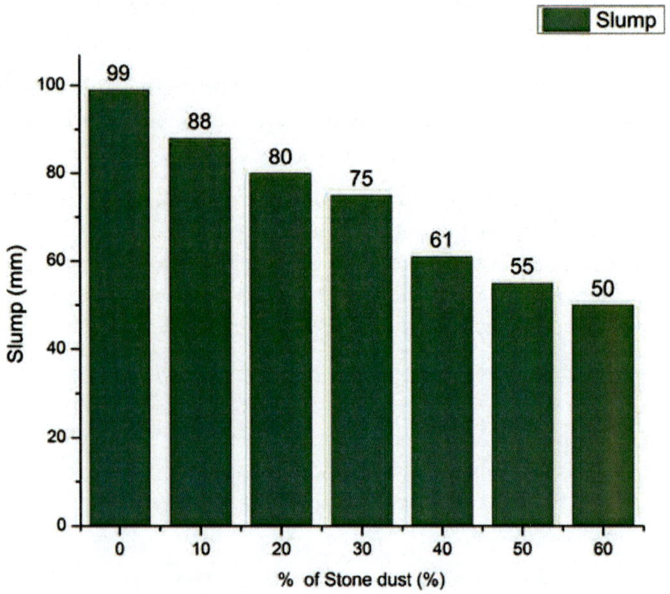

Fig. 3 Variation of a slump at the different proportion of stone dust

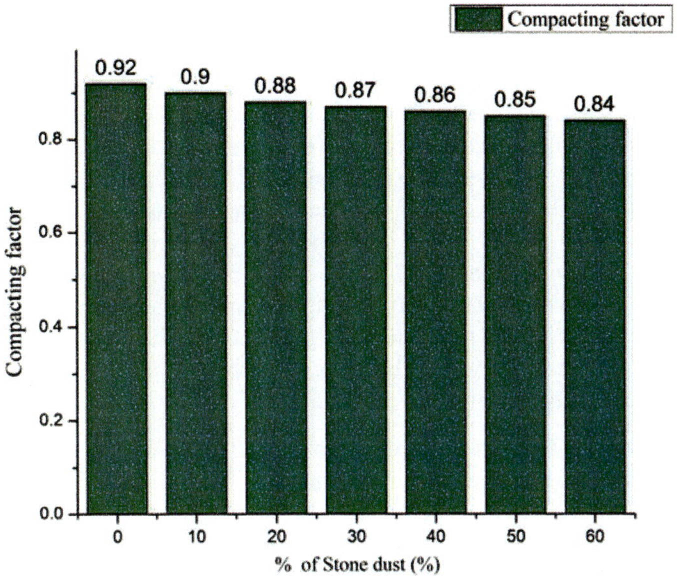

Fig. 4 Variation of CF at the different proportion of stone dust

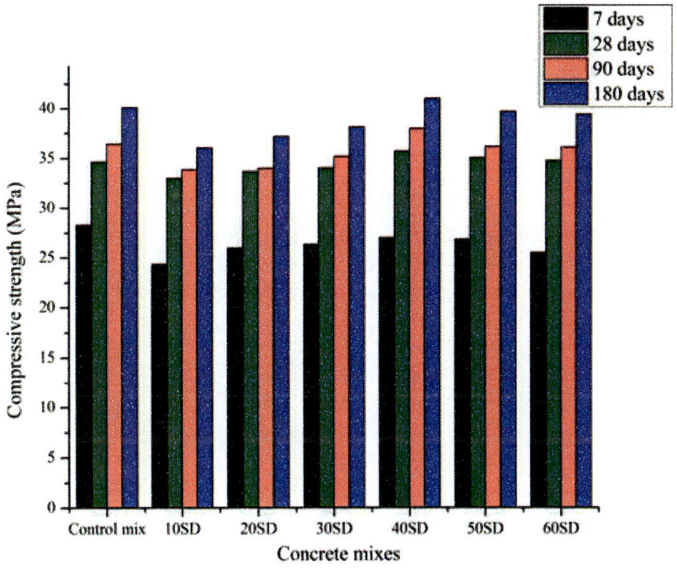

Fig. 5 Variation of CS with varying percentage of stone dust at 7, 28, 90 and 180 days

increases up to a maximum values (27, 35.69, 37.95 and 41 MPa) at 7, 28, 90 and 180 days and then decreases with the increasing percentage of SD.

In the present study, natural sand was replaced by SD with a percentage varying from 10 to 60% and the optimum replacement level of SD was observed as 40%. Variation of CS with a gradual increase in the percentage of SD as fine aggregate at an interval of 10%. The variation in CS may be due to different water absorption capacity of stone dust and sand, different dose of super plasticizer in mix, different angularity of particles etc. Various researchers have investigated similar results by replacing the sand with stone dust [7–9]. Literature reveals an optimum replacement level of natural sand by SD as 40, 50 and 60% depending upon the characteristics of SD. The optimum range may vary from 40 to 60%. Depending upon fineness of SD, the optimum replacement level was observed 40% as evident from the test results on CS, TS and durability properties of concrete.

TS of concrete The TS of all concrete mixes after 28 days and its variation shown in Fig. 6. The TS shows a similar trend as that of the CS. TS variation with increasing SD percentage were recorded as 2.50, 2.71, 2.75, 2.89, 2.78 and 2.64 MPa for 10%, 20%, 30%, 40%, 50% and 60% replacement respectively as compared to mix containing no stone dust (2.48 MPa) after 28 days.

It was observed that the tensile strength first increases up to 40% stone dust replacement, and after that, it decreases. The maximum TS at 40% stone dust replacement was 2.89 MPa.

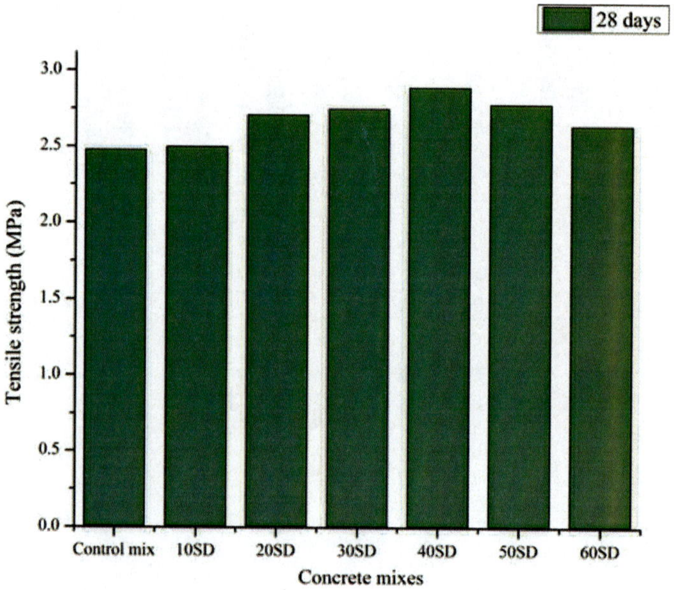

Fig.6 Variation of TS with stone dust at 28 days

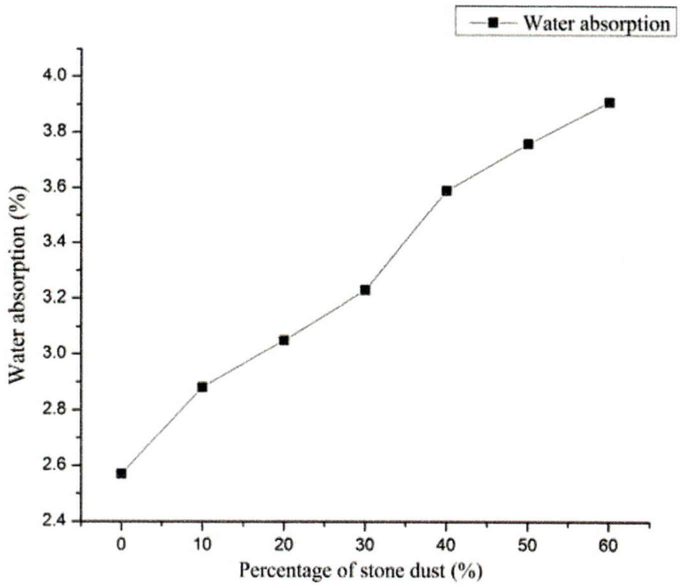

Fig.7 Water absorption test for various concrete mixes

3.2 Durability Analysis

WA The indicator of permeability of concrete is WA. Increase in percentage water absorption with increasing stone dust percentage were recorded as 2.88%, 3.05%, 3.23%, 3.59%, 3.76% and 3.91% for10%, 20%, 30%, 40%, 50% and 60% replacement respectively as compared to mix containing no stone dust (2.57%) after 28 days shown in Fig. 7.

Therefore stone dust has shown high water absorption capacity. Similar variation with the use of stone dust in concrete [10]. Concrete with stone dust has high water absorption capacity when observed at 28 days. While there are gradual decreases in WA at 90 days was observed in the later stage of experiments. Therefore, it will affect the durability of concrete in a long period. This may be due to the difference of the density and porosity between the different fine aggregates (stone dust and river sand) used in this study.

AR From Tables 5 and 6, the per cent loss in weight and per cent loss in CS percentage were determined when concrete cubes were immersed in 5% HCI solution and 5% H_2SO_4 solutions for the control mix and other concrete samples were determined. It was analyzed that there is a positive influence of replacement of sand by SD in concrete.

The percentage loss in weight and percentage loss in CS are reduced on immersion in acid solutions. Also, more significant deterioration of concrete was observed on immersion in hydrochloric acid.

Table 5 Exposure to 5% HCI solution at 90 and 180 days

S. no.	Concrete mixes	% loss in weight		% loss in CS	
		90 days	180 days	90 days	180 days
1	Control mix	4.45	4.97	17.01	25.01
2	10SD	4.27	4.80	15.23	23.46
3	20SD	4.09	4.75	13.67	21.67
4	30SD	3.98	4.69	11.89	20.45
5	**40SD**	**3.91**	**4.50**	**10.63**	**20.01**
6	50SD	3.99	4.66	10.95	21.23
7	60SD	4.06	4.73	11.25	22.13

Table 6 Exposure to 5% H_2SO_4 solution at 90 and 180 days

S. no.	Concrete mixes	% loss in weight		% loss in CS	
		90 days	180 days	90 days	180 days
1	Control mix	7.45	9.79	15.02	18.74
2	10SD	7.36	9.70	9.81	17.56
3	20SD	7.01	9.61	8.66	15.86
4	30SD	6.93	9.50	7.34	15.06
5	40SD	**6.89**	**9.30**	**6.83**	**14.12**
6	50SD	6.97	9.46	6.94	15.83
7	60SD	7.05	9.54	7.01	16.08

SR The sulphate resistance of concrete was tested at 90 and 180 days when concrete cubes were immersed in sodium sulphate solutions (10,000 ppm) by determining its compressive strength and outcomes are shown in Figs. 8 and 9. The CS at 90 days water curing was more than that CS at 90 days sulphate solution curing as shown in Fig. 8. The CS of control mix was 36.43 and 34.55 MPa when cubes cured in water and sulphate solution respectively. Highest compressive strength (37.95 MPa) was observed in 40SD sample in case of water curing and slightly lower (36.38 MPa) in case of sulphate solutions. Similarly, for 180 days, the compressive strength of control mix was 40.11 MPa (water curing), and 35.69 MPa (sulphate solution curing) and compressive strength at the optimum percentage (40SD) were 41.00 MPa (water curing) and 37.74 (sulphate solution curing) as shown in Fig. 9. From Tables 7 and 8, it has been observed that compressive strength losses were found minimum in 40SD sample at both 90 and 180 days curing. It has been revealed that 40SD sample has shown excellent durable properties than other samples.

From Fig. 10, it is evident that percentage loss in CS at 90 Days and 180 Days in sample 40SD was minimum among all concrete mixes (10SD, 20SD, 30SD, 40SD, 50SD and 60SD). The minimum loss for 10SD sample at 90 Days and 180 Days was

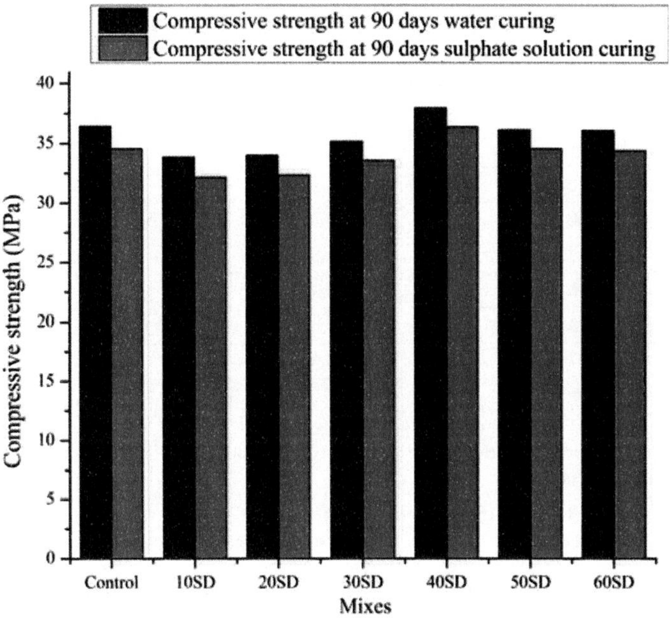

Fig.8 CS of concrete in case of water curing and sulphate solution (10000 ppm) curing at 90 days for different mixes

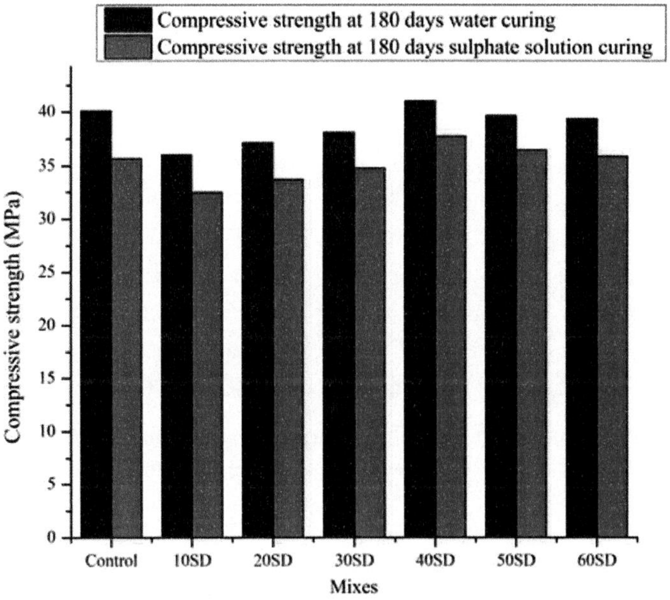

Fig. 9 CS of concrete in case of water curing and sulphate solution (10,000 ppm) curing at 180 days for different mixes

Table 7 Percentage loss in CS at 90 days in case of water curing and 10,000 ppm sulphate solutions curing

Mixes	CS (MPa) at 90 days water curing	CS (MPa) at 90 days sulphate solution	% loss in CS at 90 days
Control mix	**36.43**	**34.55**	**5.16**
10SD	33.86	32.16	5.02
20SD	34.00	32.35	4.85
30SD	35.17	33.58	4.52
40SD	**37.95**	**36.38**	**4.13**
50SD	36.13	34.55	4.37
60SD	36.07	34.39	4.66

Table 8 Percentage loss in compressive strength (CS) at 180 days when cubes cured in water and 10,000 ppm sulphate solutions

Mixes	CS (MPa) at 180 days water curing	CS (MPa) at 180 days sulphate solution	% loss in CS at 180 days
Control mix	**40.11**	**35.69**	**11.01**
10SD	36.03	32.48	9.85
20SD	37.15	33.75	9.15
30SD	38.10	34.79	8.68
40SD	**41.00**	**37.74**	**7.95**
50SD	39.66	36.47	8.04
60SD	39.36	35.88	8.84

4.13% and 7.95% respectively, which shows that 10SD sample has good resistance to sulphate attack.

4 Conclusions

Based on the research work, the following outcomes are as follows:

1. The value of slump and CF value of concrete was found decreasing with the percentage increase in the stone dust. Therefore workability decreases with an increasing percentage of stone dust.
2. The CS and TS were found to be highest at 40% stone dust replacement level, and a decreasing trend was observed beyond 40% stone dust replacement level. Optimum replacement level of stone dust is found to be 40% based on the experimental results.

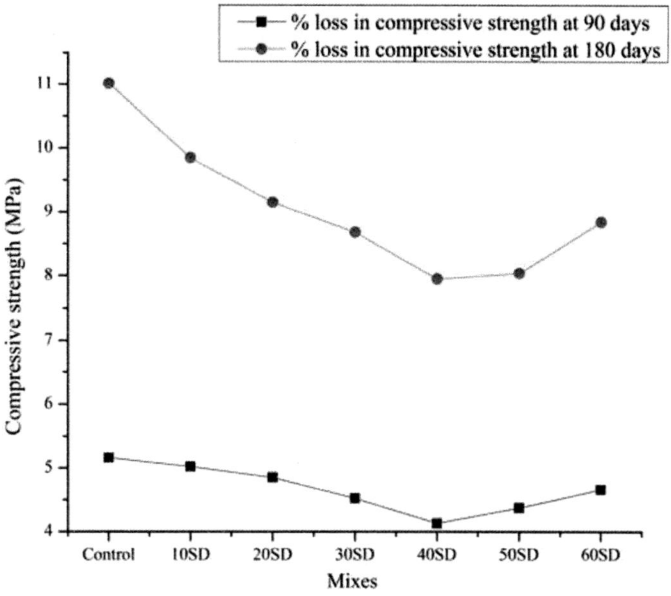

Fig. 10 Sulphate attack at 90 Days and 180 days in case of cubes cured in water and 10,000 ppm sulphate solution on different mixes

3. Stone dust has high water absorption capacity when observed at 28 days while there is a gradual decrease in WA in the later stages of experiments. Therefore it will affect the durability of concrete in the long term.
4. 40% stone dust replacement with sand provided good resistance against acid attack when cubes were cured in 5% hydrochloric acid solution and 5% sulphuric acid solution further less deterioration of concrete is observed with the sulphuric acid solution. The sulphuric acid solution proves to be advantageous against acid attack.
5. Percentage loss in compressive strength was found minimum 4.13% at 90 days and 7.95% at 180 days for 40SD mix sample which shows that resistance to sulphate attack is improved in concrete with 40% stone dust.

It has been concluded that that concrete mix with 40% replacement level of stone dust may prove advantageous in different concrete mixes without compromising the various properties of concrete.

References

1. Mahzuz, H.M.A., Ahmed, A.A.M., Yusuf, M.A.: Use of stone powder in concrete and mortar as an alternative of sand. Afr. J. Environ. Sci. Technol. **5**(5), 381–388 (2011)

2. Shyam Prakash, K., Rao, C.H.: Study on compressive strength of quarry dust as fine aggregate in concrete. Adv. Civil Eng.(2016)
3. Balamurugan, G., Perumal, P.: Behaviour of concrete on the use of quarry dust to replace sand–an experimental study. Behaviour **3**(6) (2013)
4. Kapgate, S.S., Satone, S.R.: Effect of quarry dust as partial replacement of sand in concrete. Indian Streams Res. J. **3**(5), 1–8 (2013)
5. Murthi, P., Sivakumar, V.: Studies on acid resistance of ternary blended concrete (2008)
6. Quadri, A.P.S.R.: Effective utilization of crusher dust in concrete using Portland Pozzolana cement. Int. J. Sci. Res. Publ. **415** (2013)
7. Reddy, M.V.S., Mrudula, D., Seshalalitha, M., Hariprasad, P.: The effect of crushed rock powder and Superplasticizer on the fresh and hardened properties of M30 grade Concrete. Int. J. Civil Struct. Environ. Infrastruct. (IJCSEIERD) **5**, 25–30 (2015)
8. Shukla, M., Shau, A.K., Sachan, A.K.: Performance of stone dust as fine aggregate in Portland replacing Sand on concrete and mortar. In: A National seminar on advances in special concretes, Indian Concrete Institute, Bangalore, India, pp. 241–248 (1998)
9. Bai, H.S.: Use of crushed rock powder as replacement of fine aggregate in mortar and concrete. Indian J. Sci. Technol. **4**(8), 917–9229 (2011)
10. Bonavetti, V.L., Irassar, E.F.: The effect of stone dust content in sand. Cem. Concr. Res. **24**(3), 580–590 (1994)

Water Pollution and Its Prevention Through Development of Low Cost Wastewater Treatment System

Sagar Mukundrao Gawande and Dilip D. Sarode

Abstract The development is essential part of human society which offers the opportunities in the sectors like information technology, supply chain, food-farming, infrastructure and more others. Since, last several decades universally the living standard of the human society is upgrading. Currently available and upcoming smart technologies will reduce the efforts involved in day to day activities. The development of society is endless process with respect to the time. Due to depletion in water availability will affect the rate of development and burden on water sources. This burden can be optimizing through reuse of treated wastewater by low cost Decentralized Wastewater Treatment System (DWATS). In this paper the said system is developed by using locally available materials and discussed its performance. As per guidelines of USEPA 2012 for water reuse the pH ranges with 6.5–8.4, TDS ranges from 450 to 2000 mg/L and Electrical Conductivity (EC) 0.7–3.0 μs/cm. The experimental result has pH at inlet 7.78 and 6.4 at outlet whereas The Total Dissolved Solids (TDS) is 116 mg/L at inlet to 86 mg/L at outlet of system. The Electrical conductivity (EC) observed 213 μohm/cm at inlet to 163 μohm/cm at outlet with 23.47–27.85% change in conduction and 57.69–86.41% in BOD where as 66.67–69.96% change in COD when tested at inlet and outlet.

Keywords Ausa lake · *Ipomoea carnea* · Low cost treatment · DWATS · Eco friendly

1 Introduction

The water shortage is for the most part man made because of overabundance populace development and bungle of water assets. A portion of the real explanations behind

S. M. Gawande (✉) · D. D. Sarode
Department of General Engineering, Institute of Chemical Technology, Matunga Mumbai, India
e-mail: gawande.sagar@gmail.com

D. D. Sarode
e-mail: dd.sarode@ictmumbai.edu.in

D. K. Ashish et al. (eds.), *3rd International Conference on Innovative Technologies for Clean and Sustainable Development*, RILEM Bookseries 29,
https://doi.org/10.1007/978-3-030-51485-3_35

water shortage are: Sewage and wastewater seepage into natural water bodies, inefficient utilization of water for agriculture, Lack of efficient administration in water treatment facilities and its circulation [1]. The decentralized wastewater treatment is the technology which is simpler, economical and eco friendly that utilize local living plants to clean up contaminates from soil, air, and water.

Initially the experimental setup (lab scale model) is designed and demonstrated in the engineering college campus situated in Pune, where the institutional wastewater generated by around 800 enrolled students and 100 employed staff, which generates an average 2000 L/day of wastewater during academic days this wastewater, is passed and treated through the developed model. After several trials this model offers the satisfactory results and performance which will be implementing on pilot scale in Latur district at Ausa town which is selected as study area which is one of the water stressed region of Maharashtra state.

1.1 Background and Study Area

During non rainy seasons the districts have to face difficulties of water scarcity and draught situations. Ausa town in Latur district of Marathwada region in Maharashtra state no longer has the wastewater treatment facilities. The Ausa Lake [2] in the town is taken as the study area to address the contamination level and its prevention through design of suitable decentralized wastewater treatment and to demonstrate the importance of proper sewage management among the residents in collaboration of Local governing Authorities, schools and colleges which can help to reduce overall pollution in the Ausa town [3, 4] (Fig. 1).

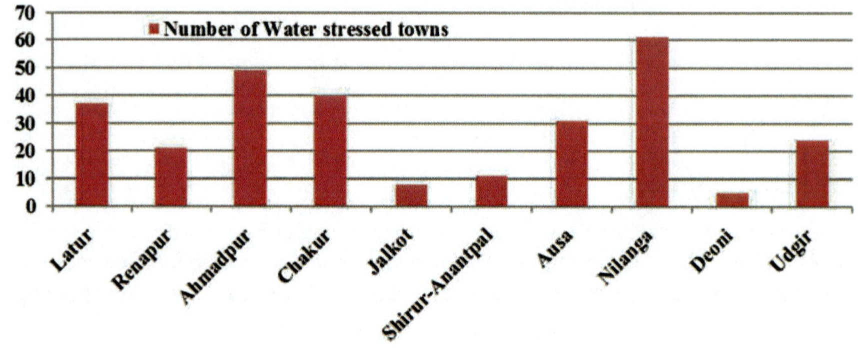

Fig. 1 Shows the water stress administrative councils in district [3]

2 Method and Materials

The wastewater is centrally collected in collection tank as shown in Fig. 2 from the experimental site at Anantrao Pawar College of Engineering and Research, Pune where the pilot designed set up is installed. The designed setup has six circular containers having volumetric capacity of 120 L.

The containers are layered with coarser porous media at bottom and finer at the top as shown in Fig. 3 with stepped slope of 3.50 m horizontal to 1.12 m vertical. Each container is divided into four compartments by using baffle walls so that each compartment has volumetric capacity as 30 L each. These baffle walls has alternate openings at bottom and middle depth along the length.

2.1 *Sampling and Testing*

Sampling and testing was done in each week to understand the performance of the developed system. The wastewater sample before treatment collected at the sewage chamber and after treatment at the outlet point of the model in sterile plastic bottles to test and analyze in laboratory.

Dissolved Oxygen (DO)- As per APHA procedure 4500-O B, the concentration of (DO) was checked. Dissolved Oxygen is an indicator of physicochemical and biochemical activities in a water bodies. It is an important parameter to define the water pollution and waste treatment process control.

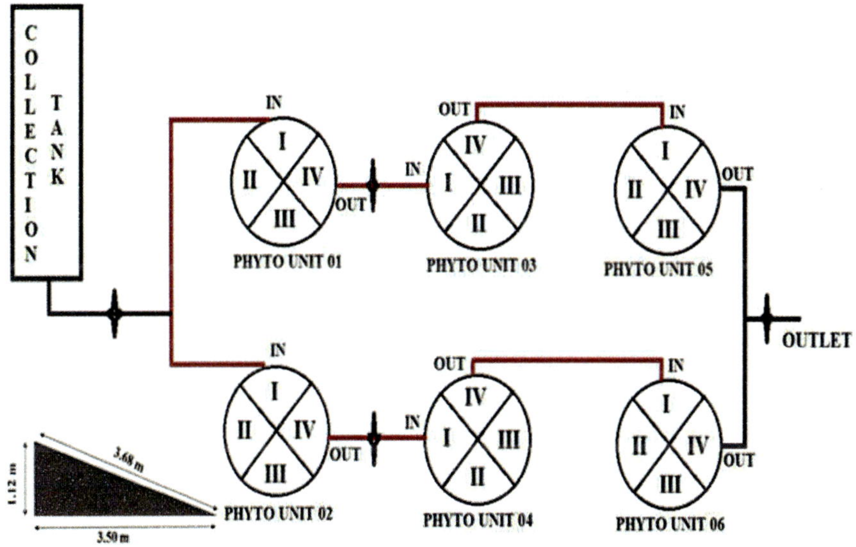

Fig. 2 Showing schematic diagram of developed model with flow control units

Details of developed phyto media and Color code used					
Media Type from bottom	Soft and hard boulders	Coarse Aggregate	Fine Aggregate	Local Soil	Free Board
Color Code					
Layer Thickness in (mm)	400 mm	120 mm	80 mm	40 mm	20 mm

Fig. 3 Showing layers of media with arrangement of baffle walls and perforated pipes to transfer of oxygen in system

Biochemical oxygen demand (BOD): This test is conducted as per APHA procedure 5210-B. It measures the alteration in dissolved Oxygen concentration caused by microorganisms as they degrade organic matter in a sample held in a BOD bottles incubated for 5 days in the at 20 °C. It is a measure of organic matter present in wastewater.

Chemical Oxygen Demand (COD): The open reflux method 5220 B is followed to determine the COD concentration. This states most types of organic matter are oxidized by a boiling mixture of chromic and sulfuric acids. The sample was refluxed in HCL solution with known quantity of potassium dichromate ($K_2Cr_2O_7$).

In combination with above the other parameters like pH, Temperature, Conductivity, solids, are also checked and analyzed by following the method of determination given in standard methods of APHA 23rd edition.

2.2 *Phyto Media*

The design of phyto-filter using locally available favorable plants and soil is used as a media. The filter media has four layers which includes the bottommost 400 mm thick layer with 40 mm soft and hard boulders, the bottom layer is covered with a

120 mm thick layer of 20–25 mm coarse aggregate. The third layer has 80 mm thick band of fine aggregate and the topmost part is covered with 40 mm thick layer of local soil.

The available voids and surface area of finer and coarser media trapped the impurities contaminates from the wastewater. The root of the plant species absorbs the settled as well accumulated as solids and translocates it to the upper part over filter media through plant root via stem [5].

2.3 *Why* Ipomoea Carnea

The conventional plant species used in remediation processes but it was observed that in most of the cases these species plucked by the visitors or eat by the animals. The *Ipomoea Carnea* (*Morning Glory*) [6] is well known for its toxicity [7] and unfavorable behavior. Mostly this species has favorable growth rate in semi arid area. The excessive growth of this species in the developed system after its maturity can be dried and utilized for biomass and construction materials.

This assessment involved the performance of *Ipomoea Carnea* [8] plant species coupled to uptake, translocation, accumulation and removal of contaminates from the supplied effluent in controlled conditions.

3 Results and Discussion

The untreated wastewater pH varies in the range from 6.2 to 6.97 and from Fig. 1 the treated wastewater varies as maximum of 7.22 and minimum of 6.82 which shows that pH of is acidic to alkaline The pH value for effluent discharge on land irrigation and into the public sewer varies from 5.5 to 9 the treated wastewater is fit for gardening purpose (Figs. 4 and 5; Table 1).

The used plants species transferred the oxygen which enhances oxygen level in the water system. The transfer of oxygen through species towards root zone has a significant role for the growth of aerobic bacteria in the root zone followed by degradation available organic matters in the wastewater. Chemical Oxygen Demand (COD) concentrations measured during experimentation for the run time of system and are plotted in the Fig. 6 the for inlet and outlet concentration. The favorable pH of water helps in reduction of COD in combination with microbial activity in the wastewater.

The Biochemical Oxygen Demand (BOD) removal in develop model was found about 57.07–80.12% hence this design is meaningful BOD removal efficiency. As shown in Fig. 7 this system gave the BOD average output as 30 mg/L which suits for gardening purpose.

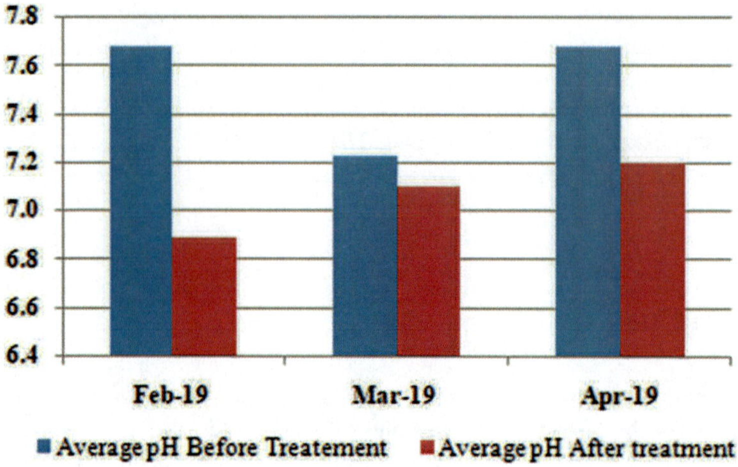

Fig. 4. pH variation of sample before and after treatment

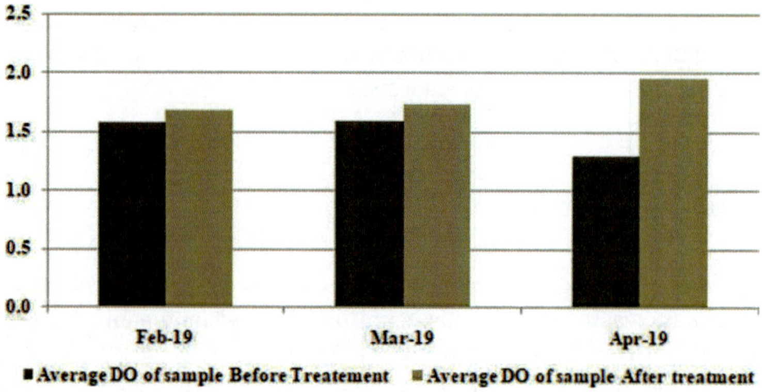

Fig. 5 Variation in dissolved oxygen in mg/L

4 Conclusion

It was found that the developed model is an efficient and economical system to treat wastewater with phyto media. The focus of the present study is recycling of wastewater for gardening and flushing purpose. The present study has some useful for applications as at source treatment of wastewater and recycling of wastewater for gardening, road washing, toilet flushing.

Table 1 Average test results of wastewater samples

S. N	Tested parameters	Set 01			Set 02		
		Tested at inlet	Tested at outlet	Percentage change	Tested at inlet	Tested at outlet	Percent-age change
1	Temperature (°C)	23.80	24.20	–	24.40	25.50	–
2	Temperature (°C)	22.40	23.30	–	24.60	25.20	–
3	pH	7.78	6.4	–	7.48	6.9	–
4	DO (mg/L)	3.95	4.89	19.22	3.62	4.12	12.14
5	COD (mg/L)	103.20	31.00	69.96	302.4	100.8	66.67
6	BOD (mg/L)	39.00	5.30	86.41	117	49.5	57.69
7	TDS (mg/L)	116	86	25.86	648	492	24.07
8	TSS (mg/L)	97	24	75.26	208	28	86.54
9	EC(μohm/cm)	213	163	23.47	1185	855	27.85
10	MPN (Coliforms)	Positive	Positive	–	Positive	Positive	–

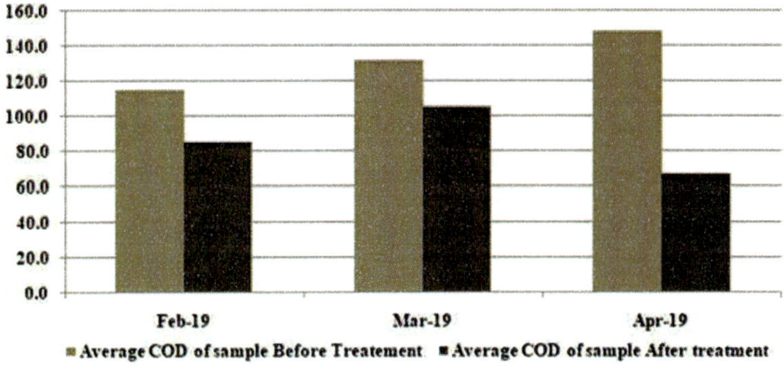

Fig. 6 Shows chemical oxygen demand (COD) in mg/L

Some highlighted point of experimental model listed as: (I) It requires low energy because naturally high quantum of oxygen transfer took place (II) As it is naturally driven no pumping required and ultimate reduction in O and M cost.(III) Applicable for rural and semi urban areas with odor less green aesthetic ambience where sludge management practices are not exits.

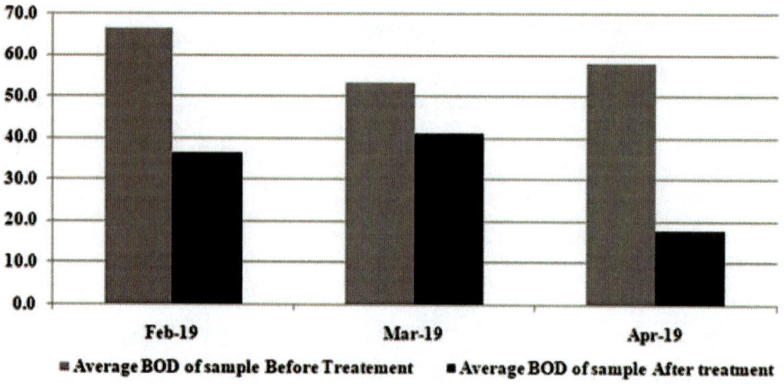

Fig. 7 Shows biochemical oxygen demand (BOD) in mg/L

Acknowledgements We are thankful to Department of Science and Technology (DST), Government of India for grant project of Rejuvenation of Ausa Lake in Latur District. Prof. Pushpito K. Ghosh for his encouragement for experimentation of low cost wastewater treatment. Dhawal Patil General Manager (Operation) of Ecosan services Foundation, Pune. Dr. Sunil Thakare Head of Institution, Anantrao Pawar College of Engineering and Research, Pune.

References

1. Phadatare, S., Gawande, S.M.: Assessment and development of water quality index in southeastern region of Satara. Imperial J. Interdisc. Res. (IJIR), 579–582 (2016)
2. Project proposal DST/WTI/2K16/306 for Mitigation of water problems in Ausa town, Latur: Wastewater management, Gaothan lake rejuvenation, Potable water production through desalination of lake water and Training of residents in matters of sanitation and water conservation.
3. Sarode, D.D., Gawande, S.M.: Quality improvement through soil stratum in non-mechanized treatment system for wastewater. In: International Conference on Reliability, Risk Maintenance and Engineering Management, 21–27 Springer Singapore (2019)
4. Gawande, S.M., Sarode, D.D.: Reuse of Wastewater to Conserve the Natural Water Resources, 353–367. Springer Nature Switzerland 2019 (2018)
5. Nemade, P.D., et al.: Wastewater renovation using constructed soil filter (CSF): a novel approach. J. Hazard. Mater., 657–665 (2009)
6. Bhalerao, S.A., Teli, N.C.: Significance of *Ipomoea Carnea* Jacq. Compreh. Rev. Asian J. Sci. Technol. **07**(08), 3371–3374 (2016)
7. Fatima, N.: A review on *Ipomoea Carnea*: pharmacology, toxicology and phytochemistry, https://doi.org/10.1515/jcim-2013-0046 (2014)
8. Eid, E.M., et. al.: Population Ecology of *Ipomoea Carnea* in Egypt. LAP LAMBERT Academic Publishing (2017)

Design of New Green Building Using Indian Green Building Council Rating System

Ashish S. Srivastava and Rajendra B. Magar

Abstract The concept of commercial green building is to minimize the wastages at all construction stages, including operation and maintenance. It requires to evaluate the building for a minimal additional cost of construction in comparison to the base case of typical building and reduction in lifetime use maintenance cost of a building. Thus, it is to ensure return on investment, not beyond the maximum of five years based on the lifecycle of a commercial building. The green building has a perspective in terms of planning and design to negate the overall impact due to construction on the universal environment and its corresponding effects on human health and wellbeing. Skillful use of water, energy, and other resources for sustainability and thereby reduce the impact on the environment, protect the health of occupants and thereby increase their productivity. The green features supplemented and considered are a minimal disturbance to landscape, site conditions available, also the effective use of recycled and recyclable environment-friendly building materials. Avoiding the use of toxic materials, like lead in interior paints, finishes and freon gas in air conditioning system, skillful use of portable water, and novel methods of recycling the used water. Effective and efficient use of energy by the use of low consumption energy efficient fittings and equipment. The use of renewable energy for augmentation of energy requirements. Efficient planning and design of quality of indoor air quality for the comfort of occupants. Increasing the overall efficiency of the building by the use of automation and building information management systems. This case study includes the overall practical evaluation of a proposed office building for the requirements as per intent. A programmed spreadsheet is prepared to comply with the intents as per the requirement of the Indian Green Building Council (IGBC) [1].

Keywords Green building · Sustainable building · Renewable energy · Recyclable material · Spreadsheet

---

A. S. Srivastava (✉) · R. B. Magar
Civil Engineering Department, Anjuman I Islam Kalsekar Technical Campus (University of Mumbai), Navi Mumbai, India
e-mail: ash@structechindia.com

D. K. Ashish et al. (eds.), *3rd International Conference on Innovative Technologies for Clean and Sustainable Development*, RILEM Bookseries 29,
https://doi.org/10.1007/978-3-030-51485-3_36

1 Introduction

Green building is also called high performance or a sustainable building that effectively uses energy water and other resources, including planning and designing, considering passive architecture. The approach of variably augmenting the efficiency of the building and its ancillary units which consume water, energy, and utilization of recyclable and recycled building materials, which also enhances human health and overall environment in the complete building life cycle. By the effective planning at the time of conceptualization, design of a building, during the construction period, operation, and maintenance of the building. Apart from renovation made during its life cycle and demolition after the satisfying complete life of the building. The concept of recycled concrete in India is not so far generically accepted, resulting in zero use of recycled concrete as aggregate for new concrete. Alternate sources for aggregates used in concrete requires new inspects, and further research and development for the use of recycled concrete as an aggregate should be carried out [2]. The concept of green building is nowadays getting attention all over the world, including India, who is a significant player in this field. The green building ensures minimal wastage in all stages of the use of buildings, including construction and operation throughout the life cycle of the building, which therein results in lower maintenance costs as compared to the base cost of a non-green building. The planning of "Green Building" includes various components like the use of rainwater harvesting techniques, innovative methods to avoid soil erosion, use of landscape to reduce heat island effect. In addition to minimization of consumption of potable water, wastewater recycling, and effective and efficient practices to save energy [3]. A sustainable solution in the field of public transportation needs urgent attention in major cities in India. Research carried out in Delhi shows alarming violations of permissible carbon oxide (CO), nitrogen dioxide (NO_2), particulate matter (PM) and other parameters. This trends to the adoption of sustainable solutions like extensive and exclusive coverage of Metro rail network, Electric vehicles, Use of CNG vehicles and other alternatives [4].

Green consultants' architects, engineers, town, and urban planners can achieve sustainable development by utilizing an integrated approach to plan design and execute any green building which efficiently uses resources and protects the environment for new construction, renovation, operation, and maintenance. Various government bodies, Environmental institutions, Municipal bodies motivate the construction industry. They further motivate them to carry out research and development in the recycling of various construction materials used in construction, specifically concrete due to significant impact by way of landfills, thereby causing environmental challenges [5]. By the use of a green building approach, one can negate the impact on our mother earth and thereby help in reducing global warming. In green building, the advent comprises of the coherence of building with the local ecosystem, for the adoption of the global environment. Green building intensifies the efficiency of energy and, in return, reduces the impact on the environment due to construction of the building and during operation and maintenance, which increases the efficiency and overall health of the occupants. Green Building concept has emerged in India from a small scale to nearly all new major constructions today. It is due to the adoption of sustainability goals by the Government of India and compulsion for all government agencies and public sector organizations to adopt and achieve green building certification. On the other side, corporates and Indian industries have adopted green construction and added more value to the sustainability community. It has resulted in extensive growth of green building professionals, the addition of research centers to carry out research and development in the area of green building, new products that are either recycled and or recyclable. It also increases the significant demand for skilled green building professionals and an uprising of the green community by various conferences and seminars all over India [6]. Green building has also become a standard for high society and marketing techniques for builders, developers, and corporates. Many a time frivolous approach is based on the adoption of western culture. Leadership in Energy and Environmental Design (LEED) is a green building rating system that ultimately represents the western approach. The Energy and Resources Institute (TERI), Energy Conservation Building Code (ECBC) and Green Rating for Integrated Habitat Assessment (GRIHA) are formulated very scientifically and efficiently for India specific use. These codes are not taken seriously and not practiced effectively as per traditional Indian architecture, the efficiency of energy and use of sustainable design is knitted skillfully as per the culture of India in GRIHA. India has a tropical climate with diversified and variable topography along with different social and cultural sentiments. In India, every part and region has its peculiar design and architecture for the climatic condition in that area which culturally has evolved as vernacular architecture. Time tested practices based on the evolution and experimentation of knowledge and practices are predominant and can be found all over the country. India is a country which has adopted sustainability in its culture and way of living. In developed and modern India, the demand for similar sustainability in the form of green building has forced significant corporates and small players to develop and construct sustainable building housing and complexes [7]. Ancient methods of construction have shown us the ways for developing modern methodologies of green building technology in line with the tradition and culture of a society [8]. The Vernacular architecture principles predominately are based on the local environmental condition, which has been entirely ignored by architects and planners in recent times. The planning techniques lessons taught by a vernacular architecture based on the conscious climate can be utilized in today's world to formulate new methods and techniques of green building [9]. The effect of improper planning and design may result in the incorrect prediction of requirements of the occupant and further results in poor performance which may include energy and or use of water effectively. The principal intent for the compliance of code of practice for building or green rating systems by the designer and architects, specifically for energy, natural light, ventilation and water used are usually counterfactual during the operation of a building Since the user of building modifies the system as per their specific requirements [10].

2 Aims and Objectives

The primary aim of this investigation is to find out the maximum savings in comparison with the base case, like the consumption of electricity, water, HVAC, and value engineering. More specifically, the research has the following objectives:

- To study the IGBC rating system for new green buildings.
- To study and find a methodology to appropriately calculate the credit calculations for the IGBC Abridge Reference Guide.
- To devise a method in a programmed spreadsheet for various options of compliances to find out the credit points.
- To prepare a case study using a spreadsheet program as per certification modalities of the green building rating system.

3 A Brief Review of the Literature

In the past 20 years, to increase sustainability as a tool and negate the impact on the environment, methods are developed that thereby add to the impact on the social, ecological, and economic aspects in the green building environment [7]. The sustainability measures are directly proportional to increasing the efficiency of green design to qualify as a green building. The minimal use of natural resources, including water and energy, give justice to the same. Also, the use of better indoor environment conditions impacts on the positive health of the occupants due to a better indoor environment. It involves educating the future occupants of the building on the correct and efficient way of using the building on its lifecycle [7, 11]. The benefits of incorporating green building practices in multiple studies demonstrated that it could help in mitigating problems, which include problems associated with the environment in an existing building and overall impacts in providing a better indoor environment and a healthy building user [6]. TERI-GRIHA, renamed as GRIHA, is an accredited national green rating system post-incorporation of numerous modifications and changes as given by architects and experts from various parts of India. To date, around 1733 building has been certified by GRIHA that comprises approximately 5.25 billion square feet of the total construction area. There has been a considerable increase in awareness for green buildings every day due to efforts imparted by IGBC, TERI and CII (Confederation of Indian Industry) [12]. Apart from savings and obtaining efficiencies in energy and water, additional sustainable improvements to be achieved by other compliances need consideration. It also will have an insignificant impact on the utilization and use of natural resource resources water consumption, reduction of waste due to the use of recycled or recyclable building construction materials [5].

Building performance improvisation by compiling sustainable policies for procurement during construction, use, and maintenance of buildings.[2, 13]. Proper and efficient green designs can achieve a green building in addition to sustainable strategies from the point of planning of a green building, construction stage, useful life and up to the demolition stage. "Whole building" integrated approach may be used for green buildings that depend on the phase of building cycle in totality which includes the stage of design, construction, operation and demolition of a green building [14]. Green buildings represent the overall responsibility for the building industry to reduce the impact on environment, economic, and social parameters on the building industry. Lesser consumption of natural resources will lead to a lesser impact on the natural environment. The use of sustainable best practices today will sustain the environment for future generations. so that they can mend their need for non-sustainable natural resources we have to use green practices today to consume lesser natural resources and lesser impact on the present environment of our mother earth [15].

Policies of government should mandate the effective use of water energy and emissions due to CO_2 [4]. As per Kohler, economic, cultural, and sociological sustainability are the three primary and vital dimensions for construction, maintenance and operation of any sustainable green building [16]. Due to much use of natural resources and thereby increase in carbon emission is due to the increase in building footprints [17]. For improving the green building management system for any construction project for sustainability, one has to consider the cost of construction at the time of planning. The pattern of green building project management for the life cycle of the building has to be on cost-effective solutions by the involvement of multi-disciplinary project team before and during the construction. The practice of motivating contracting firms shows significant results apart from achieving sustainable benchmarks. The practice of implementation of skills like public relations, hierarchical communication, and training to technical team, skilled and unskilled workforce gives better results for sustainability [18]. Around 40% of the material, including energy, 25% of the wood harvest, and around 17% of freshwater withdrawal is due to buildings [19].

As per records of the department of energy conducted in 2002 in the United States, there are 76 million residential buildings and approximately 5 million new green buildings [20]. The environmental benefits of using new green buildings have been studied by way of commission surveys, analytical reports, and thorough analysis by the U.S. federal government and alternatively by third-party organizations, including U. S. Green Building Council (USGBC). Around 70% of electrical energy consumption, 39% for allied energy use, 30% for the emissions due to greenhouse gases and 12% for the total water consumption of portable use accounts for the built environment in the United States of America [21].

A spreadsheet for Windows 10 generally is used in all fields of specialization. The use of automated calculations, preparing pivot tables, preparation of various graphs, bar charts, flow charts, tree diagrams, and gives freedom to a researcher for making their macro program using visual basic for application. The excel application used for solving various numeric problem statements and summaries and the results in the form of a spreadsheet. It also allows the end-user to furnish answers for programmed questionnaires and give output in the form of reports [22].

4 Methodology

IGBC rating program gives a direction to enable the green designers for the use of green concepts in addition to reducing environmental impacts, which can be measured. A tool to measure green concepts and environmental impacts are currently not available. Thus to ensure the final results of the credit are consistent and reliable, the methodology of the calculation is based upon a decision support system in the form of a spreadsheet as per the information of the user input and proper green measures selection. Spreadsheet prepared shows the calculated saving in operational cost and thereby reduction of carbon emission and gives the comprehensive statistics of performance which in return enables to meticulous create a business case for a green building. Only a handful of a measure planned for enhancing a positive building performance results in lesser utility cost, the longer life cycle of equipment, and lesser use of natural resources.

The basis of calculating credit points currently is as per the experience of the green building consultant. After researching a proper green building design tool, no soft computing method tools are present to calculate the credit points, as mentioned in IGBC Abridge Reference Guide Version 3.0.

5 Case Study

Given the problem mentioned above, as specified from the literature review, the following scope is outlined for the present investigation. The program prepared is intended to meet the demand for a quick, affordable, and simple tool to be utilized for the planning and assessment of the design of resource efficiency in order to escalate green building footprints in emerging markets. The details of the present study, as depicted in Table 1, are as follows:

Figure 1 shows the key-plan of the proposed AO/DO building is enclosed herewith as follows.

The above layout shows the location of the proposed office building; the natural topography has been retained and used for the green belt and other ancillary utilities. The provision of parking is made in the stilt area and also in a dense green zone. The natural pond has been retained and beautified for aesthetical purposes. Near the entrance of the building, the parking is provided and earmarked for the easy of specially-abled person. Also charging stations for the electric vehicles are provided in designated areas.

Table 1 Case study

Proposal	As desired by IOCL, SCEPL appointed as an Architect and consultant for the design and construction of a new green building located at Chithrappuzha. The work involves IGBC facilitation to achieve green building rating for the proposed development
Name of proposal	Area office and divisional office for IOCL at Survey No: 360/4, 360/5, 360/19, 360/23, 360/27, 360/28, 360/29, 360/1, 360/3 at Chithrappuzha-Ponjassery Road, Irumpanam, Thrippunithura, Ernakulam, Kerala-682 301
Name of owner	Indian Oil Corporation Limited, Kerala State office, Panampilly Nagar Avenue, Panampilly Nagar, Ernakulam, Kerala-680 063
Stage of proposal	Preliminary planning stage
Total area	1020.00 Sq. M. (Approx.)
Plot area	4162 Sq. M
Architect	Structech Consulting Engineers India Pvt. Ltd, Mumbai
General	Proposed green building with gold rating
Floor to floor height	3.2 m
General depth of slab	130 mm
Total height	22.60 m
Soil strata	Loose Clayey, with the water table at less than 1 m

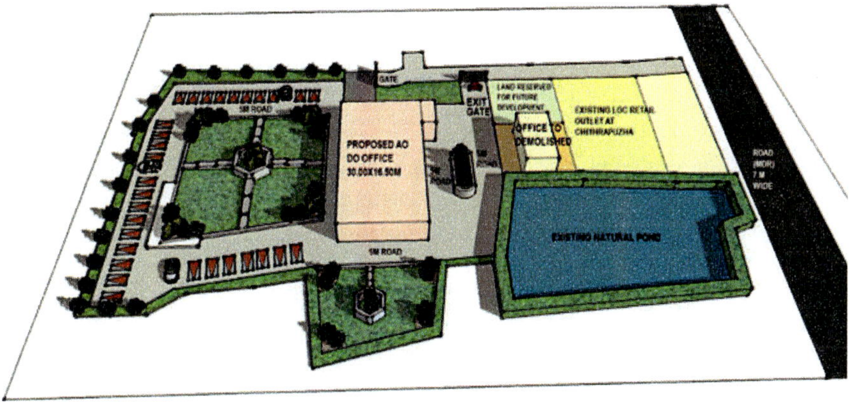

Fig. 1 Site plan showing the AO/DO office for IOCL

6 Results and Discussions

A Case study of a new commercial green building for Indian Oil Corporation Ltd, at Kochi, Kerala state for its area office and divisional office, is considered for the calculations. Based on credit no, intent and compliance total number of credits have been obtained using the programmed spreadsheet. The following are the planned

Table 2 Sustainable architecture and design

Credit no	Description	Points credited	Available points
SA credit 1	Integrated design approach	1	1
SA credit 2	Site preservation	2	2
SA credit 3	Passive architecture	2	2

activities considered for obtaining a total number of credits as per the IGBC rating system. The above is about meticulous and sustainable architecture and design of the building, proper site selection and appropriate planning thereof, innovative methods of water conservation, efficient energy planning and use of low consumption light fittings, selection of recycling and recyclable building materials preferably within the proximity of site, sustainable indoor air quality and use of non-toxic material for indoor application and innovative and sustainable development in terms of sustainability. Table 2 shows the credit points achieved for directional intent, as given by IGBC for Sustainable Architecture & Design.

The above calculations of the present case study basis are on the cost of building materials and resources. The number of occupants and annual water conservation, as shown in Figs. 2, 3, and 4.

The basis of the Building Materials and Resources (BMR) are on four criteria such as BMR Credit 1, which includes encouragement of the use of building material bearing negative impact on the environment in addition to the reuse of salvage material up to 2.5%. BMR Credit 2 is ensuring adequate management of waste by way of treating a minimum of 50% organic waste. BMR Credit 3 recommends a demonstration of a minimum of 75% of construction waste. BMR Credit 4 recommends the effective use of green building components that are certified and ensures the use of passive and active products. From Fig. 2, local building materials like metal, boulders, sand, red stone for blockwork are used widely in reinforced cement concrete, and block masonry work is highest as per BMR 1.

From Fig. 3, the ground floor is having Transient maximum occupants (since the Indian Oil public relations officer will be sitting at the ground floor in addition to reception and dispatch room). The first floor and second floor are having staff and officer work stations, where there will be very few transient occupants. The third floor is having the main conference room, training room, and mini-conference room, where the transient occupant will be maximum. Whereas the fourth and fifth floors have Area-office and Divisional-office respectively, where there are the staff and officer's work station and less transient occupants.

Further, from Fig. 4, water at the site, i.e., municipal water, groundwater, treated wastewater and stored rainwater. The water demand for the building includes demand

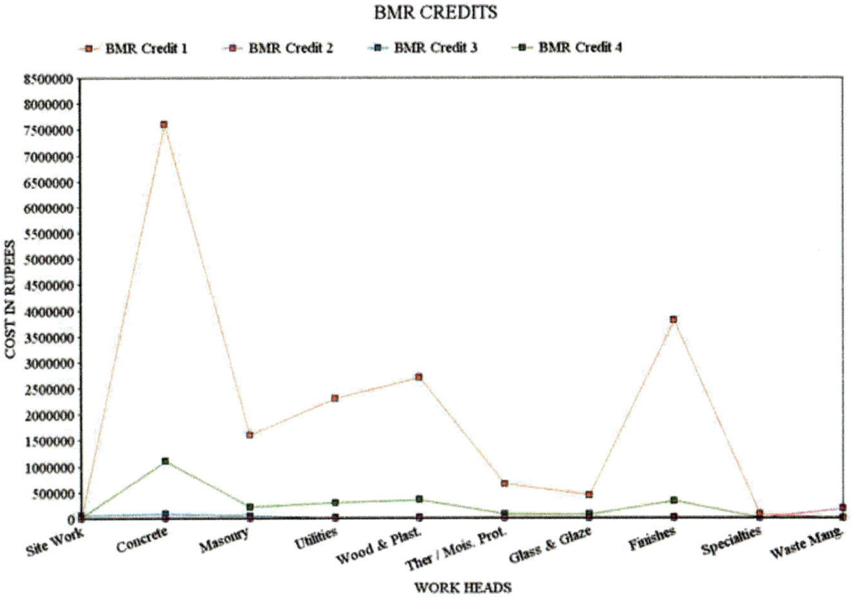

Fig. 2 Cost of building materials and resources

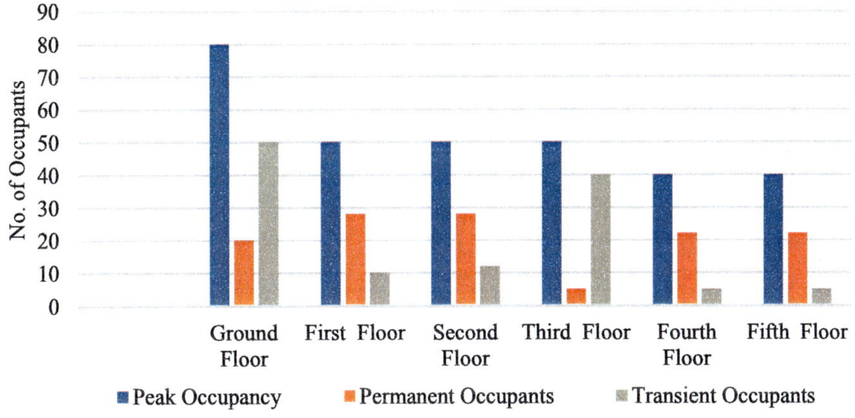

Fig. 3 Occupants living

for sanitation and plumbing, landscaping, and air conditioning, considering one full-time occupancy (F.T.E). The reuse of treated wastewater is 45.45% of the total water demand.

Table 3 shows the credit points achieved for directional intent, as given by IGBC. For Site Selection and Planning, are achieved by preparation of documentation of local building regulation compliance by enclosing proof of ownership, building plans

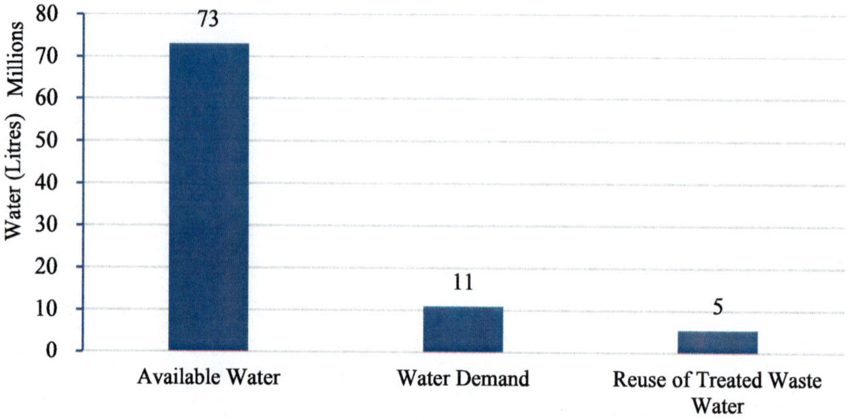

Fig. 4 Annual water conservation

Table 3 Site selection and planning

Credit no	Description	Points credited	Available points
SSP mandatory requirement 1	Local building regulations	Mandatory	Mandatory
SSP mandatory requirement 2	Soil erosion control	Mandatory	Mandatory
SSP credit 1	Basic amenities	1	1
SSP credit 2	Proximity to public transport	1	1
SSP credit 3	Low-emitting vehicles	1	1
SSP credit 4	Natural topography or vegetation	2	2
SSP credit 5	Preservation or transplantation of trees	1	1
SSP credit 6	Heat Island reduction-non-roof	2	2
SSP credit.7	Heat Island reduction-roof	2	2
SSP credit 8	Out-door light pollution reduction	1	1
SSP credit 9	Universal design	1	1
SSP credit 10	Basic	1	1
SSP credit 11	Green building guidelines	1	1

as per Floor Space Index (FSI), structural, mechanical, electrical and plumbing (MEP) drawings, also the appointment of an architect, structural engineer, MEP engineer. The other mandatory compliance of soil erosion control measures like stacking of topsoil, temporary and permanent seedings, and sedimentation control measures if found on site. Provision of a minimum of seven basic amenities. Proof of proximity to either public transport within 800 m walking distance or provision of shuttle service from the proposed building to the nearest transport hub. Encouraging electric and Compressed Natural Gas (CNG) powered vehicles for a minimum of 5% of the building occupancy. The natural topography and vegetated area to be retained a minimum of 15% of the plot area in addition to a minimum of 30% of the vegetation in the form of vertical walls and on the built area. Preservation of a minimum of 75% of existing trees on the plot in addition to the planting of a minimum of eight trees per acre, which will attract biodiversity. Reduction of heat island effect on the non-roof area by provision of the planting of saplings in an open area, provision of grass-open grid pavers in driveways and pedestrian walkways, all pavers used shall have solar reflectance index (SRI) between 29 and 64. Reduction of heat island roof by providing high SRI tiles on roof area and provision of the garden over the roof area. Reduction in outdoor night lights by providing all lights below nadir level and use of simulation approach by reduction of minimum 30% light power density (L.P.D.). Provision for convenience in building design for blinds, senior citizens and physically disabled persons. Provision of facilities for construction workers (both male and female) as per government regulation. On-site training and information of dos and don'ts for occupancy of building upon possession. The specific credit points accumulated for the proposed building.

Table 4 shows the credit points achieved for directional intent, as given by IGBC. Mandatory compliance of capturing minimum peak rainfall of a day from non-roof

Table 4 Water conservation

Credit no	Description	Points credited	Available points
WC mandatory requirement 1	Rainwater harvesting, roof and non-roof	Mandatory	Mandatory
WC mandatory requirement 2	Water efficient plumbing fixtures	Mandatory	Mandatory
WC credit 1	Landscape design	1	2
WC credit 2	Management of irrigation system	1	1
WC credit 3	Rainwater harvesting, roof and non-roof	3	4
WC credit 4	Water efficient plumbing fixtures	2	5
WC credit 5	Waste water treatment and reuse	3	5
WC credit 6	Water metering	1	1

Table 5 Energy efficiency

Credit no	Description	Points credited	Available points
EE mandatory requirement 1	Ozone depleting substances	Mandatory	Mandatory
EE mandatory requirement 2	Minimum energy efficiency	Mandatory	Mandatory
EE mandatory requirement 3	Commissioning plan for building equipment and system	Mandatory	Mandatory
EE credit 1	Eco-friendly refrigerants	1	1
EE credit 2	Enhanced energy efficiency	2	15
EE credit 3	On-site renewable energy	2	6
EE credit 4	Off-site renewable energy	0	2
EE credit 5	Commissioning, post-installation of equipment and systems	2	2
EE credit 6	Energy metering and management	2	2

areas and building roof areas. Methods of increasing the groundwater table. Other mandatory compliance of using low flow rate fixtures with aerators, since the building in consideration is public sector undertaking (PSU), considered one full-time equivalent (FTE). Landscape design compliance executed by the use of drought-tolerant trees with minimum use of turf area. Provision of a drip irrigation system for landscape areas in addition to the provision of multi water meters to control water use. Use of all plumbing fixtures with a minimum 8% lesser than base criteria. On-site installation of the wastewater treatment plant and reuse of water for flushing and landscape use. Provision of water meters at all locations like flushing, cleaning, landscape, and others.

Table 5 shows the credit points achieved for directional intent, as given by IGBC. Mandatory compliance of use of halon free fire systems and the use of chlorofluorocarbon (CFCs) free refrigerant gases in addition to the prescriptive approach to building. The next mandatory requirement of planning and placing a commissioning engineer since the construction stage to understand the systems and equipment provided in the building. Minimum 1% of the energy consumption of buildings provided by using the On-Grid Solar Plant. Off-site renewable energy and renewable energy purchase certificate not feasible due to the small size of the proposed building. All equipment and systems used are also documented and a maintenance chart for post-occupancy prepared for energy metering and energy compliance achieved by providing total automation of lighting, air conditioning, and ventilation.

Table 6 shows the credit points achieved for directional intent, as given by IGBC for Building Materials and Resources (BMR). The mandatory BMR has been complied with by the provision of centralized and separate bins for various wastages during and post-occupancy. The compliance for sustainable building material achieved is by using recycled materials during construction in addition to the

Table 6 Building materials and resources

Credit no	Description	Points credited	Available points
BMR mandatory requirement 1	Segregation of waste, post occupancy	Mandatory	Mandatory
BMR credit 1	Sustainable building materials	6	8
BMR credit 2	Organic waste management, post occupancy	0	2
BMR credit 3	Handling of waste materials, during construction	1	1
BMR credit 4	Use of certified green building materials, product & equipment	5	5

use of recyclable construction materials. The use of local material propagated along with the use of forest stewardship council (FSC). Since the building is a small office building for a PSU, thus organic waste shall be minimal and will be handed over to the municipal body. The compliance for the handling of waste material during construction to be achieved by demonstrating a minimum of 75% by way of recycling at the site. Certified green building materials conceptualized at the time of planning to be used during construction.

Table 7 shows the credit points achieved for directional intent as given by IGBC for Indoor Environmental Quality. The mandatory requirement for fresh air and

Table 7 Indoor environmental quality

Credit no	Description	Points credited	Available points
IEQ mandatory requirement 1	Minimum fresh air ventilation	Mandatory	Mandatory
IEQ mandatory requirement 2	Tobacco smoke control	Mandatory	Mandatory
IEQ credit 1	CO_2 monitoring	1	1
IEQ credit 2	Day-lighting	1	2
IEQ credit 3	Outdoor views	1	1
IEQ credit 4	Minimized indoor and out door pollutants	1	1
IEQ credit 5	Low-emitting materials	3	3
IEQ credit 6	Occupants well being facilities	1	1
IEQ credit 7	Indoor air quality testing, after construction and before occupancy	1	2
IEQ credit 8	Indoor air quality management, during construction	1	1

ventilation during construction to be documented in the quality assurance strategy for project management consultant (PMC). For controlling tobacco smoke, strict provisions made in terms and conditions of contractors, including the provision of various signages as per prevailing law. Since the anticipated building is entirely air-conditioned. Thus the CO_2 monitoring is planned by the provision of CO_2 sensors inside and outside the building to monitor differential CO_2 level (not more than 530 ppm). Unique dampeners provided in the airconditioning system allow fresh air from time to time. For the daylighting, the measurement approach adopted for a minimum of 75% of the office area provided by natural lighting. The minimum area of 75% is having access to outside views with a non-opaque glass facade. The provision of the carpet has been anticipated in the entrance canopy area to avoid outside pollution. The provision of air-curtain at the entrance of the building. Special provisions in the air-conditioning system by using a cloth-based air filter and ultraviolet (UV) lamp. Special provisions made in items of the bill of quantity (BoQ) specifying lead-free paints and finishings. Occupant wellbeing compliance made is by providing gymnasium, yoga and aerobics centers. Indoor air quality testing provision made in quality assurance strategy in PMC contract are pre and post-occupancy. The compliance for indoor air quality management during construction documented in contractors agreement by way of taking precautions like covering the materials and equipment.

Table 8 shows the credit points achieved for directional intent, as given by IGBC for Innovation and Development. The compliance for the innovation and design process considered is in architectural planning and value engineering in MEP. Optimization in a structural design carried out is for optimum and safe design. Fly ash lightweight bricks are considered, thereby reducing the dead load on the building. Compliance for water used reduction during construction achieved are for a minimum of 10% in comparison to the base case. The author is an IGBC accredited professional and working in the area of green building.

The above activities and sub-activities extracted from IGBC Green New Buildings Rating System, Version 3.0 and systematically prepared in a spreadsheet using advance visual basic for application, developed own methodology for practical calculations using an advanced spreadsheet. The certification rating of IGBC has been obtained based on seven crucial parameters, as mentioned above.

Table 9 shows the main page for a selection of credit description mentioning the credit number, points earned, and points available for the subject credit number. This spreadsheet prepared with a drop-down list for the compliance options for the intent,

Table 8 Innovation and development

Credit no	Description	Points credited	Available points
ID credit 1	Innovation in design process	1	4
ID credit 2	Optimization in structural design	1	1
ID credit 3	Water use reduction for construction	1	1
ID credit 4	IGBC accredited professional	1	1

Table 9 Visual Basic Modified Spreadsheet Showing Credit Interpretation

Module	Credit no	Description	Intent	Compliance	Points credited	Available points	Status
Sustainable architecture and design	SA credit no. 01	Integrated design approach	To demonstrate involvement of all team members from multidisciplinary field till completion stage	List of project team members with contact details, designation of roles and responsibility of all team members, documentation of more than three meetings at different stage of project with photographs	1	1	Ok
	SA credit no. 02	Site preservation	To retain the existing site features so as to minimize the site damage and associated negative environmental impact	Retain 50% minimum site contour including building area, retain 100% water bodies present, retain minimum 50% natural rock excluding building area by surface area method, retain minimum 10% topography or landscape area on plot, preserve minimum 75% of existing trees in the plot	2	2	Ok

(continued)

Table 9 (continued)

Module	Credit no	Description	Intent	Compliance	Points credited	Available points	Status
	SA credit no. 03	Passive architecture	Use of passive architecture design features to minimize negative environmental impact	Achieve minimum 2% energy saving of total annual energy consumption by adopting passive measure-simulation approach, use of climate-responsive concept and design-simulation approach	2	2	Ok

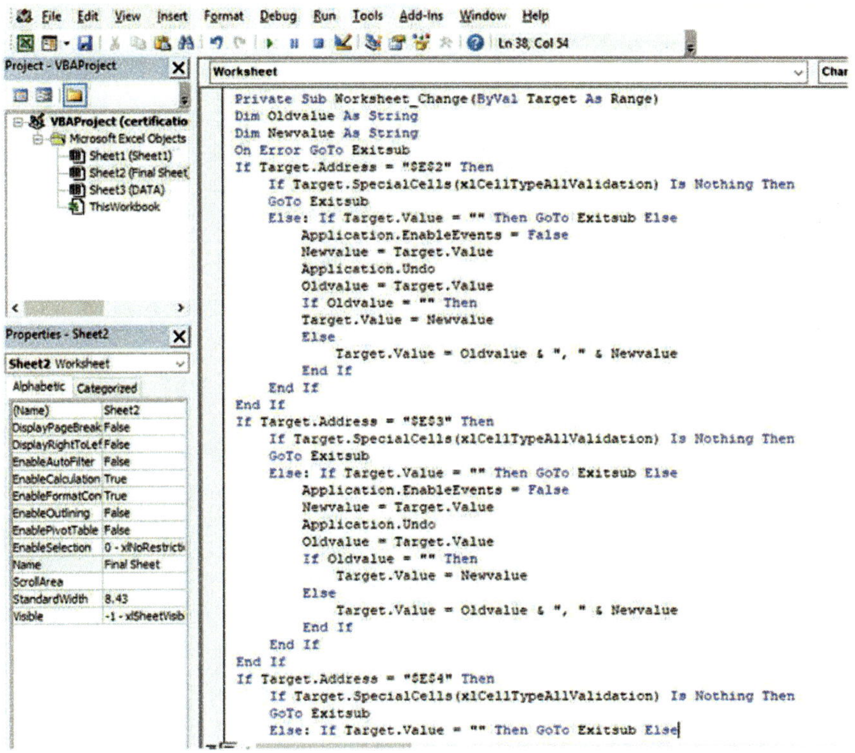

Fig. 5 Sample visual basic program

which thereby selects the credit points applicable to the compliance automatically without any human interaction.

Figure 5 shows the sample visual basic programming for the selection of compliance options for intent.

Certification is initiated at a very early design stage when the details of project IOCL, AO/DO office building entered into the spreadsheet program, and green compliance options are selected. The project has achieved the IGBC standard of a 20% savings in energy, material and water and materials as compared against base construction practices. When the project achieves this, the project registration made for gold certification. Below are the credit points achieved are as follows:

Table 10 shows the credit points achieved for the intent as given by IGBC for Credit Achievement and upon compliance of the intended points, a gold rating achieved for the IOCL, AO/DP Office Building.

As per Table 10, observed that the green building has complied with the intents as given by IGBC to achieve credit points to obtain a gold rating.

Table 10 Credit achievement

Credit points	Description
Earned 64	The IOCL, AO/DO office building, has provided the mandatory documentation which supports. Achievement of the credit requirements and associated points. For Version 3.0, the documentation of this credit is complete. Furthermore, the points considered as achieved, designated as '**Earned**'
Denied 00	The IOCL, AO/DO office building, has not applied for certification. Thus, the project has not demonstrated the achievement of these credits and designated as '**Denied**'
Final rating	'**Gold**'

7 Conclusions

As per the procedure is given and advanced spreadsheet program using a visual-basic application concluded that the computation of credits based on the compliance options selected by the users, a near to certified credit rating could be achieved ($\pm 10\%$). It is beneficial in terms of saving for Engineers/Architect or Green building consultants on their professional working day hours, thereby saving in the organizational resources and raising the output. The interpretations scheduled in the spreadsheet give an easy to use solution to the users to select the options for the compliance of the modules in a spreadsheet as given by the IGBC Green New Building Rating System checklist.

References

1. Indian Green Building Council. Web Address—https://www.igbc.in
2. Bansal, S., Singh, S.K.: Sustainable handling of construction and demolition (c & d) waste. Int. J. Sustain. Energy Environ. Res. **4**(2), 22–48 (2015)
3. Kevadiya I.G., Patil A.A., Waghmode, S.N.: Sustainable construction: green building concept—a case study. Int. J. Innov. Emerg. Res. Eng. **2**(2) (2014)
4. Gour, A.A., Singh, S.K., Tyagi, S.K., Mandal, A.: Variation in parameters of ambient air quality in National Capital Territory (NCT) of Delhi (India). Atmos. Clim. Sci. **5**, 13–22 (2015)
5. Bansal, S., Singh, S.K.: A sustainable approach towards the construction and demolition waste. Int. J. Innov. Res. Sci. Eng. Technol. **3**(2) (2014)
6. Ahn, Y.H., Pearce, A.R., Ku, K.: Paradigm shift of green buildings in the construction industry. Int. J. Sustain. Build. Technol. Urban Dev. **2**(1), 52–62 (2011)
7. Ahn, Y.H., Pearce, A.R.: Green construction: contractor experiences, expectations, and perceptions. J. Green Build. **2**(3), 106–112 (2007)
8. Harris, J., Diamond, R., Iyer, M., Paul, H.: Towards a sustainable energy balance: progressive efficiency and the return of energy conservation. Energy Effi. **1**, 175–188 (2008). doi: 10.1007/s12053-008-9011-0
9. Ar. Ganguly R.T.: Role of vernacular architecture of India in green building design—a case study of Pauni Int. J. Recent Innov. Trends Comput. Commun. **3**(2) (1993)
10. Gorgolewski, M., Brown, C., Chu, A.M., Turcato, A., Bartlett, K., Ebrahimi, G., Hodgson, M., Hill, S.M., Ouf, M., Scannell, L.: Performance of sustainable buildings in colder climates. J. Green Build. **11**(4) (2016)

11. Kibert, C.J.: Sustainable Construction: Green Building Design and Delivery, 2nd edn. Wiley , Hoboken (2008)
12. Balakrishna, B.: Gaining Ground—Sustainable Buildings Rising in India. Master Builder (2011)
13. Bansal, S., Biswas, S., Singh, S.K.: Approach of fuzzy logic for evaluation of green building rating system. Int. J. Innov. Res. Adv. Eng. (IJIRAE) **2**(3), 35–39 (2015)
14. Boecker, J., Horst, S., Keiter, T., Lau, A., Sheffer, M.B.T., Reed, B.: The Integrative Design Guide to Green Building. Wiley, Hoboken (2009)
15. Brundtland, G.H.: Protecting the Global Commons, p. 12. Earth Ethics, Fall (1989)
16. Kohler, N.: The relevance of the green building challenge: an observer's perspective. Build. Res. Inf. **27**(4/5), 309–320 (1999)
17. Montoya, M.: Green Building Fundamentals: A Practical Guide to Understanding and Applying Fundamental Sustainable Construction Practices and LEED Green building rating system. Pearson Education Publisher Edition 2 (2009)
18. Robichaud, L.B., Anantatmula, V.S.: Greening project management practices for sustainable construction. J. Manage. Eng. ASCE **27**, 48–57 (2011)
19. Roodman, D.M., Lenssen, N.: A Building Revolution: How Ecology and Health Concerns Are Transforming Construction, Worldwatch Paper 124, p. 5. Worldwatch Institute, Washington (1995)
20. U.S. Green Building Council.: Building momentum: National trends and prospects for high-performance green buildings Senate Committee on Environment and Public Works. Rep. Prepared for the U.S, Washington (2003)
21. U.S. Green Building Council.: Building momentum: National Trends and Prospects for High-Performance Green Buildings Senate Committee on Environment and Public Works. Rep. Prepared for the U.S, Washington (2006)
22. Microsoft Excel.: Stable released 16002.12325.20032.0/December 12, 2019 (2019)

Characterization and Optimization of Polyurethane Based Bituminous Waterproofing Membrane

Ashmita Rupal, Sanjay Kumar Sharma, and G. D. Tyagi

Abstract Now-a-days with the advent in technological advancements, bituminous compositions modified with polymer are being used for numerous applications such as joint filers, adhesives, protective layers, impregnating agents and roofing material The bitumen exhibits a number of mechanical properties but the conventional bitumen doesn't congregate many requirements because of its drawback of high susceptibility to temperature. Hence from a product design point of view; the most preferred form is modified bitumen. With this aim, a composition in the form of waterproofing product has been developed by modifying bitumen with polyurethane. Polyurethane because of its attractive morphology and wide range of advantageous mechanical properties is a highly promising material having the ability to alter its microstructure to suite niche applications. The developed matrix incorporates the basic useful properties of bitumen along with excellent properties of polyurethane. This modification will maximize the efficiency of product developed in different scenarios making it a resourceful product in terms of waterproofing, insulation properties and enhanced durability. This paper highlights the experimental investigations of physico–mechanical properties of the matrix developed in the form of membrane of varying thickness made from different ratios using different grades of polyurethane and bitumen. Different ratios, grades and thickness of material have been used in trail mixes to attain optimized matrix for the product. The properties such as tensile, elongation, tear, shore hardness, puncture resistance, etc. of membrane developed of various matrixes are examined with relevant ASTM codes. The research work presented here provides an experimental implication for design; development and application of membrane developed for use in engineering purpose.

A. Rupal (✉) · S. K. Sharma
Department of Civil Engineering, NITTTR, Chandigarh, India
e-mail: rupalashmitasharma@gmail.com

S. K. Sharma
e-mail: rupalashmitasharma@gmail.com

G. D. Tyagi
Shivalik Agros Poly Products Ltd., Parwanoo, HP, India

D. K. Ashish et al. (eds.), *3rd International Conference on Innovative Technologies for Clean and Sustainable Development*, RILEM Bookseries 29,
https://doi.org/10.1007/978-3-030-51485-3_37

Keywords Polyurethane (PU) · Waterproofing membrane · Mechanical characterization · Optimization · Shore hardness · Elongation · Puncture resistance · Tensile strength · Tear load

1 Introduction and Research Significance

Water and other fluids permeation in concrete results in deterioration of structures which not only affect the structure aesthetically but shorten its life by affecting the durability of structure. Most of the time water leakage or permeation is viewed as an aesthetic nuisance rather than a severe matter. Also the trapped dampness in the insulation decays the roof deck leading to premature failure of roofs. Water permeation in the structures carries chloride ions, which as a consequence result in corrosion of reinforcement steel bars causing expansion which in turn results in spalling of concrete [1]. Solution to these problems can be catered by the surface treatment or developing an impermeable barrier above/beneath the structural members of the building preventing water passing through them or simply called waterproofing [2]. By giving attention towards the cause and mechanics of water permeation, more considerate actions will be implemented for keeping the structure waterproof.

The technological changes due to advancements continue to happen with respect to design, material and in the construction industry. In present various types of waterproofing practices and products are available such as Bituminous Coating, EPDM Waterproofing Membrane; Polymer modified Liquid Membrane, Cementations Waterproofing, HDPE, Bituminous Waterproofing Membrane (APP) etc. [3]. The problems associated with the use of different innovations and new upgraded technology is keeping practitioners up to date about the benefits and attributes of utilizing them in their work area [4]. However the application and use of polymers is growing as an integral component in the construction industry. At present these types of composite building materials are being used along with application in other fields such as coatings, textiles, adhesives, automotive, furniture, nanotechnology etc. [5]. Modified bitumen in many forms is very useful as waterproofing layers, roofing and joint sealing applications and self adhering components [6, 7]. Polymer modified composite materials are being used in the manufacturing of coatings and sealing compounds [8]. Polyurethane is one of the most used products amongst polymers because it exhibits excellent physical as well as mechanical properties such as high elongation range, thermal stability, easy applicability, high energy absorption capacity, chemical resistance, high resistance in various environmental conditions and cost effectiveness [9, 10]. These specialties make polyurethane products more versatile in applicable areas.

Research Significance The multi component matrix or compositions are more efficient in various terms and those matrix account for overall higher costing of product thus implying application constraints. That's the reason the proportioning of components of the matrix to optimize the properties as per application required is to be carefully formulated. Another very important parameter is the curing as temperature variations effects significantly in the polymerisation process. The chances of foaming of polymer is also there which results in substantial changes in performance of the composite.

It can be described that the effectiveness of the composite can be represented as a complex function which depends on the components, their formation, placing, quantities or proportioning of components. So they have to be formulated to be compatible with application surface and exposure conditions. Here the composite has been formulated for making a coating with the purpose of waterproofing of concrete structure which will act as a protection layer around them. To test its efficiency of performance various tests of mechanical properties have been investigated. As the theoretical methods which are used for designing the coatings for concrete structures have achieved a very limited success till now.

This work was initiated after reviewing the literature and identifying the gaps as most of the research studies have focused on bituminous blends with different polymers for waterproofing material such as oxidized bitumen membrane, Atactic Polypropylene (APP), SBS; Polymer bitumen membrane and on other liquid applied products as fibrous acrylic, liquid rubber and liquid silicone etc. So the need was felt to identify and study the use of different materials or their blends for incorporating their properties and developing a new and more efficient product for waterproofing. Henceforth, there is a need for development of such a product for waterproofing which incorporates the basic useful properties of bitumen modified with a polymer and also utilizes properties of nanoparticles for maximizing the efficiency in different scenarios making it a versatile product in terms of waterproofing, insulation properties and enhanced durability.

Various researchers have assessed the durability of concrete structures which were coated with different coatings namely polymer emulsion, acrylic, epoxy, rubber, and polyurethane. PU was superior to all others in all the experimental tests of water absorption, permeation, chemical and chloride resistance.

2 Experimental Work

2.1 Materials Used

A liquid solvent free or low solvent hardenable polyurethane modified bitumen composition is described which is used for developing a waterproofing membrane for concrete. The various combinations of different grades of polymer (PU) and bitumen were characterized and tested for developing the polyurethane modified bitumen membrane for the desired purpose. The material and their grades used in the research work are

Bitumen: Industrial Grade Bitumen and VG30.

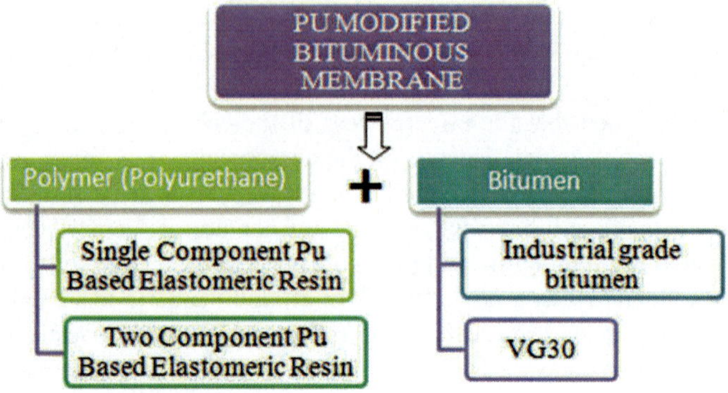

Fig. 1 Materials used in experimental work

Polymer (Polyurethane): Single and Two component Polyurethane (PU) based Elastomeric Resin (Poly-MDI) (Fig. 1).

The optimized polyurethane (PU) modified bitumen blend is selected after mechanical characterization for the product development. The purpose of developing stable mix proportion giving a uniform blend for developing membrane is to utilize the individual properties of polyurethane and bitumen by optimizing their mixing proportions accordingly [11]. The change in proportions of components vary according to usage sector or per application [12]. The combining-weight ratio of the polymer to the bitumen shall be in the range of 70:30–30:70 [13]. In the mentioned research work, four different combinations of different grades of bitumen and polyurethane were prepared, developed as membrane and testing of their mechanical properties was performed. The combination ratio of the amount of polymer (polyurethane) component to that of bitumen (depending on grade) is kept 50:50 [14]. The Polyurethane used in the experimental work has been procured from Shivalik Agro Poly Products Ltd, Parwanoo, H.P., India named as Shivabond 315 Grade PU and Shivagen 004. Table 1 shows the specifications and properties of the material considered.

The bitumen used in the research work is VG30 and Industrial Grade Bitumen (blown bitumen). Being inexpensive materials having wide applications ranging from waterproofing characteristic, adhesion, plasticity and chemical resistance. Both grades have been procured from local suppliers. Their properties and specifications are shown in Table 2.

For coating to act as a water-proofing barrier for concrete structures, it should possess the basic requirement of being watertight and having enough tensile and elongation properties to cater all the movements of the structures being elastic enough having great adhesion property with the substrate. Apart from it should have strong adhesion with the substrate. Resistance qualities to UV radiations, chemical wear tear, thermal stresses and other microbial attacks should be present.

Table 1 Specifications and properties of different grade of polymer (polyurethane)

Property	Values (Shivabond 315)	Values (Shivagen 004)
Density	1.10 ± 0.10	0.90 ± 0.10
Consistency	Medium viscosity	Low-medium viscosity
Solid content (min %)	100	100
Elongation (%)	250–350	300–400
Tensile strength (kg/cm^2)	2 ± 8	2 ± 10
Application limits (°C)	0–50	0–50
Dry time at 25°, 90% Rh (h)	4 touch dry	4 touch dry
Full cure time (days)	7	7–10 days
Resistance to industrial environment	Excellent	Excellent
Flexibility	Pass to test is: 101–1964	Pass to test is: 101–1964
Adhesion	Excellent	Excellent
Hardness, shore A (min)	5–20	20–40

Table 2 Properties and specifications of VG 30 and Industrial Grade Bitumen

Property	Values (VG30)	Values (industrial bitumen)	Test method
Specific gravity	1.01	1.02	IS: 1202-1978
Softening point (°C)	47	85	IS: 1205-1978
Penetration at (°C), 100 gm, 5 s (1/10 mm)	68	73	IS: 1203-1978
Ductility at 25° C after thin film oven test, min (mm)	40	25	IS: 1208-1978
Flash point Cleveland open cup, min (°C)	350	240	IS: 1209-1978

2.2 *Specimen Preparation and Curing*

The optimized mixed proportion is attained by selecting the sample of combination showing best suitable properties after testing. The various samples of combinations is achieved by the mixing of the different grades of bitumen and polyurethane in conventional aspects while maintaining ambient temperature. To get maximum moisture exclusion, it is more likely to carry the experimentation under inert gas atmosphere. The material mixed is then poured on metal substrate mould of particular thickness specially designed to get accuracy for formation of thin membrane/film. The composition achieved is in a liquid phase which hardens in 7–10 days through atmospheric humidity at ambient temperature (25–28 °C) and cures in a polyfunctional manner [15]. To avoid occurence of undulations because of foaming of membrane developed; mixing of the materials should be taken care of very well. The cured membrane acts as an "in situ' formed self adherent waterproofing membrane to the concrete surfaces or on the other metal substrate.

Figure 2 shows the sample of membrane developed varying thickness of 2, 3 and 4 mm on a metal substrate. The composite was made of the combination of different grades of bitumen and polyurethane. The softening point and penetration of the polyurethane bitumen blend were under the permissible limits of specification.

The samples of different combination are described as under:

Combination 1: Shivabond 315 Grade Polyurethane and VG30 grade bitumen.
Combination 2: Shivabond 315 Grade Polyurethane and Industrial Grade Liquid Bitumen.
Combination 3: Shivabond 1122 Grade Polyurethane and VG30 grade bitumen.
Combination 4: Shivabond 1122 Grade Polyurethane and Industrial Grade Liquid Bitumen.

Fig. 2 Samples of the composite membrane

3 Mechanical Characterization

Mechanical parameters test values for membrane of different composition materials blend are presented in Tables 3, 4, 5 and 6.

3.1 Tensile and Elongation Testing

Tensile tests evaluate the force that is required to break a sample whereas the elongation testing determines the amount by which the it elongates till its breaking point. These tests for coatings provide Tensile Strength and Percent Elongation (at yield and

Table 3 Test results of Combination 1: Shivabond 315 Grade Polyurethane and VG30 grade bitumen

S. no.	Mix proportion (PU: bitumen)	Thickness (mm)	Tensile strength (kg/cm^2)	Tear (kg)	Elongation (%)	Puncture resistance (kg)	Shore hardness
1	50:50	2	15.7	1.03	130	5.3	40
2	50:50	3	13.4	0.82	120	5.6	45
3	50:50	4	17.8	1.28	89	6.2	50

Table 4 Test results of Combination 2: Shivabond 315 Grade Polyurethane and Industrial Grade Bitumen

S. no.	Mix proportion (PU: bitumen)	Thickness (mm)	Tensile strength (kg/cm^2)	Tear (kg)	Elongation (%)	Puncture resistance (kg)	Shore hardness
1	50:50	2	10.6	0.15	80	3.8	35
2	50:50	3	8.5	0.18	73	4.6	40
3	50:50	4	13.5	0.55	55	4.4	45

Table 5 Test results of Combination 3: Shivabond 1122 Grade Polyurethane and VG30 grade bitumen

S. no.	Mix proportion (PU: bitumen)	Thickness (mm)	Tensile strength (kg/cm^2)	Tear (kg)	Elongation (%)	Puncture resistance (kg)	Shore hardness
1	50:50	2	12.20	0.95	320	8.40	55
2	50:50	3	10.03	1.10	305	6.50	65
3	50:50	4	9.08	0.75	375	6.90	75

Table 6 Test results of combination 4: Shivabond 1122 Grade Polyurethane and Industrial Grade Bitumen

S. no.	Mix proportion (PU: bitumen)	Thickness (mm)	Tensile strength (kg/cm^2)	Tear (kg)	Elongation (%)	Puncture resistance (kg)	Shore hardness
1	50:50	2	15.70	1.17	388	9.20	65
2	50:50	3	16.60	1.32	481	8.70	75
3	50:50	4	18.00	1.87	403	10.50	85

at break). An extensometer is used to present all the data digitally of tensile modulus and elongation and other relevant data. This test is performed in accordance with ASTM 638 [16].

3.2 *Tearing Properties*

Tear resistance determines the force required to tear the coating/membrane or thin film. It is used for quality control checks where material comparison and tear failures are possible. It is conducted in accordance with ASTM D1004.

3.3 *Puncture Resistance*

This test method finds the index value of the puncture resistance of the membranes. A Universal Testing Machine (UTM) (tensile/compressive testing) is utilized to meet the specification requirements and test is performed in accordance with ASTM D4833.

3.4 *Shore Hardness a*

Shore hardness is the determination of the term comparative hardness. The range of scales lies from 0 to 100. The test results are dependent on the specimen configuration (Fig. 3).

Fig. 3 **a**, **b** Mould for tensile, elongation and for tear load testing and Universal Tester Machine. **c** Mould for puncture resistance testing and Universal Tester Machine **d** Shore A Durometer

4 Results and Discussions

Mechanical Characterization The test results obtained from mechanical properties of samples prepared of different combinations are under the permissible limits of specification of each relevant ASTM codes.

4.1 Tensile and Elongation Properties

The tensile and elongation tests were carried out to understand the characteristics of the proportioned composition and to figure out the factors which affect the performance efficiency of the composite matrix. The fracture strain increases considerably whereas the material offers more resistance to deformation with time.

The results can be stated as following:

- The results are shown in Fig. 4 depict the values ranging from 15.7,14.7,12.2, 10.6 kg/cm^2 for 2 mm specimen thickness; values 16.6, 13.4, 10.03, 8.50 kg/cm^2 for 3 mm specimen thickness and values 18.8, 17.8, 13.5, 9.08 kg/cm^2 for 4 mm specimen thickness of combination 1–4 respectively.

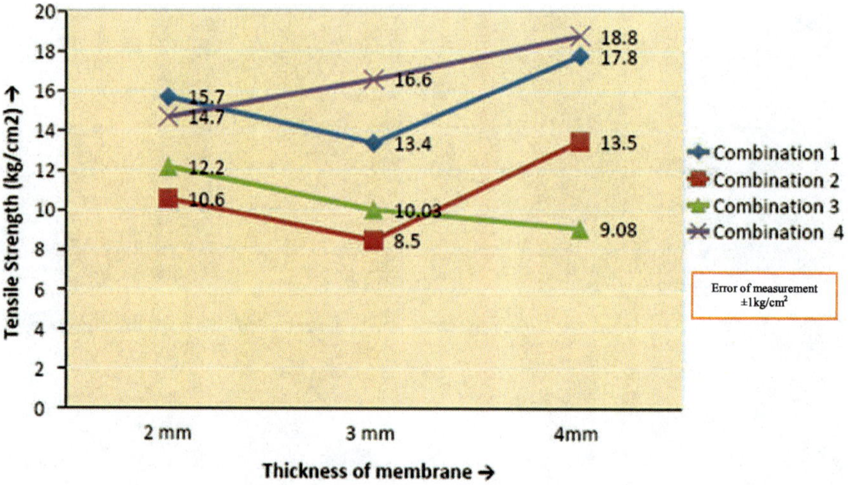

Fig. 4 Test values of tensile strength of composite membrane

- Firstly the test result states that the thinner specimens required a higher stress for relatively same deformation.
- Secondly the factor of curing **age** was observed which can be attributed to the visco elastic deformation of polymers.
- Generally, 70–80% enhancement in elongation is observed when approx 20–30% of Polyurethane is added to the base bitumen [13]. This shows that **structural morphology** can be held responsible for its elongation behavior. [17].
- Fig. 5 shows the percentage elongation test result values ranging from 388, 320, 130, 80 for 2 mm specimen thickness; values 481, 305, 120, 73 for 3 mm specimen thickness and 403, 375, 89, 55 for 4 mm specimen thickness.
- Hence analyzing the values corresponding to combination 4 specimens exhibits a large range of elongation and permissible tensile strength values after combination 1. So influencing factors are **thickness of the specimens, composition materials grade, their curing age, and strain rate**.

4.2 Tearing Properties

Tear resistance of coating or membrane is a function of its extreme resistance to rupture. Basically it tells at what load the membrane will rip apart.

It can be well analyzed from Fig. 6 that the tear load test value of the sample increases with increase in its thickness. The load value varies from 0.15 to 1.17 kg for 2 mm thickness; 0.18 to 1.32 kg for 3 mm thickness and 0.55 to 1.87 for 4 mm thickness of specimen membrane. From these results, it can be concluded that.

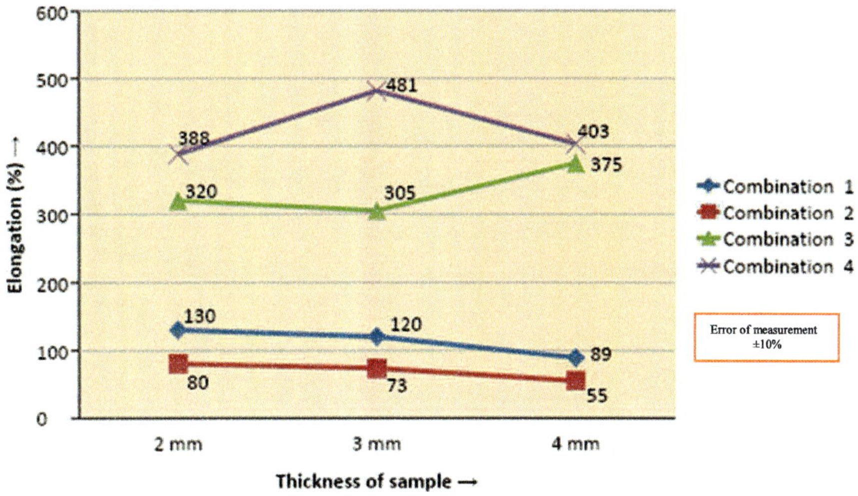

Fig. 5 Test values of elongation of composite membrane

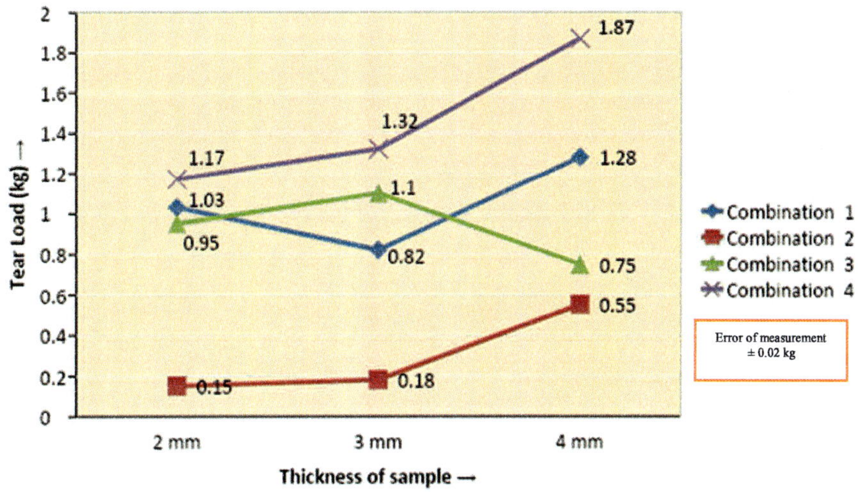

Fig. 6 Test values of tear load values of composite membrane

- Data shows that tear load greater than 1 kg is considered as required value which is therefore in accordance to the minimum required value following ASTM D1004.
- Specimen of combination 2 doesn't adhere to permissible tear load values so it is not considered for composite proportioning. It can be put in words "The load required for rupture is more when the thickness of sample is more".
- Hence values corresponding to combination 4 specimen exhibits large tear loads, which is more suitable for the product applicability.

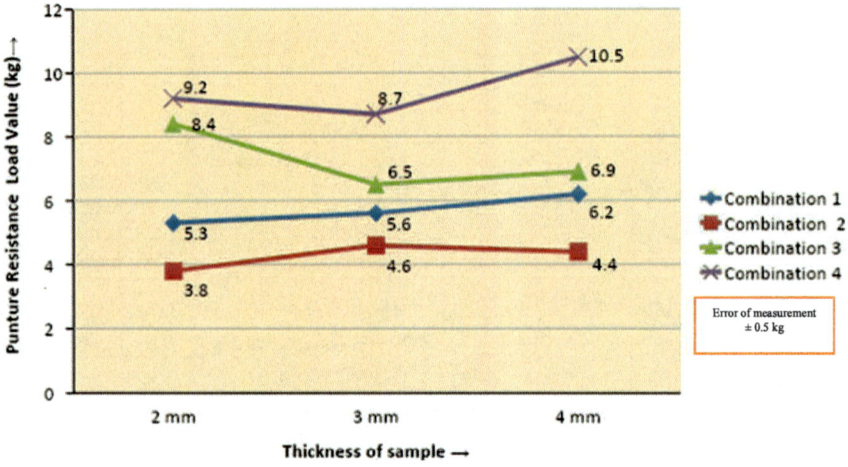

Fig. 7 Test values of puncture resistance of composite membrane

4.3 *Puncture Resistance*

The test results values of puncture resistance are shown in Fig. 7. The values range from 3.8, 5.3, 8.4, 9.2 kg for 2 mm specimen thickness; values 4.6, 5.6, 6.5, 8.7 kg for 3 mm specimen thickness and values 4.4, 6.2, 6.9, 10.5 kg for 4 mm specimen thickness. Observation can be summarized as:

- Except for combination 2, all the test values of the specimen sample of thickness of 2, 3, and 4 mm are within the required permissible limits.
- It also depends on the thickness of samples and speed of impact.
- The basic objective is to attain adequate test values of samples for puncture resistance, tear together with elongation to accommodate structural moments because of loads and sudden impacts.

4.4 *Shore Hardness*

Shore hardness gives the value of comparative hardness of the material being considered. Basically it depicts how well the membrane/coating will put through when it comes to withstand indentation. From the results obtained as shown in Fig. 7 it can be stated that:

- Shore Hardness A result values range from 35 to 65 for 2 mm specimen thickness; range 40–75 for 3 mm specimen thickness and range 45–85 for 4 mm specimen thickness.
- Higher the Shore hardness number means higher the hardness of the material.

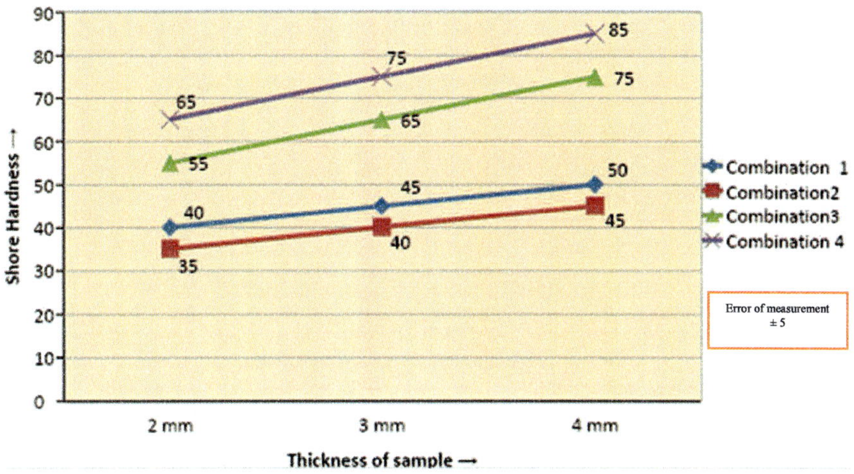

Fig. 8 Test Values of Shore Hardness of composite membrane

- A good correlation exists between asphaltenes content of the bitumen which being a domineering component for hardening mechanism and also affecting the polar functionality of polymer (polyurethane) used [18] (Fig. 8).

5 Conclusion

The composite developed performance and utility depends on the properties of matrix-physical and mechanical. Which in turn are relative to the different grades of raw material used, their quantities used for proportioning, thicknesses of coating, and also on their interaction within and how the way the product is processed. The various combinations of Polyurethane (PU) and Bitumen of different grades as raw material for development of waterproofing membrane exhibits the excellent mechanical properties in accordance to their relative ASTM codes especially offering high range of elongation percentage along with permissible values of tensile strength, puncture resistance, tearing load value and shore hardness value. From the experimental test value analysis, the combination 4 exhibit superior properties in all mechanical characterization tests. So the combination 4: Shivagen 004 and Industrial Grade Bitumen mixture is most beneficial for the development of waterproofing membrane, hence the optimized one on the basis of mechanical characterization. With the new technological developments in the area of coatings the concerned market is expected to boom in coming years. The only main point to be kept in consideration is developing such a coating that accommodates structural movement having resistance towards temperatures and other stresses.

Acknowledgements This research work was carried at Shivalik Agro Poly Products Ltd. Parwanoo, Himachal Pradesh, India. We appreciate their cooperation and permission for utilizing the resources specially R&D laboratory.

References

1. Santagata, E., Baglieri, O., Tsantilis, L., Dalmazzo, D.: Experimental evaluation of modified bituminous binders for heavy duty applications. Proc. Soc. Behav. Sci. **53**, 546–555 (2012)
2. TMR: Waterproofing Membrane Market Global Industry Analysis, Size, Share, Growth, Trends and Forecast 2016–2024. Available at: https://www.transparency.marketresearch.com/waterproofing-membrance-market.html
3. Bukowski, A., Gretkiewicz, J.: J Appl. Polym. Sci, **27**, 1197 (1982). Múnera, J.C., Ossa, E.A.: Analysis of binary bitumen-polymer mixtures. Rev. Facult. Ing. **70**, 18–15 (2014)
4. Martinez-Boza, F.J., Partal, P., Conde, B., Gallegos, C.: Linear visco-elasticity of pigmentable synthetic bitumens. In: Proceedings of the XIIth International Congress on Rheology, Quebec City, Quebec, Canada (1996)
5. Laaly, H.O.: The Science and Technology of Traditional and Modern Roofing Systems, vol. 1. Laaly, Los Angeles (1991)
6. Loeber, L., Muller, G., Morel, J., Sutton, O.: Bitumen in colloid science: a chemical, structural and rheological approach. J. Fuel Energy **77**(13), 1443–1450 (1998)
7. Navarro, F.J., Partal, P., Martínez-Boza, F.J., Gallegos, C.: Novel recycled polyethylene/ground tire rubber/bitumen blends for use in roofing applications: thermo-mechanical properties. J. Polym. Test. **29**(5), 588–595 (2010)
8. Treacy, M.M.J., Ebbesen, T.W., Gibson, J.M.: Exceptionally high Young's modulus observed for individual carbon nanotubes. Nat. Int. J. Sci. **381**, 678–680 (1996)
9. Rupal, A., Kumar Sharma, S., Tyagi, G.: Experimental investigation on mechanical properties of polyurethane modified bituminous waterproofing membrane. Mater. Today Proc. Available online 28 Dec 2019
10. Mouillet, V., Farcasb, F., Besson, S.: Ageing by UV radiation of an elastomer modified bitumen. J. Fuel Energy **87**(12), 2408–2419 (2008)
11. Singh, B., Gupta, M., Kumar, L.: Bituminous polyurethane network: preparation, properties and end use. J. Appl. Polym. Sci. **101**, 217–226 (2006)
12. Partal, P., Martínez-Boza, F.J.: Modification of bitumen using polyurethanes. In: Polymer Modified Bitumen, pp. 43–71. Woodhead Publishing Limited (2011) (Chapter)
13. Tamano, Y., Ishida, M., Okuzono, S.: Amine catalyst for producing polyurethane and process for producing polyurethane. U.S. Patent 5,374,666 (1994)
14. David, J., Vojtova, L., Bednar, K., Kucerık, J., Vavrova, M., Jancar, J.: Development of novel environmental friendly polyurethane foams. Env. Chem. Lett. **8**, 381–385 (2010)
15. Brzeziński, S., Malinowska, G., Nowak, T., Schmidt, H., Marcinkowska, D., Kaleta, A.: Structure and properties of microporous polyurethane membranes designed for textile-polymeric composite systems. Fibres Text. East. Eur. **13**(6), 54–60 (2005)
16. Goikoetxeaundia, G., González, O., Muñoz, M.E., Peña, J., Santamaría, A.: Dynamic viscoelastic characterization of bitumen/polymer roofing membranes. J. Macromol. Mater. Eng **292**(6), 685–792 (2007)
17. Sengoz, B., Topal, A., Isikyakar, G.: Morphology and image analysis of polymer modified bitumens. J. Constr. Build. Mater. **23**, 1986–1992 (2008)
18. Lesueur, D.: The colloidal structure of bitumen: consequences on the rheology and on the mechanisms of bitumen modification. Adv. Coll. Interface. Sci. **145**(2), 42–82 (2008)

An Overview on Utilization of Stone Waste in Construction Industry

Maninder Singh, Babita Saini, and H. D. Chalak

Abstract Dimensional stones are mainly used for flooring and cladding or sometimes as ornamental elements in buildings or monuments. The dimensioning process of stones accumulate very high amount of solid and slurry waste which poses serious environmental issues. Moreover, the progressive growth of construction industry consumes natural resources at large scale. Thus, the utilization of stone waste as pozzolanic and non-pozzolanic constituents in construction activities can minimize the explosive use of natural resources and promotes environmental sustainability. This paper demonstrates the review of recent research studies reported on the possible use of stone waste (SW) in cementitious products as subrogation of cement and fine aggregates. The effect of SW utilization on fresh, mechanical and durability performance such as penetration, shrinkage, porosity, carbonation and chemical immersion have been reviewed in this summary. The review of previous studies signifies that the utilization of discarded stone waste enhanced the mechanical and durability performance of the cementitious matrix due to the filling effect.

Keywords Slurry · Natural resources · Mechanical · Construction industry · Sustainability

1 Introduction

Construction industry plays an important role in the development of any nation. Mortar and concrete are the basic materials used in construction activities. The

M. Singh (✉) · B. Saini · H. D. Chalak
Civil Engineering Department, National Institute of Technology Kurukshetra, Kurukshetra, Haryana, India
e-mail: maninder_6160049@nitkkr.ac.in

B. Saini
e-mail: bsaini@nitkkr.ac.in

H. D. Chalak
e-mail: chalakhd@nitkkr.ac.in

D. K. Ashish et al. (eds.), *3rd International Conference on Innovative Technologies for Clean and Sustainable Development*, RILEM Bookseries 29,
https://doi.org/10.1007/978-3-030-51485-3_38

main constituents used in the preparation of cementitious mixture are cement and aggregates. The processing of both constituents (cement and aggregates) adversely impact the environmental ecosystem [1]. Indian construction industry approximately consumes 400 Million Tonnes (Mt) concrete per year and it is estimated to reach out 1000 Mt in less than a decade [2]. During the processing of cement a large amount of harmful greenhouse gases are emitted; approximately 1 ton of CO_2 emission is released into the surrounding air for production of 1 ton cement [3, 4]. The rapid growth in construction activities leads the demand of cement, which is very detrimental for the environment due to emissions of CO_2 and greenhouse gases. As the conventional cement-based products mainly composed of three constituents; cement, aggregate and water. Therefore, the utilization of excellent quality of constituents is necessary for making strong quality of cementitious composites. Moreover, the cementitious products comprise large amount of fine aggregates (sand). In mix design of cement products, fine aggregates cover the large part of the total constituents used. The fine aggregates are extracted from natural resources and via mining of aggregates along the river beds. The growth of construction activities also escalates the demand of fine aggregate. To complete the demand of aggregates, the extraction of natural resources has increased on high scale. The mining of sand become a challenging issue for the eco-friendly nature of environment. The excessive mining of sand erodes the riverbeds which created a natural imbalance by lowering water table and changed the river course [5]. Therefore, construction activities contribute to the degradation of environmental ecosystem.

On the other hand, the rapid growth of industrialization accumulated various types of wastes such as fly ash (FA), pond ash, blast furnace slag (BFS), plastic waste, waste foundry sand, silica fume, palm oil fuel ash, sewage sludge ash, stone dust, glass waste, marble powder, rice husk ash, red mud, jarosite, stone slurry, hyposludge etc. resulting into degradation of environment [6–11]. After observing all the possible reasons of environmental degradation; the reutilization of all these wastes in construction activities can resolve the problem of disposal, save the natural resources and protect the environment ecosystem. Numerous research studies are available on the utilization of all these wastes in construction industry. In present paper the recent studies available on the reutilization of stone waste in construction activities have been reviewed.

The dimensional stone products are mainly used for decorative purposes in buildings. The quality, type and mineralogical composition of stones depend on the type and place of origin [12]. The main source of stones production are rocks and during processing of the stones, various types of waste get accumulated. India is third largest supplier of stone products in the world. Rajasthan is the main state of marble producer in India. After mining; dressing, cutting, sawing and polishing process of stones, a huge amount of stone waste (SW) produces which results into environmental degradation in various manners. During the dimensioning of stones, the waste are generated in two forms; (a) solid stone waste (b) slurry stone waste. The solid waste is generated via mining during dressing operation and slurry stone waste are generated during cutting, sawing and finishing process [13, 14]. Disposal of these waste occupies a vast area of land which creates the problem for adjoining area living standards. The 90% particles of slurry waste are less than 200 μm in size which can be hazardous for respiratory system and eyes of human beings [15]. Therefore, the accumulation of SW is creating a problem for society and degrade the environment ecosystem. In literature, very few investigations have been available on the inclusion of SW in construction activities. The present review compiles the recent studies available on utilization of SW in construction industry and discussed the influence of SW on fresh, mechanical and durability properties of cement mortar and concrete.

2 Review on Properties of SW

As nature gifted various types of stone and its composition depends on the type and origin. The different types of stone like limestone, granite and marble etc. have been utilized in construction industry. Numerous research investigations reported that calcium oxide (CaO) and silicon di-oxide (SiO_2) are the main constituents of SW. The chemical compounds of SW depend on the type of waste generation and origin of stone. The chemical composition of some types of SW have been summarized in Table 1.

The chemical compositions summarized in Table 1 represents that the granite SW mainly consists of silicon di-oxide (SiO_2); while limestone and marble SW mainly consist of calcium oxide (CaO).

3 Utilization of SW in Mortar Mixes

The different types of SW have been utilized by various researchers in mortar mixes as cement and fine aggregates replacement. The details of some previous studies on utilization of SW in mortar mixes has been summarized in Table 2.

Bederina et al. [25] experimentally evaluated the role of limestone sand on cement mortar mixes as subrogation of sand. The experimental observations reported that the compression and flexural response of mortar mixes improved with the presence of limestone dust; whereas, decreased the mass and capillary absorption. The introduction of limestone also provides better resistance against chemical attacks. Belaidi et al. [26] evaluated the fresh and mechanical properties of self-compacting cement mortar with the incorporation of natural pozzolana (PZ) and marble powder (MP) as subrogation of cement. The experimental investigation revealed that the addition of MP (in ternary system) improved the rheological performance of the fresh matrix. Bonavetti and Irassar [27] evaluated the behaviour of mortar mix with the incorporation of quartz sand, granite and limestone waste as subrogation of sand. The observations from the study reported that the utilization of stone dust increased the strength of mortar mixes at early ages. The increase in quantity of stone dust particles negatively affects the water absorption and porosity (after 5% replacement).

Table 1 Chemical compounds of different types of SW

References	Type of stone	Chemical compounds (%)								
		CaO	SiO_2	Al_2O_3	MgO	Fe_2O_3	Na_2O	K_2O	SO_3	LOI
[16]	Limestone	45.45	7.89	2.58	1.72	1.13	0.00	0.21	0.95	42.48
[17]	Limestone	42.57	–	0.27	9.64	0.34	–	–	0.01	43.21
[18]	Limestone	49.08	5.05	1.62	1.59	0.61	0.04	0.52	–	40.88
[19]	Granite	4.90	85.5	2.10	2.50	0.40	–	–	1.80	1.10
[20]	Granite	9.10	53.30	14.10	8.30	12.30	1.20	–	–	0.30
[21]	Granite	1.85	57.58	10.42	0.58	2.06	3.29	3.42	–	20.80
[22]	Granite	–	72.57	15.63	0.83	–	4.21	6.76	–	–
[23]	Marble	54.2	1.39	0.32	0.64	0.14	<0.04	<0.06	<0.10	42.6
[20]	Marble	6.8	14.8	21.9	7.3	36.8	9.3	–	1.3	0.20
[15]	Marble	61.83	8.38	0.67	14.36	0.65	0.60	0.07	–	–
[24]	Marble	54.43	0.67	0.12	0.59	0.08	0.14	–	–	43.4

Table 2 Details of previous studies on utilization of SW in mortar mixes

References	SW type	% Used	UT	Properties evaluated									Major outcomes
				FP	CS	FR	WA	PR	CE	FT	DS	MA	
[25]	LP	25, 50, 75, 100	S	✓	✓	✓	×	×	✓	×	×	✓	Presence of LP improved SP
[26]	MP	5, 10, 15, 20, 30	C	✓	✓	×	×	×	×	×	×	×	Rheological performance improved
[27]	Quartz, GP, LP	5, 10, 15, 20	S	✓	✓	✓	×	✓	×	×	✓	×	SP improved at early ages
[28]	LP	10	S	✓	✓	✓	×	×	×	×	×	✓	Inclusion of LP produced higher thermal conductivity
[29]	LP	65% (max)	C	✓	✓	×	×	✓	✓	×	×	✓	Utilization of LP provide resistance against CE
[16]	LP	5, 7.5, 10	SCM	✓	✓	×	×	×	✓	×	×	✓	Addition of LP improved the DP
[30]	GP	10, 20, 30, 40	C	✓	✓	×	✓	✓	×	✓	×	×	20% addition of GP exhibited better performance
[31]	MP	20, 40, 60, 80, 100	S	✓	✓	×	✓	✓	×	×	✓	×	20% of river sand can be replaced by MP
[14]	BSS	5, 10, 15	FA	✓	✓	✓	×	×	×	×	×	✓	BSS improved the SP and DP

(continued)

Table 2 (continued)

References	SW type	% Used	UT	Properties evaluated									Major outcomes
				FP	CS	FR	WA	PR	CE	FT	DS	MA	
[32]	LP	4, 8, 12	FA	✓	✓	×	×	✓	×	×	×	×	Addition of LP improved the CS

UT utilization type; *BSS* burnt stone slurry; *ST* setting time; *FP* fresh properties; *CS* compressive strength; *FR* flexural response; *WA* water absorption; *CE* chemical exposure; Freeze thaw cycles; *DS* drying shrinkage; *MA* microstructural analysis; *FA* fine aggregates; *PR* porosity; *C* cement; *S* sand; *SP* strength parameters; *DP* durability performance; *GP* granite; *LP* limestone; *SCM* supplementary cementitious materials; *MP* marble powder; *SD* stone dust; *CA* coarse aggregates; *CP* chloride penetration; *TS* tensile strength; *SS* stone slurry

Corinaldesi et al. [28] investigated the strength performance and thermal conductivity of cement matrix with the utilization of rubber waste (RW) and limestone powder. The examined results depicted that the incorporation of limestone provides better thermal conductivity and strength parameters; while, the utilization of RW reduced the thermal conductivity and unit weight. Baldermann et al. [29] evaluated the performance of mortar incorporating limestone as substitution of ordinary Portland cement (OPC). The experimental observations revealed that the 50% utilization of limestone reduced porosity and gave better sulphate resistance and compressive strength than normal cement matrix. Senhadji et al. [16] studied the strength parameters, acid resistance and microstructure behaviour of cement matrix incorporating PZ, silica fume and limestone fine. The experimental results demonstrated that the limestone containing mortar mix performed better under sulphuric acid environment; while, silica fume utilized mix gets damaged in such environmental conditions. The introduction of limestone reduced the compressive strength as compared to OPC containing mortar. Mashaly et al. [30] have carried out the series of experiments on cementitious mixes containing granite sludge as subrogation of cement. The observed parameters revealed that the incorporation of granite sludge decreased the compressive strength and optimum results were found at 20% replacement. Increment in the quantity of granite sludge increased the abrasion loss, mass loss, strength loss rate and sulphate attack, minimum loss was observed at 10% substitution. Kabeer and Vyas [31] investigated the performance of cement matrix containing marble slurry as a subrogation of river sand. The experimental results reported that 20% utilization of marble slurry gave almost similar results to the control mix. Al-Akhras et al. [14] have utilized burnt stone slurry (BSS) in mortar mixes as substitution of sand. The obtained results revealed that setting time and workability reduced with an increase in the quantity of BSS; however, compressive and flexural strength increased. The microstructural analysis of mixes depicted that the BSS containing mixes were much better than plain mortar. Kwan and McKinley [32] studied the influence of limestone waste utilization on the performance of cement matrix. The obtained results revealed that the compressive strength improved with the addition of limestone waste. Keleştemur et al. [33] evaluated the performance of mortar including marble dust and glass fibers and observed that the compressive strength of matrix increased and provide resistance against high temperature. The utilization of marble dust reduced the porosity and modified the fiber-matrix interfacial bonding. Ballester et al. [34] have utilized limestone waste in mortar mix and reported that the inclusion of limestone waste enhanced the performance of cement matrix and decreased the production cost. Ramos et al. [35] investigated the various parameters of mortar included granite sludge as substitution of cement. Observations revealed that the utilization of granite sludge produces denser matrix, improved resistance to chloride without affecting workability and strength. Li et al. [36] studied the performance of cementitious mixes containing marble dust as substitution of cement and aggregates. The obtained results revealed that the inclusion of marble dust reduced the carbonation, water absorption and shrinkage strain. Kockal [37] investigated the performance of mortar with the utilization of silica fume, green ceramic powder and marble dust. Observed parameters showed that the addition of marble dust gave the highest strength and lesser permeability for all environmental conditions. Similarly, Buyuksagis et al. [38] reported that the addition of marble powder improved the strength and durability performance and contribute to making denser matrix.

4 Utilization of SW in Concrete

The different types of SW have been utilized in concrete mixes as replacement of various materials (cement and aggregates). The details of some previous studies on the utilization of SW in concrete mixes have been summarized in Table 3.

Singh et al. [4] evaluated the performance of concrete incorporating marble slurry as partial subrogation of cement up to 25%. The obtained results revealed that the strength parameters enhanced up to 15% subrogation. The incorporation of dried marble slurry improved the ultrasonic pulse velocity, durability parameter and bond strength. Rana et al. [12] studied the various parameters of concrete incorporating marble slurry as a subrogation of cement. The examined study revealed that the utilization of marble slurry reduced the workability; improved the strength parameter and provide resistance against corrosion. The optimum content of marble slurry was found 10% as cement substitution. Almeida et al. [13] examined the performance of high strength concrete containing recycled stone slurry as substitution of sand. The obtained results depicted that the utilization of stone slurry enhanced the compressive and flexural strength, modulus of elasticity and durability parameters. The enhancement was observed due to micro filling effect of stone slurry. Ashish et al. [15] checked the performance of concrete included marble powder (MP) as a subrogation of cement and sand. The experimental results revealed that the utilization of MP acts as a filler in concrete and enhanced the durability and strength parameters. The optimum level of substitution was found 10% for sand, 10% for cement and 20% for amalgam. Eren and Marar [17] studied the behaviour of concrete containing limestone crusher dust and steel fiber. The results depicted that the optimum compressive and tensile performance was found at 10% substitution of fine aggregate by limestone. The utilization of stone waste reduced water permeability due to the filing effect. Li et al. [18] evaluated the performance of two different concrete materials incorporating limestone fines as substitution of sand. The examined results depicted that the durability performance of concrete increased with the utilization of limestone fines. Abd Elmoaty [19] investigated the durability performance of concrete incorporating granite dust as a substitution of cement. The experimental observations revealed that the utilization of granite dust improved the compression behaviour, corrosion resistance of concrete and no obvious change was observed in hydration products. Similarly, Binici et al. [20], Li et al. [21] and Singh et al. [22] reported that the inclusion of granite dust in concrete enhanced the performance of cementitious matrix. Rodrigues et al. [23] studied the mechanical properties of concrete containing marble cutting sludge as a substitution of cement. It has been reported that satisfactory results were observed up to 10% replacement. André et al. [39]

Table 3 Details of previous studies on utilization of SW in concrete mixes

References	SW type	% Used	UT	Properties evaluated									Major outcomes
				FP	CS	FR	TS	WA	CP	FT	DS	MA	
[39]	MP	20, 50, 100	CA	✓	✓	×	×	✓	✓	×	×	×	MP enhanced the overall performance of the matrix
[40]	MP	20, 50, 100	FA	✓	×	×	×	✓	✓	×	✓	×	Addition of MP enhanced the DP
[18]	LP	3, 5, 7, 10, 15, 20	FA	✓	✓	✓	×	×	✓	✓	×	×	Inclusion of LP provide resistance against CE
[41]	MP	25, 50, 75, 100	FA	✓	✓	✓	✓	×	×	×	×	×	75% of aggregates can be substituted by marble aggregates
[42]	SD	5, 10, 15, 20, 25, 30	FA	✓	✓	✓	×	✓	×	×	✓	×	Addition of SD blocked the pore connectivity and improved the DP
[13]	SS	Up to 100%	FA	✓	✓	×	✓	✓	×	×	×	×	Utilization of SS enhanced the SP and DP
[43]	MP	5, 7.5, 10, 15	C	×	✓	×	✓	×	×	×	×	✓	MP depicted filling effect in concrete

(continued)

Table 3 (continued)

References	SW type	% Used	UT	Properties evaluated									Major outcomes
				FP	CS	FR	TS	WA	CP	FT	DS	MA	
[23]	MP	5, 10, 15, 20	C	✓	✓	×	✓	×	×	×	×	×	10% addition of marble sludge gave satisfactory results
[19]	GP	5, 7.5, 10, 15	C	×	✓	×	✓	×	×	×	×	✓	Mechanical properties of concrete improved up to 5% replacement
[15]	MP	10, 15; 10, 15; 20, 30	C; S; (C + S)	✓	✓	×	✓	×	×	×	×	×	MP improved the strength as well as durability performance
[44]	MP	5, 10, 20	C	✓	✓	×	✓	×	×	×	×	×	10% MP can be used as cement replacement in concrete
[45]	MP	10, 15	S	✓	✓	×	✓		×	×	×	✓	Inclusion of MP improved the SP

(continued)

Table 3 (continued)

References	SW type	% Used	UT	Properties evaluated									Major outcomes
				FP	CS	FR	TS	WA	CP	FT	DS	MA	
[4]	MP	10, 15, 20, 25	C	✓	✓	✓	✓	×	×	×	✓	✓	15% of marble slurry can be added as cement replacement

checked the performance of concrete incorporating marble waste as coarse aggregates. The examined study reported that the inclusion of marble waste aggregates improved the overall performance of matrix. Gameiro et al. [40] studied the durability performance of concrete incorporating marble waste as substitution of fine aggregates up to 100%. The examined results depicted that the flowability and absorption reduced; while, durability parameters improved. Hebhoub et al. [41] have utilized marble waste aggregates in concrete as substitution of sand. Observed parameters depicted that the addition of marble waste as sand replacement up to 75% enhanced the performance of concrete. Çelik et al. [42] investigated the effect of stone dust on concrete as a subrogation of sand. Evaluated parameters revealed that the 10% addition of stone dust provided a positive effect on compressive and flexural strength; whereas, lowest water absorption was found with 15% replacement. Aliabdo et al. [43] and Khodabakhshian et al. [44] have utilized marble industry waste in concrete as a subrogation of cement. The obtained results depicted that the incorporation of marble industry waste enhanced the durability and strength parameters and provide a positive effect on the environmental eco-system. Ashish [45] studied the behaviour of concrete incorporating marble powder waste as a subrogation of sand. The results depicted that the durability parameters of concrete enhanced up to 15% incorporation of marble waste. Similarly, Silva et al. [46] reported that the utilization of marble waste in concrete enhanced the overall performance of the concrete. Jung et al. [47] and Menadi et al. [48] studied the performance of concrete containing limestone waste. The observed results revealed that the inclusion of limestone waste in concrete solved the problem of disposal and improved the various parameters. Bacarji et al. [49] and Tennich et al. [50] reported that the utilization of granite and marble waste in concrete can save the natural resources and environmental ecosystem with acceptable strength and durability parameters.

Published literature reported that the utilized quantity and type of stone waste influence the performance of the cementitious matrix it may be due to the difference between their chemical compositions as summarized in Table 1.

5 Summary and Conclusions

The above literature review clearly states that the utilization of stone waste in construction industry is very effective in the form of fresh, mechanical and durability parameters. The properties of various types of stone waste (SW), role of SW addition as substitution of cement, sand coarse aggregates in cementitious products have been highlighted in present review. The following major outcomes can be drawn from this review:

- The addition of stone waste powder reduced the workability of the matrix. The reduction in workability enhanced with an increase in the quantity of SW powder; however, the maximum reduction was found in aggregates replacement as compared to cement replacement. Higher specific surface area and water demand of SW reduced the flowability of cementitious composites.

- The strength parameters (compressive, flexural, tensile) of cementitious products enhanced with the addition of SW. The 75% fine aggregates and 5–10% cement can be effectively replaced by SW.
- Incorporation of SW positively affect the water absorption, permeability and carbonatation resistance of cement matrix and provide resistance against chemical-rich environmental conditions. The filling effect of SW improved the interfacial transition zone characteristics and blocked the pore connectivity which alternatively enhanced the strength and durability performance of the matrix.
- Utilization of SW in construction industry solves the problem of disposal, reduce the natural resources extraction and promote the eco-friendly nature of environment.

References

1. Gautam, N., Krishna, V., Srivastava, A.: Sustainability in the concrete construction. Int. J. of Environ. Res. Dev. **4**(1), 81–90 (2014)
2. Pathak, P.P.: Inclusion of Portland and pozzolona (fly ash waste) cement in specifications. Ind. Highw. **37**, 23–29 (2009)
3. Woodson, R.D.: Concrete materials. In: Concrete Portable Handbook, pp. 5–18. Elsevier Inc., Oxford (2012)
4. Singh, M., Srivastava, A., Bhunia, D.: An investigation on effect of partial replacement of cement by waste marble slurry. Constr. Build. Mater. **134**, 471–488 (2017)
5. Bravard, J.-P., Goichot, M., Gaillot, S.: Geography of sand and gravel mining in the lower mekong river. In: First Survey and Impact Assessment. EchoGéo, vol. 26 (2013).
6. Singh, M., Saini, B., Chalak, H.D.: Performance and composition analysis of engineered cementitious composite (ECC)—a review. J. Build. Eng. **26**, 100851 (2019). https://doi.org/10.1016/j.jobe.2019.100851
7. Singh, M., Saini, B., Chalak, H.D.: Properties of engineered cementitious composites: a review. In: ICSWMD 2018: Proceedings of the 1st International Conference on Sustainable Waste Management through Design, pp 473–483. https://link.springer.com/chapter/10.1007/978-3-030-02707-0_54
8. Naik, TR., Sustainability of concrete construction. In: Practice Periodical on Structural Design and Construction, p. 13 (2008)
9. Gambhir, M.L.: Concrete Technology. The McGraw Hill Companies (2004)
10. Siddique, R.: Utilization of industrial by-products in concrete. Proc. Eng. **95**, 335–347 (2014)
11. Mehta, P.K.: Rice husk ash as a mineral admixture in concrete. In: Proceedings of the 2nd International Seminar on Durability of Concrete: Aspects of Admixtures and Industrial by Products, pp. 131–136. Gothenburg, Sweden (1989)
12. Rana, A., Kalla, P., Csetenyi, L.J.: Sustainable use of marble slurry in concrete. J. Clean. Product. **94**, 304–311 (2015)
13. Almeida, N., Branco, F., de Brito, J., Santos, J.R.: High-performance concrete with recycled stone slurry. Cem. Concr. Res. **37**, 210–220 (2007)
14. Al-Akhras, N.M., Ayman, A., Alaraji, W.A.: Using burnt stone slurry in mortar mixes. Constr. Build. Mater. **24**, 2658–2663 (2010)
15. Ashish, D.K.: Feasibility of waste marble powder in concrete as partial substitution of cement and sand amalgam for sustainable growth. J. Build. Eng. **15**, 236–242 (2018)

16. Senhadji, Y., Escadeillas, G., Mouli, M., Khelafi, H., Benosman: Influence of natural pozzolan, silica fume and limestone fine on strength, acid resistance and microstructure of mortar. Powder Technol. **254**, 314–323 (2014)
17. Özgür, E., Khaled, M.: Effects of limestone crusher dust and steel fibers on concrete. Constr. Build. Mater. **23**, 981–988 (2009)
18. Li, B., Wang, J., Zhou, M.: Effect of limestone fines content in manufactured sand on durability of low- and high-strength concretes. Constr. Build. Mater. **23**, 2846–2850 (2009)
19. Abd Elmoaty, A.E.M.: Mechanical properties and corrosion resistance of concrete modified with granite dust. Constr. Build. Mater. **47**, 743–752 (2013)
20. Binici, H., Shah, T., Aksogan, O., Kaplan, H.: Durability of concrete made with granite and marble as recycle aggregates. J. Mater. Process. Technol. **208**, 299–308 (2008)
21. Li, Y., Yu, H., Zheng, L., Wen, J., Wu, C., Tan, Y.: Compressive strength of fly ash magnesium oxychloride cement containing granite wastes. Constr. Build. Mater. **38**, 1–7 (2013)
22. Singh, S., Nagar, R., Agrawal, V., Rana, A., Tiwari, A.: Sustainable utilization of granite cutting waste in high strength concrete. J. Clean. Product. **116**, 223–235 (2016)
23. Rodrigues, R., de Brito, J., Sardinha, M.: Mechanical properties of structural concrete containing very fine aggregates from marble cutting sludge. Constr. Build. Mater. **77**, 349–356 (2015)
24. Aruntaş, H.Y., Gürü, M., Dayı, M., Tekin, İ.: Utilization of waste marble dust as an additive in cement production. Mater. Des. **31**, 4039–4042 (2010)
25. Bederina, M., Makhloufi, Z., Bounoua, A., Bouziani, T., Quéneudec, M.: Effect of partial and total replacement of siliceous river sand with limestone crushed sand on the durability of mortars exposed to chemical solutions. Constr. Build. Mater. **47**, 146–158 (2013)
26. Belaidi, A.S.E., Azzouz, L., Kadri, E., Kenai, S.: Effect of natural pozzolana and marble powder on the properties of self-compacting concrete. Constr. Build. Mater. **31**, 251–257 (2012)
27. Bonavetti, V.L., Irassar, E.F.: The effect of stone dust content in sand. Cem. Concr. Res. **24**, 580–590 (1994)
28. Corinaldesi, V., Mazzoli, A., Moriconi, G.: Mechanical behaviour and thermal conductivity of mortars containing waste rubber particles. Mater. Des. **32**(1646–1650), 0261–3069 (2011)
29. Baldermann, A., Rezvani, M., Proske, T., Grengg, C., Steindl, F., Sakoparnig, M., Baldermann, C., Galan, I., Emmerich, F., Mittermayr, F.: Effect of very high limestone content and quality on the sulfate resistance of blended cements. Constr. Build. Mater. **188**, 1065–1076 (2018)
30. Mashaly, A.O., Shalaby, B.N., Rashwan, M.A.: Performance of mortar and concrete incorporating granite sludge as cement replacement, Constr. Build. Mater. **169**, 800–818 (2018)
31. Kabeer, K.I., Ahmed, S., Ashok Kumar, V.: Utilization of marble powder as fine aggregate in mortar mixes. Constr. Build. Mater. **165**, 321–332 (2018)
32. Kwan, A.K.H., McKinley, M.: Effects of limestone fines on water film thickness, paste film thickness and performance of mortar. Powder Technol. **261**, 33–41 (2014)
33. Keleştemur, O., Arıcı, E., Yıldız, S., Gökçer, B.: Performance evaluation of cement mortars containing marble dust and glass fiber exposed to high temperature by using Taguchi method. Constr. Build. Mater. **60**, 17–24 (2014)
34. Ballester, P., Mármol, I., Morales, J., Sánchez, L.: Use of limestone obtained from waste of the mussel cannery industry for the production of mortars. Cem. Concr. Res. **37**, 559–564 (2007)
35. Ramos, T., Matos, A.M., Schmidt, B., Rio, J., Sousa-Coutinho, J.: Granitic quarry sludge waste in mortar: effect on strength and durability. Constr. Build. Mater. **47**, 1001–1009 (2013)
36. Li, L.G., Huang, Z.H., Tan, Y.P., Kwan, A.K.H., Liu, F.: Use of marble dust as paste replacement for recycling waste and improving durability and dimensional stability of mortar. Constr. Build. Mater. **166**, 423–432 (2018)
37. Kockal, N.U.: Effects of elevated temperature and re-curing on the properties of mortars containing industrial waste materials. Iran. J. Sci. Technol. Trans. Civ. Eng. **37**, 67–76 (2013)
38. Buyuksagis, I.S., Uygunoglu, T., Tatar, E.: Investigation on the usage of waste marble powder in cement-based adhesive mortar. Constr. Build. Mater. **154**, 734–742 (2017)
39. António, A., de Brito, J., Rosa, A., Pedro, D.: Durability performance of concrete incorporating coarse aggregates from marble industry waste. J. Clean. Product. **65**, 389–396 (2014)

40. Gameiro, F., de Brito, J., da Silva D.C.: Durability performance of structural concrete containing fine aggregates from waste generated by marble quarrying industry. Eng. Struct. **59**, 654–662 (2014)
41. Hebhoub, H., Aoun, H., Belachia, M., Houari, H., Ghorbel, E.: Use of waste marble aggregates in concrete. Constr. Build. Mater. **25**, 1167–1171 (2011)
42. Çelik, T., Marar, K.: Effects of crushed stone dust on some properties of concrete. Cem. Concr. Res. **26**, 1121–1130 (1996)
43. Aliabdo, A.A., Abd Elmoaty, E. Auda, E.M.: Re-use of waste marble dust in the production of cement and concrete. Constr. Build. Mater. **50**, 28–41 (2014)
44. Khodabakhshian, A., de Brito, J., Ghalehnovi, M., Shamsabadi, E.A.: Mechanical, environmental and economic performance of structural concrete containing silica fume and marble industry waste powder. Constr. Build. Mater. **169**, 237–251 (2018)
45. Ashish, D.K.: Concrete made with waste marble powder and supplementary cementitious material for sustainable development. J. Clean. Product. **211**, 716–729 (2019)
46. Silva, D., Gameiro, F., de Brito, J.: Mechanical properties of structural concrete containing fine aggregates from waste generated by the marble quarrying industry. J. Mater. Civ. Eng. **26**(6), 04014008 (2014)
47. Jung, S.-H., Saraswathy, V., Karthick, S., Kathirvel, P., Kwon, S.-J.: Microstructure characteristics of fly ash concrete with rice husk ash and lime stone powder, Int. J. Concr. Struct. Mater. https://doi.org/10.1186/s40069-018-0257-4.
48. Menadi, B., Kenai, S., Khatib, J., Aït-Mokhtar, A.: Strength and durability of concrete incorporating crushed limestone sand. Constr. Build. Mater. **23**, 625–633 (2009)
49. Bacarji E., Toledo Filho, R.D., Koenders, E.A.B., Figueiredo, E.P., Lopes J.L.M.P.: Sustainability perspective of marble and granite residues as concrete fillers. Constr. Build. Mater. **45**, 1–10 (2013)
50. Tennich, M., Kallel, A., Ouezdou, M.B.: Incorporation of fillers from marble and tile wastes in the composition of self-compacting concretes. Constr. Build. Mater. **91**, 65–70 (2015)

Author Index

D. K. Ashish et al. (eds.), *3rd International Conference on Innovative Technologies for Clean and Sustainable Development*, RILEM Bookseries 29,
https://doi.org/10.1007/978-3-030-51485-3